T0300938

THEORY OF REFLECTANCE AND EMITTANCE SPECTROSCOPY

Reflectance and emittance spectroscopy have become increasingly important tools in remote sensing, and have been employed in virtually all recent planetary spacecraft missions. They are primarily used to measure properties of disordered materials, especially in the interpretation of remote observations of the surfaces of the Earth and other terrestrial planets.

Theory of Reflectance and Emittance Spectroscopy gives a quantitative treatment of the physics of the interaction of electromagnetic radiation with particulate media, such as powders and soils. Subjects covered include electromagnetic wave propagation, single particle scattering, diffuse reflectance, thermal emittance, and polarization. This new edition has been updated and expanded to include: extension of the equivalent slab model of irregular particle scattering to include particle phase functions and coated particles; a quantitative treatment of the effects of porosity; a detailed discussion of the coherent backscatter opposition effect, including polarization; a quantitative treatment of simultaneous transport of energy within the medium by conduction and radiation; and lists of relevant databases and software.

With a strong emphasis on physical insights, this book is an essential reference for research scientists, engineers, and advanced students of planetary remote sensing.

BRUCE HAPKE is Professor Emeritus of Geology and Planetary Science at the University of Pittsburgh, where he continues to study various bodies of the solar system. He was principal investigator for the analysis of lunar samples and was associated with several other NASA missions, to Mercury, Mars, Saturn, and the outer solar system. He is a Fellow of the American Geophysical Union and was awarded the Kuiper Prize by the Division for Planetary Sciences of the American Astronomical Society for "outstanding contributions to planetary science." He has an asteroid *3549 Hapke* and a mineral *hapkeite* named in his honor.

THEORY OF REFLECTANCE AND EMITTANCE SPECTROSCOPY

Second Edition

BRUCE HAPKE

Department of Geology and Planetary Science
University of Pittsburgh

CAMBRIDGE
UNIVERSITY PRESS

University Printing House, Cambridge CB2 8BS, United Kingdom

One Liberty Plaza, 20th Floor, New York, NY 10006, USA

477 Williamstown Road, Port Melbourne, VIC 3207, Australia

314-321, 3rd Floor, Plot 3, Splendor Forum, Jasola District Centre, New Delhi - 110025, India

79 Anson Road, #06-04/06, Singapore 079906

Cambridge University Press is part of the University of Cambridge.

It furthers the University's mission by disseminating knowledge in the pursuit of education, learning and research at the highest international levels of excellence.

www.cambridge.org
Information on this title: www.cambridge.org/9780521883498

First published 2012

A catalogue record for this publication is available from the British Library

Library of Congress Cataloging in Publication data
Hapke, Bruce.
Theory of reflectance and emittance spectroscopy / Bruce Hapke. – 2nd ed.
p. cm.
Includes bibliographical references and index.
ISBN 978-0-521-88349-8
1. Reflectance spectroscopy. 2. Emission spectroscopy. 3. Moon–Surface–Spectra. I. Title.
QC454.R4H37 2012
522′.67–dc23 2011040517

ISBN 978-0-521-88349-8 Hardback

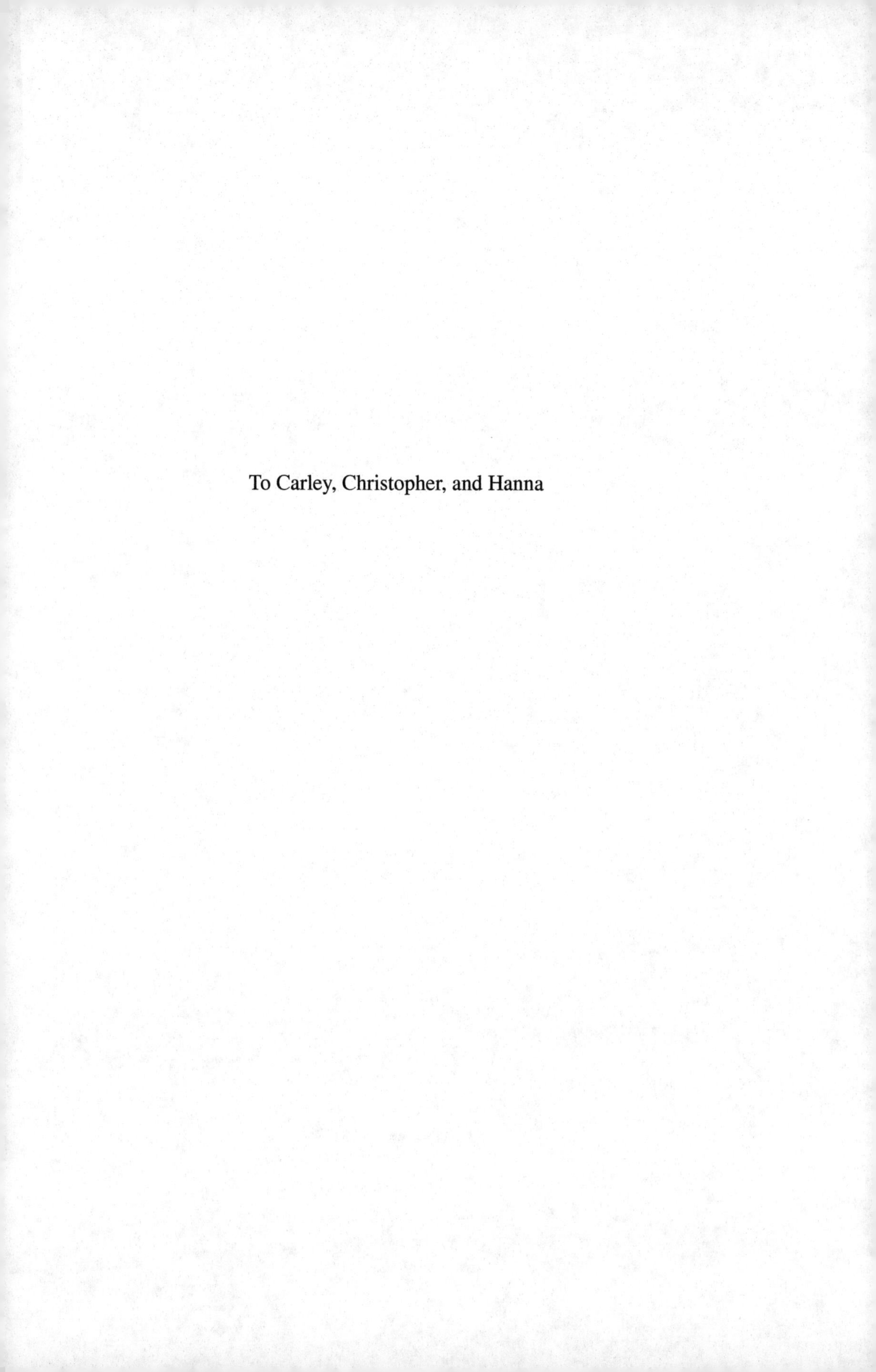

To Carley, Christopher, and Hanna

Contents

Acknowledgments *page* xi

1 Introduction 1
- 1.1 Scientific rationale 1
- 1.2 About this book 3

2 Electromagnetic wave propagation 5
- 2.1 Maxwell's equations 5
- 2.2 Electromagnetic waves in free space 6
- 2.3 Propagation in a linear nonabsorbing medium 11
- 2.4 Propagation in a linear absorbing medium 17
- 2.5 Interference 22
- 2.6 Polarization; the Stokes vector 23

3 The absorption of light 27
- 3.1 Introduction 27
- 3.2 Classical dispersion theory 27
- 3.3 Dispersion relations 33
- 3.4 Mechanisms of absorption 34
- 3.5 Band shape and temperature effects 43
- 3.6 Spectral databases 44

4 Specular reflection 45
- 4.1 Introduction 45
- 4.2 Boundary conditions in electromagnetic theory 45
- 4.3 The Fresnel equations 46
- 4.4 The Kramers–Kronig reflectivity relations 61
- 4.5 Absorption bands in reflectivity 62
- 4.6 Criterion for optical flatness 64

5 Single-particle scattering: perfect spheres 66
5.1 Introduction 66
5.2 Concepts and definitions 66
5.3 Scattering by a perfect, uniform sphere: Mie theory 72
5.4 Properties of the Mie solution 73
5.5 Other regular particles 95
5.6 The equivalent-slab approximation 95
5.7 Computer programs 99

6 Single-particle scattering: irregular particles 100
6.1 Introduction 100
6.2 Extension of definitions to nonspherical particles 101
6.3 Empirical scattering functions 101
6.4 Theoretical and experimental studies of nonspherical particles 109
6.5 The generalized equivalent-slab model 122
6.6 Computer programs and databases 144

7 Propagation in a nonuniform medium: the equation of radiative transfer 145
7.1 Introduction 145
7.2 Effective-medium theories 146
7.3 The transport of radiation in a particulate medium 148
7.4 Radiative transfer in a medium of arbitrary particle separation 158
7.5 Methods of solution of radiative-transfer problems 169
7.6 Computer programs 179

8 The bidirectional reflectance of a semi-infinite medium 180
8.1 Introduction 180
8.2 Reflectances 180
8.3 Geometry and notation 183
8.4 The radiance at a detector viewing a horizontally stratified medium 185
8.5 Empirical reflectance expressions 187
8.6 The diffusive reflectance 189
8.7 The bidirectional reflectance 195
8.8 Comparison of the IMSA model with measurements 210
8.9 Bidirectional reflectance of a medium of arbitrary filling factor 216

9 The opposition effect 221
9.1 Introduction 221
9.2 The shadow-hiding opposition effect (SHOE) 224
9.3 The coherent backscatter opposition effect (CBOE) 237
9.4 Combined SHOE, CBOE, and IMSA models 260

10 A miscellany of bidirectional reflectances and related quantities 263
10.1 Introduction 263
10.2 Some commonly encountered bidirectional reflectance quantities 263
10.3 Reciprocity 264
10.4 Diffuse reflectance from a medium with a specularly reflecting surface 266
10.5 Oriented scatterers: applications to vegetation canopies 268
10.6 Reflectance of a layered medium 272
10.7 Mixing formulas 282

11 Integrated reflectances and planetary photometry 287
11.1 Introduction 287
11.2 Integrated reflectances 287
11.3 Planetary photometry 295

12 Photometric effects of large-scale roughness 303
12.1 Introduction 303
12.2 Derivation 307
12.3 Applications to planetary photometry 323
12.4 Summary of the roughness correction model 331
12.5 Other planetary photometric models 335

13 Polarization of light scattered by a particulate medium 339
13.1 Introduction 339
13.2 Linear polarization of particulate media 340
13.3 The positive branch of polarization 344
13.4 The negative branch of polarization 354
13.5 Summary 367

14 Reflectance spectroscopy 369
14.1 Introduction 369
14.2 Measurement of reflectances 370
14.3 Inverting the reflectance to find the scattering parameters 372
14.4 Absorption bands in reflectance 378
14.5 The reflectance spectra of intimate mixtures 388
14.6 Absorption bands in layered media 392
14.7 Retrieving the absorption coefficient from the single-scattering albedo 395
14.8 Other methodologies 400
14.9 Particulate media with $X \ll 1$ 406
14.10 Planetary applications 407

15 Thermal emission and emittance spectroscopy 412
15.1 Introduction 412
15.2 Black-body thermal radiation 413
15.3 Emissivity 415
15.4 Kirchhoff's law 425
15.5 Combined reflectance and emittance 427
15.6 Emittance spectroscopy 428
15.7 The thermal shadow-hiding opposition effect: thermal beaming 435

16 Simultaneous transport of energy by radiation and thermal conduction 440
16.1 Introduction 440
16.2 Equations 440
16.3 Some time-independent applications of the equations 449
16.4 Time-dependent radiative and conductive models 460

Appendix A A brief review of vector calculus 463
Appendix B Functions of a complex variable 467
Appendix C The wave equation in spherical coordinates 470
Appendix D Fraunhofer diffraction by a circular hole 478
Appendix E Table of symbols 482

Bibliography 488
Index 509

Acknowledgments

From the first edition

Many persons have contributed to this book. Foremost is my wife, Joyce, to whom this book is dedicated. Without her continuing support, to say nothing of pleas, cajoleries, and sometimes even threats, this book would not have been written. My children, Kevin, Jeff, and Cheryl, all managed to launch themselves successfully while this work was under way. In spite of several anxious moments, they have been a joy and an inspiration.

Next are my former students. Their suggestions, criticisms, measurements, and computations made important contributions. Bob Nelson built the goniometric photopolarimeter that took much of the data used in this book. He will be happy to know the instrument is still functional. I am especially grateful to Eddie Wells for his careful measurements and suggestions, and also to Jeff Wagner, Deborah Domingue, and Audrey McGuire.

Over the years I have benefited from conversations with many other persons, particularly Jack Salisbury, Paul Helfenstein, Carle Pieters, Joe Veverka, Roger Clark, Marcia Nelson, Jim Pollack, Ted Bowell, and Kaari Lumme. I also wish to thank Sophia Prybylski of Cambridge University Press. Her careful attention to the manuscript caught many errors of both typography and grammar.

My father was fond of quoting Albert Einstein to the effect that a scientist never really understands his own theories unless he can satisfactorily explain them to an average person. I never was able to verify whether or not Einstein actually said this, but it seemed like a good principle to follow. My colleague and friend Bill Cassidy is an explorer, finder of meteorites, and raconteur extraordinaire, but he has never claimed to be a mathematician, and thus he was an ideal man-on-the-street for my ideas. Many times he watched while I covered a blackboard with equations, and then asked me to explain in English what I had just written. I hope his patience has had a positive effect on the clarity of this book.

The major impetus behind the development of the reflectance models described here has been the desire to provide a tool that will enable planetary scientists to better understand the surfaces of the various bodies we study. I am grateful to the Planetary Geology and Geophysics Program, Solar System Exploration Division, Office of Space Science and Applications of the National Aeronautics and Space Administration (NASA) for their continuing support. I especially wish to thank former NASA program manager Steve Dwornik, who continued to approve my grant proposals even though at times he was not quite sure what I was attempting to do. I also thank the National Research Council of the National Academy of Sciences for a senior research fellowship at NASA's Ames Research Center that supported me while I was working out some of these ideas.

I would be remiss if I did not especially acknowledge Thomas Gold, who may be said to have started it all. I had just finished my graduate studies at Cornell University when President Kennedy announced that we were going to the Moon. As a result, planetary science was suddenly revitalized. I thought that this would be a much more exciting field in which to do research than neutron physics, which was the subject of my doctoral dissertation, and Tommy agreed to accept me for postdoctoral research.

No one knew what the surface of the Moon was like, but Tommy thought that it was covered with a very fine-grained soil, which he referred to by the generic term "dust." At the time, that idea was at odds with the prevailing wisdom, which was divided between those who expected to find volcanic extrusive rock similar to Hawaiian aa and those who expected cobbly gravel, thought to have been generated by meteorite impacts. Tommy returned from a conference at which astronomers from the then Soviet Union had emphasized the strongly backscattering character of the lunar bidirectional refectance function. He was sure that "dust" could have this property and suggested that I build a goniometric photometer to investigate the diffuse reflectances of particulate media.

He assigned a young graduate student, Hugh Van Horn, as my research assistant. Hugh has since gone on to study brighter and denser objects than the Moon. We built the photometer and proceeded to measure the bidirectional reflectance functions of everything we could lay our hands on, including pulverized rocks, but the only material that was as backscattering as the Moon was reindeer moss – hardly a likely candidate. Somewhat in desperation, we began referring to the mysterious shapes that would produce a lunar type of photometric function as "fairy castle structures."

One day we discovered that very fine SiC abrasive powder was strongly backscattering, but that coarse SiC powder was not. There was no obvious reason for that difference in scattering properties, so I went off in search of microscope to see if a magnified inspection of the surface would give me a clue. It was late on Friday afternoon, and most of my colleagues had gone home, so that the only instrument

I could borrow was a low-power, stereoscopic microscope. This turned out to be serendipitous. First, I looked through the microscope at the coarse-grained powder. It resembled a pile of gravel and was not very interesting. Then I placed the fine-grained powder on the stage, and there were the fairy castles.

The name we had given the structures turned out not to be facetious at all, but is in fact a rather accurate description of the morphology of a powder in which the surface forces that act between grains exceed the gravitational forces exerted on them by the Earth. As I looked through the microscope, I saw a miniature world of deep, mysterious valleys and soaring towers leaning at crazy angles atop rugged, porous hills, with flying buttresses, and all connected by lacy bridges. Readers can easily verify these features for themselves. The complexity of such a texture is impossible to perceive with a monocular microscope, but it is just what is necessary to produce a lunar-type reflectance function. This discovery, along with Lyot's polarization data, turned out to be among the strongest pre-Apollo evidences that the lunar surface consists of a fine-grained regolith.

After we had solved the problem experimentally, I thought I would see if I could describe it mathematically, and I have been thinking about reflectances, off and on, ever since.

Finally, I wish to thank Ted Bowell for proposing that an asteroid that he discovered be named 3549 Hapke after me. I hope that someday my granddaughter Carley will land on it.

Preface to the second edition

In the years since the first edition of this book appeared reflectance and emitance spectroscopy have evolved into mature techniques for the remote study of surfaces of bodies of the solar system. The first edition had been generously received by my colleagues, so Cambridge University Press suggested that I write a second edition. Interactions with many colleagues, including those cited above and also Paul Lucey, Jack Mustard, and Mark Robinson, continue to help my understanding of reflectance. I am particularly indebted to my former students Jennifer Piatek and Amy Snyder Hale. The many discussions with them and their research while at the University of Pittsburgh made important contributions to this book. Jen's wizardy with MAC computers and her general programming skills were astonishing to someone who finished his formal schooling with only a slide rule. Bob Nelson has grown from a former graduate student into a respected planetary scientist and personal friend. We continue to collaborate on measurements of the reflectance and polarization of particulate media, and these measurements and his insights have been invaluable.

Finally, I thank Larry Taylor for naming the mineral hapkeite (Fe_2Si) after me.

1
Introduction

Then we shall rise and view ourselves with clearer eyes.

Henry King, bishop of Chichester (1592–1669)

1.1 Scientific rationale

All models are wrong, but some are useful.

George E. P. Box

The subject of this book is remote sensing, that is, seeing "with clearer eyes." In particular, it is concerned with how light is emitted and scattered by media composed of discrete particles and what can be learned about such a medium from its scattering properties.

If you stop reading now and look around, you will notice that most of the surfaces you see consist of particulate materials. Sometimes the particles are loose, as in soils or clouds. Sometimes they are embedded in a transparent matrix, as in paint, which consists of white particles in a colored binder. Or they may be fused together, as in rocks, or tiles which consist of sintered ceramic powder. Even vegetation is a kind of particulate medium in which the "particles" are leaves and stems. These examples show that if we wish to quantitatively interpret the electromagnetic radiation that reaches us, rather than simply form an image from it, it is necessary to consider the scattering and propagation of light within nonuniform media.

One of the first persons to use remote sensing to learn about the surface of a planet was Galileo Galilei. Galileo (1638) noticed that the full Moon, as it rose over his garden wall opposite the setting Sun, was darker than the sunlit wall. He also noted that the bidirectional reflectance function (of course, he did not use that term) of the lunar surface was diffuse in nature, rather than specular. From those observations he argued that the Moon was not a smooth, perfectly reflecting, crystalline sphere, as the prevailing wisdom of the day supposed, but was a planet not unlike the Earth.

In the intervening years since Galileo, remote sensing of planetary surfaces has become a quantitative science that has been used by two major groups of scientists. One is the community of planetary scientists. Virtually everything we know about the surfaces of the other bodies of the solar system comes to us through remote sensing. Most spacecraft missions are flybys or orbiters, and even landers, manned or unmanned, can sample only tiny portions of the surfaces of the bodies on which they set down. The second group consists of those scientists who are concerned with processes on the surface of the Earth, including geologists, meteorologists, geographers, and agronomists, who use remote sensing to study the Earth from balloons, aircraft, and satellites. However, the book is sufficiently general that it should be useful to any scientist or engineer who uses reflectance or emittance as an analytical tool.

In addition, reflectance spectroscopy turns out to be a powerful method for quantitatively measuring the characteristic absorption spectrum of a material. It has several advantages over more conventional methods. The dynamic range of the technique is large, typically four or more orders of magnitude in the absorption coefficient. The method is effective when the imaginary component of the refractive index is in the range of 10^{-3} to 10^{-1}, where both transmission and specular reflection techniques are difficult. Finally, sample preparation is convenient and simply requires grinding and sieving the material. Thus, this book should also be of interest to anyone who measures absorption coefficients.

The purpose of this book is to present quantitative models that describe the diffuse scattering and thermal emission of electromagnetic radiation from particulate media, such as planetary regoliths or powders in the laboratory. There are two general approaches to this problem. The first is to start from Maxwell's electromagnetic equations and attempt to find exact solutions. However, even with the use of modern high-speed compters this turns out to be impossible except for systems that are so simple as to have little resemblance to actual media, and the amount of computer time required is so long as to be impractical for most applications. Such solutions are useful primarily for illuminating some of the physical processes involved in light scattering by particulate media. However, at the present stage of our computational abilities they cannot justify the rather astonishing claim made by some persons that these processes are understood perfectly.

The second method, and the one used in this book, is based primarily on the equation of radiative transfer. This infamous equation is so notoriously intractable that most students have tried to stay as far away from it as possible. However, it will be seen that, by making appropriate approximations, solutions that are mathematically simple but surprisingly accurate can be obtained. (For example, see Figure 8.10.) Essentially, the radiative transfer equation assumes that photons of light can be treated as a gas diffusing between the particles of the medium. Even though the mathematical basis of this assumption has not been fully established, and some of the solutions presented in this book are not exact, this approach can be justified on several grounds:

(1) In most remote-sensing measurements the accuracy with which *absolute* reflectances can be measured is usually relatively low, typical uncertainties being of the order of $\pm 10\%$. Even in the laboratory, considerable effort must be expended to improve this figure substantially.
(2) Most applications deal with large, irregular particles whose scattering properties are poorly known, so that a detailed numerical calculation may give a misleading impression of high accuracy.
(3) Often in applications to planetary science a first-order estimate is sufficient to explain an observation or evaluate a hypothesis, and greater precision is unwarranted by the data.
(4) To evaluate the effect of varying a parameter, a numerical solution must be repeated many times, whereas if an analytic expression is available the effect can often be ascertained by inspection.

Thus, in most cases, exact numerical solutions are no more useful and are much less convenient than approximate analytic ones. The philosophy of this book is to be as rigorous as possible but, where necessary, to make approximations that retain the essential physics and at the same time are sufficiently simple that solutions to the problems of interest can be given by closed analytic functions.

1.2 About this book

The book is aimed at advanced undergraduate and beginning graduate students in the physical sciences. It is assumed that the reader has had at least an introductory course in physics, plus calculus through differential equations, and has a slight familiarity with vectors and complex variables. For those whose knowledge of mathematics is limited or rusty, brief reviews of those aspects of vector calculus, complex variables, and standard solutions of the wave equation that will be needed in this book have been included in Appendixes A–C.

I have tried to use notations that are in wide use, but this is not always possible in a treatise that covers a large range of topics. A list of symbols is included as Appendix E. Occasionally I have had to use the same symbol for two different quantities; in that case the meaning should be obvious from the context.

The book is divided into four parts. The first part, consisting of Chapters 2–4, introduces the general subjects of wave propagation and absorption in continuous media, polarization, and specular reflection from boundaries. The second part, Chapters 5 and 6, describes the scattering of light by single particles. The third part, Chapters 7–14, treats diffuse reflectance and polarization by particulate media and applications to reflectance spectroscopy. Thermal emittance is treated in Chapters 15 and 16. The discussion of thermal emittance spectroscopy can be brief because Kirchhoff's laws show that the processes of emittance and reflectance are complementary. Hence, the results of the earlier chapters on reflectance can be carried over directly to emittance.

Finally, an important warning must be issued to those who might want to use equations printed in this book to interpret reflectance or emittance data. I have tried to ensure that the important equations are correct, but I doubt that I have been completely successful. All practitioners of science know that Murphy's law is real, and in the context of this book the law can be phrased, "The equation you need the most contains a typographical error." I have tried to give sufficient detail that a reader can readily reproduce most of the derivations. Always rederive a critical equation!

2
Electromagnetic wave propagation

2.1 Maxwell's equations

Because reflectance spectroscopy uses electromagnetic radiation to probe matter, this book begins with Maxwell's electromagnetic equations and the solutions to them that describe propagating plane waves. For those whose knowledge of vectors may be a bit rusty, a brief review of vector notation is provided in Appendix A.1.

In their general form, Maxwell's equations can be written as follows:

$$\text{div } \mathbf{D}_e = \rho_e, \tag{2.1}$$

$$\text{div } \mathbf{B}_m = 0, \tag{2.2}$$

$$\text{curl } \mathbf{E}_e = -\partial \mathbf{B}_m / \partial t, \tag{2.3}$$

$$\text{curl } \mathbf{H}_m = \mathbf{j}_e + \partial \mathbf{D}_e / \partial t. \tag{2.4}$$

In these equations, $\mathbf{E}_e$ is the electric field, $\mathbf{D}_e$ is the electric displacement, $\mathbf{B}_m$ is the magnetic-induction field, $\mathbf{H}_m$ is the magnetic intensity, ρ_e is the electric-charge density, $\mathbf{j}_e$ is the electric current density, t is the time, and "div" and "curl" are, respectively, the vector divergence and curl operators. The reader is referred to the many excellent textbooks on electromagnetic theory for detailed derivations and more rigorous discussions of these equations, including Stratton (1941), Panofsky and Phillips (1962), Marion (1965), Elliott (1966), Landau and Lifschitz (1975), and Jackson (1999).

Equation (2.1) states that electric charges can generate electric fields and that the field lines diverge from or converge toward the charges. Equation (2.2) states that there are no sources of magnetic fields that are analogous to electric charges; that is, magnetic monopoles do not exist. According to equation (2.3), electric fields can also be generated by magnetic fields that change with time, and electric fields generated in this manner tend to coil or curl around the magnetic-field lines. Similarly, according to equation (2.4), magnetic fields can be generated by both electric currents and time-varying electric fields, and the magnetic lines of force tend to curl around these sources.

Equations (2.1) – (2.4) are called the *field equations*. In order to solve them, additional relations, known as *constitutive equations*, that connect two or more of the quantities in the field equations are needed, along with appropriate boundary conditions. The constitutive relations describe the way that microscopic charges and currents generated in the medium by the applied field alter those fields. In the remainder of this chapter, solutions of the field equations that describe propagating plane waves will be discussed for several different types of constitutive equations.

2.2 Electromagnetic waves in free space

2.2.1 The wave equation

The simplest case of electromagnetic radiation is wave propagation in a vacuum. In free space there are no charges or currents, so the constitutive equations have the simple form

$$\rho_e = 0, \tag{2.5}$$

$$\mathbf{j}_e = 0, \tag{2.6}$$

$$\mathbf{D}_e = \varepsilon_{e0}\mathbf{E}_e, \tag{2.7}$$

$$\mathbf{B}_m = \mu_{m0}\mathbf{H}_m, \tag{2.8}$$

where ε_{e0} is the permittivity of free space ($\varepsilon_{e0} = 8.85 \times 10^{-12}$ coulomb^2/newton-meter^2), and μ_{m0} is the permeability of free space (($\mu_{m0} = 12.57 \times 10^{-7}$ amperes/m^2). Then the field equations become

$$\text{div } \mathbf{E}_e = 0, \tag{2.9}$$

$$\text{div } \mathbf{B}_m = 0, \tag{2.10}$$

$$\text{curl } \mathbf{E}_e = -\partial \mathbf{B}_m/\partial t, \tag{2.11}$$

$$\text{curl } \mathbf{B}_m = \mu_{m0}\varepsilon_{e0}\partial \mathbf{E}_e/\partial t. \tag{2.12}$$

These equations may be combined to yield relations that each contain only one field quantity, as follows. Take the curl of both sides of (2.11) and use vector identity (A.9) from Appendix A:

$$\text{curl}(\text{curl}\,\mathbf{E}_e) = \text{grad}(\text{div}\,\mathbf{E}_e) - (\text{div}\cdot\text{grad})\mathbf{E}_e = -\partial(\text{curl}\,\mathbf{B}_m)/\partial t. \tag{2.13}$$

Using (2.9) and (2.12) in (2.13) gives

$$\nabla^2\mathbf{E}_e - \mu_{m0}\varepsilon_{e0}\partial^2\mathbf{E}_e/\partial t^2 = 0, \tag{2.14}$$

where $\nabla^2 = \text{div} \cdot \text{grad}$ is the Laplacian operator (often called "del-squared"). Applying a similar analysis starting with the magnetic field, equation (2.12), yields

$$\nabla^2 \mathbf{B}_m - \mu_{m0}\varepsilon_{e0}\partial^2 \mathbf{B}_m/\partial t^2 = 0. \tag{2.15}$$

Equations (2.13) and (2.14) are of the general form

$$\nabla^2 f = (1/v^2)\partial^2 f/\partial t^2. \tag{2.16}$$

Equation (2.16) is known as the wave equation and describes a disturbance at position $\boldsymbol{r}$ of arbitrary shape propagating with velocity v. If the geometry of the situation has plane-parallel symmetry, the general solution of (2.16) is $f = f(K\varnothing)$, where K is any constant, $\varnothing = \mathbf{u}_p \cdot \mathbf{r} \pm vt$ is the phase, $\mathbf{u}_p$ is a unit vector parallel to the direction of propagation, and f is an arbitrary function. All points satisfying $\mathbf{u}_p \cdot \mathbf{r} = \text{constant}$ lie on a plane perpendicular to $\mathbf{u}_p$ and passing through the point $\mathbf{r}$ (Figure 2.1).

In the remainder of this chapter, only plane waves will be considered. Without loss of generality, the direction of propagation may be taken to lie along the z-axis. In this case, f is independent of x and y, the differential vector operator "del" is $\nabla = \mathbf{u}_z\partial/\partial z$, where $\mathbf{u}_z$ is a unit vector pointing in the z direction, and $\varnothing = z \pm vt$. Then (2.16) becomes

$$\partial^2 f/\partial z^2 - (1/v^2)\partial^2 f/\partial t^2 = 0. \tag{2.17}$$

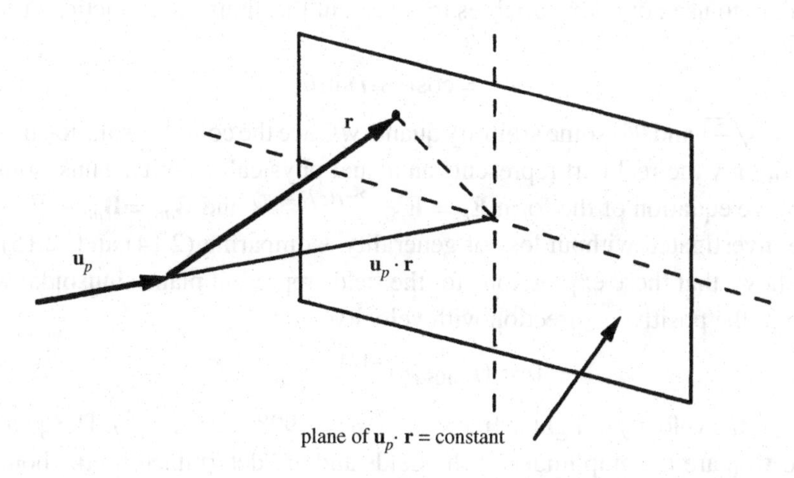

Figure 2.1

The proof that this equation is satisfied by any arbitrary function $f(K\varnothing)$ is almost trivial:

$$\frac{\partial^2 f}{\partial z^2} = \frac{d}{d\varnothing}\left(\frac{df}{d\varnothing}\frac{\partial\varnothing}{\partial z}\right)\frac{\partial\varnothing}{\partial z} = K^2\frac{d^2 f}{d\varnothing^2}, \tag{2.18}$$

$$\frac{1}{\upsilon^2}\frac{\partial^2 f}{\partial t^2} = \frac{1}{\upsilon^2}\frac{d}{d\varnothing}\left(\frac{df}{d\varnothing}\frac{\partial\varnothing}{\partial t}\right)\frac{\partial\varnothing}{\partial t} = K^2\frac{d^2 f}{d\varnothing^2}. \tag{2.19}$$

If $\varnothing = z - \upsilon t$, f represents a pattern whose shape is described by $f(K\varnothing)$ moving in the positive z direction with velocity υ; if $\varnothing = z + \upsilon t$, the pattern is moving toward negative z.

If the propagating pattern is periodic and repeats over a distance λ, then at any point z as the wave passes its pattern will repeat over a time interval $t = \lambda/\upsilon$. This interval is called the period. The reciprocal of the period is the frequency ν, and λ is the wavelength. It is convenient to let $K = 2\pi/\lambda$, so that $f(K\varnothing) = f(2\pi z/\lambda \pm 2\pi\nu t)$. Thus, υ and ν are related by

$$\upsilon = \nu\lambda. \tag{2.20}$$

A large class of problems can be solved by representing the waves by sinusoidally varying functions of the form $f = f_0 \sin 2\pi(z/\lambda \pm \nu t)$ or $f = f_0 \cos 2\pi(z/\lambda \pm \nu t)$. The reason sinusoidal solutions are so useful is that the principles of Fourier analysis and synthesis allow an arbitrary function to be described mathematically by sums of sinusoidal waves of appropriate amplitude and phase, provided the function is mathematically well behaved. In practice this qualification is not greatly restrictive, because virtually any function that describes a physically real quantity will be well behaved.

Because exponential functions are easy to manipulate mathematically, it is often convenient to use complex variables to represent the sinusoidal functions through the relation

$$e^{i\theta} = \cos\theta + i\sin\theta, \tag{2.21}$$

where $i = \sqrt{-1}$ and θ is some arbitrary quantity. Once the complex solution has been obtained, only the real part represents an actual physical quantity. Thus, solutions of the wave equation of the form $\mathbf{E}_e = \mathbf{E}_{e0}e^{2\pi i(z/\lambda - \nu t)}$ and $\mathbf{B}_m = \mathbf{B}_{m0}e^{2\pi i(z/\lambda - \nu t)}$ may be investigated without loss of generality. Comparing (2.14) and (2.15) with (2.16) shows that these expressions for the fields represent plane sinusoidal waves moving in the positive z direction with velocity

$$\upsilon = (\mu_{m0}\varepsilon_{e0})^{-1/2} = c_0, \tag{2.22}$$

where c_0 is the velocity of light in free space ($c_0 = 2.998 \times 10^8\,\mathrm{m\,s^{-1}}$). The quantities $\mathbf{E}_{e0}$ and $\mathbf{B}_{m0}$ are the amplitudes of the fields and are determined by the boundary conditions.

The electric and magnetic fields are not independent, but are related through (2.11), which for a plane sinusoidal wave propagating in the z direction is

$$\text{curl}\,\mathbf{E}_e = \mathbf{u}_z \times \partial \mathbf{E}_e/\partial z = (2\pi i/\lambda)\mathbf{u}_z \times \mathbf{E}_e = -\partial \mathbf{B}_m/\partial t = 2\pi i\nu \mathbf{B}_m. \quad (2.23)$$

Thus,

$$\mathbf{B}_m = \mathbf{u}_z \times \mathbf{E}_e/c_0. \quad (2.24)$$

Similarly (2.12) is

$$\mathbf{u}_z \times \partial \mathbf{B}_m/\partial z = i\,(2\pi/\lambda)\,\mathbf{u}_z \times \mathbf{B}_m = \mu_{m0}\varepsilon_{e0}\partial \mathbf{E}_e/\partial t = -2\pi i\nu\mu_{m0}\varepsilon_{e0}\mathbf{E}_e, \quad (2.25)$$

so that

$$\mathbf{E}_e = -(1/\lambda\nu\mu_{m0}\varepsilon_{e0})\mathbf{u}_z \times \mathbf{B}_m = -c_0\mathbf{u}_z \times \mathbf{B}_m. \quad (2.26)$$

Equations (2.24) and (2.26) show that $\mathbf{E}_e$ and $\mathbf{B}_m$ are perpendicular to each other and to the direction of propagation (Figure 2.2), and also that the amplitudes of the fields are related by

$$B_{m0} = E_{e0}/c_0. \quad (2.27)$$

Two independent orthogonal solutions to the wave equation are possible in which the electric vectors are perpendicular to each other. If the positive x direction is chosen to be parallel to $\mathbf{E}_e$, then the component of $\mathbf{B}_m$ corresponding to this solution points in the positive y direction. Figure 2.2 illustrates this solution. If the positive y direction is chosen to be parallel to the electric vector, then the accompanying magnetic vector points in the negative x direction. In free space, neither $\mathbf{E}_e$ nor $\mathbf{B}_m$ has a component parallel to z.

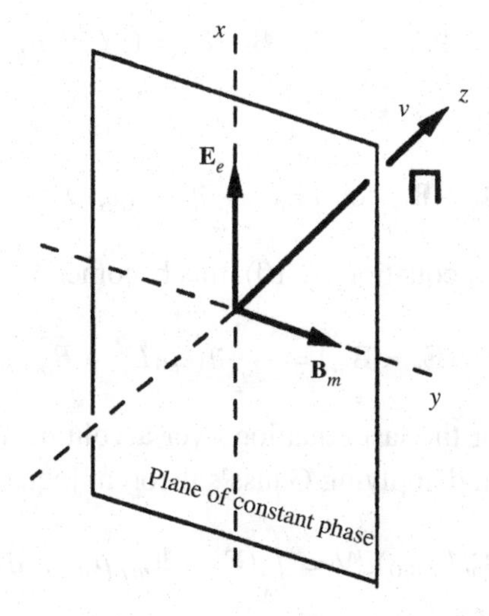

Figure 2.2 Relation between the fields and the propagation velocity vector in a plane electromagnetic wave.

2.2.2 Huygens's principle

Three hundred years ago when it was realized that light had a wave nature, the question arose as to what medium the waves were propagating through. It was postulated that all space was filled with an invisible fluid called the "aether." However, in the modern view of an electromagnetic wave no aether is required. According to Maxwell's equations, a changing electric field generates a magnetic field, and a changing magnetic field generates an electric field. Thus, a propagating electromagnetic wave may be regarded as generating itself; that is, the changing fields on the wave front continually regenerate the wave. This concept leads to *Huygens's principle*, in which each point on a wave front may be considered to be a source of spherical wavelets that travel radially outward and combine coherently with wavelets from all the other points to produce a new wave front. If the wave front is plane and infinite in lateral extent, this process simply produces another plane wave front. However, if part of the wave front is obstructed, Huygens's principle predicts that fields still exist behind the obstructing object. This phenomenon is called *diffraction*, and Huygen's principle can be used to calculate the resultant fields in the vicinity of the object.

2.2.3 The Poynting vector and the irradiance

An important quantity, the power contained in the wave, can be obtained by forming the vector dot products of **E** with equation (2.12) and of **B** with (2.13),

$$\mathbf{E}_e \cdot \operatorname{curl} \mathbf{B}_m = \mu_{m0}\varepsilon_e \mathbf{E}_e \cdot \partial \mathbf{E}_e/\partial t = (\mu_{m0}\varepsilon_{e0}/2)\partial \mathbf{E}_e^2/\partial t, \tag{2.28}$$

$$\mathbf{B}_m \cdot \operatorname{curl} \mathbf{E}_e = -\mathbf{B} \cdot \partial \mathbf{B}_m/\partial t = (1/2)\partial B_m^2/\partial t; \tag{2.29}$$

subtracting gives

$$\mathbf{E}_e \cdot \operatorname{curl} \mathbf{B}_m - \mathbf{B}_m \cdot \operatorname{curl} \mathbf{E}_e = \frac{1}{2}\partial(\mu_{m0}\varepsilon_{e0}E_e^2 + B_m^2)/\partial t. \tag{2.30}$$

Using the vector identity, equation (A.10), this becomes

$$(1/\mu_{m0})\operatorname{div}(\mathbf{E}_e \times \mathbf{B}_m) = -\frac{1}{2}\partial(\varepsilon_{e0}E_e^2 + B_m^2/\mu_{m0})/\partial t. \tag{2.31}$$

Integrating both sides of the last equation over a volume V bounded by a closed surface A (Figure 2.3), and applying Gauss's theorem [equation (A.11)], gives

$$\begin{aligned}\int_V \operatorname{div}(\mathbf{E}_e \times \mathbf{B}_m/\mu_{m0})dV &= \int_A (\mathbf{E}_e \times \mathbf{B}_m/\mu_{m0}) \cdot dA \\ &= -\frac{\partial}{\partial t}\int_V (\varepsilon_{e0}E_e^3/2 + B_m^2/2\mu_{m0})dV.\end{aligned} \tag{2.32}$$

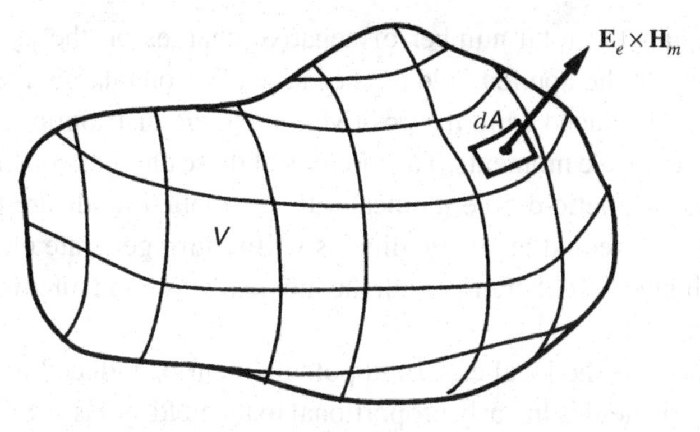

Figure 2.3

Now, $\varepsilon_{e0}E_e^2/2$ and $B_m^2/2\mu_{m0}$ are, respectively, the energy densities in the electric and magnetic fields, so that the right-hand side of (2.32) describes the rate of decrease of energy density in V. Thus, the left-hand side must represent the outward movement of energy through the surface A bounding V. The quantity $\mathbf{E}_e \times \mathbf{B}_m/\mu_{m0}$ is the local flux of energy per unit area per unit time crossing surface element and is known as the *Poynting vector* Π,

$$\Pi = \mathbf{E}_e \times \mathbf{B}_m/\mu_{m0} = \mathbf{E}_e \times \mathbf{H}_m = \mathbf{u}_z \mathbf{E}_e^2/\mu_{m0}c_0. \tag{2.33}$$

The Poynting vector oscillates with time. The *irradiance* J of the wave is the average energy crossing unit area perpendicular to the direction of propagation per unit time. This quantity is often called the intensity, although the latter term is usually reserved for the power flux per unit solid angle. The irradiance is the time average (denoted by angular brackets) over one cycle of the product of the real parts of the fields in the Poynting vector,

$$\begin{aligned} \mathbf{u}_z J &= \langle \mathrm{Re}(\mathbf{E}_e) \times \mathrm{Re}(\mathbf{B}_m)/\mu_{m0} \rangle \\ &= \mathbf{u}_z (\nu/\mu_{m0}) \int_0^{1/\nu} E_{e0} B_{m0} \cos^2 2\pi(z\lambda - \nu t) dt. \end{aligned} \tag{2.34}$$

Using the relation $\cos^2 x = (1 + \cos 2x)/2$, this integral is readily evaluated to give

$$J = E_{e0} B_{m0}/2\mu_{m0} = E_{e0}^2/2\mu_{m0}c_0 = \sqrt{\varepsilon_{e0}/\mu_{m0}} E_{e0}^2/2. \tag{2.35}$$

2.3 Propagation in a linear nonabsorbing medium

2.3.1 Induced electric dipoles

When fields are present in a medium, they act on the electrons and ions that make up the medium to induce electric and magnetic-dipole moments. The net electric charge ρ_e usually is zero because on the average the total number of positive charges on the ions equals the total number of negative charges on the electrons. However, in response to the applied field the negative electron charge distribution may become slightly displaced from the positive ion charge distribution, thus generating local electric-dipole moments. The motions of these charges constitute currents, which generate magnetic-dipole moments. If the applied fields are time-varying, these induced electric and magnetic dipoles will in turn generate electromagnetic waves that will interact coherently with the incident wave and alter its propagation characteristics.

In many materials the local electric-dipole moment $\mathbf{p}_e$ induced in a medium by an applied electric field is linearly proportional to the field and is parallel to it. Thus, there is often a relation of the form

$$\mathbf{p}_e = \alpha_e \varepsilon_{e0} \mathbf{E}_{\text{loc}}, \tag{2.36}$$

where $\mathbf{E}_{\text{loc}}$ is the local electric field, and α_e is a constant of proportionality called the electric *polarizability*. The local field may be different from the general macroscopic field $\mathbf{E}_e$ in the medium.

The *electric polarization* $\mathbf{P}_e$ is the dipole moment per unit volume,

$$\mathbf{P}_e = N \mathbf{p}_e, \tag{2.37}$$

where N is the number of dipoles per unit volume. Often $\mathbf{P}_e$ is linearly proportional and parallel to the applied field $\mathbf{E}_e$, so that it can be described by a relation of the form

$$\mathbf{P}_e = \chi_e \varepsilon_{e0} \mathbf{E}_e, \tag{2.38}$$

where χ_e is the electric susceptibility. The electric displacement $\mathbf{D}_e$ is defined as

$$\mathbf{D}_e = \varepsilon_{e0} \mathbf{E}_e + \mathbf{P}_e = \varepsilon_{e0}(1 + \chi_e) \mathbf{E}_e. \tag{2.39}$$

The quantity in parentheses is the *dielectric constant* or specific inductive capacity,

$$K_e = 1 + \chi_e. \tag{2.40}$$

Thus,

$$\mathbf{P}_e = \varepsilon_{e0}(K_e - 1) \mathbf{E}_e, \tag{2.41}$$

and

$$\mathbf{D}_e = \varepsilon_e \mathbf{E}_e, \tag{2.42}$$

where

$$\varepsilon_e = \varepsilon_{e0} K_e \tag{2.43}$$

is the permittivity of the medium.

In general, ε_e may vary from place to place, as in an inhomogeneous or discontinuous medium. It may depend on the direction of $\mathbf{E}_e$, as in birefringent crystals.

In some materials $\mathbf{D}_e$ is not parallel to $\mathbf{E}_e$, in which case ε_e is a tensor and (2.42) is a matrix equation. The permittivity may be a function of $\mathbf{E}_e$, as often occurs when a substance is illuminated by an intense laser beam. However, in this chapter it will be assumed that ε_e is a homogeneous, isotropic, scalar constant. The problem of propagation in a discontinuous medium will be considered later in the book.

2.3.2 The Clausius–Mossotti / Lorentz–Lorenz relation

A question of considerable interest is the relation between the local field $\mathbf{E}_{\text{loc}}$ at the molecule, which defines α_e and $\mathbf{p}_e$, and the applied field $\mathbf{E}_e$, which determines K_e, ε_e, and $\mathbf{P}_e$. This is not a trivial question, and it has never been answered satisfactorily in general. It is discussed extensively in many advanced textbooks on electrostatics and electrodynamics.

Often it is simply assumed that $\mathbf{E}_{\text{loc}} = \mathbf{E}_e$. In that case, $\mathbf{P}_e = N\alpha_e\varepsilon_{e0}\mathbf{E}_{\text{loc}}$, $\chi_e = N\alpha_e$, and

$$K_e = 1 + N\alpha_e. \tag{2.44}$$

Another approach, which has considerable experimental support, is the following. Assume that at an arbitrary point the medium can be divided into two regions separated by a spherical surface of radius R and area $a = 4\pi R^2$ centered on some molecule having an induced dipole moment $\mathbf{p}_e$. The geometry of the situation is shown in Figure 2.4. The distance R represents the range of influence of the induced dipole. Outside the sphere the electrical forces may be represented by the macroscopic field and polarization, whereas inside, the detailed fields due to the other dipoles must be taken into account. Assume that the sphere is small compared with the wavelength in the medium, so that the macroscopic field is uniform across the sphere.

The field at the center of the sphere is the vector sum of three fields:

(1) The applied field $\mathbf{E}_e$.
(2) The field $\mathbf{E}_{ep}$ due to charges on the surface a induced by $\mathbf{E}_e$. The induced charge dq_e in an incremental surface area da is $dq_e = P_e \cos\vartheta' da$, where $da = 2\pi R^2 \sin\vartheta' d\vartheta'$, and ϑ' is the angle between $\mathbf{P}_e$ (parallel to $\mathbf{E}_e$) and the radius vector to da. The component of the field at the center parallel to $\mathbf{E}_e$, due to the charge in da is $dq_e \cos\vartheta'/4\pi\varepsilon_{e0}R^2 = P_e \cos^2\vartheta' da/4\pi\varepsilon_{e0}R^2$. By symmetry, the components perpendicular to $\mathbf{E}_e$ vanish. Integrating this field over the surface of the sphere gives the contribution due to the induced surface polarization charges:

$$E_{ep} = \int_0^{\pi} \frac{P_e \cos^2\vartheta' 2\pi R^2 \sin\vartheta' d\vartheta'}{4\pi\varepsilon_{e0}R^2} = \frac{P_e}{3\varepsilon_{e0}}. \tag{2.45}$$

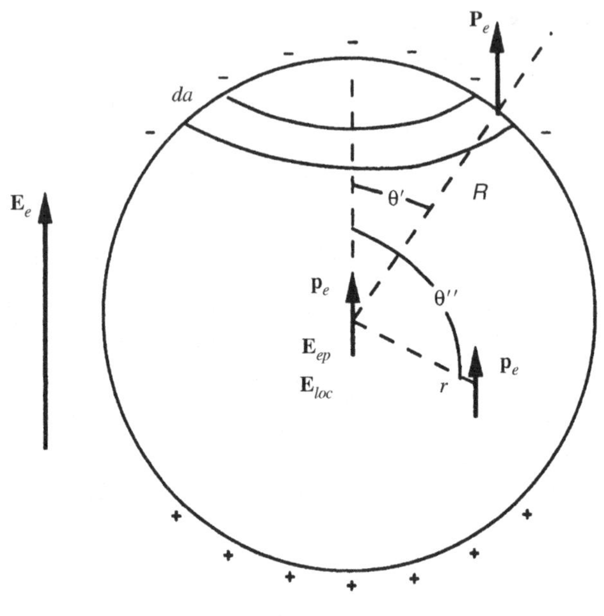

Figure 2.4

(3) The field due to the sum of the other individual dipoles in the sphere. Assume that these are statistically distributed isotropically about the center of the cavity. Then the average component at the center due to the other dipoles may be found as follows. The potential U of a dipole of moment p_e aligned with the x-axis, at a point at distance r making an angle ϑ'' with the x-axis, is given by

$$U = p_e \cos\vartheta''/4\pi\varepsilon_{e0}r^2 = p_e x/4\pi\varepsilon_{e0}(x^2+y^2+z^2)^{3/2}.$$

The x component of the field is

$$\begin{aligned} E_{ex} &= -\partial U/\partial x \\ &= -[p_e/4\pi\varepsilon_{e0}]\left[(x^2+y^2+z^2)^{-3/2} - 3x^2(x^2+y^2+z^2)^{-5/2}\right] \\ &= -p_e(1-3\cos^2\vartheta'')/4\pi\varepsilon_{e0}r^3. \end{aligned}$$

The total value of the x component of the field at the center of a sphere of radius $r < R$ due to an ensemble of such dipoles uniformly distributed over the surface of that sphere may be found by integrating the foregoing expression for $\mathbf{E}_{ex}$ over the surface, giving

$$\langle E_{ex}\rangle = -\int_0^{\pi} (p_e/4\pi\varepsilon_{e0}r^3)(1-3\cos^2\vartheta'')2\pi r^2 \sin\vartheta'' d\vartheta'' = 0. \qquad (2.46)$$

Hence, the field at the center due to the other dipoles in the sphere vanishes.

The total field at the center is the sum of the three contributions,

$$\mathbf{E}_{\rm loc} = \mathbf{E}_e + \mathbf{P}_e/3\varepsilon_{e0}. \tag{2.47}$$

Substituting from (2.36) and (2.37),

$$\mathbf{P}_e = \mathrm{N}\alpha_e\varepsilon_{e0}(\mathbf{E}_e + \mathbf{P}_e/3\varepsilon_{e0}).$$

Solving for $\mathbf{P}_e$,

$$\mathbf{P}_e = N\alpha_{e0}\mathbf{E}_e/(1 - N\alpha_e/3) = \varepsilon_{e0}(K_e - 1)\mathbf{E}_e,$$

from (2.41). Solving for K_e,

$$K_e = (1 + 2N\alpha_e/3)/(1 - N\alpha_e/3). \tag{2.48}$$

If $N\alpha_e \ll 1$, then to first order $K_e \simeq 1 + N\alpha_e$, so that $\mathbf{E}_{\rm loc} = \mathbf{E}_e$. Equation (2.48) may be written in the alternative form

$$(K_e - 1)/(K_e + 2) = N\alpha_e/3 = N_0\rho\alpha_e/3W, \tag{2.49}$$

where N_0 is Avogadro's number ($N_0 = 6.023 \times 10^{23}$ molecules/mole), ρ is the mass density, and W is the molecular weight.

Equation (2.49) is known variously as the Clausius–Mossotti or Lorentz–Lorenz relation. Apparently it was first derived semi-empirically by R. Clausius and O. Mossotti for static fields, and later more rigorously by H. Lorentz and L. Lorenz. A similar relation was derived by Rayleigh. It describes the dependence of the dielectric constant of a gas on density reasonably well. It also gives fair agreement for many solids, as illustrated by Figure 2.5. Obviously, however, this relation is not completely general, because it predicts that K_e becomes infinite when $N\alpha_e = 3$. Further discussion of the Clausius–Mossotti relation may be found in many places, including Elliott (1966), Frohlich (1958), and Bottcher (1952).

2.3.3 Induced magnetic dipoles

In a manner analogous to the treatment of the effects of induced electric dipoles, the magnetization $\mathbf{M}_m$ is defined as the induced magnetic-dipole moment per unit volume. It will be assumed that $\mathbf{M}_m$ is linearly proportional to $\mathbf{B}_m$. The magnetic intensity $\mathbf{H}_m$ is also assumed to be proportional to $\mathbf{B}_m$ and thus to $\mathbf{M}_m$. The relationships among these quantities are

$$\mathbf{M}_m = \mu_{m0}\chi_m\mathbf{M}_m, \tag{2.50}$$

where χ_m is the magnetic susceptibility, and

$$\mathbf{B}_m = \mu_{m0}\mathbf{H}_m + \mathbf{M}_m. \tag{2.51}$$

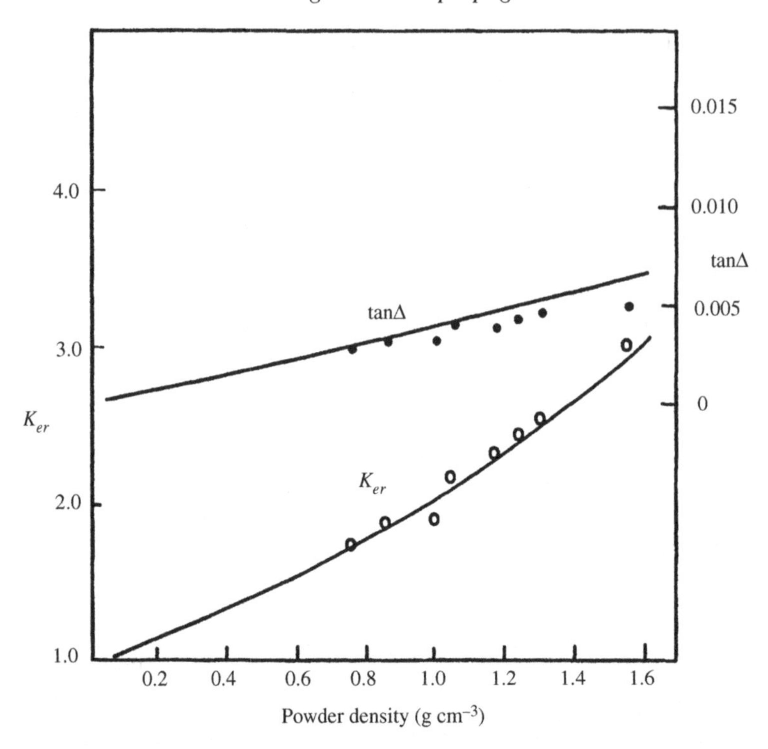

Figure 2.5 Permittivity and loss tangent as a function of density of a basalt powder finer than 37 μm at $\nu = 450$ MHz. The dots show the measured values; the lines show the Clausius–Mossotti/Lorentz–Lorenz relation normalized to the measured values of the solid material. (Reproduced from Campbell and Ulrichs [1969], copyright 1969 by the American Geophysical Union.)

Thus,

$$\mathbf{B}_m = \mu_{m0}(1 + \chi_m)\mathbf{H}_m. \tag{2.52}$$

The quantity in parentheses is the relative permeability K_m,

$$K_m = 1 + \chi_m. \tag{2.53}$$

Then

$$\mathbf{B}_m = \mu_m \mathbf{H}_m, \tag{2.54}$$

where

$$\mu_m = \mu_{m0} K_m \tag{2.55}$$

is the *magnetic permeability.*

2.3.4 The wave equation

Equations (2.42) and (2.54), in which ε_e and μ_m are assumed to be constants for a particular material, along with $\rho_e = 0$ and $j_e = 0$, are the constitutive equations for a linear nonabsorbing medium. In such a material, Maxwell's equations, (2.1) – (2.4), reduce to the same form as (2.9) – (2.12) for free space, except that ε_{e0} and μ_{m0} are replaced by ε_e and μ_m, respectively. Reasoning identical with that in Section 2.2 then leads to equations of the form

$$\nabla^2 \mathbf{E}_e - \mu_m \varepsilon_e \partial^2 \mathbf{E}_e / \partial t^2 \tag{2.56}$$

and

$$\nabla^2 \mathbf{B}_m - \mu_m \varepsilon_e \partial^2 \mathbf{B}_m / \partial t^2. \tag{2.57}$$

These describe waves propagating with velocity

$$\upsilon = 1/\sqrt{\mu_m \varepsilon_e} = c_0/\sqrt{K_e K_m} = c_0/n \tag{2.58}$$

where n is the refractive index

$$n = \sqrt{K_e K_m}. \tag{2.59}$$

For sinusoidal waves, the frequency is determined by the source of the radiation and hence is the same in the medium and in free space. However, because $\upsilon = \lambda\nu = c/n$, the wavelength is decreased by a factor $1/n$ from its free-space value. Thus, if λ is the wavelength in free space, the field solutions are of the form

$$\mathbf{E}_e = \mathbf{E}_{e0} \exp\left[2\pi i \left(\frac{nz}{\lambda} - \nu t\right)\right] \tag{2.60}$$

and

$$\mathbf{B}_m = \mathbf{B}_{m0} \exp\left[2\pi i \left(\frac{nz}{\lambda} - \nu t\right)\right] \tag{2.61}$$

The fields $\mathbf{E}_e$ and $\mathbf{B}_m$ are related by

$$\mathbf{B}_m = \mathbf{u}_Z \times \mathbf{E}_e/\upsilon \text{ and } \mathbf{E}_e = -\upsilon \mathbf{u}_Z \times \mathbf{B}_m, \tag{2.62}$$

so that

$$B_{m0} = E_{e0}/\upsilon = n E_{e0}/c_0. \tag{2.63}$$

Then the irradiance is

$$J = \sqrt{\varepsilon_e/\mu_m} E_{e0}^2/2. \tag{2.64}$$

2.4 Propagation in a linear absorbing medium

2.4.1 The wave equation with finite conductivity

Absorption of radiation may be introduced into the electromagnetic equations in a number of ways. The most intuitively direct way is to assume that the medium has a finite electrical conductivity, so that the currents induced by the incident radiation suffer losses that are described by a constitutive relation having the form of Ohm's law,

$$\mathbf{j}_e = \sigma_e \mathbf{E}_e, \tag{2.65}$$

where σ_e is the electrical conductivity.

Assume that there are no free charges, so that $\rho_e = 0$, and that the other constitutive relations are (2.42), (2.54), and (2.65). Then Maxwell's equations (2.1) – (2.4) become

$$\text{div } \mathbf{E}_e = 0, \tag{2.66}$$

$$\text{div } \mathbf{B}_m = 0, \tag{2.67}$$

$$\text{curl } \mathbf{E}_e = -\partial \mathbf{B}_m / \partial t, \tag{2.68}$$

$$\text{curl } \mathbf{B}_m = \mu_m (\sigma_e \mathbf{E}_e + \varepsilon_e \partial \mathbf{E}_e \partial t). \tag{2.69}$$

As previously, taking the curl of (2.68), using (2.66), and substituting (2.69) gives

$$\nabla^2 \mathbf{E}_e - \mu_m \sigma_e \partial \mathbf{E}_e / \partial t - \mu_m \varepsilon_e \partial^2 \mathbf{E}_e / \partial t^2 = 0. \tag{2.70}$$

The nature of the solutions to (2.70) depends on whether the second or third term on the left-hand side dominates. The second term is of the order of $\mu_m \sigma_e \Delta E_e / \Delta t$, where ΔE_e is the change in $\mathbf{E}_e$ occurring in time Δt, while the third term is of order $\mu_m \varepsilon_e \Delta E_e / \Delta t^2$.

For periodic solutions the controlling time interval is $\Delta t \simeq 1/\nu$. Thus, if $\mu_m \sigma_e \Delta E_e \nu \ll \mu_m \varepsilon_e \Delta E_e \nu^2$, or $\sigma_e / \nu \varepsilon_e \ll 1$, the second term of (2.70) is negligible compared to the third, and the solution is essentially the wave equation. On the other hand, if $\sigma_e / \nu \varepsilon_e \gg 1$, then the second term dominates, and (2.70) is essentially the diffusion equation. Equations of the latter type describe such phenomena as the diffusion of one gas into another, the density of neutrons in a fission reactor, and heat flow. If σ_e is large, then when an external field is applied to the medium, induced charges and currents generate internal fields that are of opposite polarity to the incident field and tend to cancel it. The motion of the field appears to be retarded, and the field acts as if it were diffusing slowly into the medium rather than propagating as a wave.

2.4.2 Representations of absorption of radiation

In many types of absorptions there are no macroscopic currents, so that conductivity is an unsatisfactory way to describe the losses. Hence, absorption of electromagnetic radiation is described in the literature by a variety of parameters besides conductivity; these include imaginary component of the dielectric constant, imaginary component of the refractive index, loss tangent, optical conductivity, and the absorption coefficient. All of these quantities are equivalent, and the choice of which particular one to use is largely a matter of taste and convention. Conductivity is usually reserved for use with metals in the frequency range where $\nu \ll \sigma_e/\varepsilon_e$, complex dielectric constant and loss tangent in radio-frequency applications, and complex refractive index, optical conductivity or absorption coefficient in the ultraviolet (UV), visible, and infrared (IR) wavelength ranges.

Consider a periodically varying field of the form

$$\mathbf{E}_e(z,t) = E_e(z)e^{-2\pi i\nu t}, \tag{2.71}$$

where E_e is the amplitude of $\mathbf{E}_e$. Substituting (2.71) into (2.70) gives

$$(d^2E_e/dz^2)e^{-2\pi i\nu t} + 2\pi i\nu t\mu_m\sigma_e E_e e^{-2\pi i\nu t} + 4\pi^2\nu^2\mu_m\varepsilon_e E_e e^{-2\pi i\nu t}. \tag{2.72}$$

It is of interest to ask if a solution of the form of (2.71) may be constructed that will satisfy both equations (2.56) and (2.70). The answer is affirmative provided that the dielectric constant is allowed to be complex. That is, K_e, K_m, and σ_e are not used explicitly; instead, their effects are simply lumped together into an effective *complex dielectric constant* K_e, and the wave equation is forced to be of the form of (2.56),

$$\nabla^2 E_e - K_e\mu_{m0}\varepsilon_{e0}\partial^2 E_e/\partial t^2 = 0, \tag{2.73}$$

where

$$K_e = K_{er} + iK_{ei} = K_{er}(1 + i\tan\Delta); \tag{2.74}$$

K_{er} and K_{ei} are, respectively, the real and imaginary parts of K_e, and

$$\tan\Delta = K_{ei}/K_{er} \tag{2.75}$$

is the *loss tangent.* (In many references, K_e is defined as $K_e = K_{er}–iK_{ei}$. The sign of the imaginary component makes no practical difference as long as the same convention is maintained throughout.)

Putting (2.71) into (2.72) gives

$$(d^2E_e/dz^2)e^{-2\pi i\nu t} + 4\pi^2\nu^2 K_e\mu_{m0}\varepsilon_{e0}E_e e^{-2\pi i\nu t}. \tag{2.76}$$

Comparing this last equation with (2.72) shows that in order for the two equations to be equivalent, the following must hold:

$$K_{er} = K_e K_m, \tag{2.77}$$

$$K_{ei} = K_m\sigma_e/2\pi\nu\varepsilon_{e0}. \tag{2.78}$$

Solving for σ_e gives

$$\sigma_e = 2\pi\nu\varepsilon_{e0}K_{ei}/K_m. \tag{2.79}$$

Equation (2.79) can be used to calculate an effective σ_e from measured values of K_{ei} and K_m. The conductivity defined in this way is called the *optical conductivity*. Except for the special case of ferromagnetic materials, μ_m is usually not significantly different from μ_0, and $K_m = 1$.

If the dielectric constant is complex, then the refractive index is also. From (2.45),

$$n = \sqrt{K_e} = n_r + in_i, \tag{2.80}$$

where n_r and n_i are the real and imaginary parts, respectively, of the *complex refractive index n*. (If K_e is defined to be $K_e = K_{er} - iK_{ei}$, then m must be defined as $m = n - ik$.)

Combining (2.74) and (2.80),

$$K_e = K_{er} + iK_{ei} = n^2 = (n_r + in_i)^2 = n_r^2 - n_i^2 + 2in_r n_i. \tag{2.81}$$

Thus,

$$K_{er} = n_r^2 - n_i^2, \tag{2.82}$$

and

$$K_{ei} = 2n_r n_i. \tag{2.83}$$

Solving for n_r and n_i,

$$n_r = \left\{\tfrac{1}{2}\left[\left(K_{er}^2 + K_{ei}^2\right)^{1/2} + K_{er}\right]\right\}^{1/2}, \tag{2.84}$$

and

$$n_i = \left\{\tfrac{1}{2}\left[\left(K_{er}^2 + K_{ei}^2\right)^{1/2} - K_{er}\right]\right\}^{1/2}. \tag{2.85}$$

If $\tan\Delta \ll 1$, these become

$$n_r \approx \sqrt{K_{er}}, \tag{2.86}$$

$$n_i \approx K_{ei}/2\sqrt{K_{er}}. \tag{2.87}$$

At this point the reader may well be wondering how a refractive index can be complex, because this also would seem to imply a complex wavelength and propagation velocity. One answer is that this is simply a convenient mathematical representation; the imaginary part can always be recast into a form that describes a real quantity. To see the physical meaning and implications of a complex dielectric constant or refractive index, write the solution to the wave equation explicitly in the form of (2.60), which describes a plane wave propagating in the positive z direction in a medium of complex refractive index n:

$$\mathbf{E}_e = \mathbf{E}_{e0} e^{2\pi i(nz/\lambda - \nu t)} = \mathbf{E}_{e0} e^{2\pi i[(n_r + in_i)z/\lambda - \nu t]} = \mathbf{E}_{e0} e^{-2\pi n_i z/\lambda} e^{2\pi i(n_r z/\lambda - \nu t)}. \tag{2.88}$$

Equation (2.88) represents a sinusoidal wave with frequency ν, wavelength λ/n_r and velocity $\upsilon = c_0/n_r$, where λ is the wavelength in free space, propagating through the medium with exponentially decreasing amplitude. The magnetic field has the same form,

$$\mathbf{B}_m = \mathbf{B}_{m0} e^{-2\pi n_i z/\lambda} e^{2\pi i (n_r z/\lambda - \nu t)}, \tag{2.89}$$

where, from (2.63),

$$\mathbf{B}_{m0} = (n_r + i n_i) E_{e0}/c_0. \tag{2.90}$$

Now, the actual electric field is

$$\mathrm{Re}(\mathbf{E}_e) = E_{e0} e^{-2\pi n_i z/\lambda} \cos[2\pi(n_r z/\lambda - \nu t)], \tag{2.91}$$

and the actual magnetic field is

$$\begin{aligned} \mathrm{Re}(\mathbf{B}_m) &= \mathrm{Re}\left[(n_r + i n_i)(E_{e0}/c_0) e^{-2\pi n_i z/\lambda} e^{2\pi i (n_r z/\lambda - \nu t)}\right] \\ &= (E_{e0}/c_0) e^{-2\pi n_i z/\lambda} \{n_r \cos[2\pi (n_r z/\lambda - \nu t)] \\ &\quad - n_i \sin[2\pi (n_r z/\lambda - \nu t)]\} \end{aligned} \tag{2.92}$$

This can also be written

$$\mathrm{Re}(\mathbf{B_m}) = (E_{e0}/c_0) e^{2\pi n_i/\lambda} n_r \cos[2\pi(n_r z/\lambda - \nu t + \psi)] \tag{2.93}$$

where $\tan\psi = n_i/n_r$. Comparing (2.91) and (2.93) shows that in an absorbing medium the phases of the electric and magnetic fields are no longer the same, but differ by angle ψ.

The irradiance is given by $J = \langle \mathrm{Re}(\mathbf{E}_e) \times \mathrm{Re}(\mathbf{B}_m)/\mu_{m0} \rangle$, where μ_m has been set equal to μ_{m0} in the spirit of lumping all the propagation constants into K_e, and the angular brackets denote a time average over one period. Then the integration over time is readily carried out to give

$$J = \mathrm{Re}\left(n|\mathbf{E}_e|^2/2\mu_{m0}\right) = \tfrac{1}{2}\sqrt{\varepsilon_{e0}/\mu_{m0}} n_r E_{e0}^2 e^{-4\pi n_i z/\lambda}. \tag{2.94}$$

Note that only the portion of $\mathbf{B}_m$ that is in phase with $\mathbf{E}_e$ contributes to the irradiance.

Expression (2.94) for J is of the form $J(z) = J(0)e^{-4\pi n_i z/\lambda}$. This has the same form as Beer's law,

$$J(z) = J(0)e^{-\alpha z}, \tag{2.95a}$$

where α is the *absorption coefficient*,

$$\alpha = 4\pi n_i/\lambda. \tag{2.95b}$$

Equation (2.95b) is often called the *dispersion relation*, although, as will be seen in Chapter 3, it is but one of several dispersion relations. It shows that the irradiance is reduced by a factor of e when the wave has propagated a distance $l_e = \alpha^{-1} = \lambda/4\pi n_i$. This distance is called the *absorption length.*

2.5 Interference

Suppose that two electromagnetic waves $E_1 = E_{10}\cos 2\pi(z/\lambda_1 - \nu_1 t + \varnothing_1)$ and $E_2 = E_{20}\cos 2\pi(z/\lambda_2 - \nu_2 t + \varnothing_2)$ are propagating in the same direction, where $\varnothing_1$ and $\varnothing_2$ are the phases of the two waves at point z and time t, and that the fields are parallel to each other. We have seen that irradiance is proportional to the time average of the square of the electric field, $<(E_1 + E_2)^2>$, which is

$$
\begin{aligned}
&\frac{1}{T}\int_0^T E_{10}^2\cos^2 2\pi\left(\frac{z}{\lambda_2} - \nu_2 t + \varnothing_2\right)dt + \frac{1}{T}\int_0^T E_{20}^2\cos^2 2\pi\left(\frac{z}{\lambda_2} - \nu_2 t + \varnothing_2\right)dt \\
&\quad + \frac{1}{T}\int_0^T 2E_{10}E_{20}\cos 2\pi\left(\frac{z}{\lambda_1} - \nu_1 t + \varnothing_1\right)\cos 2\pi\left(\frac{z}{\lambda_2} - \nu_2 t + \varnothing_2\right)dt \\
&= \frac{E_{10}^2}{2T}\int_0^T\left[1 + \cos 4\pi\left(\frac{z}{\lambda_1} - \nu_1 t + \varnothing_1\right)\right]dt + \frac{E_{20}^2}{2T}\int_0^T \\
&\quad \times\left[1 + \cos 4\pi\left(\frac{z}{\lambda_2} - \nu_2 t + \varnothing_2\right)\right]dt \\
&\quad + \frac{E_{10}E_{20}}{T}\int_0^T\left[\cos 2\pi\left(\frac{z}{\lambda_1} - \nu_1 t + \varnothing_1 + \frac{z}{\lambda_2} - \nu_2 t + \varnothing_2\right)\right. \\
&\quad \left. + \cos 2\pi\left(\frac{z}{\lambda_1} - \nu_1 t + \varnothing_1 - \frac{z}{\lambda_2} + \nu_2 t - \varnothing_2\right)\right]dt
\end{aligned}
\tag{2.96}
$$

Now, all of the cosine functions in the integrands of (2.69) are oscillating functions of t with average values of zero. If $\nu_1 \neq \nu_2$ their integrals become vanishingly small over times T long compared with $1/\nu_1$ and $1/\nu_2$, so that the total integral is proportional to $E_{10}^2 + E_{20}^2$. Thus the total irradiance is equal to the sum of the individual irradiances. However, if $\nu_1 = \nu_2$ $(\lambda_1 = \lambda_2)$, then the last integral becomes

$$
\frac{E_{10}E_{20}}{T}\int_0^T\left[\cos 2\pi\left(\frac{2z}{\lambda_1} - 2\nu_1 t + \varnothing_1 + \varnothing_2\right) + \cos 2\pi(\varnothing_1 - \varnothing_2)\right]dt. \tag{2.97}
$$

The integral over the first, but not the second, cosine vanishes, so that now the integral over time is proportional to $E_{10}^2 + 2E_{10}E_{20}\cos 2\pi(\varnothing_1 - \varnothing_2) + E_{20}^2$. Hence the total irradiance now depends on the difference in phase between the two waves. If the two waves are in phase, $\varnothing_1 = \varnothing_2$, the integral over time is proportional to $(E_{10} + E_{20})^2$, and the total irradiance is proportional to the square of the sum of the amplitudes of the individual waves. The waves are said to *interfere positively*. However, if $2\pi(\varnothing_1 - \varnothing_2) = \pi$, the integral over time is proportional to $(E_{10} - E_{20})^2$, the total irradiance is proportional to the square of the difference between the amplitudes of the individual waves, and the waves are said to *interfere negatively*. In that case if $E_{10} = E_{20}$ the two waves exactly cancel each other out and the irradiance is zero.

2.6 Polarization; the Stokes vector

2.6.1 Linear and circular polarization

An electromagnetic wave propagating in the z direction can be resolved into two waves with electric vectors parallel to the x and y axes: $\boldsymbol{E}_x = \boldsymbol{u}_x E_{x0}\cos(2\pi x/\lambda - 2\pi \nu t + \varnothing_x)$ and $\boldsymbol{E}_y = \boldsymbol{u}_y E_{y0}\cos(2\pi x/\lambda - 2\pi \nu t + \varnothing_y)$, where $\mathbf{u}_x$ and $\mathbf{u}_y$ are unit vectors, and $\varnothing_x$ and $\varnothing_y$ are the phases of the fields at point $x = 0$ and time $t = 0$. These axes are perpendicular to each other, but their directions are otherwise arbitrary and can be chosen to fit the situation of interest. Suppose $\varnothing_x = \varnothing_y = 0$. Then at any point z, say $z = 0$, the two components will oscillate in phase; that is, both will increase, go through maximum, decrease, reverse, and so on, at the same time. The resultant combination of the two vectors will be an electric field that always oscillates in the same plane making an angle with the x and y axes that is determined by the ratio of E_{y0} to E_{x0}. The light is said to be *linearly polarized*.

However, suppose the two waves are not in phase. Suppose $\varnothing_y = 0$, and $\varnothing_x = -\pi/2$; that is, the x field lags behind the y field by 90°. Then at $t = 0$ the y field points in the positive y direction and is at its maximum, but the x field is zero. As time proceeds the y field decreases while the x field grows in the positive x direction until one quarter of a period later the y field is zero while the x field is maximum. If the resultant of the combined fields is represented by a vector, the vector has rotated in the clockwise direction and the tip has traced out one quarter of an ellipse. During one period the tip traces out a complete ellipse, so the light is said to be *elliptically polarized*.

If $E_{y0} = E_{x0}$ the tip of the electric vector traces out a circle and the light is said to be *circularly polarized*. A little thought will show that an elliptically polarized wave can be represented by the sum of two circularly polarized waves rotating in opposite directions. If $\varnothing_y - \varnothing_x < 0$ the electric vector rotates in the clockwise (CW) direction and the light is said to be *right-handed circularly polarized* or have *right-handed helicity*. If $\varnothing_y - \varnothing_x > 0$ the vector rotates in the counterclockwise (CCW) direction and the light is said to be *left-handed circularly polarized* or have *left-handed helicity*.

Unfortunately there is an ambiguity over the definition of *handedness* or *helicity*. Historically, workers in the optical region of the spectrum have defined the helicity as that when the observer is looking in the direction from which the radiation is coming, but persons working at radio frequencies have defined the helicity as that when looking in the direction into which the radiation is propagating. In this book we will follow the latter convention and define CW (right-handed) and CCW (left-handed) rotation as that seen by an observer looking in the direction into which the radiation is propagating. Thus, a circularly polarized wave has right-handed helicity if a screw with a right-handed thread would advance in the same direction as the direction of propagation of the wave if turned CW. When consulting a reference the convention used should always be ascertained.

2.6.2 The Jones and Stokes vectors

A propagating electromagnetic wave may be conveniently described in matrix form by the *Jones vector.* Suppose the wave is propagating along the z-axis in the positive z direction. Then in matrix notation the general form of the Jones vector for the electric field is

$$\mathbf{E} = \begin{pmatrix} E_{x0}\exp[2\pi i(z/\lambda - \nu t) + i\varnothing_x] \\ E_{y0}\exp[2\pi i(z/\lambda - \nu t) + i\varnothing_y] \end{pmatrix}. \tag{2.98a}$$

Then the intensity of the wave is proportional to $|\mathbf{E}|^2 = E_{x0}^2 + E_{y0}^2$. Since the absolute phase is usually not of interest, the Jones vector is often simplified by dividing by the common factor $\exp[2\pi i(z/\lambda - \nu t)]$, so that the simplified Jones vector contains only amplitude and phase difference:

$$\mathbf{E} = \begin{pmatrix} E_{x0}\exp(i\varnothing_x) \\ E_{y0}\exp(i\varnothing_y) \end{pmatrix}. \tag{2.98b}$$

Finally it is common practice to normalize the vector by dividing by $\sqrt{E_{x0}^2 + E_{y0}^2}\exp(i\varnothing_x)$, so that the *normalized Jones vector* is

$$\mathbf{E} = \begin{pmatrix} \dfrac{E_{x0}}{\sqrt{E_{x0}{}^2 + E_{y0}{}^2}} \\ \\ \dfrac{E_{y0}}{\sqrt{E_{x0}{}^2 + E_{y0}{}^2}} e^{i(\varnothing_y - \varnothing_x)} \end{pmatrix}. \tag{2.98c}$$

Examples of the normalized Jones vector are given in Table 2.1.

Light may be *unpolarized*, in which case its electric vectors point randomly in all directions perpendicular to the direction of propagation, or it may be a combination of unpolarized, linearly polarized, or circularly polarized radiation. The difficulty with the Jones vector is that it cannot describe unpolarized light. If the light is partially polarized it may be described by a quantity known as the *Stokes vector*, which has four components and is usually denoted by

$$S = \begin{pmatrix} I \\ Q \\ U \\ V \end{pmatrix}. \tag{2.99}$$

Table 2.1. *Some examples of the normalized Jones and Stokes vectors*
Polarization state Jones vector Stokes vector

Polarization state	Jones vector	Stokes vector
Unpolarized		$I\begin{pmatrix}1\\0\\0\\0\end{pmatrix}$
Linearly polarized along x-axis	$\begin{pmatrix}1\\0\end{pmatrix}$	$I\begin{pmatrix}1\\1\\0\\0\end{pmatrix}$
Linearly polarized along y-axis	$\begin{pmatrix}0\\1\end{pmatrix}$	$I\begin{pmatrix}1\\-1\\0\\0\end{pmatrix}$
Linearly polarized 45° CCW from x-axis	$\begin{pmatrix}1/\sqrt{2}\\1/\sqrt{2}\end{pmatrix}$	$I\begin{pmatrix}1\\0\\+1\\0\end{pmatrix}$
Linearly polarized 45° CW from x-axis	$\begin{pmatrix}1/\sqrt{2}\\-1/\sqrt{2}\end{pmatrix}$	$I\begin{pmatrix}1\\0\\-1\\0\end{pmatrix}$
Left-handed circularly polarized	$\begin{pmatrix}1/\sqrt{2}\\-i/\sqrt{2}\end{pmatrix}$	$I\begin{pmatrix}1\\0\\0\\+1\end{pmatrix}$
Right-handed circularly polarized	$\begin{pmatrix}1/\sqrt{2}\\i/\sqrt{2}\end{pmatrix}$	$I\begin{pmatrix}1\\0\\0\\-1\end{pmatrix}$

Denoting the time average of a quantity by angular brackets, the components of the Stokes vector are defined as follows:

$$I = C(<\mathbf{E}_x{}^2> + <\mathbf{E}_y{}^2>), \tag{2.100a}$$

$$Q = C(<\mathbf{E}_x{}^2> - <\mathbf{E}_y{}^2>), \tag{2.100b}$$

$$U = C(<2\mathbf{E}_x\mathbf{E}_y\cos(\varnothing_y - \varnothing_x)>), \tag{2.100c}$$

$$V = C(<2\mathbf{E}_x\mathbf{E}_y\sin(\varnothing_y - \varnothing_x)>), \tag{2.100d}$$

where $C = \frac{1}{2}\sqrt{\varepsilon_{e0}/\mu_{m0}}n_r$. Physically, I is the total intensity; Q is the linearly polarized intensity parallel to the x-axis minus that parallel to the y-axis; U is the linearly polarized intensity along a direction rotated 45° CCW from the x-axis minus that along a direction rotated 45° CW from the x-axis; V is the left-handed polarized intensity minus the right-handed polarized intensity. Often the Stokes vector is normalized by dividing all quantities by I, in which case it is written

$$S = I \begin{pmatrix} 1 \\ Q \\ U \\ V \end{pmatrix}, \tag{2.101}$$

where Q, U, and V have values between -1 and $+1$. Since I is the total intensity the components must satisfy the relation: $Q + U + V \leq 1$. Examples of the normalized Stokes vector are given in Table 2.1.

2.6.3 Polarization

The *degree of polarization*, also often called the *linear polarization ratio*, and just plain *polarization*, is defined as

$$P = -Q/I, \tag{2.102}$$

and is the difference between the linearly polarized intensity along two perpendicular axes divided by the total intensity.

3

The absorption of light

3.1 Introduction

The differential reflection and scattering of light as a function of wavelength form the basis of the science of reflectance spectroscopy. This chapter discusses the absorption of electromagnetic radiation by solids and liquids. The classical descriptions of absorption and dispersion are derived first, followed by a brief discussion of these processes from the point of view of quantum mechanics and modern physics. Finally, the various types of mechanisms by which light is absorbed are summarized.

3.2 Classical dispersion theory

3.2.1 Conductors: the drude model

The simplest model for absorption and dispersion by a solid is that of Drude (1959). This model assumes that some of the electrons are free to move within the lattice, while the ions are assumed to remain fixed. These approximate the conditions within a metal. The average electric-charge density associated with the semifree electrons is equal to the average of that associated with the lattice ions, so that the total electric-charge density $\rho_e = 0$. Because the quantum-mechanical wave functions of the conduction electrons are not localized in a metal, the local field E_{loc} seen by the electrons is equal to the macroscopic field $\mathbf{E}_e$. Thus, the force on each electron is $-e_0\mathbf{E}_e$,where e_0 is the charge of an electron. Assume that $\mathbf{E}_e$ is parallel to the x-axis.

In addition to the electric field, there is a force due to collisions of each electron with the lattice, resulting in nonradiative loss of energy. This force, which is proportional to the velocity $d\mathbf{x}/dt$ of the electron, but opposite in direction, may be characterized by a parameter Ξ, defined such that the average collisional force on the electron is given by $-2\pi\,\Xi m_e d\mathbf{x}/dt$, where m_e is the mass of the electron, and $\mathbf{x}$ is the displacement of the electron relative to the lattice. Physically, it can be shown that $\Xi = 1/2\pi\bar{t}$, where $\bar{t}$ is the mean time between collisions of the electron with the lattice; Ξ is called the *collision frequency*. The equation of motion of an electron is then

$$m_e d^2\mathbf{x}/dt^2 + 2\pi \Xi m_e d\mathbf{x}/dt = -e_0\mathbf{E}_e. \tag{3.1}$$

Assume that $\mathbf{E}_e$ is a periodic wave described by $\mathbf{E}_e = \mathbf{E}_{e0}e^{-2\pi i\nu t}$, where ν is the frequency, and we seek periodic solutions of (3.1) of similar form, $\mathbf{x} = \mathbf{x}_0 e^{-2\pi i\nu t}$. Substituting this form of $\mathbf{x}$ into (3.1) gives

$$-4\pi^2\nu^2 m_e\mathbf{x}_0 - 4\pi^2 i m_e \Xi \nu \mathbf{x}_0 = -e_0\mathbf{E}_{e0}, \tag{3.2}$$

or

$$\mathbf{x}_0 = \frac{e_0/4\pi^2 m_e}{\nu^2 + i\Xi\nu}\mathbf{E}_{e0}. \tag{3.3}$$

The amount $\mathbf{x}_0$ by which the electron is offset from the ion lattice induces an electric dipole moment

$$\mathbf{p}_e = -e_0\mathbf{x}_0 = -\frac{e_0^2/4\pi^2 m_e}{\nu^2 + i\Xi\nu}\mathbf{E}_e = \alpha_e\varepsilon_{e0}\mathbf{E}_{e0}, \tag{3.4}$$

where α_e is the electric polarizability. Thus,

$$\alpha_e = -\frac{e_0^2}{4\pi^2 m_e\varepsilon_{e0}}\frac{1}{\nu^2 + i\Xi\nu}. \tag{3.5}$$

The electric susceptibility is $\chi_e = N_e\alpha_e$, where N_e is the free-electron density. Hence, the dielectric constant is

$$K_e = 1 + \chi_e = 1 - \frac{N_e e_0^2}{4\pi^2 m_e\varepsilon_{e0}}\frac{1}{\nu^2 + i\Xi\nu}. \tag{3.6}$$

The factor $N_e e_0^2/4\pi^2 m_e\varepsilon_{e0}$ has the dimensions of reciprocal time squared, and the square root of this quantity is called the *plasma frequency* ν_p ,

$$\nu_p = (N_e e_0^2/4\pi^2 m_e\varepsilon_{e0})^{1/2}. \tag{3.7}$$

Thus the dielectric constant may be written

$$K_e = 1 - \nu_p^2/(\nu^2 + i\Xi\nu). \tag{3.8}$$

The real and imaginary parts of K_e are, respectively,

$$K_{er} = n_r^2 - n_i^2 = 1 - \nu_p^2/(\nu^2 + \Xi^2) \tag{3.9}$$

and

$$K_{ei} = 2n_r n_i = \frac{\nu_p^2}{\nu^2 + \Xi^2}\frac{\Xi}{\nu}, \tag{3.10}$$

where $n = n_r + in_i = \sqrt{K_e}$ is the refractive index and can be found from equations (2.84) and (2.85).

In most metals, $\Xi \ll \nu_p$, so that the real part of K_e becomes zero at or close to the plasma frequency. When this happens, the phase velocity becomes infinite. The physical meaning of a dielectric constant of zero is that true electromagnetic waves cannot propagate. The only disturbances that can occur in the medium are standing waves; these are called *plasma oscillations*.

If $\nu \gg \nu_p$, then $K_{er} \to 1$ and $K_{ei} \ll 1$, so that the metal becomes transparent. Conversely, when $\nu \ll \nu_p$, both K_{er} and K_{ei} are large, and the material is opaque. In the limit of small ν, $K_{ei} \simeq (\nu_p^2/\Xi\nu) = N_e e_0^2/4\pi^2 m_e \varepsilon_{e0} \Xi \nu$. From equation (2.78), K_{ei} is related to the electrical conductivity σ_e by $K_{ei} = \sigma_e/2\pi\nu\varepsilon_{e0}$. When ν is very small, σ_e is the direct-current (DC) conductivity σ_{DC}. Thus,

$$\Xi = 2\pi\nu_p^2\varepsilon_{e0}/\sigma_{\mathrm{DC}} = N_e e_0^2/2\pi m_e \sigma_{\mathrm{DC}}. \tag{3.11}$$

The plasma frequency ν_p can be measured by noting where the metal becomes transparent, or it can be calculated by estimating N_e as the number of atoms per unit volume times the number of free electrons per atom, and using (3.7). The collision frequency Ξ can be measured from the DC conductivity, using (3.11).

3.2.2 Absorption in insulators: the Lorentz model

In nonconducting solids the electrons are not free to move through the lattice, but are localized by being bound to individual atoms. Because the wave functions of the electrons are localized in insulators, $\mathbf{E}_{\mathrm{loc}}$ must be used rather than $\mathbf{E}_e$. The equations of motion of the electrons contain all the terms of (3.1), plus a force describing the bonding of the electrons to lattice sites. The Lorentz model (Lorentz, 1952) assumes that this force is proportional to the displacement $\mathbf{x}$ and is characterized by a parameter ν_0, defined such that the restoring force is $-4\pi^2 m_e \nu_0^2 \mathbf{x}$. Then the equation of motion of an electron is

$$m_e d^2\mathbf{x}/dt^2 + 2\pi m_e \Xi d\mathbf{x}/dt + 4\pi^2 m_e \nu_0^2 \mathbf{x} = -e_0 \mathbf{E}_{\mathrm{loc}}. \tag{3.12}$$

This equation is identical with that of a mass moving in a potential well of parabolic shape, and the Lorentz model is sometimes referred to as the parabolic-well model of electron oscillations.

The periodic solution to (3.9) with both $\mathbf{E}_{\mathrm{loc}}$ and $\mathbf{x}$ proportional to $e^{-2\pi i\nu t}$ is

$$-4\pi^2\nu^2 m_e \mathbf{x}_0 - 4\pi^2 i m_e \Xi \nu \mathbf{x}_0 + 4\pi^2 m_e \nu_0^2 \mathbf{x}_0 = -e_0 \mathbf{E}_{\mathrm{loc0}}, \tag{3.13}$$

which gives

$$\mathbf{x}_0 = \frac{e_0}{4\pi^2 m_e} \frac{1}{\nu_0^2 - \nu^2 - i\Xi\nu} \mathbf{E}_{\mathrm{loc0}}. \tag{3.14}$$

As in the Drude model, this gives

$$X_e = N_e\alpha_e = \frac{N_e e_0^2}{4\pi^2 m_e \varepsilon_{e0}} \frac{1}{\nu_0^2 - \nu^2 - i\,\Xi\nu} = \frac{\nu_p^2}{\nu_0^2 - \nu^2 - i\,\Xi\nu}. \tag{3.15}$$

If $\mathbf{E}_{\text{loc}} \neq \mathbf{E}_e$, but the Clausius–Mossotti relation is valid,

$$(K_e - 1)/(K_e + 2) = N_e\alpha_e/3 = (\nu_p^2/3)/(\nu_0^2 - \nu^2 - i\,\Xi\nu), \tag{3.16}$$

or

$$K_e = [1 + (2\nu_p^2/3)/(\nu_0^2 - \nu^2 - i\,\Xi\nu)]/[1 - (\nu_p^2/3)/(\nu_0^2 - \nu^2 - i\,\Xi\nu)]. \tag{3.17}$$

Usually in the Lorentz model it is assumed, without particular justification, that $\mathbf{E}_{\text{loc}} = \mathbf{E}_e$. In that case,

$$K_e = 1 + \chi_e = 1 + N_e\alpha_e = 1 + \frac{N_e e_0^2}{4\pi^2 m_e \varepsilon_{e0}} \frac{1}{\nu_0^2 - \nu^2 - i\,\Xi\nu} = 1 + \frac{\nu_p^2}{\nu_0^2 - \nu^2 - i\,\Xi\nu}. \tag{3.18}$$

Then

$$K_{er} = n_r^2 - n_i^2 = 1 + \frac{\nu_p^2(\nu_0^2 - \nu^2)}{(\nu_0^2 - \nu^2)^2 + \Xi^2\nu^2}, \tag{3.19}$$

and

$$K_{ei} = 2n_r n_i = \frac{\nu_p^2 \Xi\nu}{(\nu_0^2 - \nu^2)^2 + \Xi^2\nu^2}. \tag{3.20}$$

Equations (3.19) and (3.20) also describe absorption in solids by ion vibrations instead of electronic oscillations, except that the electron mass m_e, which occurs in the expression for the plasma frequency, must be replaced by the reduced mass of the negative ions with respect to the positive ions in the molecule. Usually an absorption band is superimposed on a continuum real dielectric constant K_{ec}. In that case the term equal to 1 in equation (3.19) is replaced by K_{ec} or $n_c^2 = K_{ec}$.

Figures 3.1 and 3.2 give an example of the variations of K_{er}, K_{ei}, n_r, and n_i with ν for a solid that obeys the Lorentz model. The dielectric constant and refractive index have a resonance at $\nu = \nu_0$. Note that over most of the range of ν, K_{er} increases as ν increases; this is called *normal dispersion*. However, near ν_0, K_{er} decreases as ν increases; this is called *anomalous dispersion*. Note that K_{er} can be negative in this region, but n_r and n_i are positive.

If the absorption band is strong enough, n_r can be <1 on the high-frequency side of the absorption band. This means that the wave velocity is $c_0/n_r > c_0$, which would appear to violate the theory of special relativity. However, c_0/n_r is the phase velocity, the speed with which the peaks of the wave move. It can be shown that the power and information content conveyed by the wave always move with velocity $\leq c_0$..

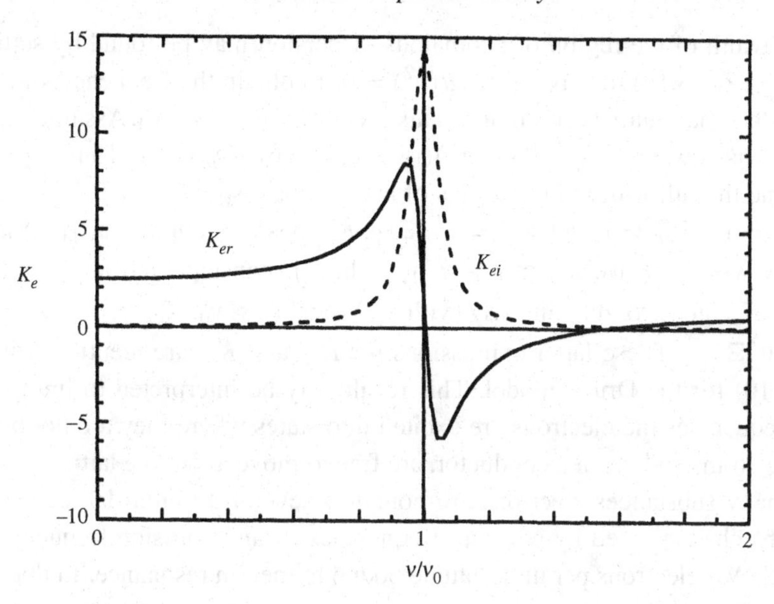

Figure 3.1 Real and imaginary parts of the dielectric constant in the vicinity of a strong absorption band, according to the Lorentz model. This example is calculated for $\nu_p/\nu_0 = 1.2$ and $\Xi/\nu_0 = 0.1$.

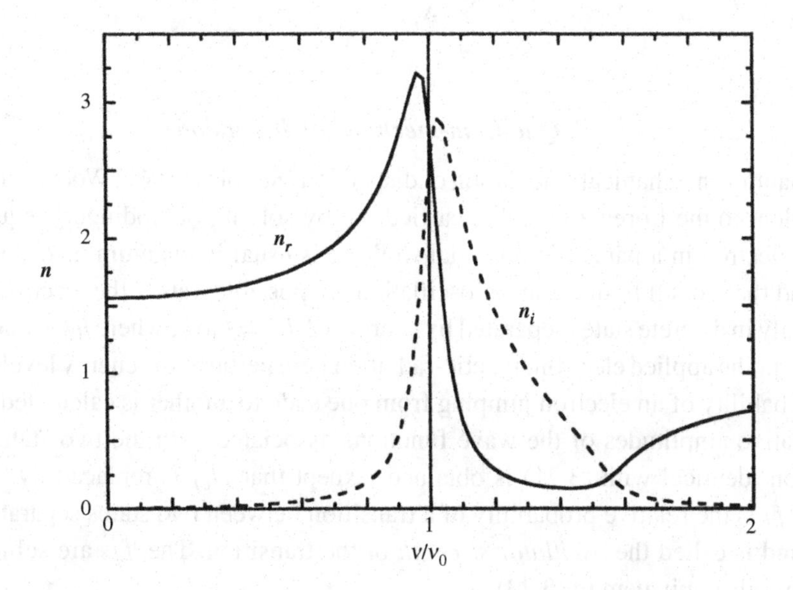

Figure 3.2 Real and imaginary parts of the refractive index in the vicinity of a strong absorption band for a material with the dielectric constant of Figure 3.1.

The width of the region of anomalous dispersion may be found by setting the derivative of (3.19) to zero, $dK_{er}/d(\nu^2) = 0$, to obtain the frequencies ν_m where K_e is either maximum or minimum. This gives $\nu_0^2 - \nu_m^2 = \pm \Xi \nu_0$. Assuming that the region is narrow, $\nu_0^2 - \nu_m^2 = (\nu_0 - \nu_m)(\nu_0 + \nu_m) \simeq (\nu_0 - \nu_m) \cdot 2\nu_0$. Thus, $|\nu_0 - \nu_m| = \Xi/2$, and the full width of the region of anomalous dispersion is Ξ.

At $\nu = \nu_0$, $K_{er} = 1$ and $K_{ei} = \nu_p^2/\Xi \nu_0$; K_{ei} has a maximum at ν_0. Assuming $\Xi \ll \nu_0$, when $\nu \ll \nu_0$, $K_{er} \simeq 1 + \nu_p^2/\nu_0^2$, which is constant, and $K_{ei} \simeq \nu_p^2 \Xi \nu/\nu_0^4$, so that K_{ei} goes to zero linearly with ν. When $\nu \gg \nu_0$, $K_{er} \simeq 1 - \nu_p^2/\nu^2$, and $K_{ei} \simeq \nu_p^2 \Xi/\nu^3$. These latter expressions for K_{er} and K_{ei} are identical with (3.9) and (3.10) for the Drude model. This result may be interpreted to imply that at high frequencies the electrons are excited into states where they are not bound to discrete atoms and, as in a conductor, are free to move about the lattice.

In many substances electrons are bound to several different lattice sites with strengths characterized by resonant frequencies ν_j and collision frequencies Ξ_j, and with N_{ej} electrons per unit volume bound to the jth resonance. In that case,

$$K_e = 1 + \frac{e_0^2}{4\pi^2 m_e \varepsilon_{e0}} \sum_j \frac{N_{ej}}{\nu_j^2 - \nu^2 - i\Xi_j \nu}, \tag{3.21}$$

where theN_{ej}s are subject to the sum rule

$$\sum_j N_{ej} = N_e. \tag{3.22}$$

3.2.3 Quantum-mechanical dispersion

The quantum-mechanical calculation of dispersion (Sokolov, 1967; Wooten, 1972), equivalent to the Lorentz model is carried out by solving Schrödinger's equation for an electron in a parabolic potential well. As is usual in quantum mechanics, it is found that not all frequencies of oscillation are possible; rather the electrons can exist only in discrete states separated by energies $\Delta E_j = h_0 \nu_j$, where h_0 is Planck's constant. The applied electromagnetic-radiation field perturbs the energy levels, and the probability of an electron jumping from one state to another is calculated from the relative amplitudes of the wave functions associated with the two states. An equation identical with (3.21) is obtained, except that N_{ej} is replaced by $N_e f_j$, where f_j is the relative probability of a transition between two states separated by ΔE_j and is called the *oscillator strength* of the transition. The f_js are subject to the sum rule equivalent to (3.23),

$$\sum_j f_j = 1. \tag{3.23}$$

3.3 Dispersion relations

A *dispersion relation* is an equation that describes the frequency or wavelength dependence of an optical quantity. One example is equation (2.95), which relates the absorption coefficient to the imaginary part of the reflective index and the wavelength. Although this equation is often called *the* dispersion relation, it is only one of many. Another example is (3.19) and (3.20) for K_{er} and K_{ei} versus ν, which, together with (2.84) and (2.85), can be used to calculate the angular dispersion of white light as it passes through a prism. The latter is, of course, the original meaning of "dispersion."

An important relation may be derived from equation (3.15) for K_e by mathematically allowing ν to be a complex quantity. Then $K_e - 1$ has a singularity (that is, it becomes infinite) only when $\nu^2 + i\Xi\nu - \nu_0^2 = 0$ or $\nu = [-i\Xi \pm (-\Xi^2 + 4\nu_0^2)^{1/2}]/2$. If $\Xi \ll \nu_0$, $\nu \approx i\Xi/2 \pm \nu_0$. Because these singularities lie below the real axis, $K_e - 1$ is analytic in the upper half of the complex frequency plane. Also, $|K_e - 1| \to 0$ as $\nu \to \infty$. Hence, relation (B.11) from Appendix B is applicable, which gives

$$K_e - 1 = \frac{1}{i\pi}\int_{-\infty}^{\infty} \frac{K_e(\nu') - 1}{\nu' - \nu} d\nu'. \tag{3.24}$$

In addition, K_e has crossing symmetry, so that $K_{er}(-\nu) = K_{er}(\nu)$ and $K_{ei}(-\nu) = -K_{ei}(\nu)$; hence, equations (B.14) and (B.15) from Appendix B are also applicable, so that

$$K_{er}(\nu) - 1 = \frac{2}{\pi}\int_{0}^{\infty} \frac{\nu' K_{ei}(\nu')}{\nu'^2 - \nu^2} d\nu', \tag{3.25}$$

and

$$K_{ei}(\nu) = -\frac{2\nu}{\pi}\int_{0}^{\infty} \frac{K_{er}(\nu') - 1}{\nu'^2 - \nu^2} d\nu'. \tag{3.26}$$

Equations (3.25) and (3.26) are known as the Kramers–Kronig dispersion relations. They show that the real and imaginary parts of the dielectric constant are not independent, but that one can be calculated from the other.

Although we have used the Lorentz model to derive these dispersion relations, it can be shown (e.g., Wooten, 1972) that they are much more general and are independent of the particular model for the dielectric constant. In fact, these dispersion relations follow from the requirement of causality, that is, that the system cannot have a response before the event that excites the response occurs.

Numerous other dispersion relations and sum rules exist. They have been discussed by many authors, including Wooten (1972) and Smith (1985).

3.4 Mechanisms of absorption

3.4.1 Introduction

The mechanisms by which electromagnetic radiation interacts with condensed matter may be classified into four broad categories: rotational, vibrational, electron excitation, and free carrier. Each will be discussed qualitatively in this section. More detailed, rigorous descriptions of the modern theory of the solid state, on which this section is based, can be found in books on modern physics, solid-state physics, and optical physics, such as those by Sproull and Phillips (1980), Kittel (1976), and Garbuny (1965). A succinct and clear summary of absorption mechanisms, particularly in the visible and infrared, has been provided by Hunt (1980). Further discussion and references may be found in the work of Salisbury (1993). See also Egan and Hilgeman (1979) and Gaffey *et al.* (1989). Although the emphasis here will be on absorption by solids, many of the mechanisms are also applicable to liquids. Tabulated values of optical constants for a variety of materials have been provided by Palik (1991).

3.4.2 Molecular rotation

Molecules possessing permanent electric-dipole moments contribute to the complex dielectric constant by changing their orientations when acted on by an electric field. The molecules in gases and liquids are able to rotate in an applied field, and their permanent dipole moments can cause a large electric polarizability. In a solid, the ability of a molecule to rotate depends on its shape and the strength of its interactions with its environment. In general, the more nearly spherical the molecule and the smaller its dipole moment, the more easily it rotates. Molecules often have several stable orientations in the solid, and they may flip from one orientation to another during an interval called the *relaxation time*.

If the period of the applied field is long compared with the relaxation time, the alignments saturate, and the dielectric constant is large. Conversely, if the ratio of the period to the relaxation time is small, the molecules do not have time to rotate, and the dielectric constant is small. Thermal agitation tends to randomize the orientations of the molecules, so that the portion of the dielectric constant associated with rotational transitions decreases with increasing temperature.

A substance of considerable interest in planetary remote sensing is water H_2O. In liquid water at room temperature, the relaxation time is on the order of 3×10^{-11} sec, corresponding to a wavelength of 1 cm. Consequently, the real part of the dielectric constant of water for wavelengths shorter than 1 cm is around 80, whereas the value in the visible part of the spectrum is 1.77. However, in ice at $-20\,^{\circ}$C the relaxation time is around 1 msec, so that the microwave dielectric constant of ice is

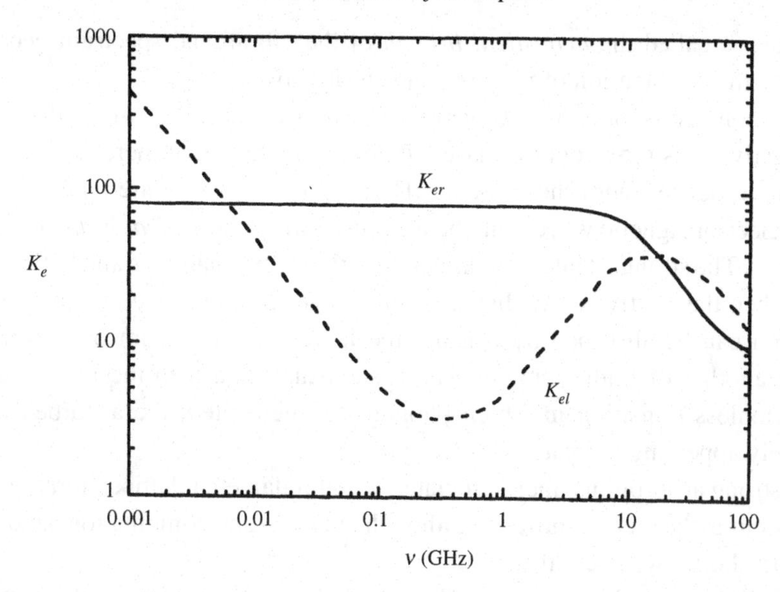

Figure 3.3 Dielectric constant of fresh water at 20 °C. Data from Fung and Ulaby (1983).

1.78. Figure 3.3 shows the dielectric constant of liquid water in the microwave region.

The ubiquitous occurrence of water on the surface of the Earth and its high dielectric constant effectively prevent the penetration of microwaves very far into the surface. However, because the Earth appears to be the only body in the solar system where H_2O occurs in the liquid state on the surface, this limitation does not apply to microwave sensing of other planets.

3.4.3 Lattice vibrations

A solid can be considered to be made up of positive and negative ions arranged in a periodic array, with each ion able to vibrate about its equilibrium position. The mathematical description of lattice vibrations is found by setting up the equations of motion of the ions and seeking solutions having the form of periodic waves propagating through the array. In the quantum-mechanical description the lattice vibrations are considered to be made up of waves, called *phonons*, with discrete, quantized energy levels.

In general, in both classical and quantum-mechanical physics two classes of solutions are found. One type describes motions in which the positive and negative ions move together so that there is no net dipole moment associated with the vibrations, and hence they are not able to interact with an electromagnetic wave.

This class is called the *acoustical branch* of the vibrational spectrum because it describes the propagation of pressure and shear waves.

The second class, or branch, of waves describes vibrations in which the positive and negative ions move out of phase with each other, so that there is a net dipole moment associated with the motions. These types of waves are able to emit or absorb electromagnetic waves, and hence this is called the *optical branch* of lattice vibrations. The strongest interactions between the electromagnetic and lattice waves occur when the relative ionic displacements result in transverse dipole moments. Radiation can be absorbed by causing the lattice to jump from one vibrational state to another of higher energy. The system may return to the lower state by a radiationless transition in which phonons are generated, the absorbed energy ultimately appearing as heat.

The strong absorption bands associated with fundamental lattice vibrations generally occur in the thermal infrared, although overtone and combination bands often extend to shorter wavelengths.

Figure 3.4 shows the measured vibrational absorption spectrum of quartz. The figure also shows the fit of classical dispersion theory, equation (3.21), to the data.

3.4.4 Electronic transitions

3.4.4.1 The band model of electrons in a solid

Absorption in the near ultraviolet, visible, and near infrared can occur by several different mechanisms, but most can be classed as transitions in which single electrons are induced by the radiation to jump from a state of lower energy to one of higher energy. These transitions can be described in terms of the band model of electrons in a solid.

A classical oscillating system has a well-defined resonant frequency. An example is the pendulum, in which the resonant frequency is determined by the length. Two isolated, identical pendulums will have the same resonant frequency, but if the pendulums are coupled, the resonances will split, so that the system as a whole will have two resonant frequencies. The stronger the coupling, the greater the splitting.

Similarly, electrons in isolated atoms or ions have well-defined, discrete energy states, which (in principle) can be calculated from the Schrödinger equation. As the ions are brought together in a solid, the wave functions of the outer electrons overlap and perturb one another. The result of the perturbations is to split the energy levels, so that the system of many atoms has a large number of states that are closely spaced in energy. The quantum-mechanical energy levels overlap because they have a finite width. As a result, the system has several wide, continuous bands in which the electrons can exist, separated by gaps in which no solution of the wave equation is possible.

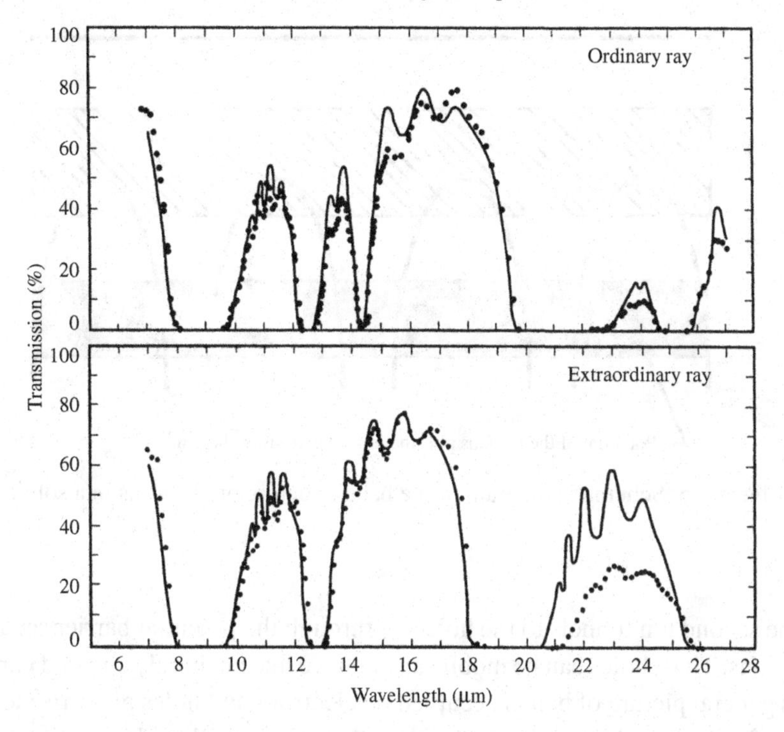

Figure 3.4 Transmission spectrum of quartz for the ordinary and extraordinary rays. The lines show classical dispersion theory fitted to the data. (Reproduced from Spitzer and Kleinman [1961], copyright 1961 by the American Physical Society.)

The highest energy band, in which all electron states are occupied, is called the *valence band*. The wave functions of electrons in the valence band generally are localized on discrete ions or atoms. The lowest energy band in which all states are not occupied, is called the *conduction band*, which may be partially filled or completely empty. (To be sure, states of lower energy than the valence band and of higher energy than the conduction band exist and may also properly be called valence and conduction bands. However, these other bands usually play no role in absorption and so may be ignored.) Electrons in the conduction band are not bound to any specific atom or ion, but are free to move about the solid lattice, where they contribute to both the electrical and thermal conductivities. The width of the gap between the valence and conduction bands typically is several electronvolts. The bands are illustrated schematically in Figure 3.5.

If an electron is excited from the valence band to the conduction band, it leaves an ion with an excess of positive charge behind it. This positively charged electron vacancy is called a *hole*. A hole can jump from one ion to another when an electron

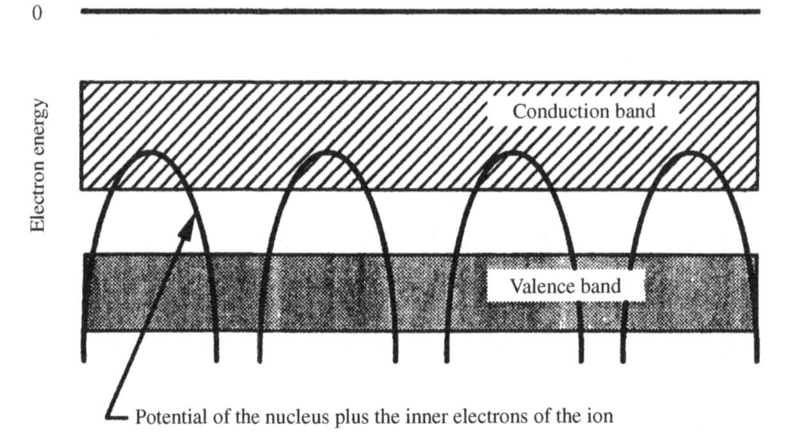

Figure 3.5 Schematic diagram of the band structure of electrons in a solid.

from the second ion tunnels to the hole site through the potential barrier separating the two ions. Thus, holes can be mobile and behave like positively charged particles.

This general picture of bands occupied by electrons and holes gives rise to a rich variety of ways by which light can be absorbed. These will be described next.

3.4.4.2 Color centers

All real solids have many different types of lattice imperfections, including *vacancies*, where an ion is missing from the lattice, *interstitials*, where an extra ion is forced into the space between ions of the lattice, and *substitutional* impurities, where a lattice ion is replaced by one of a different species. Generally an excess charge of one sign accompanies the imperfection, so that to preserve overall electrical neutrality a second imperfection having a charge of the opposite sign also exists near the first.

An electron can orbit a positively charged imperfection and a hole a negatively charged one to form systems analogous to the hydrogen atom. Such a system is called a *color center*, and absorption can be regarded as taking place when the electron or hole jumps to a higher energy level in this hydrogen-like system. The wavelengths of absorption bands due to color centers usually are in the visible and near infrared. Figure 3.6 shows color centers induced in soda-silica glass by x-ray irradiation.

Color centers have been extensively studied in alkali halides and a number of metal oxides. Detailed discussions can be found in the works of Schulman and Compton (1962) and Fowler (1968).

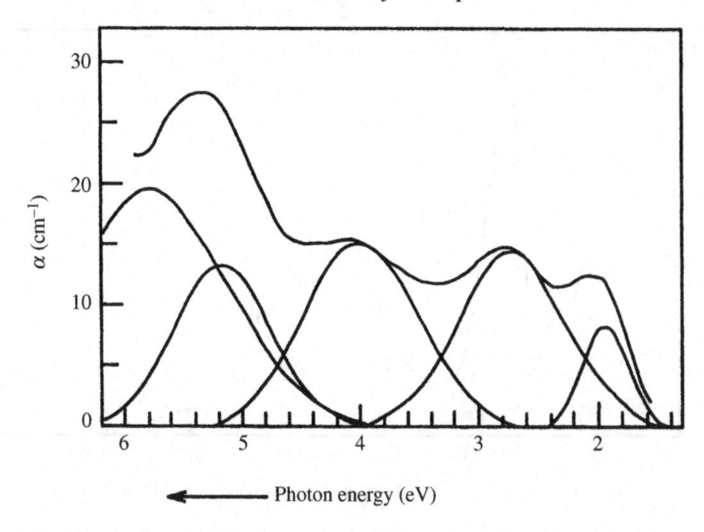

Figure 3.6 Absorption bands in a soda-silica glass irradiated by x-rays. The upper curve is the absorption spectrum; the lower curves show the spectrum deconvolved into individual bands of Gaussian shapes. (Reproduced from Cohen and Janezic [1983], copyright 1983 with permission of Akademie-Verlag GmbH.)

3.4.4.3 Crystal-field effects in transition elements

Transition elements have partly filled d or f shells in both the neutral atoms and their common oxidation states. In an isolated ion the orbital states in the partly filled shell are degenerate, so that electrons have equal probabilities of residing in any state. However, when the ion is in a solid, the electric field due to the surrounding ions is not isotropic. This anisotropy removes the degeneracies, so that the orbitals in the shell have different energy levels, an effect known as *crystal-field splitting*. Light can be absorbed when an electron jumps from an orbital of lower energy to one of higher energy in the same shell. This mechanism is called *crystal-field absorption.*

The probability of a transition between the orbital energy levels is subject to certain selection rules. One is the spin-multiplicity rule, which states that transitions may take place only between states that have the same number of unpaired electrons. A second is the Laporte rule, which forbids transitions between orbitals of the same type and quantum number. Thus, in the isolated ion, crystal-field transitions would be strictly forbidden even if the energy levels were not degenerate. In a crystal lattice these selection rules are relaxed by various effects, including spin-orbit coupling and departure of the environment of the ion from perfect symmetry. However, the transition probabilities are relatively low, so that the absorptions are weak. Transitions within the d shell of an ion in octahedral coordination are

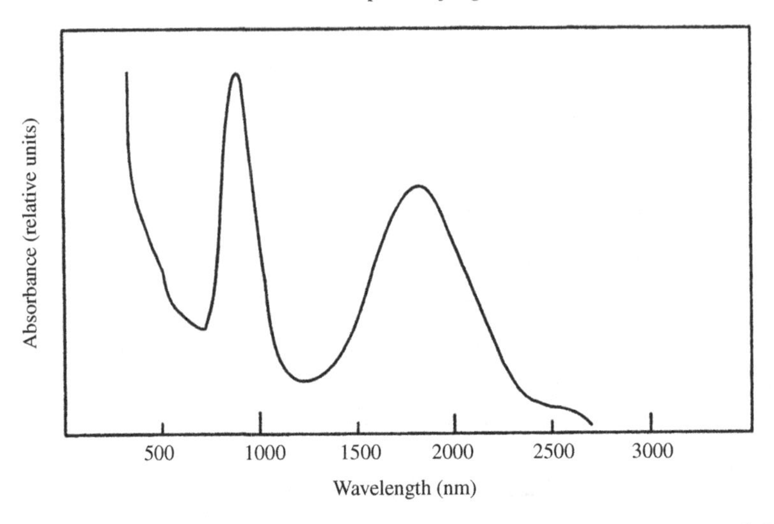

Figure 3.7 Absorption spectrum of enstatite ([Mg, Fe]SiO_3) showing crystal-field bands of Fe^{2+} at 900 and 1800 nm. (Reproduced from White and Keester [1966], copyright 1966 with permission of the Mineralogical Society of America.)

Laporte-forbidden and thus are very weak. However, if the ion is in tetrahedral coordination, which lacks a center of symmetry, the transition probability is roughly 100 times larger.

Absorption bands due to crystal-field transitions typically occur in the visible and near infrared. The crystal-field absorption spectrum of Fe^{2+} in octahedral coordination with O^{2-} in enstatite is shown in Figure 3.7. An excellent discussion of crystal-field absorption has been provided by Burns (1970, 1993). See also Adams (1975) and Vaughan (1990). The Fe^{2+} bands near 1000 and 2000 nm and the Fe^{3+} band near 860 nm have important applications in geochemical remote sensing.

3.4.4.4 Charge transfer

Absorption by charge transfer occurs when an electron jumps from a state localized on one ion to another on a nearby ion. Because the transition is not spin- or Laporte-forbidden, charge-transfer bands are strong. Theoretical prediction of the wavelength at which charge-transfer absorption will occur is difficult because both ions and their neighbors must all be considered as one system in calculating the electronic energy levels.

Charge-transfer bands generally are in the near ultraviolet, visible, and near infrared spectral region. Figure 3.8 shows the $Fe^{2+} - Ti^{4+}$ charge-transfer bands

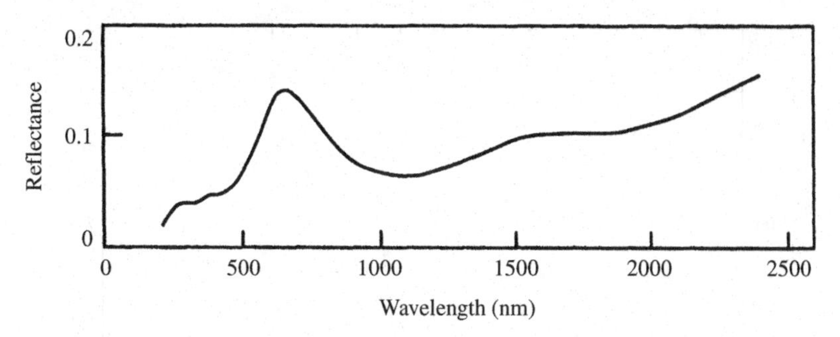

Figure 3.8 Reflectance spectrum of powdered silicate glass. The spectrum shows the $Fe^{2+} - Ti^{4+}$ charge-transfer bands at 340 and 420 nm, and the weak crystal-field bands of Fe^{2+} near 1000 and 2000 nm.

at 340 and 420 nm in a silicate glass (Wells and Hapke, 1977). This spectrum also exhibits the Fe^{2+} crystal-field bands.

In certain minerals, such as magnetite (Fe_3O_4) and the ilmenite–hematite ($FeTiO_3 - Fe_2O_3$) series, charge-transfer mechanisms are thought to cause electron delocalization (Burns *et al.*, 1980). The electrons behave as nearly free carriers, with the result that the minerals are opaque over a wide spectral region.

3.4.4.5 Excitons

When an electron is excited into the conduction band, a hole is left in the valence band. The positively charged hole can attract an unbound electron and form a hydrogen-like system, called an *exciton*, which can absorb light by transitions between levels within the system. The energies of excitonic transitions are slightly smaller than the valence–conduction band gap, so that the wavelengths corresponding to them usually lie in the far ultraviolet. An exciton transition in forsterite is illustrated in Figure 3.9.

3.4.4.6 Interband and impurity transitions

If the wavelength is short enough, absorption may occur when the radiation excites an electron directly from the valence band to the conduction band. Generally this type of transition occurs in the far ultraviolet, but for a few materials, called intrinsic semiconductors, thermal phonons are sufficient. Elemental Ge and Si are examples of intrinsic semiconductors. Absorption associated with interband transitions is extremely strong. Figure 3.9 shows the interband absorption spectrum of forsterite.

Often electrons of impurity ions have energy levels within the forbidden gap between the valence and conduction bands. In that case, transitions may occur in

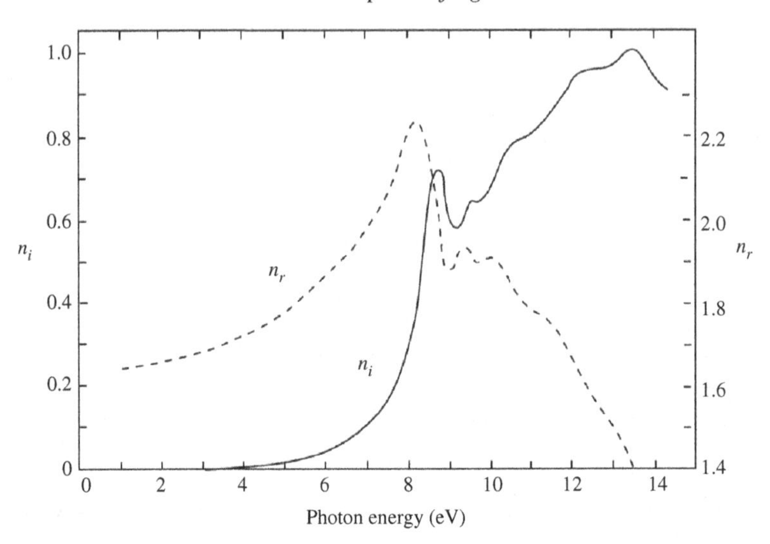

Figure 3.9 Refractive index of forsterite ($Mg_2\ SiO_4$) showing the exciton band at about 9 eV. (Reproduced from Nitsan and Shankland [1976], copyright 1976 with permission of the Royal Astronomical Society.)

which an electron is excited from the valence band to an impurity level, or from impurity level to conduction band.

3.4.5 Free carriers

Free carriers of charge, including both electrons and holes, can absorb light in a solid when they are stimulated to move by the electric field of the incident radiation and then collide with the lattice, converting some of the energy of the radiation into lattice vibrations. According to quantum theory, an electron can propagate without loss through a perfect periodic structure; hence, the absorption actually takes place by collisions with lattice imperfections. The imperfections can be of many types, including dislocations, grain boundaries, vacancies, interstitials, impurities, and random displacements of the ions from their average positions due to thermal motions. The latter can be described quantum-mechanically as phonons, and absorptions associated with this type of imperfection are sometimes referred to as electron-phonon losses.

Free-carrier loss is the main source of absorption in the visible and near infrared in metals and semiconductors. Unlike the other mechanisms discussed in this section, free-carrier absorption occurs over a broad range of wavelengths and not in well-defined bands. It is described approximately by the Drude theory at long wavelengths, but not at short, where other absorption mechanisms dominate.

Electromagnetic-wave propagation and losses in metals have been discussed by Abeles (1966) and Sokolov (1967).

3.5 Band shape and temperature effects

It has been shown in this chapter that an absorption band due to charged particles oscillating in a parabolic potential well has a Lorentzian shape described by equations (3.19) and (3.20). For all of the loss mechanisms considered in the preceding section the assumption of parabolic shapes is a reasonable first approximation, and measured infrared absorption bands associated with lattice vibrations can be described adequately by Lorentzian shapes (Figure 3.4).

However, the shorter-wavelength electronic bands are better described empirically by Gaussian shapes of the form $\alpha = A \exp[-(\nu - \nu_0)^2/B]$, where A and B are constants (Figure 3.6). The reasons for this have been discussed by Lax (1954) and Dexter (1956). Essentially it occurs because the widths of the electronic potential wells are affected by the distances to the nearest neighbor ions. These distances differ from their averages because the lattice may be distorted by a variety of effects, including dislocations, impurities, and interstitials and because the ions are in constant thermal vibration. The effect of these distortions averaged over a large

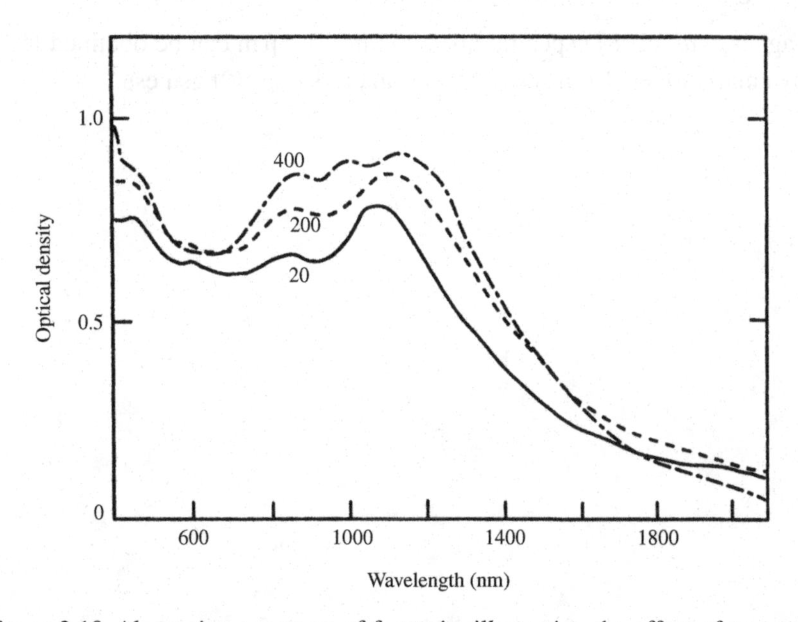

Figure 3.10 Absorption spectrum of forsterite illustrating the effect of temperature on the crystal-field band. The number next to each curve is the temperature in degrees Celsius. (Reproduced from Sung *et al.* [1977], copyright 1977 with permission of Pergamon Press Ltd.)

number of systems in the solid is to smear out the band shapes from Lorentzians to Gaussians and to cause the bands to become wider with increasing temperature.

Raising the temperature usually causes a solid to expand, increasing the average distance between the ions in the lattice and the widths of the electronic potential wells. The wavelength of an electron wave function varies directly with the width of the potential well, so that the energies of the electron states become less negative. The lower energy levels are raised more than the higher ones, so that the transition energy and thus the frequency of the center of an absorption band decrease with increasing temperature. This effect is illustrated in Figure 3.10, which shows the increase in the central wavelength of the Fe^{2+} crystal-field band in olivine with increasing temperature.

3.6 Spectral databases

Reflectance and emittance spectra of substances of interest in the geological and planetary sciences can be found on the following websites.

speclib.jpl.nasa.gov The Jet Propulsion Laboratory, US Geological Survey and Johns Hopkins University spectral libraries, vis-midIR.
speclab.asu.edu The Arizona State University spectral library, mid-IR.
lf314-rlds.geo.brown.edu also planetary.brown.edu/relabdata - vis/nearIR/midIR.

The Wagner *et al.* (1987) spectral atlas 90 nm – 1.8 μm can be obtained in digital form by email request from Jeffrey K. Wagner, jkwagn@bgsu.esu.

4
Specular reflection

4.1 Introduction

In this chapter the specular or mirror-like reflection that occurs when a plane electromagnetic wave encounters a plane surface separating two regions with different refractive indices is discussed quantitatively, along with the accompanying transmission, or refraction, through the interface. Specular reflection is important to the topic of this book for several reasons. First, it is an important tool for investigating properties of materials in the laboratory. Second, it occurs in remote-sensing applications when light is reflected from smooth parts of a planetary surface, such as the ocean. Third, it is one of the mechanisms by which light is scattered from a particle whose size is large compared with the wavelength, so that an understanding of this phenomenon is necessary to an understanding of diffuse reflectance from planetary regoliths.

4.2 Boundary conditions in electromagnetic theory

Whenever a volume contains a boundary separating regions of differing electric or magnetic constants, the components of $\mathbf{D}_e$ and $\mathbf{B}_m$ perpendicular to the surface and the components of $\mathbf{E}_e$ and $\mathbf{H}_m$ tangential to the surface must be continuous across the boundary. If the fields constitute an electromagnetic wave propagating through the surface from one medium to another, the amplitudes of the fields are different within the two regions. Therefore, the continuity conditions cannot be satisfied unless there is another wave propagating backward from the surface into the first medium, in addition to the wave propagating forward from the surface into the second medium. That is, the boundary partly transmits and partly reflects the incident wave. These two additional waves may be thought of as arising because the incident wave induces charges and currents on the surface and in the interior that in turn generate new waves that are coherent with the original wave and travel in both directions away from the interface. The fractions of the energy reflected, transmitted, or refracted, can be found from solutions of the electromagnetic wave equation that satisfy these boundary conditions.

4.3 The Fresnel equations

4.3.1 Introduction

The equations describing the reflection and transmission of a wave by a plane boundary were first derived by A. J. Fresnel and are known as the Fresnel equations.

It was shown in Chapter 2 that at any point $\mathbf{r}$ the fields of a plane electromagnetic wave traveling in a direction parallel to a unit propagation vector $\mathbf{u}_p$ can be described by

$$\mathbf{E}_e = \mathbf{E}_{e0} \exp[2\pi i(n\mathbf{u}_p \cdot \mathbf{r}/\lambda - \nu \mathbf{t})] \tag{4.1}$$

and

$$\mathbf{H}_m = \mathbf{H}_{m0} \exp[2\pi i(n\mathbf{u}_p \cdot \mathbf{r}/\lambda - \nu t)], \tag{4.2}$$

where ν is the frequency, λ is the wavelength in free space, and n is the refractive index, which may be complex. From (2.62), $\mathbf{E}_e$ and $\mathbf{H}_m$ are related by

$$\mathbf{H}_m = n\mathbf{u}_p \times \mathbf{E}_e/\mu_m c_0 \text{ or } \mathbf{E}_e = -\mu_m c_0 \mathbf{u}_p \times \mathbf{H}_m/n, \tag{4.3}$$

where μ_m is the permeability, and $c_0 = \lambda v$ is the speed of light in vacuum. In the spirit of the preceding chapters, in which all propagation parameters were lumped together into a complex index of refraction, it will be assumed in the remainder of this chapter that $\mu_m = \mu_{m0}$ everywhere and that the media on opposite sides of the boundary differ only in n.

The geometry and notation relevant to the processes of reflection and transmission are shown in Figure 4.1. The boundary separates a medium with refractive index $n_1 = n_{1r} + in_{1i}$ from one with $n_2 = n_{2r} + in_{2i}$. A plane wave is incident on the surface from medium 1 into medium 2. The normal to the surface pointing into the first medium is parallel to the unit vector $\mathbf{u}_n$. The incident wave has electric field $\mathbf{E}_e$ and magnetic intensity $\mathbf{H}_m$ and propagates in a direction parallel to the unit vector $\mathbf{u}_p$, which makes an angle ϑ with $\mathbf{u}_n$. Similar quantities with single primes denote the transmitted wave, and those with double primes the reflected wave.

Let n denote the refractive index of the second medium relative to the first,

$$n = n_2/n_1. \tag{4.4}$$

If either or both n_1 and n_2 are complex, the relative refractive index is given by

$$n = n_r + n_i = \frac{n_{2r} + in_{2i}}{n_{1r} + n_{1i}} = \frac{n_{1r}n_{2r} + n_{1i}n_{2i} + i(n_{1r}n_{2i} - n_{2r}n_{1i})}{{n_{1r}}^2 + {n_{1i}}^2},$$

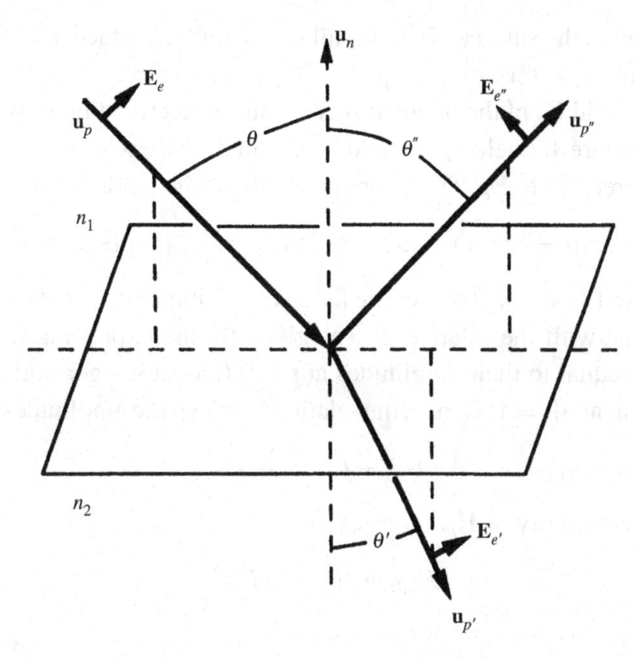

Figure 4.1 Geometry of specular reflection and transmission.

so that

$$n_r = (n_{1r}n_{2r} + n_{1i}n_{2i})/({n_{1r}}^2 + {n_{1i}}^2), \tag{4.5}$$

$$n_i = (n_{1r}n_{2i} - n_{2r}n_{1i})/({n_{1r}}^2 + {n_{1i}}^2). \tag{4.6}$$

The electric fields in the three waves are described by the following equations:

$$\text{incident}: \quad \mathbf{E}_e = \mathbf{E}_{e0}\exp[2\pi i(n_1\mathbf{u}_p \cdot \mathbf{r}/\lambda - \nu t)], \tag{4.7}$$

$$\text{refracted}: \quad \mathbf{E}'_e = \mathbf{E}'_{e0}\exp[2\pi i(n_2\mathbf{u}'_p \cdot \mathbf{r}/\lambda - \nu t)], \tag{4.8}$$

$$\text{reflected}: \quad \mathbf{E}''_e = \mathbf{E}''_{e0}\exp[2\pi i(n_1\mathbf{u}''_p \cdot \mathbf{r}/\lambda + \nu t)]. \tag{4.9}$$

The associated magnetic fields are given by (4.3). The coefficients $E_{e0} = |\mathbf{E}_{e0}|$, $E'_{e0} = |\mathbf{E}'_{e0}|$, and $E''_{e0} = |\mathbf{E}''_{e0}|$ are the amplitudes of the waves, and the exponents are the phases.

4.3.2 Reflection and transmission at normal incidence

The case when the incident wave is traveling perpendicular to the boundary surface will be considered first as a prelude to the more complicated case of incidence from an arbitrary direction. In this case the reflected and refracted waves also travel perpendicularly to the surface, which means that all of the electric and magnetic fields must be parallel to the surface. Then for all points on the surface, $\mathbf{r}$ is perpendicular to the propagation vectors so that $\mathbf{u}_p \cdot \mathbf{r} = \mathbf{u}'_p \cdot \mathbf{r} = \mathbf{u}''_p \cdot \mathbf{r} = 0$.

The electric field $\mathbf{E}_e$ of the incident wave induces electric dipoles on the surface. As shown in Figure 4.2 below, the fields generated by these dipoles are pointed in the opposite direction to $\mathbf{E}_e$. Thus, for continuity of the fields across the surface,

$$E_{e0}\exp(-2\pi i\nu t) - E''_{e0}\exp(2\pi i\nu t) = E'_{e0}\exp(-2\pi i\nu t).$$

This equation indicates that the electric field vectors increase, decrease, and reverse with period $1/\nu$ with the relative phases given by the exponents. The fields are maximum and equal to their amplitudes at $t = 0$ (and at integer multiples of $1/\nu$) when the exponentials $= 1$. Hence, the relation between the amplitudes are given by

$$E_{e0} - E''_{e0} = E'_{e0}. \tag{4.10}$$

Similarly, the continuity of $\mathbf{H}_m$ requires

$$H_{m0} + H''_{m0} = H'_{m0},$$

or

$$n_1 E_{e0}/\mu_{m0}c_0 + n_1 E''_{e0}/\mu_{m0}c_0 = n_2 E'_{e0}/\mu_{m0}c_0. \tag{4.11}$$

Solving the simultaneous equations (4.10) and (4.11) gives

$$E'_{e0} = \frac{2}{n_2/n_1 + 1}E_{e0} = \frac{2}{n+1}E_{e0} \tag{4.12}$$

and

$$E''_{e0} = \frac{1 - n_2/n_1}{n_2/n_1 + 1}E_{e0} = \frac{n-1}{n+1}E_{e0}. \tag{4.13}$$

Note that the electric vector of the reflected wave points oppositely to that of the incident wave, as shown in the figure. The sign reversal may also be thought of as the reflection causing a phase shift of 180°.

Figures 4.2a and 4.2b illustrate the case when the light is traveling from a medium with a lower refractive index to one of higher. Then the electric vector of the reflected field points oppositely to the incident and refracted vectors. For the opposite case where $n < 1$ the quantity $(n - 1)/(n + 1)$ is negative, indicating that the reflected electric field vector is opposite to that shown in Figure 4.2. Thus the field points in the same direction as the other two vectors.

If the incident light is circularly polarized it induces a separation of positive and negative charges on the surface, and the direction of the line separating the charges rotates along with the incident field. Thus, if the incident light is right-handed the charge-separation vector rotates CW when seen looking into the same direction as the incident light is propagating, as shown in the top of Figure 4.3. This generates a reflected wave whose electric vector is rotating in the same direction. However,

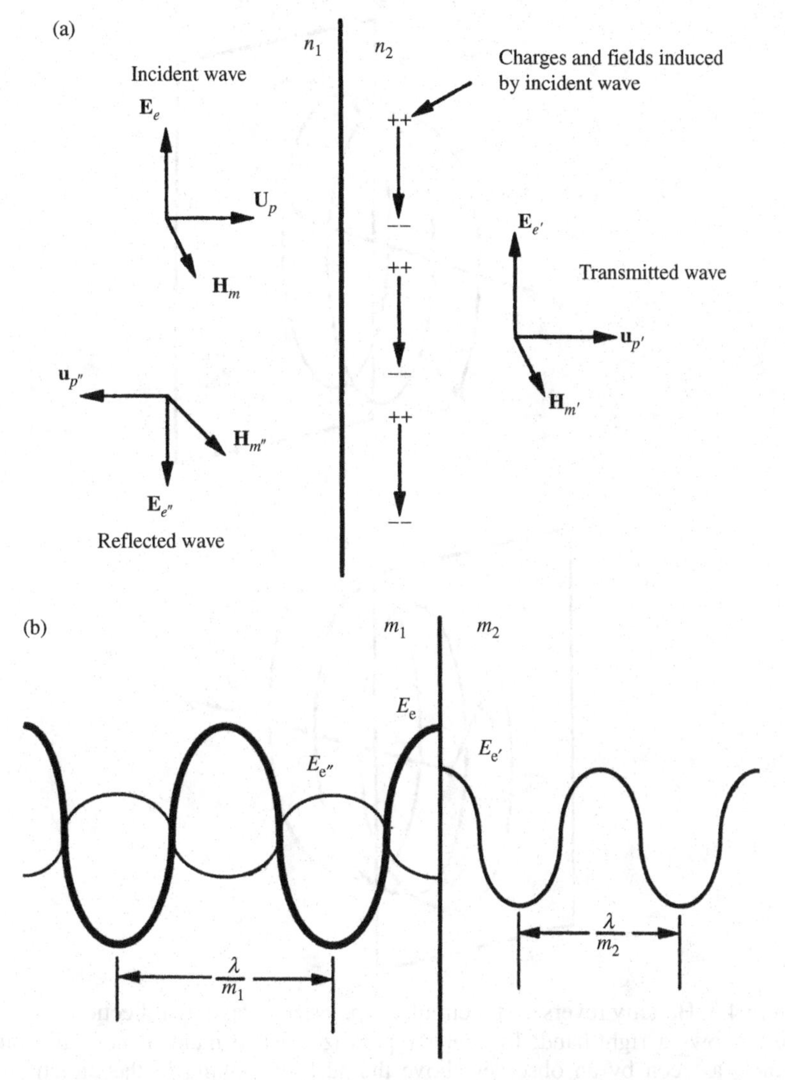

Figure 4.2 (a) Schematic diagram of the charges and fields induced on the surface by the incident wave and the fields in the incident, transmitted, and reflected waves. (b) Reflected and transmitted sinusoidal waves.

when seen looking in the direction into which the reflected wave is propagating, as shown in the bottom of Figure 4.3, the electric vector appears to rotate CCW so that the reflected wave is left-handed. Thus, reflectance reverses the helicity of a circularly polarized wave. This is true regardless of whether n_r is smaller or larger than 1.

It was shown in Chapter 2 (equation [2.94]) that the irradiance is proportional to the square of the absolute value of the electric field. Hence, the reflection coefficient,

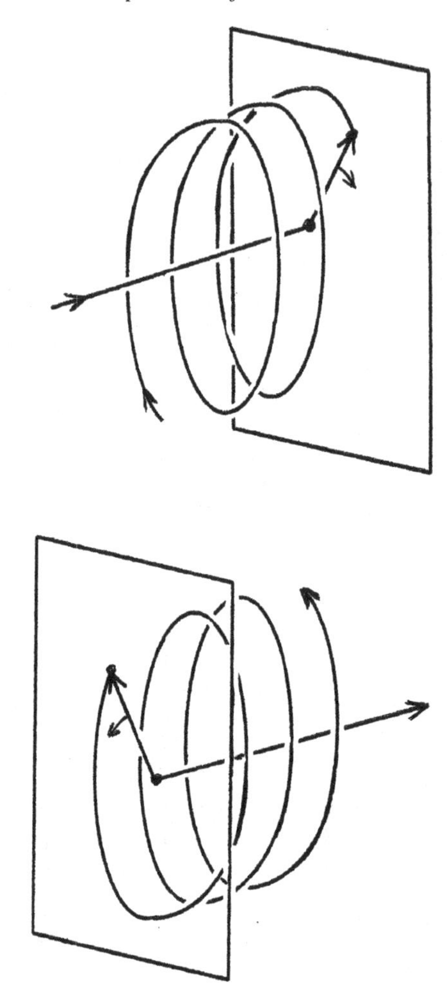

Figure 4.3 Helicity reversal of a circularly polarized wave by reflection. The top figure shows a right-handed circularly polarized wave incident normally on a surface, as seen by an observer above the surface looking in the direction of propagation. The bottom figure shows the reflected wave as seen by an observer located just below the surface looking in the direction of propagation.

the ratio of reflected to incident intensity, is

$$R = \left|\frac{E''_{e0}}{E_{e0}}\right|^2 = \left|-\frac{n-1}{n+1}\right|^2 = \frac{(n_r-1)^2+n_i^2}{(n_r+1)^2+n_i^2}. \tag{4.14}$$

The transmission coefficient, the ratio of transmitted to incident light, is

$$T = \mathrm{Re}\left[n * \left|\frac{E''_{e0}}{E_{e0}}\right|^2\right] = n_r\left|\frac{2}{n+1}\right|^2 = \frac{4n_r}{(n_r+1)^2+n_i^2}. \tag{4.15}$$

It may be readily verified that $T = 1 - R$, as it must, because the boundary does not absorb or create energy. If $n_2 = n_1$ then $R = 0$ and $T = 1$; thus, if there is no change in refractive index, the boundary is invisible.

If the first medium is a vacuum, so that $n_1 = 1 + i0$; then $n_r = n_{2r}$ and $n_i = n_{2i}$. If $n_{2i} = 0$, then

$$R = \left(\frac{n_r - 1}{n_r + 1}\right)^2, \tag{4.16}$$

an equation familiar to all students in introductory mineralogy courses. However, if either medium has losses, then

$$R = \frac{(n_r - 1)^2 + n_i^2}{(n_r + 1)^2 + n_i^2}. \tag{4.17}$$

This equation increases monotonically with n_i from a value of $R = [(n_r - 1)/(n_r + 1)]^2$ at $n_i = 0$ to $R \to 1$ when $n_i \gg 1$. Thus, an imaginary component of the refractive index increases the reflectivity of the boundary. This is why metals appear bright when seen in reflected light.

At first sight it may appear contradictory that an imaginary component of the refractive index that is associated with absorption of radiation (as we saw in Chapter 2) also causes an increase in the reflected light. The incident fields may be thought of as inducing dipoles and currents at the surface and in the interior, and they generate radiation that propagates in both directions from the surface. The radiation traveling backward constitutes the reflected light; that traveling forward combines coherently with the incident fields and reduces the intensity of the transmitted light. If $n_i = 0$, the induced dipole moments are relatively small in most substances, so that the radiation generated by them is weak. Hence R is small and T is large. In order to affect the reflectance appreciably, n_i must be of the order of 1. However, an imaginary component this large implies a large conductivity (equation [2.79]), which means that the charges are highly mobile. The induced dipoles and the radiation they generate are strong, resulting in a high reflectivity. Most of the radiation is reflected, rather than being absorbed, as a consequence of the ability of the charges to rearrange themselves in such a manner as to inhibit the fields from entering the second medium.

4.3.3 Reflection and transmission at arbitrary angles

4.3.3.1 Snell's law

We now consider the more general case when $\theta \neq 0$. Let the point described by $\mathbf{r}$ lie in the boundary surface. The tangential components of both $\mathbf{E}_e$ and $\mathbf{H}_m$ must be continuous across this surface. This can be true only if the spatial parts of the phases are equal,

$$n_1 \mathbf{u}_p \cdot \mathbf{r}/\lambda = n_2 \mathbf{u}'_p \cdot \mathbf{r}/\lambda = n_1 \mathbf{u}''_p \cdot \mathbf{r}/\lambda. \qquad (4.18)$$

This equation implies that $\mathbf{u}_n$, $\mathbf{u}_p$, $\mathbf{u}'_p$, and $\mathbf{u}''_p$ are coplanar. To see this, suppose that $\mathbf{r}$ is a vector in the surface and perpendicular to the plane containing both $\mathbf{u}_p$ and $\mathbf{u}_n$, so that $\mathbf{u}_p \cdot \mathbf{r} = 0 = \mathbf{u}_n \cdot \mathbf{r}$. But the only nontrivial way that $\mathbf{u}'_p \cdot \mathbf{r}$ and $\mathbf{u}''_p \cdot \mathbf{r}$ can also be zero is if all three propagation vectors and $\mathbf{u}_n$ lie in the same plane.

By definition, for any vector $\mathbf{r}$ in the surface, $\mathbf{u}_n \cdot \mathbf{r} = 0$. Thus, using vector identity (A.5),

$$\mathbf{u}_n \times (\mathbf{u}_n \times \mathbf{r}) = (\mathbf{u}_n \cdot \mathbf{r})\, \mathbf{u}_n - (\mathbf{u}_n \cdot \mathbf{u}_n)\, r = -\mathbf{r},$$

so that, using vector identity (A.6),

$$\mathbf{u}_p \cdot \mathbf{r} = -\mathbf{u}_p \cdot [\mathbf{u}_n \times (\mathbf{u}_n \times \mathbf{r})] = -\left(\mathbf{u}_p \times \mathbf{u}_n\right) \cdot (\mathbf{u}_n \times \mathbf{r}).$$

Similar expressions hold for $\mathbf{u}'_p$ and $\mathbf{u}''_p$. Hence, equation (4.9) can be put in the forms

$$(n_1/\lambda)\left(\mathbf{u}_p \times \mathbf{u}_n\right) \cdot (\mathbf{u}_n \times \mathbf{r}) = (n_1/\lambda)\left(\mathbf{u}''_p \times \mathbf{u}_n\right) \cdot (\mathbf{u}_n \times \mathbf{r}) \qquad (4.19)$$

and

$$(n_1/\lambda)\left(\mathbf{u}_p \times \mathbf{u}_n\right) \cdot (\mathbf{u}_n \times \mathbf{r}) = (n_2/\lambda)\left(\mathbf{u}'_p \times \mathbf{u}_n\right) \cdot (\mathbf{u}_n \times \mathbf{r}). \qquad (4.20)$$

Equation (4.19) shows that $\theta'' = \theta$. This is most easily seen by supposing that $\mathbf{r}$ is a vector in the surface and also in the common plane of the propagation vectors. Then $\mathbf{u}_n \times \mathbf{r}$ is perpendicular to the propagation plane and parallel to $\mathbf{u}_p \times \mathbf{u}_n$ and $\mathbf{u}''_p \times \mathbf{u}_n$. In this case, (4.19) becomes $(n_1 r/\lambda)\sin\theta = (n_1 r/\lambda)\sin\theta''$, from which it follows that $\theta'' = \theta$.

Similarly, equation (4.20) shows that θ and θ' are related by $n_1 \sin\theta = n_2 \sin\theta'$, or

$$\sin\theta = n \sin\theta', \qquad (4.21)$$

which is Snell's law.

To proceed further, it is convenient to resolve the field vectors into two components parallel and perpendicular to the plane of propagation and consider them separately. Let subscript "$\perp$" denote the component with $\mathbf{E}$ perpendicular to the plane containing the propagation vectors, in which case $\mathbf{H}$ is in that plane; let subscript "$\|$" denote the component with $\mathbf{E}$ in the same plane as the propagation vectors, in which case $\mathbf{H}$ is perpendicular to that plane. The first case is called the *transverse electric* case and the second the *transverse magnetic* case.

4.3.3.2 $\mathbf{E}_e$ perpendicular to the plane of propagation

In the transverse electric case the electric vectors are all perpendicular to $\mathbf{u}_n$ and parallel to the boundary surface. The conditions that the tangential components of $\mathbf{E}_e$ and $\mathbf{H}_m$ be continuous across the boundary can be written, respectively,

$$\mathbf{u}_n \times \left(\mathbf{E}_e + \mathbf{E}_e''\right) = \mathbf{u}_n \times \mathbf{E}_e' \tag{4.22}$$

and, from (4.2),

$$n_1 \mathbf{u}_n \times \left(\mathbf{u}_p \times \mathbf{E}_e/\mu_{m0}c_0 + n_1 \mathbf{u}_p'' \times \mathbf{E}_e''/\mu_{m0}c_0\right) = n_2 \mathbf{u}_n \times (\mathbf{u}_p' \times \mathbf{E}_e'/\mu_{m0}c_0), \tag{4.23}$$

where the fields are given by (4.4), and $\mathbf{r}$ lies in the surface. Because spatial parts of the phases must all be equal on the surface, (4.22) becomes

$$E_{e0} + E_{e0}'' = E_{e0}'. \tag{4.24}$$

Using vector relation (A.5) gives

$$\mathbf{u}_n \times \left(\mathbf{u}_p \times \mathbf{E}_e\right) = \left(\mathbf{u}_n \cdot \mathbf{E}_e\right)\mathbf{u}_p - \left(\mathbf{u}_n \cdot \mathbf{u}_p\right)\mathbf{E}_e = -\left(\mathbf{u}_n \cdot \mathbf{u}_p\right)\mathbf{E}_e,$$

since $\mathbf{u}_n \cdot \mathbf{E}_e = 0$. Similar expressions hold for $\mathbf{E}_e'$ and $\mathbf{E}_e''$. Hence, (4.23) is

$$n_1\left(\mathbf{u}_n \cdot \mathbf{u}_p\right)\mathbf{E}_e + n_1\left(\mathbf{u}_n \cdot \mathbf{u}_p''\right)\mathbf{E}_e'' = n_2\left(\mathbf{u}_n \cdot \mathbf{u}_p'\right)\mathbf{E}_e',$$

or

$$-n_1 E_{e0}\cos\theta + n_1 E_{e0}''\cos\theta'' = -n_2 E_{e0}'\cos\theta'.$$

Because $\theta'' = \theta$, and $n = n_2/n_1$, this becomes

$$E_{e0}\cos\theta - E_{e0}''\cos\theta = nE_{e0}'\cos\theta'. \tag{4.25}$$

Solving the simultaneous equations (4.24) and (4.25) gives

$$\frac{E_{e0}'}{E_{e0}} = \frac{2\cos\theta}{\cos\theta + n\cos\theta'} \tag{4.26}$$

and

$$\frac{E_{e0}''}{E_{e0}} = \frac{\cos\theta - n\cos\theta'}{\cos\theta + n\cos\theta'}. \tag{4.27}$$

4.3.3.3 $\mathbf{E}_e$ parallel to the plane of propagation

In the transverse magnetic case the magnetic vectors are perpendicular to $\mathbf{u}_n$ and parallel to the surface. Hence, the boundary condition that the magnetic intensity be continuous across the surface is

$$\mathbf{u}_n \times \left(\mathbf{H}_m + \mathbf{H}_m''\right) = \mathbf{u}_n \times \mathbf{H}_m',$$

so that

$$H_{m0} + H''_{m0} = H'_{m0}. \tag{4.28}$$

The condition that the tangential component of the electric field be continuous is, from (4.2),

$$\mathbf{u}_n \times \left(\mu_{m0}c_0\mathbf{u}_p \times \mathbf{H}_m/n_1\right) + \mathbf{u}_n \times \left(\mu_{m0}c_0\mathbf{u}''_p \times \mathbf{H}''_m/n_1\right) = \mathbf{u}_n \times \left(\mu_{m0}c_0\mathbf{u}'_p \times \mathbf{H}'_m/n_2\right).$$

Because $\mathbf{u}_n \cdot \mathbf{H}_m = 0$,

$$\mathbf{u}_n \times (\mathbf{u}_n \times \mathbf{H}_m) = (\mathbf{u}_n \cdot \mathbf{H}_m)\,\mathbf{u}_p - (\mathbf{u}_n \cdot \mathbf{u}_p)\,\mathbf{H}_m = -(\mathbf{u}_n \cdot \mathbf{u}_p)\,\mathbf{H}_m;$$

thus,

$$\mu_{m0}c_0\left(\mathbf{u}_n \cdot \mathbf{u}_p\right)\mathbf{H}_m/n_1 + \mu_{m0}c_0\left(\mathbf{u}_n \cdot \mathbf{u}''_p\right)\mathbf{H}''_m/n_1 = \mu_{m0}c_0\left(\mathbf{u}_n \cdot \mathbf{u}'_p\right)\mathbf{H}'_m/n_2,$$

or

$$mH_{m0}\cos\theta - mH''_{m0}\cos\theta = H'_{m0}\cos\theta'. \tag{4.29}$$

Combining (4.28) and (4.29) gives

$$\frac{H'_{m0}}{H_{m0}} = \frac{2n\cos\theta}{n\cos\theta + \cos\theta'}$$

and

$$\frac{H''_{m0}}{H_{m0}} = \frac{n\cos\theta - \cos\theta'}{n\cos\theta + \cos\theta'},$$

or, using (2.63),

$$\frac{E'_{e0}}{E_{e0}} = \frac{2\cos\theta}{n\cos\theta + \cos\theta'}, \tag{4.30}$$

$$\frac{E''_{e0}}{E_{e0}} = \frac{n\cos\theta - \cos\theta'}{n\cos\theta + \cos\theta'}. \tag{4.31}$$

These results have been obtained using only the tangential components of $\mathbf{E}_e$ or $\mathbf{H}_m$, as appropriate. Using a similar procedure, it can be shown that the continuity conditions on the normal components of $\mathbf{D}_e$ and $\mathbf{B}_m$ have also been satisfied.

4.3.4 *Reflection and transmission when* n *is real*

The reflection and transmission coefficients and their properties when n_1 and n_2 are real will be discussed first. According to (2.94) the irradiance is proportional to $\mathrm{Re}(n|E_e|^2)$. Hence, the reflectivities $R_\perp$ and $R_\parallel$ when $\mathbf{E}_e$ is perpendicular and parallel, respectively, to the plane of propagation are given by

$$R_\perp = \left[\frac{\cos\theta - n\cos\theta'}{\cos\theta + n\cos\theta'}\right]^2, \tag{4.32}$$

$$R_\mathrm{p} = \left[\frac{n\cos\theta - \cos\theta'}{n\cos\theta + \cos\theta'}\right]^2, \tag{4.33}$$

where θ' is given by Snell's law

$$\sin\theta = n\sin\theta'. \tag{4.34}$$

The corresponding transmission coefficients are given by

$$T_\perp = \frac{4n\cos^2\theta}{(\cos\theta + n\cos\theta')^2} = (1 - R_\perp)\frac{\cos\theta}{\cos\theta'}, \tag{4.35}$$

$$T_\parallel = \frac{4n\cos^2\theta}{(n\cos\theta + \cos\theta')^2} = \left(1 - R_\parallel\right)\frac{\cos\theta}{\cos\theta'} \tag{4.36}$$

Expressions (4.32) – (4.36) are the Fresnel equations for n real. The factor $\cos\theta/\cos\theta'$ in the transmission coefficients arises because the flux of energy crossing an area A perpendicular to $\mathbf{u}_p$ in the incident wave illuminates an area $A\sec\theta$ on the surface; the light transmitted through this area is contained in area $A\sec\theta\cos\theta'$ in the second medium (Figure 4.4).

It is left as an exercise for the reader to show that (4.32) and (4.33) can be put in the following equivalent forms:

$$R_\perp = \left[\frac{\cos\theta - \sqrt{n^2 - \sin^2\theta}}{\cos\theta + \sqrt{n^2 - \sin^2\theta}}\right]^2 = \left[\frac{\sin(\theta - \theta')}{\sin(\theta + \theta')}\right]^2, \tag{4.37}$$

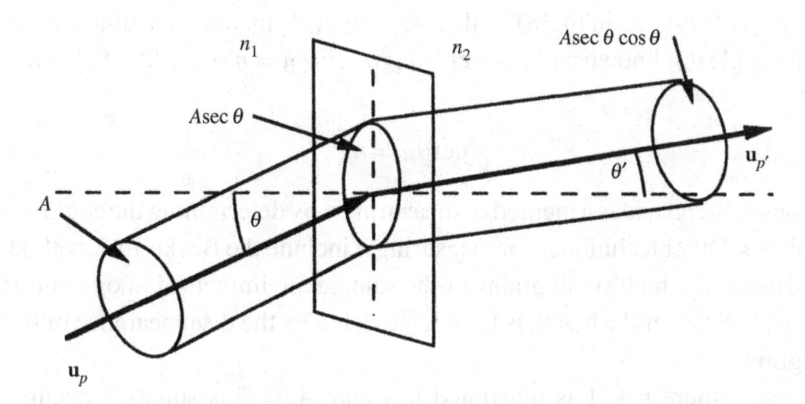

Figure 4.4 Continuity of the flux through the surface.

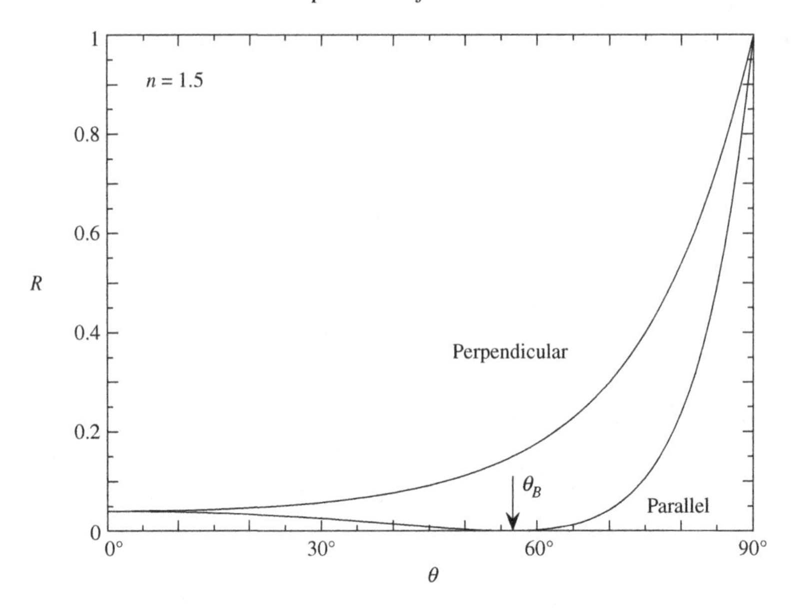

Figure 4.5 Fresnel reflection coefficients for $n = 1.50 + i0$ versus the angle of incidence.

and

$$R_{\parallel} = \left[\frac{n^2 \cos\theta - \sqrt{n^2 - \sin^2\theta}}{n^2 \cos\theta + \sqrt{n^2 - \sin^2\theta}} \right]^2 = \left[\frac{\tan\left(\theta - \theta'\right)}{\tan\left(\theta + \theta'\right)} \right]^2 . \tag{4.38}$$

The reflectivities $R_{\perp}$ and $R_{\parallel}$ versus θ are plotted in Figure 4.5 for a case where $n > 1$. Equation (4.21) shows that in this case $\theta > \theta'$. When $\theta = 0$, $R_{\perp}(0) = R_{\parallel}(0) = [(n-1)/(n+1)]^2$, which is identical with (4.16). When $\theta = \pi/2$, $R_{\perp}(\pi/2) = R_{\parallel}(\pi/2) = 1$. As θ increases from zero, $R_{\perp}$ increases monotonically. However, $R_{\parallel}$ first decreases and becomes zero at the angle $\theta = \theta_B$ where $\theta_B + \theta' = \pi/2$ so that $\tan(\theta_B + \theta') \rightarrow \infty$ in (4.38), following which $R_{\parallel}$ increases to unity at $\theta = \pi/2$.

At the angle θ_B, known as *Brewster's angle,* $\sin\theta_{\mathrm{B}} = n\sin(\pi/2 - \theta_{\mathrm{B}}) = n\cos\theta_{\mathrm{B}}$, so that

$$\tan\theta_B = n \tag{4.39}$$

Equation (4.39) provides a method of measuring n by determining the angle at which R_{p} vanishes. Other techniques for measuring n include the Becke-line method (e.g., Bloss, 1961), in which small grains of the sample are immersed in oils of different refractive indices until a match is found, indicated by the disappearance of the edge of the grain.

The case where $n < 1$ is illustrated in Figure 4.6. This situation occurs when $n_1 > n_2$. For example, the interface may be between two materials of different

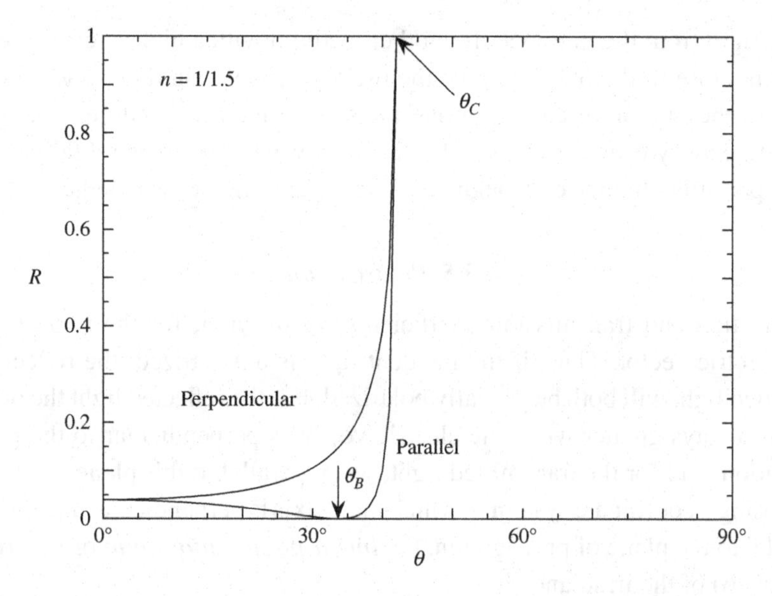

Figure 4.6 Fresnel reflection coefficients for $n = 1/1.50 + i0$ versus the angle of incidence.

refractive index, or between water and air, or between air and a medium with $n_r < 1$ in the anomalous dispersion region of a strong absorption band. When $\theta = 0$,

$$\frac{E'_{e0}}{E_{e0}} = \frac{1 - 1/n}{1 + 1/n} = \frac{n-1}{n+1}.$$

Thus, at small angles the reflectivities are equal to the case where $n_r > 1$; that is, they have the value $R_\perp(0) = R_\parallel(0) = |(n-1)/(n+1)|^2$, and $R_\perp$ increases monotonically, while $R_\parallel$ first decreases to zero at the Brewster angle, given by (4.39), and then increases. However, θ' increases more rapidly than θ. When $\theta' = \pi/2$, θ is equal to the *critical angle* θ_c. At the critical angle Snell's law is

$$\sin\theta_c = n, \tag{4.40}$$

and $R_\perp(\theta_c) = R_\parallel(\theta_c) = 1$. Because θ' cannot exceed $\pi/2$, the reflectivities are unity for all θ between θ_c and $\pi/2$.

When $\theta > \theta_c$, $R_\perp = R_\mathrm{p} = 1$ and $T_\perp = T_\parallel = 0$. For these angles the boundary completely reflects all incident radiation, and the system is said to exhibit *total internal reflection.* A detailed analysis shows that the fields actually penetrate a short distance into the medium of lower refractive index, but that waves propagating away from the surface do not exist. Instead a system of moving fields, called an *evanescent wave,* is generated that propagates along the surface, and whose amplitude is appreciable in the lower refractive index medium only within about a wavelength from the interface. If another medium with a higher refractive index is brought close to the first interface the oscillating fields in the evanescent wave will cause the motion of charges along the second interface and generate a wave propagating away from the surface. The incident wave is no longer totally reflected and has partially "tunneled through" the gap separating the two media.

4.3.5 Polarization

The reflection and transmission coefficients are different for the two directions of the electric vector. Thus, if the incident light is unpolarized, the reflected and transmitted light will both be partially polarized. For the reflected light the reflected power is always greater when the electric vector is perpendicular to the plane of propagation, and for the transmitted light, when parallel to this plane.

Choosing a coordinate system in which the x-axis is perpendicular and the y-axis is parallel to the plane of propagation, the *linear polarization ratio* or *polarization* (Section 2.6) of the irradiance is

$$P = (J_\perp - J_\parallel)/(J_\perp + J_\parallel), \tag{4.41}$$

where, as usual, the subscript "$\perp$" denotes the electric vector perpendicular to the plane of propagation, and the subscript "$\parallel$" denotes the electric vector parallel to this plane; P is a scalar. If the incident radiation J is unpolarized, then $J_\perp = J_\parallel = J/2$. Hence the polarization of the reflected light is

$$P = (R_\perp - R_\parallel)/(R_\perp + R_\parallel),$$

and that of the transmitted light is

$$P = (T_\perp - T_\parallel)/(T_\perp + T_\parallel) = -(R_\perp - R_\parallel)/(2 - R_\perp - R_\parallel).$$

The polarization is shown in Figure 4.7. Note that it is always positive for reflected light and becomes $+100\%$ at the Brewster angle. The polarization is always negative for transmitted light but nowhere reaches –100%.

If the incident light is linearly polarized in an arbitrary direction it can be resolved into transverse electric and transverse magnetic components vibrating in phase with each other. As the angle of incidence increases from zero the unequal reflection coefficients of the two components cause the plane of polarization to rotate until at the Brewster angle the plane of polarization of the reflected light is entirely perpendicular to the scattering plane.

If the incident light is circularly polarized the two components are 90° out of phase. As the angle of incidence increases the unequal reflection coefficients cause the light to become elliptically polarized. At the Brewster angle the ellipse is perfectly flat and the reflected light is linearly polarized.

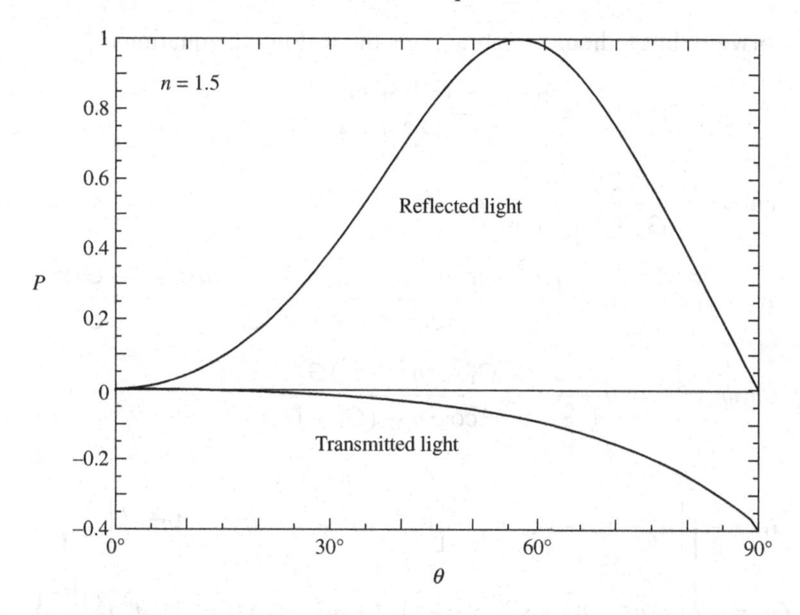

Figure 4.7 Polarization of the reflected and transmitted light for $n = 1.5 + i0$ versus the angle of incidence.

4.3.6 Reflection and refraction when n is complex

When n_1 or n_2 is complex, the amplitude ratios E''_{e0}/E_{e0}, given by (4.27) and (4.31) for the directions of **E**, are complex also. Denote the amplitude reflection coefficients, that is, the ratio of the amplitudes of the fields, by

$$r_\perp = \frac{\cos\theta - n\cos\theta'}{\cos\theta + n\cos\theta'} = \eta_\perp e^{i\phi_\perp}, \tag{4.42a}$$

$$r_\parallel = \frac{n\cos\theta - \cos\theta'}{n\cos\theta + \cos\theta'} = \eta_\parallel e^{i\phi_p}. \tag{4.42b}$$

where the ηs are the amplitudes and the ϕs are the phases of the reflection coefficients. The physical meaning of $\phi_\perp$ and $\phi_\parallel$ is that shifts of this amount in the phase of the wave occur upon reflection. Transmission is similarly accompanied by phase shifts.

The bending of a ray of light upon refraction is caused by the change in the velocity of propagation, c_0/n_r (equation [2.88]). Hence, in media with complex refractive indices Snell's law is

$$n_{1r}\sin\theta = n_{2r}\sin\theta'. \tag{4.43}$$

Straightforward, but tedious, algebra gives the following equations:

$$R_\perp = |r_\perp|^2 = \eta_1^2 = \frac{[\cos\theta - G_1]^2 + G_2^2}{[\cos\theta + G_1]^2 + G_2^2}, \tag{4.44}$$

$$\tan\phi_\perp = \frac{2G_2\cos\theta}{G_1^2 + G_2^2 - \cos^2\vartheta}, \tag{4.45}$$

$$R_\parallel = |r_\parallel|^2 = \eta_2^2 = \frac{[(n_r^2 - n_i^2)\cos\theta - G_1]^2 + [2n_r n_i \cos\theta - G_2]^2}{[(n_r^2 - n_i^2)\cos\theta + G_1]^2 + [2n_r n_i \cos\theta + G^2]^2}, \tag{4.46}$$

$$\tan\phi_\parallel = 2\cos\theta \frac{2n_r n_i G_1 - (n_r^2 - n_i^2)G_2}{(n_r^2 + n_i^2)^2\cos^2\theta - (G_1^2 + G_2^2)}, \tag{4.47}$$

where

$$G_1^2 = \frac{1}{2}\left\{\left[n_r^2 - n_i^2 - \sin^2\theta\right] + \left[(n_r^2 - n_i^2 - \sin^2\theta)^2 + 4n_r^2 n_i^2\right]^{1/2}\right\}, \tag{4.48}$$

$$G_2^2 = \frac{1}{2}\left\{-\left[n_r^2 - n_i^2 - \sin^2\theta\right] + \left[(n_r^2 - n_i^2 - \sin^2\theta)^2 + 4n_r^2 n_i^2\right]^{1/2}\right\}. \tag{4.49}$$

At $\theta = 0$ the reflectivities are $R_\perp(0) = R_\parallel(0) = \left[(n_r - 1)^2 + n_i^2\right]/\left[(n_r + 1)^2 + n_i^2\right]$, which is the same as (4.17). As θ increases, the component of the reflectivity perpendicular to the plane of propagation increases monotonically to 1 at $\theta = \pi/2$. The parallel component first decreases to a minimum and then increases to 1 at $\theta = \pi/2$; however, the minimum never becomes zero, as it does at the Brewster angle when $n_i = 0$. The reflectivities are plotted versus θ in Figure 4.8 for a medium with a complex refractive index.

In two special cases equations (4.46) – (4.49) can be simplified somewhat. When $n_i \ll 1$,

$$n_r^2 - n_i^2 \simeq n_r^2, \quad G_1 \simeq \left(n_r^2 - \sin^2\theta\right)^{1/2}, \quad G_2 \simeq n_r n_i/(n_r^2 - \sin^2\theta)^{1/2}. \tag{4.50}$$

At the opposite extreme, when $|n_r^2 - n_i^2| \gg 1$,

$$G_1 \simeq n_r, \quad G_2 \simeq n_i, \tag{4.51}$$

and, after a little algebra, the reflectivities become

$$R_\perp = \frac{(n_r - \cos\theta)^2 + n_i^2}{(n_r + \cos\theta)^2 + n_i^2}, \tag{4.52}$$

$$R_\parallel = \frac{(n_r - 1/\cos\theta)^2 + n_i^2}{(n_r + 1/\cos\theta)^2 + n_i^2}. \tag{4.53}$$

When $n_i \neq 0$ and the incident light is unpolarized, the different phase shifts for the two reflection coefficients cause the reflected light to be partially circularly polarized.

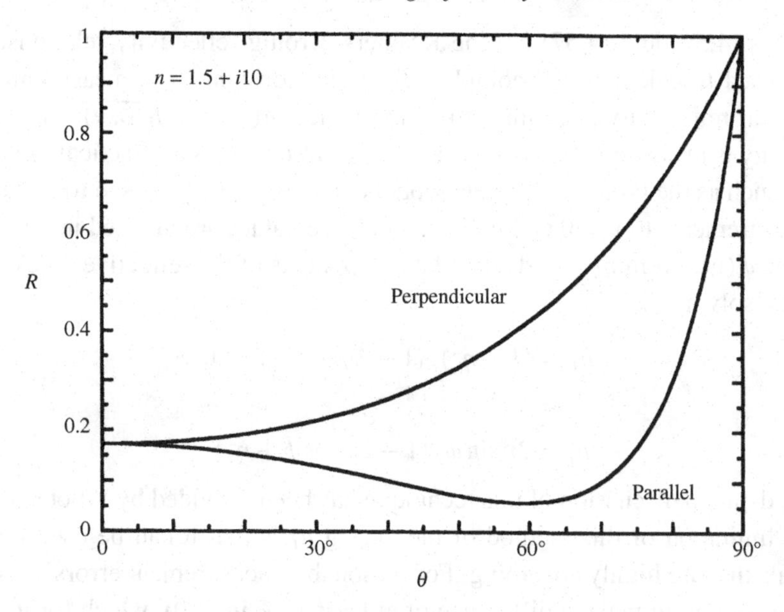

Figure 4.8 Fresnel reflection coefficients for $n = 1.50 + i1.0$ versus the angle of incidence.

4.4 The Kramers–Kronig reflectivity relations

At $\theta = 0$ the complex amplitude reflection coefficient for the two directions of polarization ([4.42a] and [4.42b]) are equal and can be written

$$r(0) = r_{\perp}(0) = r_{\parallel}(0) = (n_r - 1 + in_i)/(n_r + 1 + in_i) = \eta e^{i\phi}, \tag{4.54}$$

where η is the amplitude of the reflection coefficient and ϕ is its phase.

Because n_r and n_i are both functions of frequency ν, η and ϕ are also functions of ν. Taking the logarithm of (4.54) gives

$$\ln[\mathbf{r}(0, \nu)] = \ln[\eta(\nu)] + i\phi(\nu). \tag{4.55}$$

The reflectivity must obey causality. Thus, as discussed in Chapter 3, the expression for the amplitude reflectivity possesses crossing symmetry; that is, $r(0, -\nu) = r^*(0, \nu)$. Hence, according to the discussion in AppendixB, $\eta(\nu)$ and $\phi(\nu)$ must obey relations of the form of (B.14) and (B.15):

$$\ln[\eta(\nu)] = \frac{2}{\pi}\int_0^{\infty} \frac{\nu'\phi(\nu')}{\nu'^2 - \nu^2} d\nu', \tag{4.56}$$

$$\phi(\nu) = -\frac{2\nu}{\pi}\int_0^{\infty} \frac{\ln[\eta(\nu')]}{\nu'^2 - \nu^2} d\nu'. \tag{4.57}$$

Equations (4.56) and (4.57) are the Kramers–Kronig reflectivity relations. They can be used to calculate the complex dielectric constant from measurements of the normal reflectivity as a function of frequency, $\eta(\nu) = \sqrt{R(0,\nu)}$. The normal reflectivity is measured over as wide a range of frequencies as practical. A suitable theory, such as the Drude or Lorentz model (Chapter 3), is then used to extrapolate the measurements to $\nu = 0$ and $\nu = \infty$, and these values are inserted into (4.57) to calculate $\phi(\nu)$. From $\eta(\nu)$ and $\phi(\nu)$ the components of the refractive index can be found by solving (4.54):

$$n_r = (1-\eta^2)/(1-2\eta\cos\phi+\eta^2). \tag{4.58}$$

and

$$n_i = 2\eta\sin\phi/(1-2\eta\cos\phi+\eta^2). \tag{4.59}$$

A more detailed discussion of this technique has been provided by Wooten (1972).

One limitation of this method of measuring n_i is that it can be used only for materials that are highly absorbing. For reasonable measurement errors, n_i should be large enough to make a difference of at least 10% in $R(0)$, which for $n_r = 1.5$ means $n_i \geq 0.15$. From (2.95) it may be seen that this requires that the absorption coefficient be so large that the irradiance propagating through the medium is reduced by a factor of e over a distance of less than half a wavelength.

4.5 Absorption bands in reflectivity

The spectral reflectivity at normal incidence of a material having an absorption band given by a classical Lorentz oscillator with the same parameters as in Figures 3.1 and 3.2 is shown in Figure 4.9. The material is assumed to be in air, so that $n = n_2 = n_r + in_i$. Note that the peak of the reflectivity occurs at a higher frequency than the band center ν_0 and that the band is broadened. The reason the frequency of the band center shifts is the anomalous dispersion of the real part of the refractive index, which, as can be seen in Figure 3.2, decreases with frequency throughout the main region of the absorption band.

If the material has only one absorption band, and is in vacuum, then Figure 3.2 shows that $n_r = 1$ only at ν_0, the band center. However, if the continuum refractive index n_c is > 1, n_r will also $= 1$ at a second frequency on the high-frequency side of ν_0. This second frequency at which $n_r = 1$ is known as the *Christiansen frequency*, denoted by ν_C, and the corresponding wavelength is the Christiansen wavelength λ_C.

The Christiansen wavelength provides a method for making a narrow-bandwidth transmission filter. Small crystals of suitable material are embedded in a second medium whose refractive index is chosen to make $n_r = 1$ at the desired frequency, which also must be away from the center of any strong absorption bands in either

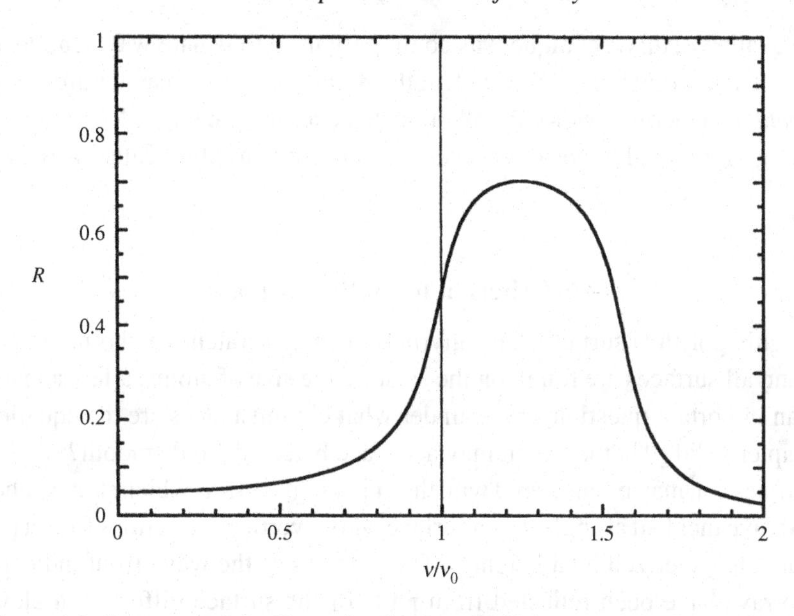

Figure 4.9 Spectrum of the Fresnel reflection coefficient at normal incidence for a material having the Lorentz dielectric constant and refractive index shown in Figures 3.1 and 3.2. Note that the reflectivity peak is much broader than, and is displaced to the high-frequency side of the dielectric-constant peak.

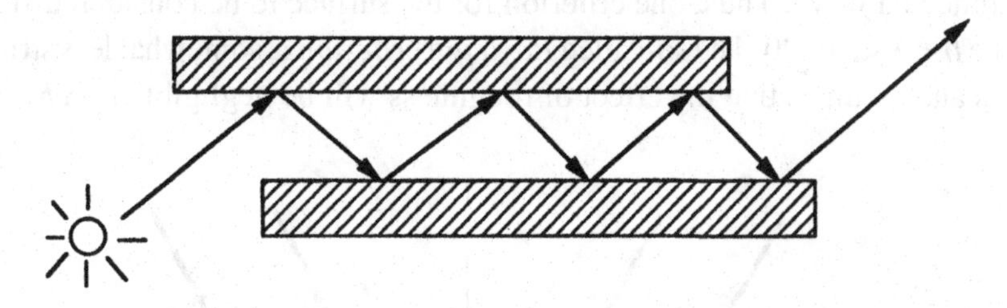

Figure 4.10 Isolating the frequencies in a restrahlen band by multiple reflections.

material. Rays of light whose frequencies are close to the Christiansen frequency will pass undeviated through the surfaces between the two materials, but rays of other frequencies will be bent out of the beam by refraction.

The peak in reflectivity near the center of a strong absorption band provides another method of obtaining a nearly monochromatic beam of radiation without using a prism or diffraction grating. Light from a source giving off a broad range of wavelengths is arranged to be reflected many times from the polished surfaces of an appropriate material, as shown in Figure 4.10. After several reflections the beam consists almost entirely of light of a narrow range of wavelengths around the reflectivity maximum. Before long-wavelength diffraction gratings were widely available, this technique was commonly used to isolate wavelengths in the infrared. The narrow bands of wavelengths produced in this manner are known as *Reststrahlen*, a German word literally translated as "residual rays." For this reason the strong vibrational bands of solids in the infrared are often called *Reststrahlen bands*.

4.6 Criterion for optical flatness

Even highly polished surfaces have myriads of small scratches and other imperfections, and all surfaces are rough on the scale of the size of atoms, a few angstroms. Thus, an important question arises: under what circumstances are the equations of this chapter valid? That is, when may a surface be considered smooth?

This question may be answered with the aid of Figure 4.11, which shows schematically a wave incident at angle θ on a surface whose vertical deviations from a perfect plane are characterized by a height h. The portions of the wave front indicated by the two rays have been reflected from parts of the surface differing in elevation by h. As indicated in the figure, the path difference over which the two rays have traveled is $\Delta L = 2h\cos\theta$, so that the portions of the wave front they represent differ in phase by $\Delta\phi = 2\pi\,\Delta L/\lambda = 4\pi h\cos\theta/\lambda$.

If $\Delta\phi \ll 2\pi$, say $\Delta\phi < 2\pi/10$, then the roughness will have negligible effect on the reflected wave. Thus, one criterion for the surface to be considered smooth would be $h < \lambda\sec\theta/20$. The so-called *Rayleigh criterion* is somewhat less stringent than this and assumes that the effect of roughness will be negligible if $\Delta\phi < \pi/2$,

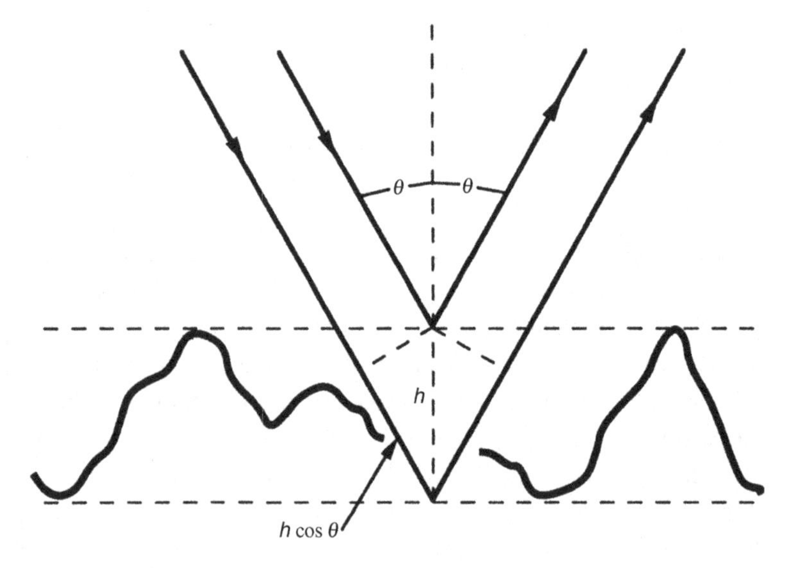

Figure 4.11 Schematic diagram of two parts of a wave reflected from a rough surface.

leading to

$$h < \lambda \sec\theta/8 \tag{4.60}$$

as the criterion for a surface to be considered smooth.

On the other hand, if the two portions of the wave front differ in phase by π or more, then they will certainly interfere with each other, either constructively or destructively, so that the surface will act as a rough scatterer if

$$h > \lambda \sec\theta/4. \tag{4.61}$$

Note that the roughness criteria depend on θ. A surface may be rough for normal incidence but smooth for glancing incidence.

The transition between rough- and smooth-surface scattering has been studied experimentally by Schaber, Berlin, and Brown (1976) as a method for measuring the surface relief of a terrain using radar.

5

Single-particle scattering: perfect spheres

5.1 Introduction

One of the fundamental interactions of electromagnetic radiation with a particulate medium is scattering by individual particles, and many of the properties of the light diffusely reflected from a particulate surface can be understood, at least qualitatively, in terms of single-particle scattering. This chapter considers scattering by a sphere. Although perfectly spherical particles are rarely encountered in the laboratory and never in planetary soils, they are found in nature in clouds composed of liquid droplets. For this reason alone, spheres are worth discussing. Even more important, however, is the fact that a sphere is the simplest three-dimensional object whose interaction with a plane electromagnetic wave can be calculated by exact solution of Maxwell's equations. Therefore, in developing various approximate methods for handling scattering by nonuniform, nonspherical particles, the insights afforded by uniform spheres are invaluable.

In the first part of this chapter some of the quantities in general use in treatments of diffuse scattering are defined. Next, the theory of scattering by a spherical particle is described qualitatively, and conclusions from the theory are discussed in detail. Finally, an analytic approximation to the scattering efficiency that is valid when the radius is large compared with the wavelength is derived.

5.2 Concepts and definitions

5.2.1 Radiance

In a radiation field where the light is uncollimated, the amount of power at position $\mathbf{r}$ crossing unit area perpendicular to the direction of propagation Ω, traveling into unit solid angle about Ω, is called the *radiance* and will be denoted by $I(\mathbf{r}, \Omega)$. Radiance is often also called *specific intensity,* or simply *intensity,* or *brightness.*

Note the difference between irradiance $\boldsymbol{J}$, which refers to power per unit area of a collimated beam, and radiance $\boldsymbol{I}$, which is the uncollimated power per unit

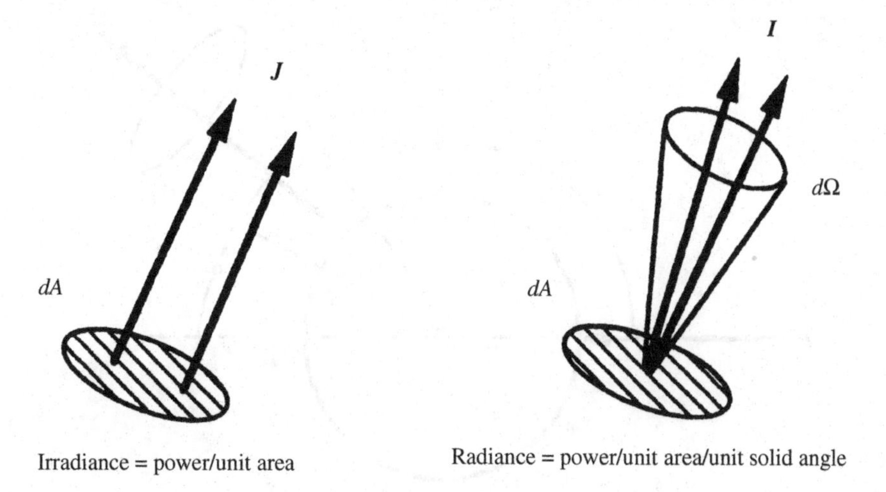

Figure 5.1 Irradiance and radiance.

area per unit solid angle (Figure 5.1). The power per unit area traveling into a small range of solid angles $\Delta\Omega$ about some direction Ω, is $\mathbf{I}(\mathbf{r}, \Omega)\Delta\Omega$. The irradiance may be considered to be the limit of a highly collimated radiance, $\boldsymbol{J}(\boldsymbol{r}, \Omega_0) = \lim_{\Delta\Omega\to 0}[\boldsymbol{I}(\boldsymbol{r}, \Omega)\Delta\Omega] = \boldsymbol{I}(\mathbf{r}, \Omega)\delta(\Omega - \Omega_0)$,where $\delta(x)$ is the delta function.

The delta function, which was introduced by P. A. M. Dirac, is an extremely useful mathematical tool. It has the following properties: $\delta(x) = 0$ everywhere except at $x = 0$, where $\delta(0)$ is infinite in such a manner that its integral over an interval $\overline{ab}$ containing $x = 0$ is $\int_a^b \delta(x)dx = 1$. Then for any function $f(x)$, $\int_a^b f(x)\delta(x)dx = f(0)$ if $\overline{ab}$ contains $x = 0$, but if not, the integral has the value zero. Thus, if a radiation field consisting of a collimated irradiance $\boldsymbol{J}(\Omega_0)$ is examined by a directional detector with a solid angle of admittance $\Delta\Omega$, the response will be proportional to $\int_{\Delta\Omega} \boldsymbol{J}(\Omega_0)\delta(\Omega - \Omega_0)d\Omega$ which is equal to $\boldsymbol{J}$ if $\Delta\Omega$ includes the direction from the source Q and is zero otherwise.

In this chapter a number of particle scattering parameters will be defined for the case where the particles are spherically symmetric. The definitions will be extended to nonspherical particles in Chapter 6, and to assemblages of irregular, randomly oriented particles in Chapter 7. The geometry of the single-particle scattering problem is shown schematically in Figure 5.2. In the discussions and definitions in this chapter the radiation incident on the particle will be referred to as if it is an irradiance $\boldsymbol{J}$; however, it may also be considered to be an increment of radiance $I\Delta\Omega$. This incident radiation interacts with a particle whose refractive index relative to the surrounding medium is $n = n_r + in_i$.

When an electromagnetic wave is incident on a particle, a certain amount of power is removed, or extinguished, from the incident wave and either is absorbed

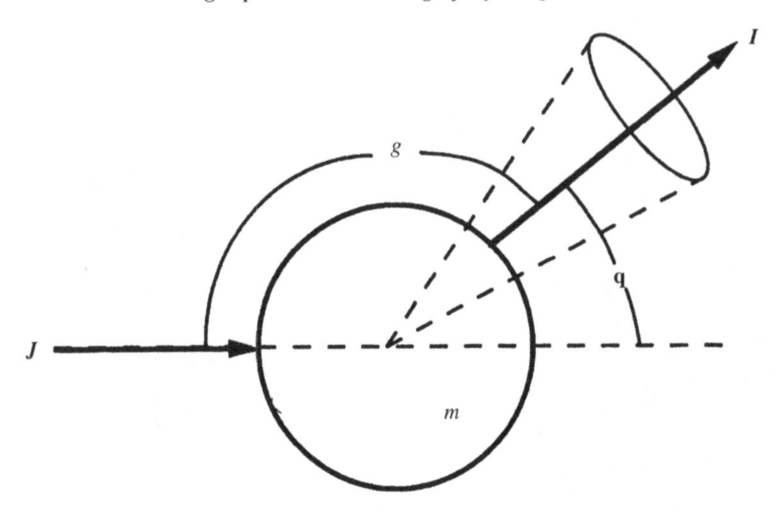

Figure 5.2 Scattering by a single particle. The plane containing $\boldsymbol{J}$ and $\boldsymbol{I}$ is the scattering plane.

or has its direction of propagation altered. In this book it will be assumed that the energy removed from the beam can be only scattered or absorbed; phenomena such as fluorescence and Raman scattering will be ignored. However, the energy may be re-emitted at a different wavelength as thermal radiation; this topic will be treated in Chapter 15. The scattered radiance is the power per unit area deflected into unit solid angle traveling radially away from the center of the particle and measured at a distance large compared with both the particle size and wavelength.

5.2.2 Cross sections

Let $\boldsymbol{J} = \boldsymbol{u}_p J$, where $\boldsymbol{u}_p$ is a unit vector pointing in the direction of propagation of $\boldsymbol{J}$. Let the total amount of power of $\boldsymbol{J}$ that is affected by the particle be P_E. Then the *extinction cross section* is

$$\sigma_E = P_E/J. \tag{5.1}$$

An amount P_S of P_E is *scattered* into all directions, and the remainder P_A is absorbed by the particle. The *scattering cross section* is

$$\sigma_S = P_S/J, \tag{5.2}$$

and the *absorption cross section* is

$$\sigma_A = P_A/J. \tag{5.3}$$

The cross sections have the units of area. Because $P_E = P_S + P_A$, the cross sections are not independent, but must be related by $\sigma_E = \sigma_S + \sigma_A$.

5.2.3 *Efficiencies and size parameter*

Denote the geometrical cross-sectional area of the particle by $\sigma = \pi a^2$, where a is the particle radius. The ratios of the various cross sections to the geometrical cross-sectional area are the corresponding efficiencies. Thus, the *extinction, scattering,* and *absorption efficiencies* are, respectively,

$$Q_E = \sigma_E/\sigma, \tag{5.4}$$

$$Q_S = \sigma_S/\sigma, \tag{5.5}$$

$$Q_A = \sigma_A/\sigma, \tag{5.6}$$

where these quantities must satisfy

$$Q_E = Q_S + Q_A. \tag{5.7}$$

The ratio of the circumference of the particle to the wavelength is called the *size parameter*,

$$X = 2\pi a/\lambda = \pi D/\lambda, \tag{5.8}$$

where $D = 2a$ is the particle diameter.

5.2.4 *Particle single-scattering albedo and espat function*

The ratio of the total amount of power scattered to the total power removed from the wave is the *particle single-scattering albedo*, ϖ. From the definitions of the cross sections and efficiencies,

$$\varpi = P_S/P_E = \sigma_S/\sigma_E = Q_S/Q_E. \tag{5.9}$$

A related parameter, which will be studied and found to be useful later, is the *effective single-particle absorption thickness*, or *espat function*,

$$W = Q_A/Q_S = \frac{1-\varpi}{\varpi}. \tag{5.10}$$

In general, the efficiencies and the single-scattering albedo are functions of wavelength.

5.2.5 *Scattering and phase angles*

Consider the radiance scattered by a particle into some direction making an angle θ with the original direction of propagation of the incident irradiance; θ is the *scattering angle,* and the plane containing the incident and scattered propagation vectors is the *scattering plane.*

The complement of the scattering angle is called the *phase angle,* $g = \pi - \theta$. Historically the term comes from the astronomical lunar phase angle, which is the angle subtended by the Earth and the Sun as seen from the Moon. The phase angle is the angle between the directions to the source and detector of radiation as measured from the center of the particle. In the literature on scattering by a single particle, the scattering angle θ is usually used. However, in discussions of diffuse scattering by particulate media, including laboratory and remote-sensing applications, the phase angle g is often more convenient. The phase and scattering angles are shown in Figure 5.2.

5.2.6 Particle phase function

Suppose the power described by irradiance $\boldsymbol{J}(\mathbf{r}, \Omega_0)$ traveling into a direction Ω_0 is scattered by a particle at position $\mathbf{r}$ into a radiance pattern $I(\mathbf{r}, \Omega)$. The *particle phase function* $\Pi(\theta)$ describes the angular pattern into which the power $P_S = J\sigma Q_S$ is scattered. Let $(dP_S/d\Omega)(\Omega_0, \Omega)$ be the power scattered by the particle from direction Ω_0 into unit solid angle centered about direction Ω, and let θ be the scattering angle between Ω_0 and Ω. Then $\Pi(\theta)$ is defined by the equation

$$\frac{dP_S}{d\Omega}(\Omega_0, \Omega) = J(\Omega_0)\sigma Q_S \frac{\Pi(\theta)}{4\pi}. \tag{5.11}$$

The particle phase function may be described equivalently by $\Pi(g) = \Pi(\pi - \theta)$. Because the area perpendicular to the incident radiation affected by the particle is σQ_E, the scattered radiance, or power per unit area scattered into unit solid angle, is

$$I(\Omega) = J(\Omega_0)\frac{\sigma Q_S \Pi(g)}{\sigma Q_E 4\pi} = J\varpi \frac{\Pi(g)}{4\pi}. \tag{5.12}$$

If the particle scatters isotropically, $\Pi(g) = 1$.

The total power scattered in all directions is $\int_{4\pi} \frac{dP_S}{d\Omega} d\Omega = P_S$; thus $\Pi(g)$ is required to satisfy the normalization condition: $(1/4\pi)\int_{4\pi} \Pi(g)d\Omega = 1$. Because the particle is spherically symmetric, the scattered power is independent of azimuth; hence, the integration may be carried out over azimuth giving $d\Omega = 2\pi \sin g\, dg$ or $d\Omega = 2\pi \sin\theta d\theta$ as appropriate, and the normalization condition becomes

$$\frac{1}{2}\int_0^\pi \Pi(g)\sin g\, dg = \frac{1}{2}\int_0^\pi \Pi(\theta)\sin\theta d\theta = 1. \tag{5.13}$$

In some references the factor 4π is omitted in the definition (5.11), in which case $\Pi(g)$ is normalized so that $\int_{4\pi} \Pi(g)d\Omega = 1$. The particle phase function is sometimes also known by the unwieldy name of *single-particle angular scattering function.*

5.2.7 The asymmetry factor

The particle angular scattering function may be characterized by various parameters, usually referred to as asymmetry factors. One such parameter is the *cosine asymmetry factor* ξ, which is the average value of the cosine of the scattering angle θ weighted by the particle phase function,

$$\xi = < \cos\theta > = [\int_{4\pi} \cos\theta \Pi(\theta) d\Omega] / [\int_{4\pi} \Pi(\theta) d\Omega] = \frac{1}{2} \int_0^{\pi} \cos\theta \Pi(\theta) \sin\theta d\theta$$

$$= -\langle \cos g \rangle = -\frac{1}{2} \int_0^{\pi} \cos g \Pi(g) \sin g \, dg. \quad (5.14)$$

A positive value of ξ implies that most of the light is scattered into the forward hemisphere, while a negative value of ξ means that the particle is predominantly backscattering. For example, if $\Pi(g) = 1 + b\cos g$, then $\xi = -b/3$. If the particle scatters light isotropically, $\xi = 0$. Note, however, that the inverse may not be true: $\xi = 0$ implies only that the particle scatters symmetrically about the $g = \pi/2$ plane, but $\Pi(g)$ is not necessarily isotropic.

5.2.8 The scattering (Mueller) matrix

The particle phase function $\Pi(g)$ defined by equation (5.12) refers to situations in which we are interested only in unpolarized light. If, instead, the incident irradiance and scattered radiance are partially polarized they can be described, respectively by the 1×4 column Stokes vectors **J** and **I** defined in Chapter 2. In this case the quantity analagous to the particle phase function is the *scattering matrix* **M** defined by

$$\mathbf{I} = \frac{\varpi}{4\pi} \mathbf{MJ}. \quad (5.15)$$

In general, the scattering matrix, also known as the *Mueller matrix*, is a 4×4 matrix. However, if either the particles are spherically symmetric (that is, consist of concentric spherical shells), or are ensembles of randomly oriented nonspherical particles, each of which has a mirror-symmetric twin (which probably describes most laboratory and regolith powders reasonably well), then only 8 of the 16 elements of the scattering matrix are non-zero, and only 5 of these 8 are independent. In this case **M** has the form

$$\mathbf{M} = \begin{bmatrix} M_{11} & M_{12} & 0 & 0 \\ M_{12} & M_{22} & 0 & 0 \\ 0 & 0 & M_{33} & M_{34} \\ 0 & 0 & -M_{34} & M_{44} \end{bmatrix}, \quad (5.16)$$

All elements of the matrix are functions of g. In particular, $\Pi(g) = M_{11}$, and $P(g) = -M_{12}/M_{11}$.

5.3 Scattering by a perfect, uniform sphere: Mie theory

The simplest case of single-particle scattering is that of a plane electromagnetic wave scattered by a uniform spherical particle. This problem was independently solved by several persons, including Gustav Mie, and the result has come to be known as Mie theory; however, other individuals, especially Debye and Lorenz, also have legitimate claim to having been the first to obtain a solution. Bohren and Huffman (1983) discuss its history. Frequently in the remote-sensing literature the term *Mie theory* or *Mie scattering* is used to refer to scattering by a particle large compared with the wavelength of any shape. However, this is an incorrect and corrupt usage of the term, which refers *only* to scattering by an *isolated spherical* particle of *any* size. Mie theory is treated in greater or lesser detail in a number of books, including the work of Born and Wolf (1980), Stratton (1941), and Van de Hulst (1957). The most readable, especially for the novice, are the books by Bohren and Huffman (1983) and Kerker (1969).

The solution to the Mie problem is lengthy and complicated, and its details are not particularly instructive nor insightful. Hence, rather than force the reader to, in the words of Bohren and Huffman, "acquire virtue through suffering," I will simply outline the mathematical procedure, referring the interested reader to the excellent treatments in the books mentioned earlier. Following this summary, the properties of the solution will be discussed in some detail.

The same procedure is followed as in the derivation of the Fresnel equations in Chapter 4. First the wave equation is solved in a coordinate system appropriate to the problem, in this case spherical, and then the electric and magnetic fields are required to satisfy continuity conditions on the surface of the sphere. The resulting expressions give internal and external fields as a function of distance from the center of the sphere and scattering angle, along with the extinction and scattering efficiencies and the particle phase function.

Because the particle is a sphere, a spherical coordinate system is chosen whose origin is at the center of the particle and whose polar axis is parallel to the direction of the incident radiation. The general solution to the wave equation in spherical coordinates is derived in Appendix C and is of the form

$$F(r, \vartheta, \psi, t) = \sum_{l=0}^{\infty} \left[C_l h_l^{(1)}(kr) + D_l h_l^{(2)}(kr) \right] \left[A_{lm} \Psi_{lm}^{(e)}(\vartheta, \psi) + B_{lm} \Psi_{lm}^{(o)}(\vartheta, \psi) \right] e^{-2\pi i \nu t} \tag{5.17}$$

where $F(r, \vartheta, \psi, t)$ can refer to either the electric or magnetic field, $k = 2\pi/\lambda$, λ is the wavelength, r is the radial coordinate, ϑ is the polar coordinate, Ψ is the azimuthal coordinate, $h_l^{(1)}(kr)$ and $h_l^{(2)}(kr)$ are spherical Hankel functions of the first and second kind, respectively, $\Psi_{lm}^{(e)}(\vartheta, \psi)$ and $\Psi_{lm}^{(0)}(\vartheta, \psi)$ are even and odd spherical harmonic functions, respectively, l is an integer ($l \geq 0$), m is an integer ($0 \leq m \leq l$), and the *A*s, *B*s, *C*s, and *D*s are constants that are determined by the specific problem. Hankel and spherical harmonic functions are tabulated, and their properties have been exhaustively studied. They are discussed in Appendix C. However, because the electric and magnetic fields are vectors the wave equations in vector form must be solved.

Equation (5.17) describes a series of spherical waves propagating radially inward and outward with respect to the particle. In order to complete the solution, the incident plane wave must also be expressed in a form similar to (5.17) as an infinite series of spherical wave functions. This can be done because the spherical Hankel and harmonic functions form complete, orthogonal sets (see Appendix C). The incident and scattered fields must satisfy various conditions, including that the fields be finite everywhere, that the only field traveling inward far from the particle be the incident irradiance, and that the usual conditions of continuity on the transverse and perpendicular components of the electric and magnetic fields (Chapter 4) at the surface of the particle be satisfied. If the distance r from the sphere is large the amplitude of the outwardly propagating wave is proportional to $1/r$, which means that the power falls off as $1/r^2$, and the strength of this wave varies with direction. Within a wavelength or so of the surface other fields also exist that fall off faster than $1/r^2$, somewhat analgous to the evanescent waves associated with total internal reflection from a plane surface discussed in Chapter 4. However, these fields, known as *near fields,* are not propagating waves and do not carry energy away from the surface.

The resulting expressions for the extinction, scattering, and absorption efficiencies and the particle phase function consist of slowly converging infinite series. Except when the particle is small compared with the wavelength a large number ($\sim X$) of terms are necessary to evaluate the expressions, and a computer is virtually a necessity for the solution. Because these expressions do not yield particularly useful insights, they will not be written down explicitly here. Again, the interested reader is referred to the references cited earlier.

5.4 Properties of the Mie solution

5.4.1 General properties

The Mie solutions turn out to depend on only two parameters: the refractive index n of the particle relative to the surrounding medium and the *size parameter X*.

The solutions have different properties depending on whether X is small, comparable to, or large compared with unity. Each of these regimes will be discussed in detail in this section. The solutions also depend on whether the electric vector in the incident wave is perpendicular to (denoted by subscript "⊥") or parallel to (denoted by subscript "‖") the scattering plane. It is found that the scattering process does not change the direction of polarization of the wave. Thus, if the incident light is polarized perpendicular to the scattering plane, the scattered light will be perpendicularly polarized also, and similarly for parallel-polarized incidence. However, the scattering coefficients are different for the two directions of polarization.

The general behavior of the extinction efficiency as the size of the sphere is varied is illustrated by the solid line in Figure 5.3 for the case of a real index of refraction, $n = n_r = 1.50$.

In this figure, Q_E is plotted versus the parameter $(n_r - 1)X$. Because $n_i = 0$, $Q_S = Q_E$, and $Q_A = 0$. For small values of X the efficiency increases nonlinearly proportionally to X^4 to a value near 2 at $(n_r - 1)X \simeq 1$. As $(n_r - 1)X$ increases, Q_E develops a series of large oscillations on which small ripples are superimposed. The first maximum of the large oscillations is roughly at $Q_E \simeq 4$ and $(n_r - 1)X \simeq 2$,

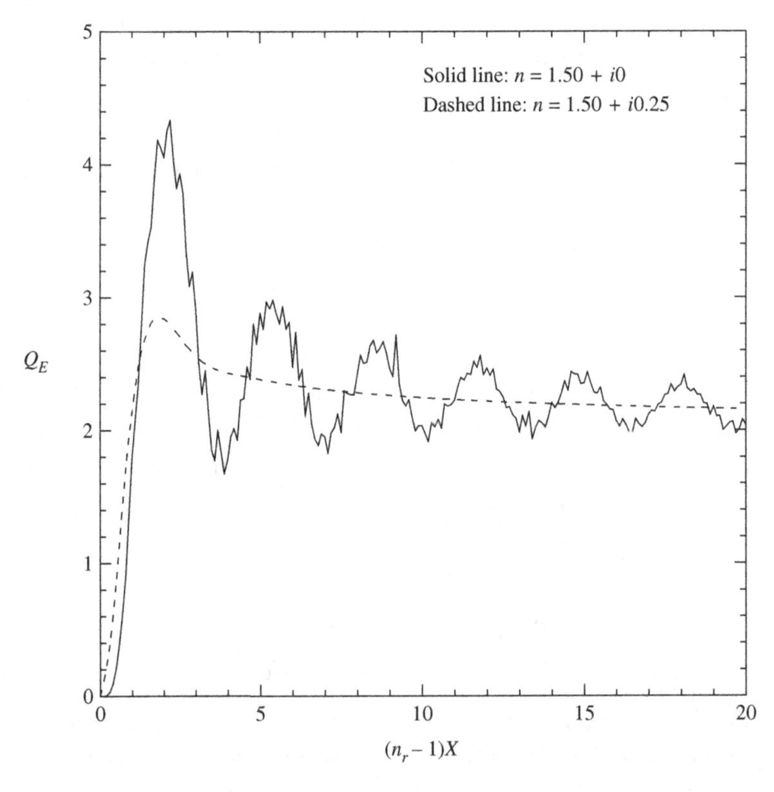

Figure 5.3 Solid line: extinction efficiency of a sphere of refractive index $n = 1.50 + i0$ and size parameter X vs. $(n_r - 1)X$. Dashed line: extinction efficiency of a sphere of refractive index $n = 1.50 + i0.25$ vs. $(n_r - 1)X$.

followed by a minimum of somewhat less than 2 at $(n_r - 1)X \simeq 3.8$. Thereafter, Q_E oscillates about the value 2 with decreasing amplitude and with successive extrema separated by values of $(n_r - 1)X$ differing by π.

The dashed line in Figure 5.3 shows the extinction efficiency of an absorbing sphere with $n = 1.50 + 0.25i$. This value of n_i corresponds to an absorption length $\alpha^{-1} = \lambda/4\pi n_i$ (Chapter 2) equal to the diameter of a sphere with $X = 1$. For small values of X, Q_E is proportional to X. As X increases, Q_E goes through a maximum at $(n_r - 1)X \simeq 2$ that is lower than the maximum for the nonabsorbing sphere and then slowly approaches $Q_E = 2$ with no oscillations or ripples.

The fact that the efficiencies can be greater than 1 may be surprising at first sight. Physically, this means that the particle affects a larger portion of the wave front than is obstructed by the geometrical cross-sectional area of the sphere.

From the figures and discussion it is apparent that the scattering properties can be divided into three size regions: $X << 1$, $X \sim 1$, and $X >> 1$. These will be discussed next.

5.4.2 $X = 1$: *the Rayleigh region*

When $X = 1$ it is useful to expand the Mie expressions for the scattering parameters in powers of X. It is found that to terms of order X^4,

$$Q_E = 4X\mathrm{Im}\left\{\frac{n^2-1}{n^2+2}\left[1 + \frac{X^2}{15}\frac{n^2-1}{n^2+2}\frac{n^4+27n^2+38}{2n^2+3}\right],\right\} + \frac{8}{3}X^4\mathrm{Re}\left\{\left[\frac{n^2-1}{n^2+2}\right]^2\right\}, \tag{5.18}$$

$$Q_S = \frac{8}{3}X^4\left|\frac{n^2-1}{n^2+2}\right|^2, \tag{5.19}$$

and $Q_A = Q_E - Q_S$. If $n_i = 0$,

$$Q_S = Q_E = \frac{8}{3}X^4\left|\frac{n_r^2-1}{n_r^2+2}\right|^2, \tag{5.20}$$

$Q_A = 0$, and $\varpi = 1$. In this region the scattered radiance is proportional to $1/\lambda^4$, a result that was first obtained by Rayleigh (1871) using dimensional arguments. Thus, particles small compared with the wavelength are known as *Rayleigh scatterers*.

If n_i is not negligible, but $|n|X = 1$, so that higher powers of X can be ignored, then

$$Q_A \simeq Q_E \simeq \frac{24 n_r n_i}{(n_r^2+n_i^2)^2+4(n_r^2-n_i^2)+4} X, \tag{5.21}$$

$$Q_S \simeq \frac{8}{3} \frac{[(n_r^2+n_i^2)^2+n_r^2-n_i^2-2]^2+36 n_r^2 n_i^2}{[(n_r^2+n_i^2)^2+4(n_r^2-n_i^2)+4]^2} X^4. \tag{5.22}$$

In this case $\varpi = Q_S/(Q_S+Q_A) \simeq Q_S/Q_A \propto X^3$, so that if the particle is absorbing, its single-scattering albedo is very small. Small absorbing particles are called *Rayleigh absorbers.* The absorption efficiency is proportional to X, so that the absorption cross section $\sigma_A \propto \pi a^2 X$, and the power absorbed is proportional to the volume or mass of the particle.

The intensity of the radiance scattered with the electric vector perpendicular to the scattering plane is proportional to 1, and parallel to the scattering plane is proportional to $\cos^2 g$. If the incident light is unpolarized, the particle phase function is

$$\Pi(g) = \frac{3}{4}(1+\cos^2 g). \tag{5.23}$$

Note from equations (5.18) and (5.19) that the denominators of Q_E and Q_S both contain the factor n^2+2. Thus, the efficiencies can become very large when $n^2 = K_e \simeq -2$, which requires $K_{ei} = 1$ and $K_{er} \simeq -2$. Figure 3.1 shows that this can happen in the region of anomalous dispersion on the high-frequency side of a strong absorption band. This phenomenon is known as a *plasma resonance,* and it can be important in small particles, or asperities that behave as small particles on the surfaces of larger particles, in the infrared.

Refering to Chapter 4, the *linear polarization* of the radiance is defined relative to some special plane. In Mie theory the special plane is the scattering plane, so that the polarization is defined as

$$P = (I_\perp - I_\parallel)/(I_\perp + I_\parallel), \tag{5.24}$$

where the subscript "$\perp$" denotes the electric vector perpendicular to the scattering plane, and the subscript "$\parallel$" denotes the electric vector parallel to this plane. Thus, the polarization of the scattered radiance is

$$P(g) = (1-\cos^2 g)/(1+\cos^2 g) = \sin^2 g/(1+\cos^2 g). \tag{5.25}$$

The phase function and polarization are plotted in Figures 5.4a and 5.4b The polarization is always positive, and $P = 1$ at $g = 90°$.

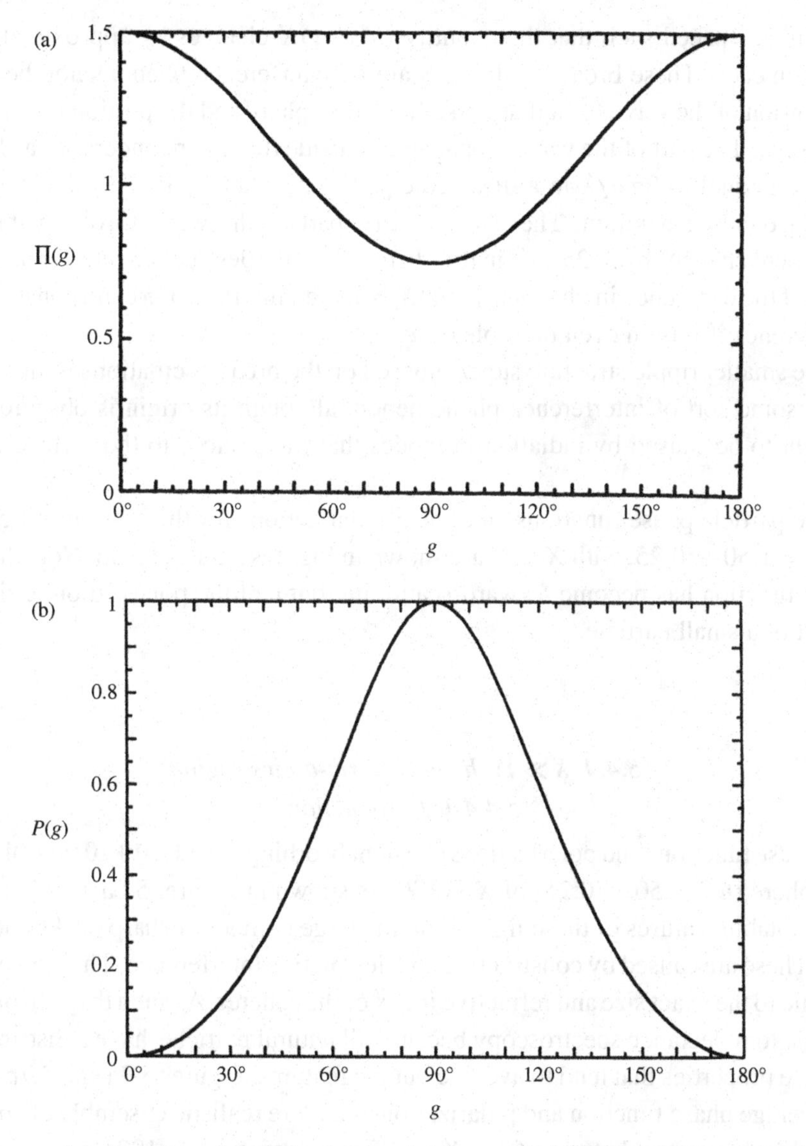

Figure 5.4 (a) Phase function of a Rayleigh particle. (b) Polarization of light scattered by a Rayleigh particle.

5.4.3 $X \sim 1$: *the resonance region*

When the particle size is of the order of the wavelength, the behaviors of the efficiencies and phase function are complicated and vary from case to case. If the particle is not too absorbing, the extinction and scattering efficiencies tend to oscillate about the value 2, with decreasing amplitude and successive maxima or minima spaced such that the quantity $(n_r - 1)X$ changes by approximately π between each. These broad oscillations are an interference phenomenon between the portion of the wave front that passes near the sphere and the portion transmitted through it. The part of the wave propagating outside the sphere undergoes a change in phase equal to $2\pi a/\lambda$ in a distance equal to a, where λ is the wavelength in the surrounding medium. The phase of that part of the wave traveling through the sphere changes by $n_r 2\pi a/\lambda$ in this distance. Thus, destructive interference will occur if the difference in phases $(n_r - 1)X$ is an odd multiple of π, and constructive interference if it is an even multiple of π.

The smaller ripple structure superimposed on the broad oscillations is also obviously some sort of interference phenomenon, although its origin is obscure. It is thought to be caused by radiation in modes that travel close to the surface of the particle.

The particle phase functions and linear polarizations for the cases $n = 1.5 + i0$ and $n = 1.50 + i0.25$, with $X = 1$, are shown in Figures 5.5a and 5.5b. Note that the phase function has become forward-scattering, but that the polarization is similar to that of a small particle.

5.4.4 $X \gg 1$: *the geometric-optics region*

5.4.4.1 *Introduction*

The phase functions and polarizations of a nonabsorbing ($n = 1.50 + i0$) and absorbing sphere ($n = 1.50 + i0.25$) of $X = 100$ are shown in Figures 5.6a and 5.6b. The most notable features of these figures are the large number of sharp peaks and valleys. These are caused by constructive and destructive interference. They are highly specific to the exact size and refractive index of the spheres. As such they are of little interest to reflectance spectroscopy because all natural particles have a distribution of these properties that tend to average out the features. Figures 5.7a and 5.7b show the average phase function and polarization of a more realistic ensemble of spheres with size parameters between $67 < X < 133$ and average $X = 100$.

The major features of Figure 5.7a are an extremely high, narrow peak centered around $g = 180°$ in the forward direction, a strong broad forward-scattering lobe, a weaker narrower lobe in the backscattering direction, and a low valley between the forward and backward lobes. When the sphere is large compared with the wavelength, the parts of the wave front separated by distances of the order of the radius are nearly independent, so that what happens to one part has very little effect on the other parts. That is, to a large extent the propagating wave can be treated as if it were made up of independent bundles of rays that are refracted and reflected by the sphere. Except for the high, narrow forward peak the features of Figure 5.7a can

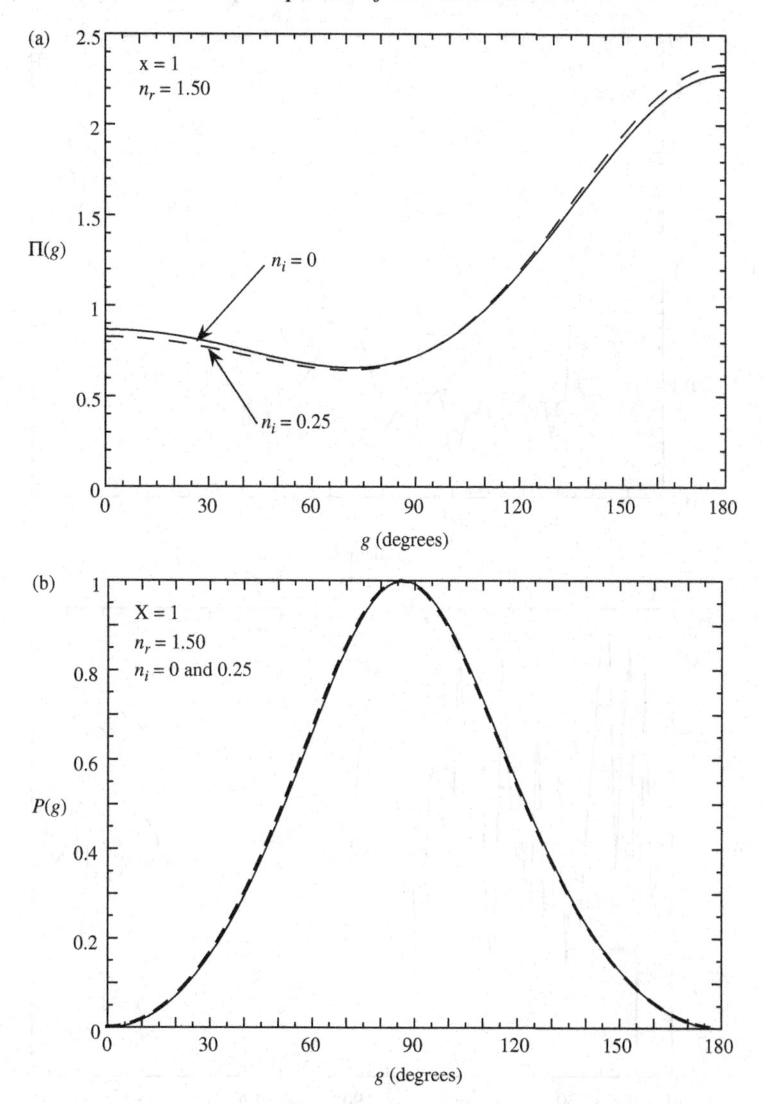

Figure 5.5 (a) Phase functions of particles of $X = 1$ with $n = 1.50 + i0$ (solid line) and $n = 1.50 + i0.25$ (dashed line). The single-scattering albedo of the particle with $n_i = 0$ is $\varpi = 1.00$ and that of the particle with $n_i = 0.25$ is $\varpi = 0.27$. (b) Polarization of the particles of Figure 5.5a.

be understood in terms of geometric optics. The various phenomena that control different portions of the curves of this figure will now be discussed in detail.

5.4.4.2 Diffraction

At large values of X the extinction efficiency of an isolated sphere approaches the asymptotic value of 2, not 1, indicating that the sphere affects a portion of

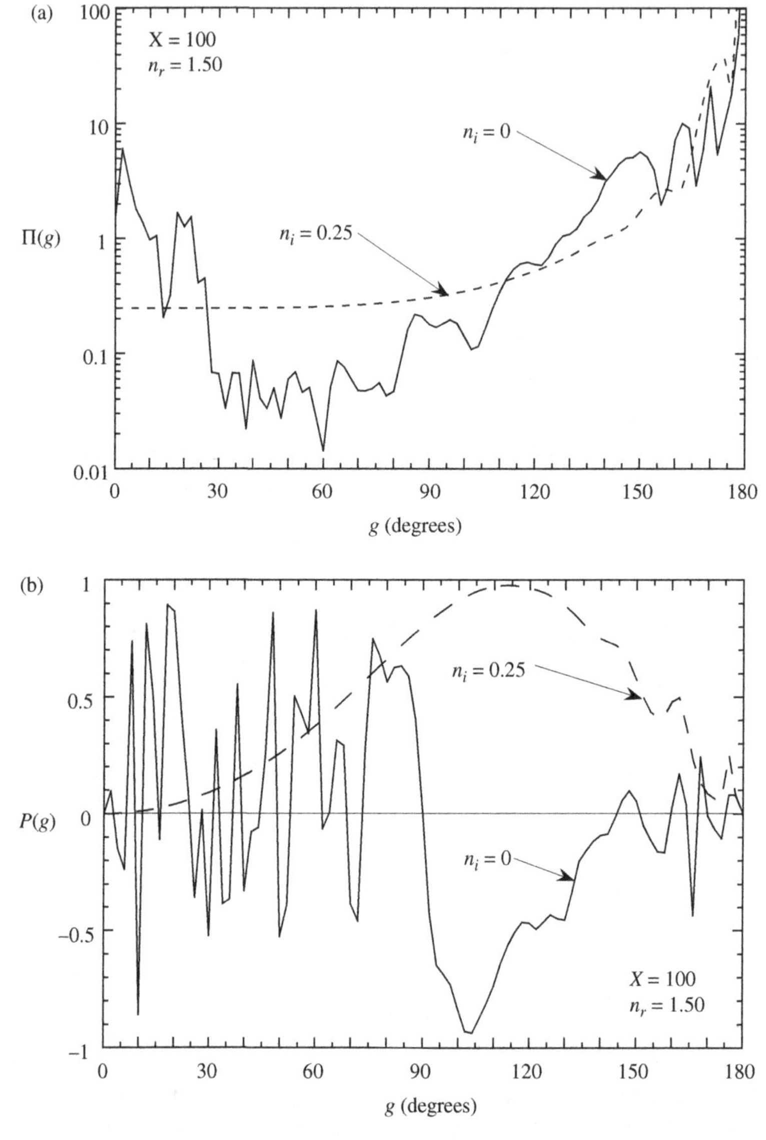

Figure 5.6 (a) Phase functions of a particle of $X = 100$ with $n = 1.50 + i0$ (solid line) and $n = 1.50 + i0.25$ (dashed line) calculated using Mie theory. The single-scattering albedo of the particle with $n_i = 0$ is $\varpi = 1.00$, and that of the particle with $n_i = 0.25$ is $\varpi = 0.54$. (b) Polarization of the particles of Figure 5.6a.

the incident wave front equal to twice its geometric cross section. Also, the phase curves of both the clear and absorbing particles exhibit a very strong, narrow peak in the forward-scattering direction. Both of these effects are caused by diffraction and occur because of the wave nature of light.

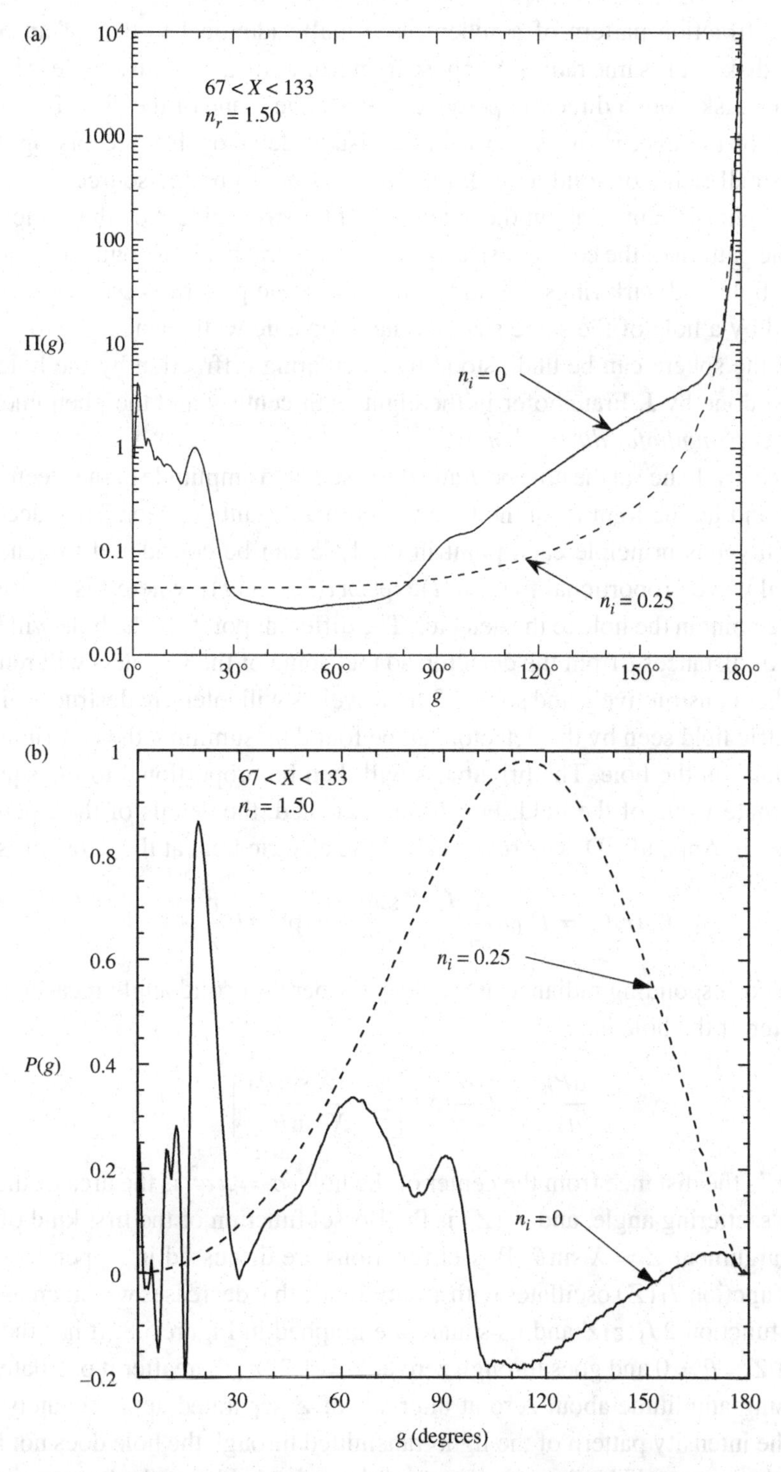

Figure 5.7 (a) Mean phase functions of emsembles of particles with $63 < X < 133$ having $n = 1.50 + i0$ (solid line) and $n = 1.50 + i0.25$ (dashed line) calculated using Mie theory. (b) Polarization of the particles of Figure 5.7a.

The diffraction pattern of a sphere is virtually identical to that of an opaque circular disk of the same radius. Suppose light from a distant point source is incident on such a disk from a direction perpendicular to the plane of the disk. If the phase function of the disk is measured using a distant detector that accepts light from only a small range of solid angles, the disk appears to be the source of a narrow cone of light concentric about the extension of the ray incident on the center of the disk. The pattern of the cone consists of a bright central peak surrounded by a series of faint light and dark rings. Paradoxically, the same pattern is seen if the disk is replaced by a hole of the same size in a large opaque wall. Hence, the diffraction peak of the sphere can be understood by calculating diffraction by the hole. This was first done by J. Fraunhofer in the eighteenth century and the phenomenon is known *as Fraunhofer diffraction.*

Let the irradiance at the hole be J and the associated amplitude of the electric field be E_{e0}, and let the light from the hole fall onto a distant detector. Then according to the Huygens principle each point in the hole can be considered to generate a spherical wave proportional to $(E_{e0}/l)\exp[2\pi\iota(l/\lambda - \nu t)]$, where l is the distance from the point in the hole to the detector. The different points in the hole will be at a variety of distances l from the detector, so that some of the wavelets will reinforce each other constructively and some of the wavelets will interfere destructively. The net electric field seen by the detector can be found by summing the contribution of every point in the hole. The brightness will then be proportional to the square of the absolute value of the field. For those interested, the details of the calculation are given in Appendix D. The result is that the electric field at the detector is

$$E_e(r, \theta) = E_{e0}a\frac{X}{r}\frac{J_1(X\sin\theta)}{X\sin\theta}\exp[2\pi i(\frac{r}{\lambda} - \nu t)], \tag{5.26}$$

and the corresponding radiance at the detector per unit solid angle measured from the center of the hole is

$$\frac{dP_d}{d\Omega} = J\frac{\sigma}{4\pi}X^2\left[2\frac{J_1(X\sin\theta)}{X\sin\theta}\right]^2, \tag{5.27}$$

where r is the distance from the center of the hole, $\sigma = \pi a^2$ is the area of the hole, θ is the scattering angle, and $J_1(Z)$ is the Bessel function of the first kind of order 1 with argument $Z = X\sin\theta$. Bessel functions are discussed in Appendix C. The Bessel function $J_1(Z)$ oscillates with an amplitude that decreases with increasing Z.

The function $2J_1(z)Z$ and its square are graphed in Figure 5.8. It has the value unity at $Z = \theta = 0$ and goes through zero at $Z = 1.22\pi$, thereafter it oscillates with decreasing amplitude about zero at intervals of Z separated approximately by π. Thus, the intensity pattern of the light transmitted through the hole does not have a sharp edge, as would be the case if geometric optics held exactly, but consists of a

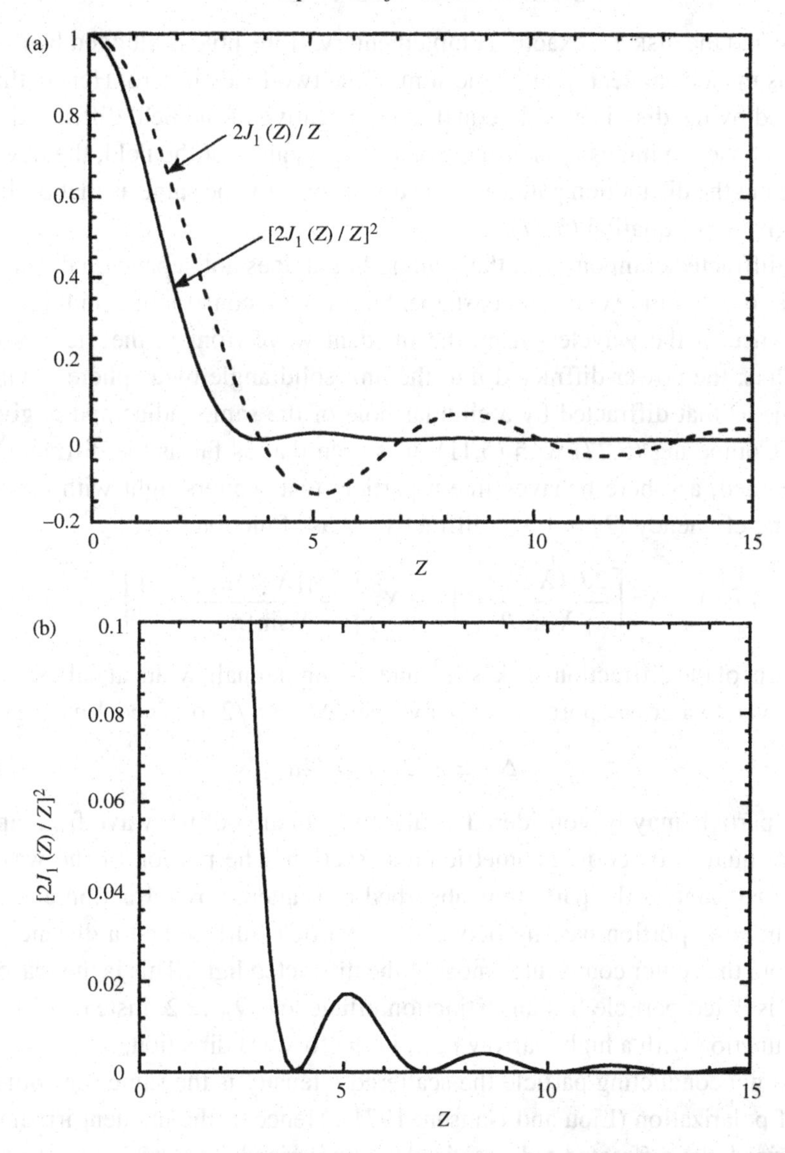

Figure 5.8 (a) Electric field and intensity in the Fraunhofer diffraction pattern of a hole in an opaque wall, normalized to the center of the pattern. (b) Enlargement of the intensity pattern of Figure 5.7a.

bright central peak surrounded by a series of alternating dark and light rings. The angular width of the central peak is inversely proportional to the radius of the hole.

Now consider the complementary situation. Remove the wall and replace it by an isolated opaque disk of the same size and at the same position as the hole. The field in the diffraction pattern of the disk can be found from *Babinet's principle:*

the hole and the disk are exactly complementary. If the hole is plugged by the disk, all fields must disappear; that is, the sum of the two fields is zero. Hence, the field diffracted by the disk is exactly equal to the negative of the field diffracted by the hole. Because the intensity is proportional to the square of the field, the power per unit area in the diffraction pattern of the disk is exactly the same as that of the hole and is given by equation (5.27).

The diffracted component of the scattered light does not depend on the nature of the disk, or even the details of its shape, but only on constructive and destructive interference of the wavelets from the incident wave front in the vicinity of the disk. Thus, the power diffracted into the unit solid angle by a sphere is virtually the same as that diffracted by a circular hole of the same radius and is given by (5.27). Comparing (5.27) with (5.11), it is seen that as far as the diffracted light is concerned, a sphere behaves like a particle that scatters light with diffractive scattering efficiency $Q_d = 1$ and diffractive phase function

$$\Pi_d(g) = X^2 \left[\frac{2J_1(X\sin\theta)}{X\sin\theta} \right]^2 = X^2 \left\{ \frac{2J_1[X\sin(\pi - g)]}{X\sin(\pi - g)} \right\}^2. \tag{5.28}$$

The height of the diffraction peak is X^2, and its angular half-width at half-maximum $\Delta\theta$ is given to a good approximation by $X \sin \Delta\theta \approx \pi/2$, or since X is large,

$$\Delta\theta \simeq \pi/2X = \lambda/4a. \tag{5.29}$$

A large particle may be considered as affecting an area of the wave front approximately equal to twice its geometric cross section. The portion of the wave that actually encounters the particle is absorbed or scattered by refraction and reflection, while the portion passing between the particle surface and a distance about $\sqrt{2}a$ from the center contributes most of the diffracted light. This is the reason that a large isolated particle has an extinction efficiency $Q_E \simeq 2$, instead of 1, and a phase function with a high, narrow peak in the forward direction.

For a nonconducting particle the scattered intensity is the same for both directions of polarization (Liou and Hansen, 1971). Hence, if the incident irradiance is unpolarized, the diffracted radiance is also unpolarized.

Although the diffracted light appears to emanate from the particle we have seen that it actually comes from an annulus around the particle. Thus, if the space near a particle is partially blocked by other particles, as would be the case if it is part of a powder or soil, the diffraction pattern will be changed both quantitatively and qualitatively: quantitatively because some of the light that would have contributed to the diffraction has been removed, and qualitatively because the diffracting object now consists of several particles and the spaces between them instead of a simple sphere. We will return to this topic in Chapter 7, where it will be shown that the Fraunhofer diffraction of a single particle does not exist in a close packed powder

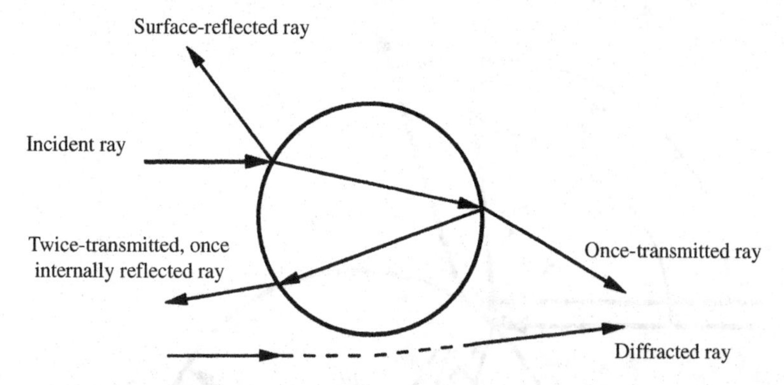

Figure 5.9 Schematic diagram of the scattering processes in a sphere with $X \gg 1$.

or regolith. Hence, in such media $Q_d = 0$. Any model of scattering by powder that ignores this blocking violates conservation of energy and is inherently incorrect.

5.4.4.3 Scattering not caused by diffraction

It will be useful to explicitly separate the part of the scattering that is associated with diffraction from that due to other effects. Denote the nondiffractive portion of the scattering efficiency by Q with a lowercase subscript s and define it by

$$Q_s = Q_S - Q_d, \tag{5.30}$$

where $Q_d \simeq 1$. The nondiffractive interactions of the sphere with the incident wave can be described reasonably well by geometric optics. These interactions consist of rays that are specularly reflected from the surface and rays that are refracted and transmitted through the sphere. The latter may be internally reflected one or more times from the inside of the surface. The major phenomena are shown schematically in Figure 5.9. It is important to note that because of the symmetry of the sphere, the rays that are reflected or refracted by the sphere always remain in the same plane as that formed by the incident ray of interest and the axis of the sphere parallel to the incident rays.

5.4.4.4 Surface reflection

If the sphere is large, the portion of the surface encountered by a ray can be considered as locally flat and effectively infinite, and the Fresnel equations for reflection by a plane interface may be used to calculate the amounts and directions of the light reflected and transmitted. The geometry of the surface-reflected light is shown in Figure 5.10.

Let a sphere of radius a be illuminated by irradiance $J_{\perp\parallel}$, where the subscript "$\perp\ \parallel$" is to be interpreted as "$\perp$" if the incident irradiance is polarized with the

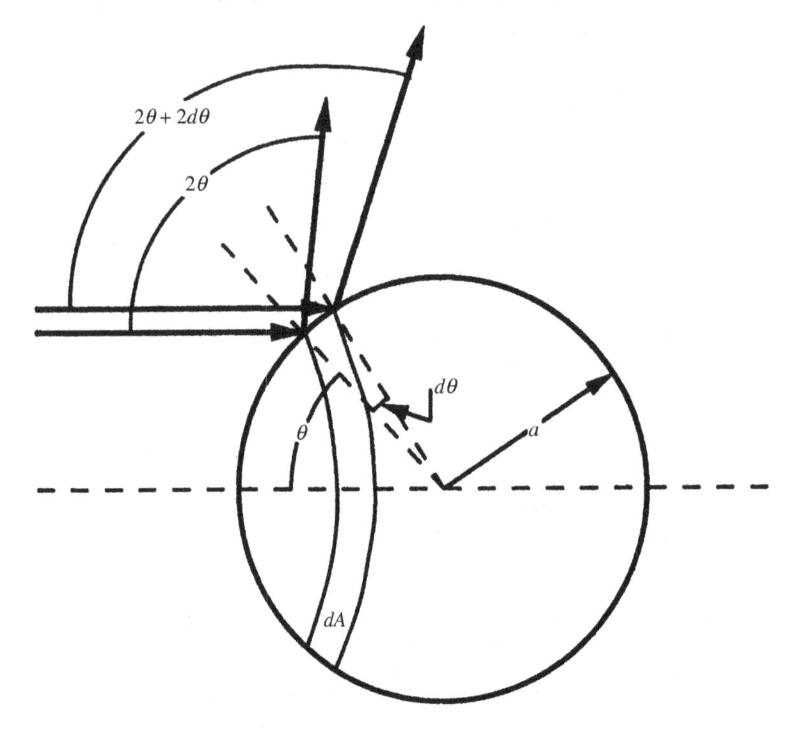

Figure 5.10 Specular reflection from the surface of a large sphere.

electric vector perpendicular to the scattering plane, and as "||" if the electric vector is parallel to the plane. The rays are incident at points where the radius vector makes an angle ϑ with the direction of propagation. (For clarity only one such ray is shown; however, the rays are distributed symmetrically in azimuth all around the axis of the sphere.) These rays are then reflected through a phase angle $g = 2\vartheta$.

Consider a second set of rays incident at points making angles $\vartheta + d\vartheta$ with the radius vector and reflected through phase angles $g + dg = 2\vartheta + 2d\vartheta$. The area of surface between all those rays whose intersections with the sphere are bounded by the angles ϑ and $\vartheta + d\vartheta$ is $dA = 2\pi a \sin\vartheta \cdot a d\vartheta$. Because the projection of dA onto the plane perpendicular to the direction of incident irradiance is $dA \cos\vartheta$, the power dP_R specularly reflected by dA is

$$dP_{R\perp\|} = J_{\perp\|} dA \cos\vartheta\, R_{\perp\|}(\vartheta) = 2\pi a^2 J_{\perp\|} R_{\perp\|}(\vartheta) \sin\vartheta \cos\vartheta d\vartheta, \qquad (5.31)$$

where $R_\perp(\vartheta)$ and $R_\|(\vartheta)$ are the Fresnel reflectivities derived in Chapter 4.

At a large distance from the sphere this light passes through an incremental solid angle $d\Omega = 2\pi \sin(2\vartheta) d(2\vartheta) = 8\pi \sin\vartheta \cos\vartheta d\vartheta$. Hence, the power per unit solid angle specularly reflected from the surface of the sphere through an angle g is

$$\frac{dP_{R\perp\parallel}}{d\Omega} = \frac{J_{\perp\parallel} 2\pi a^2 R_{\perp\parallel}(\vartheta)\sin\vartheta\cos\vartheta d\vartheta}{8\pi\sin\vartheta\cos\vartheta d\vartheta} = J_{\perp\parallel}\sigma\frac{R_{\perp\parallel}(g/2)}{4\pi}. \tag{5.32}$$

The curvature of the spherical surface has a defocusing effect, which causes the radiation incident on the sphere through a range of angles $d\vartheta$ to be reflected through a range of angles $2d\vartheta$. Because $R_\perp \geq R_\parallel$, if the incident irradiance is unpolarized, the light reflected from the surface always has a net positive polarization.

The total amount of light reflected from the surface of the sphere is, from (5.32),

$$P_{R\perp\parallel} = \int_{4\pi}\frac{dP_{R\perp\parallel}}{d\Omega}d\Omega = \int_{\vartheta=0}^{\pi/2} dP_{R\perp\parallel} = J_{\perp\parallel}2\pi a^2\int_0^{\pi/2} R_{\perp\parallel}(\vartheta)\sin\vartheta\cos\vartheta d\vartheta. \tag{5.33}$$

If the incident light is unpolarized, so that $J_\perp = J_\parallel = J/2$, then the contribution to Q_s and $\Pi(g)$ from surface reflection is

$$Q_R\Pi_R(g) = [R_\perp(g/2) + R_\parallel(g/2)]/2, \tag{5.34}$$

where

$$Q_R = S_e, \tag{5.35}$$

and S_e is the total fraction of the light externally incident on the surface of the particle that is specularly reflected

$$S_e = \int_0^{\pi/2}[R_\perp(\vartheta) + R_\parallel(\vartheta)]\cos\vartheta\sin\vartheta d\vartheta = 2\int_0^{\pi/2} R(\vartheta)\cos\vartheta\sin\vartheta d\vartheta, \tag{5.36}$$

where $R(\vartheta) = [R_\perp(\vartheta) + R_P(\vartheta)]/2$ is the average of the polarized reflectivities.

Values of S_e have been calculated numerically and tabulated by Kerker (1969) for $1.2 \leq n_r \leq 4$ and $0 \leq n_i \leq 4$. They are plotted in Figure 5.11. A useful approximate expression for S_e can be derived by plotting S_e against $R(0)$, where

$$R(0) = \frac{(n_r - 1)^2 + n_i^2}{(n_r + 1)^2 + n_i^2}$$

is the specular reflection coefficient for normal incidence. Thus is done in Figure 5.12, in which the dots are the values calculated by Kerker. Except very close to $R(0) = 0$, the points can all be well represented by the second-order polynomial

$$S_e = 0.0587 + 0.8543R(0) + 0.0870R(0)^2, \tag{5.37}$$

where the coefficients were chosen as the best fit that is forced to go through the point ($R(0) = 1$, $S_e = 1$). Equation (5.37) is plotted as the line in Figure 5.12 and provides a convenient approximation for S_e.

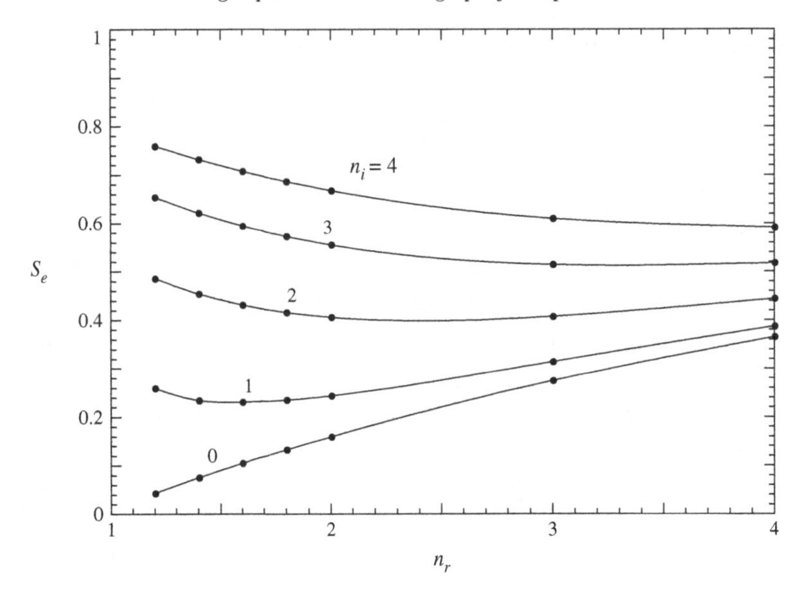

Figure 5.11 Surface reflection coefficient of a large sphere for various values of the complex refractive index. (The dots are the values from Kerker [1969]; the lines are interpolated using a polynomial fit.)

5.4.4.5 Refracted rays

The remainder of the light scattered by the sphere consists of rays that are refracted into the sphere and back out after making one or more internal traverses of the particle, and are illustrated in Figure 5.13. For these rays it may be assumed that $n_i \ll 1$, because otherwise the particle would be opaque and the contribution of the refracted rays would be negligible.

Consider first a ray of irradiance $J_{\perp\|}$ that makes only one pass through the sphere after encountering the surface at the point where the radius vector makes an angle ϑ with the direction of propagation. A fraction $R_{\perp\|}(\vartheta)$ is reflected by the first surface, and the remainder enters the particle as $[1 - R_{\perp\|}(\vartheta)]\cos\vartheta/\cos\vartheta'$, making an angle ϑ' with the radius vector, where ϑ' is given by Snell's law (Chapter 4). As the ray traverses the sphere it is attenuated by absorption according to $\exp(-\alpha x) = \exp(-4an_i\cos\vartheta')$, where $\alpha = 4\pi n_i/\lambda$ is the absorption coefficient and $x = 2a\cos\vartheta'$ is the path length through the sphere. Because $n_i \ll 1$, equations (4.32) and (4.33) may be used to calculate $R_{\perp\|}(\vartheta)$.

At the far side of the sphere a fraction $R_{\perp\|}(\vartheta')$ of the ray is reflected. It was shown in Chapter 4 that the reflection coefficient for rays incident internally on a surface at angle ϑ' is the same as the reflection coefficient for rays incident externally at angle ϑ. Hence, at the forward surface, a fraction $R_{\perp\|}(\vartheta)$ of the energy is reflected, and

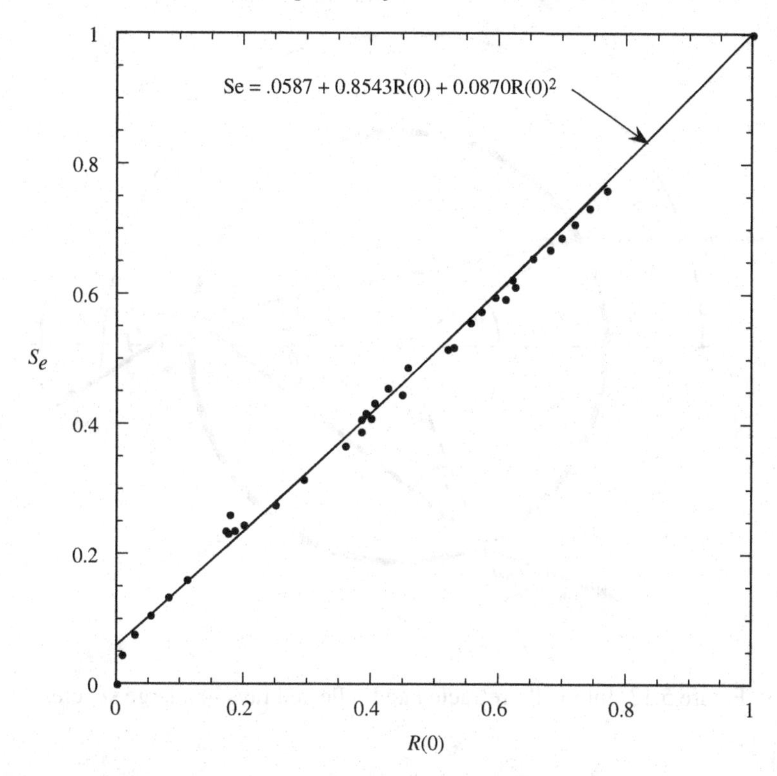

Figure 5.12 Surface reflection coefficient of a large sphere with $n_r > 1$ versus the specular reflectivity of a plane interface at normal incidence. (The dots are the calculated values from Kerker [1969]; the line is equation [5.37].)

fraction $[1 - R_{\perp\|}(\vartheta)]\cos\vartheta'/\cos\vartheta$ is transmitted. At all positions the ray remains in the same plane.

Figure 5.13 shows that the once-transmitted ray exits the sphere in a direction making an angle $\Upsilon = \pi + 2\vartheta - 2\vartheta'$ with the direction to the source. The phase angle of the exiting ray is $g = 2\pi - \Upsilon = \pi - 2(\vartheta - \vartheta')$. The minimum value of g occurs when $\vartheta = \pi/2$, so that $\vartheta' = \vartheta_c = \sin^{-1}(1/n_r)$, or $g = \pi - 2(\pi/2 - \vartheta_c) = 2\vartheta_c$. Hence, the angular width of the forward-refracted peak is $\pi - 2\vartheta_c$, where ϑ_c is the critical angle for total internal reflection.

Of the power incident on the sphere at all points on the surface between angles ϑ and $\vartheta + d\vartheta$ through the area $dA\cos\vartheta = 2\pi a \sin\vartheta a d\vartheta \cos\vartheta$, an amount

$$dP_{1\perp\|} = J_{\perp\|} dA \cos\vartheta \left[1 - R_{\perp\|}(\vartheta)\right]^2 e^{-\alpha x}$$

emerges between angles g and $g + dg$, where the subscript 1 on dP indicates the power emerging after one transit of the sphere. At a large distance from the particle this light passes through an incremental solid angle $d\Omega = 2\pi \sin g dg$. Hence, the

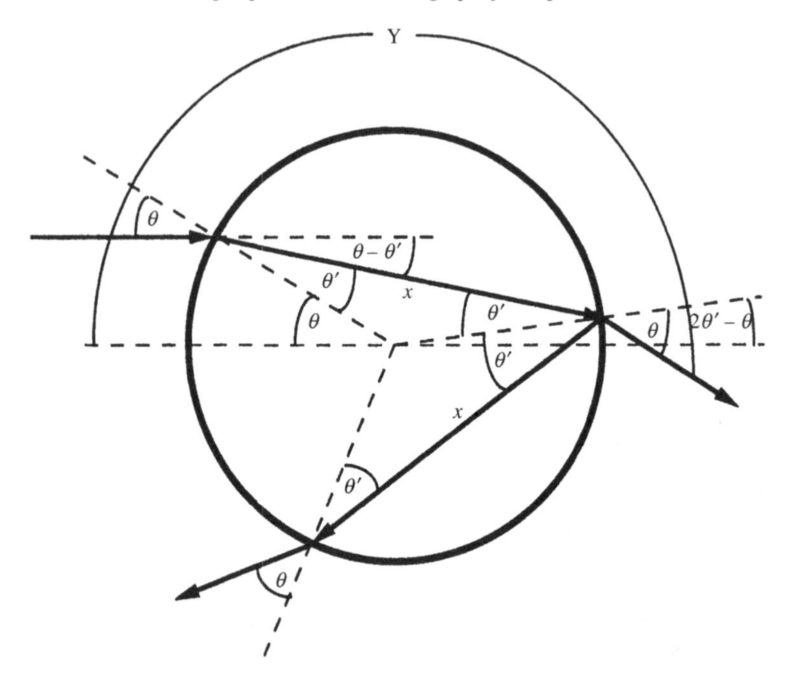

Figure 5.13 Internally refracted and reflected rays in a large sphere.

power per unit solid angle of light transmitted once through the sphere is

$$\frac{dP_{1\perp\|}}{d\Omega}=\frac{J_{\perp\|}2\pi a^2\left[1-R_{\perp\|}(\vartheta)\right]^2 e^{-4n_i X\cos\vartheta'}\sin\vartheta\cos\vartheta d\vartheta}{2\pi\sin g\,dg}. \tag{5.38}$$

The remaining power is internally reflected, passes through the sphere a second time, and is partially transmitted out of the sphere; the rest is again internally reflected, and the process continues. Because of the particle symmetry, at each reflection ϑ' and x are the same, and the ray stays in the same plane. From Figure 5.13 it is seen that light emerging from the sphere after each subsequent internal reflection and transmission differs from $dP_{1\perp\|}/d\Omega$ in three ways. (1) Each internal reflection attenuates the intensity by $R_{\perp\|}(\vartheta)$. (2) Each transmission attenuates the intensity by $\exp(-4n_i X\cos\vartheta')$. (3) Each internal reflection increases the deflection angle Υ by an additional amount $\pi-2\vartheta'$. Hence, rays that have passed through the particle q times and scattered into phase angle g have power per unit solid angle

$$\frac{dP_{q\perp\|}}{d\Omega}=J_{\perp\|}\frac{\alpha}{4\pi}[1-R_{\perp\|}(\vartheta)]^2 R_{\perp\|}(\vartheta)^{q-1}e^{-4qn_i X\cos\vartheta'}F_q(g,\vartheta,\vartheta'), \tag{5.39}$$

where $F_q = (g, \vartheta, \vartheta')$ is the focusing factor,

$$F_q(g,\vartheta,\vartheta') = \left|\frac{4\sin\vartheta\cos\vartheta}{\sin g(dg/d\vartheta)}\right| = \left|\frac{4\sin\vartheta\cos\vartheta}{\sin\Upsilon(d\Upsilon/d\vartheta)}\right|, \tag{5.40}$$

$$\Upsilon = q\pi + 2\vartheta - 2q\vartheta', \tag{5.41}$$

$$g = |\Upsilon - 2N\pi|, \tag{5.42}$$

where N is the integer ($N \geq 0$) that makes $0 \leq g \leq \pi$, ϑ and ϑ' are related by Snell's law, $\sin\vartheta = n_r \sin\vartheta'$, so that

$$d\Upsilon/d\vartheta = 2(1 - q d\vartheta'/d\vartheta) = 2\left\{1 - q\left[(1-\sin^2\vartheta)/(n^2-\sin^2\vartheta)\right]^{1/2}\right\}. \tag{5.43}$$

Equation (5.40) predicts that the intensity becomes infinite at certain angles where rays incident on the sphere over a range of entrance angles ϑ are focused into the same direction Υ. These resonances are another situation where geometric optics breaks down. Although the particle phase function is sharply peaked at these angles, the actual intensity remains finite, and its value must be calculated by Mie theory. However, ray theory does correctly predict the angles at which the peaks occur.

According to equation (5.41), a peak occurs when $d\Upsilon/d\vartheta = 0$, or

$$\sin\vartheta = [(q^2 - n^2)/(q^2 - 1)]^{1/2}. \tag{5.44}$$

The ray corresponding to $q = 2$ (two transits plus one internal reflection) is responsible for the primary *rainbow* in water droplets. Taking the refractive index of water in visible light as $n = 1.33$, these equations show that the rainbow ray enters the sphere at $\vartheta = 59.4°$, with $\vartheta' = 40.2°$, and exits at deflection angle $\Upsilon = 318°$ or phase angle $g = 42°$. Fainter rainbows that are produced by rays making more than two transits of the sphere may also be observed. For substances other than water the generic term *cloudbow* is used instead of rainbow.

Another situation in which a singularity in $F_q(g, \vartheta, \vartheta')$ occurs is if $g = 0$, but $\vartheta \neq 0$ or $\pi/2$. The resulting peak is called the *glory*. (A glory can frequently be seen from an airplane flying above a cloud of water droplets as a bright glow around the shadow of the aircraft on the cloud. However, in water the refractive index is too small for the $q = 2$ ray to cause a glory, which in this case is thought to arise from rays with $q \gg 2$ that enter the sphere at nearly glancing incidence and are multiply internally reflected or guided around the surface of the particle.)

Hence, the total contribution of the refracted rays to the scattering is the sum of all terms of the form of (5.34),

$$\frac{dP_{r\perp\|}}{d\Omega}=\sum_{q=1}^{\infty}\frac{dP_{q\perp\|}}{d\Omega}=J_{\perp\|}\frac{\sigma}{4\pi}[1-R_{\perp\|}(\vartheta)]^2$$
$$\times\sum_{q=1}^{\infty}R_{\perp\|}(\vartheta)^{q-1}e^{-4qn_i X\cos\vartheta'}F_q(g,\vartheta,\vartheta'). \qquad (5.45)$$

In order to illustrate the behavior of the refracted rays, the $q=1$ and $q=2$ terms will be discussed in detail. When $q=1$, $\Upsilon=\pi+2(\vartheta-\vartheta')$. For the axial ray, $\vartheta=0$, $\Upsilon=g=\pi$, and $F_1(\pi,0,0)=(1-1/n_r)^{-2}$. As ϑ increases, Υ increases to a maximum value of $\Upsilon=2\pi-2\vartheta'_c$, where $\vartheta'_c=\sin^{-1}(1/n_r)$ is the critical angle for total internal reflection, corresponding to the ray that enters the sphere at $\vartheta=\pi/2$. At the same time, g decreases from π to $2\vartheta'_c$. The result is a bright lobe of forward-refracted light. The sphere acts as a thick lens to focus the incident light into a beam of angular half-width $2\vartheta'_c$.

When $q=2$, $\Upsilon=2\pi+2\vartheta-4\vartheta'$. For the axial ray, $\vartheta=0$, $\Upsilon=2\pi$, $g=0$, and $F_2(0,0,0)=(1-2/n_r)^{-2}$. As ϑ increases, Υ first decreases, and g increases to the cloudbow angle. As ϑ continues to increase, the rays are prevented by Snell's law and by the curvature of the sphere from exiting at any lesser deflection angle Υ, but instead pile up, causing the cloudbow. Then Υ increases to a maximum angle of $3\pi-4\sin^{-1}(1/n_r)$ corresponding to $\vartheta=\pi/2$ and $\sin\vartheta'=1/n_r$. If n_r is large enough, this maximum Υ may exceed 2π. In this case there will be a glory when $\Upsilon=2\pi$; according to (5.41), this occurs when $\vartheta=2\vartheta'$, or $\sin\vartheta=n_r(1-n_r^2/4)^{1/2}$.

5.4.4.6 *Total light scattered by a large sphere*

Combining the equations developed in this section, the total power, including diffraction, scattered into unit solid angle by an isolated sphere, large compared with the wavelength, is

$$dP_{\perp\|}/d\Omega=dP_d/d\Omega+dP_{R\perp\|}/d\Omega+dP_{r\perp\|}/d\Omega,$$

where the terms on the right-hand side of this equation are given by (5.27), (5.32), and (5.45). Hence,

$$[Q_S\Pi_S(g)]_{\perp\|}=\frac{1}{J_{\perp\|}\sigma}\frac{dP_{\perp\|}}{d\Omega}=X^2\left\{\frac{2J_1[X\sin(\pi-g)]}{X\sin(\pi-g)}\right\}^2+R_{\perp\|}(g/2)$$
$$+[1-R_{\perp\|}(\vartheta)]^2\times\sum_{q=1}^{\infty}R_{\perp\|}(\vartheta)^{q-1}e^{-4qn_i X\cos\vartheta'}F_q(g,\vartheta,\vartheta'). \qquad (5.46)$$

The scattering efficiency can be found by integrating $[Q_S\Pi(g)]_{\perp\parallel}$ over all solid angles:

$$Q_{S\perp\parallel} = \frac{1}{4\pi}\int_{4\pi}[Q_S\Pi_s(g)]_{\perp\parallel}/d\Omega, \tag{5.47}$$

and the particle phase function from

$$\Pi_{s\perp\parallel}(g) = [Q_S\Pi_s(g)]_{\perp\parallel}/Q_S. \tag{5.48}$$

As a check on computations, Q_S must equal 2 when $n_i = 0$, because the only loss occurs by absorption of the refracted rays.

Approximately 99% of the light scattered by a large nonabsorbing particle is included in the terms up to and including two transits of the particle. Hence, the infinite sums in (5.45) may usually be truncated at $q=2$ or 3.

If the incident irradiance is unpolarized, so that $J_\perp = J_\parallel = J/2$, then

$$\begin{aligned}[Q_s\Pi_s(g)] = {} & X^2\left\{\frac{2J_1[X\sin(\pi-g)]}{X\sin(\pi-g)}\right\}^2 + \frac{R_\perp(g/2)+R_\parallel(g/2)}{2} \\ & + \frac{1}{2}\sum_{q=1}^{\infty}\left\{[1-R_\perp(\vartheta)]^2 R_\perp(\vartheta)^{q-1} + [1-R_\parallel(\vartheta)]^2 R_\parallel(\vartheta)^{q-1}\right\} \\ & \times e^{-4qn_iX\cos\vartheta'} F_q(g,\vartheta,\vartheta').\end{aligned} \tag{5.49}$$

The phase function of an isolated sphere with $m = 1.50 + 0i$ and $X = 100$ calculated from geometric optics is shown in Figure 5.14a, where the different contributions are shown explicitly.

The sum has been carried out to three transits. Comparing this figure with the exact Mie-theory calculation shown in Figure 5.7, it is seen that the two results are qualitatively similar.

Hodkinson and Greenleaves (1963), and Ungut *et al.* (1981) have shown that in the forward-scattering portion of the phase diagram where the scattering angle is smaller than 20° there is good agreement between Mie theory and geometric optics for X as small as 6. However, according to Liou and Hansen (1971), satisfactory agreement over the complete range of phase angles is not reached until X exceeds about 400.

We are now in a position to understand the various parts of Figures 5.7a and 5.14a. The sharp, narrow peak at $g = \pi$ is due to diffraction. The broad forward-scattered lobe is primarily due to the refracted ray that has made one traverse of the sphere. The backscattered lobe, the rainbow at 21°, and the glory are all due to the internally scattered rays. Note that there is a gap between the maximum value of g at which the internally reflected ray emerges and the minimum phase angle at which the forward-refracted ray emerges. In this gap, the only contribution to the scattered light is from surface Fresnel reflection, so that $\Pi(g)$ is small there.

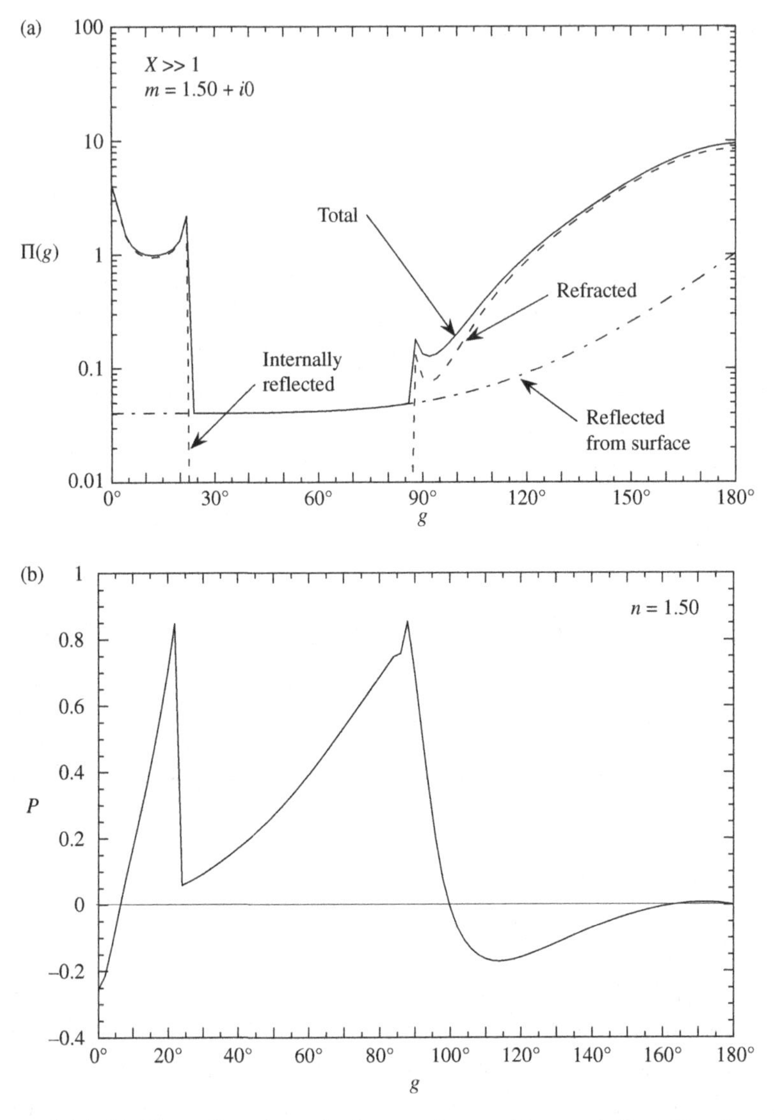

Figure 5.14 (a) Phase function of a large sphere with $n = 1.50 + i0$ calculated according to geometric optics. Compare with the solid curve of Figure 5.7a calculated using Mie theory. (b) Polarization of a large sphere with $n = 1.50 + i0$ calculated according to geometric optics. Compare with the solid curve of Figure 5.7b calculated using Mie theory.

The diffracted light is independent of n_i, and if $n_i << 1$, the surface reflectivity is nearly independent of n_i. However, the intensities of the refracted rays depend exponentially on n_i. As n_i increases, the refracted component decreases, and the twice-transmitted lobe scattered in the backward direction decreases more rapidly than the once-transmitted forward-scattered light. If $n_i X >> 1$, but $n_i << 1$, the particle is opaque, but the surface reflectivity remains small, so that $Q_s \simeq S_e = 1$. When $n_i >\sim 0.1$, the surface reflectivity behaves like $[(n_r-1)^2+n_i^2]/[(n_r+1)^2+n_i^2]$, and Q_s increases with n_i.

5.4.4.7 *Polarization*

The linear polarization for unpolarized incident irradiance may be calculated from

$$P(g) = \frac{[Q_s \Pi_s(g)]_\perp - [Q_s \Pi_s(g)]_\|}{[Q_s \Pi_s(g)]_\| + [Q_s \Pi_s(g)]_\|}, \tag{5.50}$$

where the $[Q_s \Pi_s(g)]_{\perp\|}$ are given by (5.46). The polarization for a sphere with $n = 1.50 + i0$ and $X = 100$ calculated from geometric optics is shown in Figure 5.14b. The forward-scattered lobe involves two refractions and is negatively polarized. At intermediate angles the major contribution is light reflected from the surface, and thus is positively polarized. The backward-scattered lobe consists of a mixture of light that enters the sphere at small ($\vartheta << \pi/2$) angles, which tends to be positively polarized, and large ($\vartheta \approx \pi/2$) angles, which tends to be negatively polarized, so that the polarization is complicated. The cloudbow is positively polarized, while the glory may be positively or negatively polarized. The polarization tends to be negative close to $g = 0$.

5.5 Other regular particles

The general principles on which Mie theory is based can be extended to other relatively simple geometries, including layered spheres, right circular cylinders of infinite length, and ellipsoids of revolution. For cylinders, the size parameter refers to the radius of the cylinder. The equations for cylinders and coated spheres are derived in the standard references, including those by Van de Hulst (1957), Kerker (1969), and Bohren and Huffman (1983). The Bohren and Huffman book contains FORTRAN programs for calculating scattering by a coated sphere and infinite circular cylinder. Scattering by oblate and prolate ellipsoids of revolution is treated in Asano and Yamamoto (1975).

The properties of the solutions are qualitatively similar to those of a sphere with similar size parameter X. Although different in detail, the scattering exhibits Rayleigh behavior when X is small, and diffraction, interference, and cloudbow-like resonance phenomena when$X \gg 1$.

5.6 The equivalent-slab approximation

When the particle is large compared with the wavelength and the incident light is unpolarized, an analytic approximation for the nondiffractive component of the

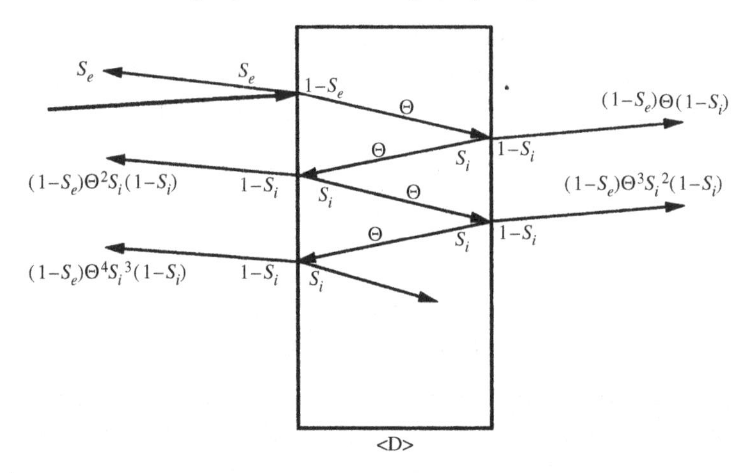

Figure 5.15 Schematic diagram of the equivalent-slab model for Q_s.

scattering and absorption efficiencies that is sufficiently accurate for many applications may be derived by replacing the sphere by a slab with appropriate optical properties. The equivalent-slab model is illustrated schematically in Figure 5.15.

The total fraction of incident light specularly reflected into all directions from the surface of the equivalent slab is S_e, where S_e is the integral of the Fresnel reflection coefficients, equation (5.36). As shown in Section 5.4.4, this is the same as the fraction of light reflected from the outer surface of a sphere. A fraction $1-S_e$ of the incident light then enters the slab and is attenuated by a factor Θ by absorption, where Θ is the *internal-transmission factor*, defined as the total fraction of light entering the particle that reaches another surface after one transit. Following the first passage through the particle, a fraction S_i is internally reflected, and the remainder $1 - S_i$ is refracted through the surface, where S_i is the reflection factor for light internally incident on the surface of the particle. The process continues, as indicated schematically in Figure 5.15.

The total fraction of incident light that is reflected or emerges from the back surface of the slab (the surface facing the source) is the backscattering efficiency

$$Q_{sB} = S_e + (1 - S_e)(1 - S_i)S_i\Theta^2 + (1 - S_e)(1 - S_i)S_i^2\Theta^4 + - - -$$
$$= S_e + (1 - S_e)(1 - S_i)\frac{S_i\Theta^2}{1 - S_i^2\Theta^2}, \quad (5.51a)$$

and that emerging from the forward surface of the slab (the surface facing away from the source) is the forward-scattering efficiency

$$Q_{sF} = (1 - S_e)(1 - S_i)\Theta + (1 - S_e)(1 - S_i)S_i^2\Theta^3 + (1 - S_e)(1 - S_i)S_i^4\Theta^5$$
$$+ - - - = (1 - S_e)(1 - S_i)\frac{\Theta}{1 - S_i^2\Theta^2}. \quad (5.51b)$$

The sum of the two efficiencies is the nondiffractive scattering efficiency. The back- and forward-scattering efficiencies can be conveniently expressed by two quantities, the scattering efficiency Q_s and the scattering difference Δ_s, where

$$Q_s = Q_{sB} + Q_{sF} = S_e + (1 - S_e)(1 - S_i)\frac{\Theta + S_i\Theta^2}{1 - S_i^2\Theta^2} = S_e + (1 - S_e)(1 - S_i)\frac{\Theta}{1 - S_i\Theta} \tag{5.52a}$$

and

$$\Delta Q_s = Q_{sB} - Q_{sF} = S_e + (1 - S_e)(1 - S_i)\frac{S_i\Theta^2 - \Theta}{1 - S_i^2\Theta^2} = S_e + (1 - S_e)(1 - S_i)\frac{-\Theta}{1 + S_i\Theta}. \tag{5.52b}$$

Then

$$Q_{sB} = \frac{Q_s + \Delta Q_s}{2}, \tag{5.53a}$$

and

$$Q_{sF} = \frac{Q_s - \Delta Q_s}{2}. \tag{5.53b}$$

It remains to find S_i and Θ. Because of the spherical symmetry of the particle, a ray encounters each internal surface at the same angle at which it entered the interior; hence, the reflection coefficients are the same for each order of internal reflection and $S_i = S_e$. From the discussion in Section 5.4.4.5, the factor Θ is given by $\Theta = 2\int_{\vartheta=0}^{\pi/2} e^{-4n_i X \cos\vartheta'} \sin\vartheta \cos\vartheta \, d\vartheta$. Differentiating Snell's law gives $\sin\vartheta \cos\vartheta \, d\vartheta = n_r^2 \sin\vartheta' \cos\vartheta' d\vartheta'$. Also, $4n_i X = \alpha D$, hence, $\Theta = 2n^2 \int_{\vartheta'=0}^{\vartheta'_c} e^{-\alpha D \cos\vartheta'} \sin\vartheta' \cos\vartheta' d\vartheta'$, where $\vartheta'_c = \sin^{-1}(1/n_r)$ is the critical angle for total internal reflection. This integral is readily evaluated to give

$$\Theta = \frac{2n_r^2}{(\alpha D)^2}\left\{ e^{-\alpha D(1-1/n_r^2)^{1/2}} \left[1 + \alpha D\left(1 - 1/n_r^2\right)^{1/2} \right] - e^{-\alpha D}[1 + \alpha D] \right\}. \tag{5.54}$$

Define the mean ray path length through the interior $\langle D \rangle$ as the thickness of a slab that will have the same value of Θ as (5.54) as $\alpha \to 0$. For the slab,$\Theta = e^{-\alpha\langle D \rangle} \simeq 1 - \alpha\langle D \rangle$. Expanding (5.54) in powers of αD gives $\Theta \simeq 1 - \frac{2}{3}[n_r^2 - (1/n_r)(n_r^2 - 1)^{3/2}]\alpha D$. Hence,

$$\langle D \rangle = \frac{2}{3}\left[n_r^2 - \frac{1}{n_r}\left(n_r^2 - 1\right)^{3/2} \right] D. \tag{5.55}$$

This quantity is the average distance traveled by all rays during a single transit of the particle. The value of $\langle D \rangle / D$ ranges from 0.85 to 0.93 as n_r increases from 1.3 to 2.0.

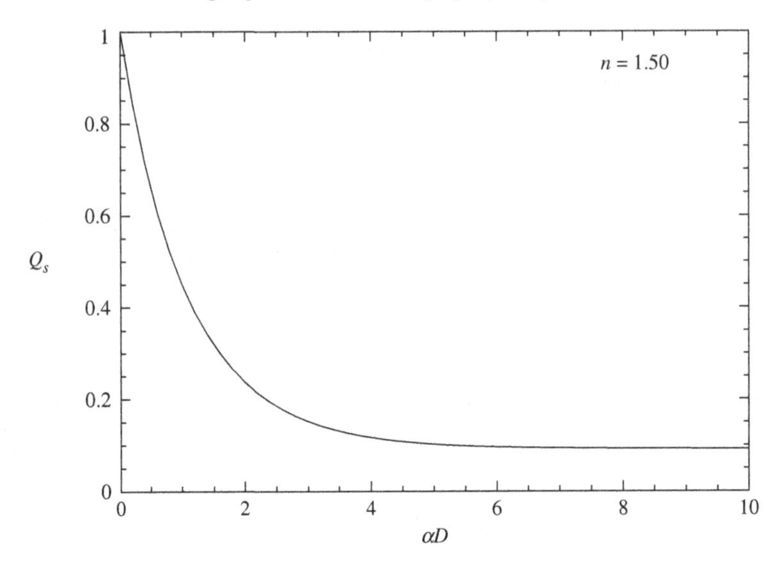

Figure 5.16 Scattering efficiency of a large sphere vs. αD calculated according to geometric optics and from the equivalent-slab model, equation (5.55); $n = 1.50 + i0$. The two curves are indistinguishable on the scale of this figure. The maximum error is 0.002.

When $\alpha\langle D\rangle \ll 1, \Theta = 1 - \alpha\langle D\rangle \simeq \exp(-\alpha < D >)$. If $\exp(-\alpha < D >)$ is compared with equation (5.54) for n_r between 1.3 and 2.0, the two quantities are found to be within 0.002 of each other for all values of $\alpha\langle D\rangle$. Hence, to a good approximation,

$$\Theta \simeq e^{-\alpha\langle D\rangle}, \tag{5.56}$$

where $\langle D\rangle$ is given by (5.55), so that the thickness of the equivalent slab is set equal to $\langle D\rangle$.

Inserting these values into (5.52), the equivalent-slab approximation for the scattering efficiency of a spherical particle much larger than the wavelength is

$$Q_s = S_e + (1 - S_e)\frac{1 - S_e}{1 - S_e\Theta}\Theta, \tag{5.57a}$$

and for the scattering difference,

$$\Delta Q_s = S_e - (1 - S_e)\frac{1 - S_e}{1 - S_e(-\Theta)}(\Theta), \tag{5.57b}$$

where S_e is given by (5.35) and Θ by (5.56).

The scattering efficiency versus $\alpha\langle D\rangle$ calculated using the exact geometric-optics expressions and the equivalent-slab approximation is shown in Figure 5.16, where the approximation is seen to be excellent.

Since $X >> 1$, we assume that $Q_E = 2$, while $Q_d = 1$, so $Q_A + Q_s = 1$. Then to the same accuracy as (5.57a) the absorption efficiency is

$$Q_A = 1 - Q_S \simeq (1 - S_e)\frac{1 - \Theta}{1 - S_e\Theta}. \tag{5.58}$$

Note that when $\alpha\langle D\rangle \ll 1$, $Q_A \simeq \alpha\langle D\rangle \propto \alpha D$, so that the total amount of light absorbed $J\sigma Q_A$ is proportional to the volume of the particle. However, as αD increases, Q_A approaches the lower limit $1 - S_e$, and the power absorbed is proportional to the cross-sectional area of the particle.

5.7 Computer programs

Bohren and Huffman (1983) contains a FORTRAN computer program for numerical calculation of the intensity and polarization of light scattered from a sphere. Michael Mishchenko maintains a website from which a computer program that calculates the average scattering parameters for spheres with a distribution of sizes may be downloaded at www.giss.nasa.gov/~crimm.

6

Single-particle scattering: irregular particles

6.1 Introduction

The scattering of electromagnetic radiation by perfect, uniform, spherical particles was described in Chapter 5. However, such particles are rarely found in nature. Most pulverized materials, including planetary regoliths, volcanic ash, laboratory samples, and industrial substances, have particles that are irregular in shape, have rough surfaces, and are not uniform in either structure or composition. Even the liquid droplets in clouds are not perfectly spherical, and they contain inclusions of submicroscopic particles around which the liquid has condensed, so that they are not perfectly uniform. At the present state of our computational and analytical capabilities it is possible to find exact solutions of scattering by such particles only by the expenditure of considerable computer time and memory (to say nothing of the effort of writing detailed programs), so that approximate models remain extremely useful.

The objective of any model of single-particle scattering is to relate the microscopic properties of the particle (its structure and complex refractive index) to the macroscopic properties (the scattering and extinction efficiencies and the phase function) that, in principle, can be measured by an appropriate scattering experiment. This chapter describes a variety of models that have been proposed to describe the scattering of light by irregular particles. This is not an exhaustive survey; rather, it is a commentary on those models that are most often encountered in remote-sensing applications or that offer some particular insight into the problem.

The organization of this chapter is as follows. First, the quantities defined in Chapter 5 for spherical particles are extended to particles that are nonspherical in shape. Next, some empirical models that are widely used in remote sensing are introduced. Theoretical approaches to the formidable problem of scattering by irregular particles are summarized. Laboratory measurements of the efficiencies and phase functions are described. The experimental and theoretical results are then summarized. Finally, the equivalent-slab approximation introduced in Chapter 5 for spheres is generalized to irregular particles and further extended to coated particles.

6.2 Extension of definitions to nonspherical particles

The physical meanings of the particle scattering parameters defined in Chapter 5, including the cross sections and phase functions, are intuitively clear for a particle that is spherically symmetric. However, if a particle is not spherical, the proper interpretation of these quantities is not obvious. Unless explicitly stated otherwise, in this book it will always be assumed that we are dealing with ensembles of particles whose orientations are random. Thus, the power extinguished by a particle of arbitrary shape is defined to be the average of the power removed from the beam of light as the particle is randomly oriented in all directions. Similar definitions apply for the power scattered and absorbed. The geometric cross section σ of a particle is the average area of the geometric shadow cast by the particle as it is oriented at random in all directions. The equivalent particle radius and diameter are defined as

$$a = \sqrt{\sigma/\pi}, \tag{6.1a}$$

$$D = 2a = 2\sqrt{\sigma/\pi}. \tag{6.1b}$$

With this understanding of implicitly averaging over all orientations, the quantities defined in Section 5.2 have physical meanings for nonspherical particles as well as spherical ones. These meanings are intuitively clear and internally consistent. Note that the scattering properties defined in this way are azimuthally symmetric, so that the normalization condition, equation (5.13) on $\Pi(g)$ holds.

Some workers define the effective particle radius a to be the radius of a sphere of the same volume as the particle under consideration. With such a definition the power absorbed by the particle is the same as that absorbed by the equivalent sphere if the particle is optically thin, but not if the particle is opaque. In this book the definition of the equivalent radius a in terms of a sphere of the same mean geometric cross-sectional area is preferred, because this leads to unambiguous meanings for the efficiencies and phase functions. Also, the power absorbed by an optically thick irregular particle and that absorbed by its equivalent sphere are equal.

6.3 Empirical scattering functions

6.3.1 The Allen approximation for Fraunhofer diffraction

It was shown in Chapter 5 that the light scattered by Fraunhofer diffraction around a sphere is identical with that diffracted through a circular aperture of the same radius in an opaque screen, and consists of a strong central peak surrounded by a series of weaker fringes. A noncircular opening has a qualitatively similar diffraction pattern, but the main peak and the fringes are not azimuthally symmetric. For instance, the angular size and spacing of the diffraction pattern of a rectangular slit is wider along the direction perpendicular to the long edge of the slit than along the direction perpendicular to the short edge of the slit. Jenkins and White (1950) give a number of examples of the diffraction patterns of openings of various shapes. Dauger has created a computer program that calculates the Fraunhofer diffraction pattern of an object of arbitrary cross section (see end of chapter).

Ensembles of large, irregular particles have narrow, forward-scattered main diffraction peaks, but the fringes and asymmetries tend to be averaged out, so that the diffracted intensity decreases monotonically from the center of the pattern and approximates the envelope of the diffraction pattern of a sphere. Allen (1946) has pointed out that a good empirical fit to the envelope of the diffraction fringes is provided by the function.

$$\Pi_d(g) = \frac{X^2}{1+0.470X^3(\pi - g)^3}, \tag{6.2}$$

where the coefficient of X^3 in the denominator comes from normalizing $\Pi_d(g)$. This function is mathematically more convenient than equation (5.26). The exact and approximate diffraction functions are compared in Figure 6.1.

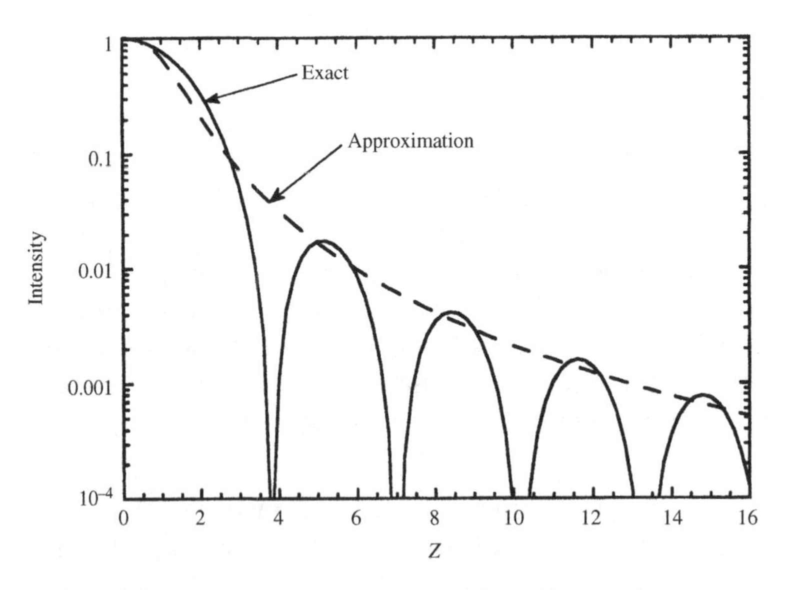

Figure 6.1 Allen approximation to the Fraunhofer diffraction function (dashed line) compared with the exact expression (solid line).

6.3.2 Legendre polynomial representation of $\Pi(g)$

It is often convenient to represent $\Pi(g)$ as a series of Legendre polynomials:

$$\Pi(g) = \sum_{j=0}^{\infty} b_j P_j(g), \tag{6.3}$$

where the b_js are constants, and the $P_j\ (g)$s are Legendre polynomials of order j. The properties of these functions are reviewed and illustrated in Appendix C. This representation of Π(g) is most useful when the departures from isotropic scattering are not very large, so that only a few terms are necessary. Thus (6.3) is most often used to represent the nondiffractive portion of Π(g). It has been used by many authors, including Hapke (1981) and Mustard and Pieters (1989).

From the normalization condition on $\Pi(g)$, equation (5.13), the series must satisfy

$$\frac{1}{4\pi}\int_{4\pi} \sum_j b_j P_j(g) d\Omega = \frac{1}{4\pi}\int_0^{\pi} \sum_j b_j P_j(g) 2\pi \sin g dg = 1.$$

But, from the orthogonality of the Legendre polynomials, the only nonvanishing integral is that for $j = 0$, which equals $4\pi b_0$. Hence, $b_0 = 1$. No general model exists to specify the other coefficients. A constraint on the b_j's is that $\Pi(g)$ cannot be negative anywhere. For example, if it is desired to represent the phase function by a first-order expansion, $\Pi(g) = 1 + b_1 \cos g$, then b_1 is restricted to the range $-1 \leq b_1 \leq +1$. A first-order expansion is adequate if the phase function is single-lobed. However, the phase functions of most real, nonopaque particles are double-lobed, which requires at least a second-order expansion: $\Pi(g) = 1 + b_1 P_1(g) + b_2 P_2(g)$.

The cosine asymmetry factor (equation 5.14) is

$$\xi = -<\cos g> = -\frac{1}{4\pi}\int_0^{\pi} \sum_{j=0}^{\infty} P_j(g) \cos g 2\pi \sin g dg$$

$$= -\frac{1}{2}\int_0^{\pi} b_1 P_1(g) \cos g \sin g dg = -\tfrac{1}{2}\textstyle\int_0^{\pi} b_1 \cos^2 g \sin g\, dg = -b_1/3, \tag{6.4}$$

where we have again used the orthogonality property, which implies that all terms except $j = 1$ in the sum vanish. Hence, $b_1 = -3\xi$.

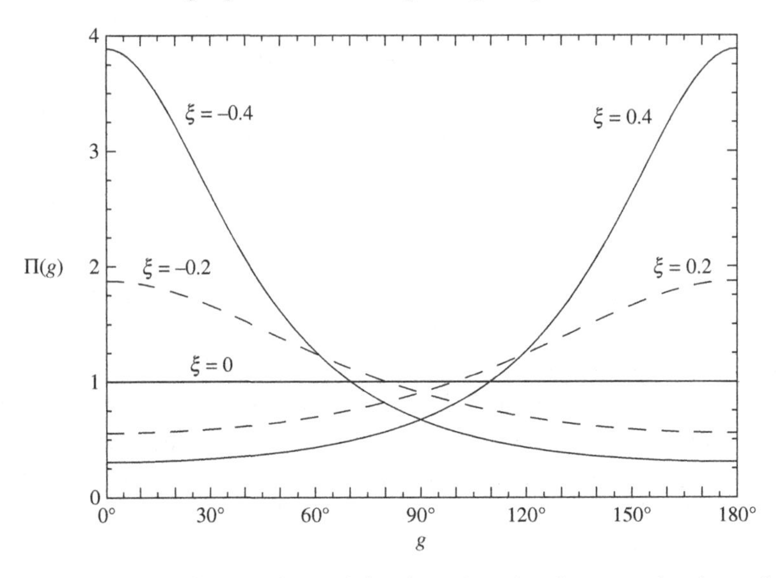

Figure 6.2 Henyey–Greenstein particle phase function for several values of the asymmetry parameter ξ .

6.3.3 Henyey–Greenstein function representation of $\Pi(\mathbf{g})$

Henyey and Greenstein (1941) introduced the single-parameter empirical phase function

$$\Pi_{HG1}(\theta) = \frac{1-\xi^2}{(1-2\xi\cos\theta+\xi^2)^{3/2}},$$

or equivalently,

$$\Pi_{HG1}(g) = \frac{1-\xi^2}{(1+2\xi\cos g+\xi^2)^{3/2}}. \tag{6.5}$$

The function is illustrated for several values of ξ in Figure 6.2. It is widely used in planetary reflectance work (e.g., Lumme and Bowell, 1981b; Buratti, 1985; Helfenstein *et al.*, 1988).

This versatile function has the following useful properties, which are left as an exercise for the reader to prove. It is normalized: $(1/4\pi)\int_{4\pi}\Pi_{HG1}(g)d\Omega = 1$. The Henyey–Greenstein parameter ξ is the cosine asymmetry factor, $\xi = \langle\cos\theta\rangle = -\langle\cos g\rangle$. The function is isotropic ($\Pi_{HG1}(g) = 1$) when $\xi = 0$. At $g = 0$ and π, $\Pi_{HG1}(g)$ has the values $(1-\xi)/(1+\xi)^2$ and $(1+\xi)/(1-\xi)^2$, respectively. If $\xi > 0$, $\Pi_{HG1}(g)$ increases monotonically between 0 and π, and decreases monotonically if $\xi < 0$. The shape of the peak depends on ξ, becoming higher and narrower as $|\xi|$ increases. In the limiting case, as $\xi \to 1(\text{or} -1)$, $\Pi_{HG1}(g) \to 0$ everywhere except at $g = \pi$ (or 0), where it becomes infinite in such a way that its integral equals 4π. Hence, this function is often used to represent the diffraction peak.

The expansion of the Henyey-Greenstein function in Legendre polynomials is (Kattawar, 1975)

$$\Pi_{HG1}(g) = \sum_{j=0}^{\infty} (2j+1)(-\xi)^j P_j(g). \tag{6.6}$$

However, equation (6.5) has only one lobe, whereas the phase functions of most particles are double-lobed. In order to represent a double-lobed phase function, two Henyey-Greenstein functions of opposite symmetry are required. The double-lobed Henyey-Greenstein function can be formulated using either two or three parameters; the two-parameter function is:

$$\Pi_{HG2}(g) = \frac{1+c}{2}\frac{1-b^2}{(1-2b\cos g+b^2)^{3/2}} + \frac{1-c}{2}\frac{1-b^2}{(1+2b\cos g+b^2)^{3/2}}; \tag{6.7a}$$

and the three-parameter function is:

$$\Pi_{HG3}(g) = \frac{1+c}{2}\frac{1-b_1^2}{(1-2b_1\cos g+b_1^2)^{3/2}} + \frac{1-c}{2}\frac{1-b_2^2}{(1+2b_2\cos g+b_2^2)^{3/2}}. \tag{6.7b}$$

In these experessions the first term describes the backward lobe and the second the forward lobe; the relative strengths of the lobes are determined by c, and their shapes by b, or b_1 and b_2. The b-parameters are constrained to lie in the range $0 \leq b$, b_1, $b_2 \leq 1$; there is no constraint on c except that $\Pi_{HG}(g) \geq 0$ everywhere. The mean cosines of the two functions are:

$$\text{two-parameter } \xi = -bc; \tag{6.8a}$$

$$\text{three-parameter } \xi = -\frac{1+c}{2}b_1 + \frac{1-c}{2}b_2. \tag{6.8b}$$

The disadvantage of the two-parameter function is that the shapes of the lobes of the particle being described cannot be too different; the advantage is that it has only two parameters to be fitted instead of three. Domingue and Verbiscer (1997) and Hartman and Domingue (1998) investigated the relative ability of the two- and three-parameter functions to fit phase angle measurements of several types of real materials. They concluded that, although slightly better fits were obtained with the three-parameter function, the improvement was marginal and not sufficient to justify a third parameter.

6.3.4 *Lambert and Lommel–Seeliger sphere phase functions for nondiffractive scattering*

Suppose an element of area dA on the surface of a sphere is illuminated by collimated light of irradiance J making an angle i with the normal to dA and observed from a direction making an angle e with the surface normal. The phase angle g is the angle between the directions to the source and detector as seen from the particle. Suppose each element of the surface scatters the incident light into unit solid angle described by a surface reflectance function $dI = JY(i, e, g)dA$. Then the nondiffractive scattering efficiency Q_s and phase function of an opaque sphere, each of whose surface elements scatters light as described by $Y(i, e, g)$, may be calculated as follows.

It is assumed that the particle is sufficiently absorbing that internally transmitted light can be neglected. Choose a spherical coordinate system with the origin at the center of the particle and such that the great circle through the sub-source point and the sub-observer point forms the equator of the sphere, with the central meridian of longitude passing through the sub-observer point. The sub-observer point and sub-source point are separated in longitude by an angle equal to the phase angle g. The geometry is shown in Figure 6.3. The element of area $dA = a^2 \cos L dL d\Lambda$ is located on the surface of the sphere at longitude Λ and latitude L, where a is the radius of the particle. The outward normal to the surface makes an angle i with the incident illumination, and an angle e with the direction to the observer.

From the law of cosines for spherical triangles, $\cos i = \cos(\Lambda + g) \cos L$, and $\cos e = \cos \Lambda \cos L$. Then the radiance scattered from the sphere into the direction toward the observer is the integral of $Y(i, e, g)$ over all areas of the surface that are

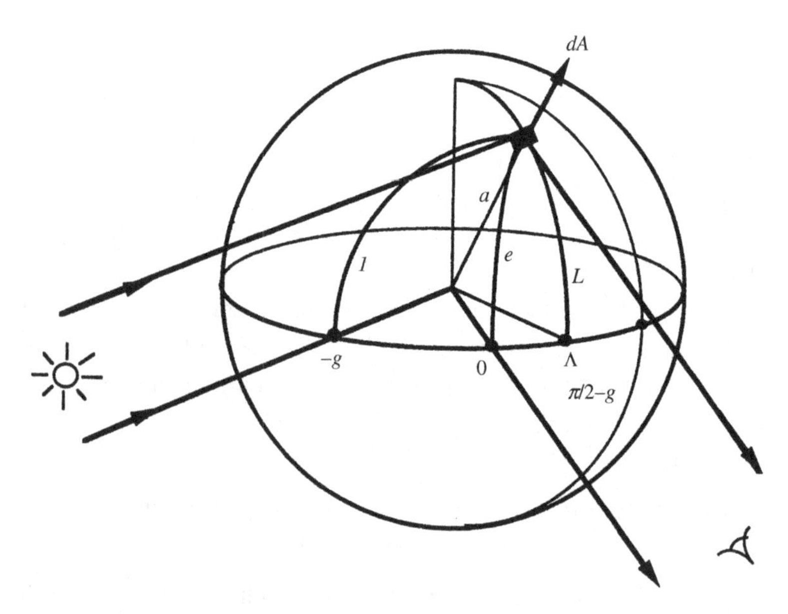

Figure 6.3

both visible and illuminated:

$$I = \int_{\Lambda=-\pi/2}^{\pi/2-g} \int_{L=-\pi/2}^{\pi/2} JY(i, e, g) dA. \tag{6.9}$$

Two widely used surface reflectance functions are Lambert's law,

$$Y_{\mathrm{L}}(i, e, g) = \frac{1}{\pi} \cos i \cos e, \tag{6.10}$$

and the Lommel–Seeliger law,

$$Y_{\mathrm{LS}}(i, e, g) = \frac{1}{4\pi} \frac{\cos i \cos e}{\cos i + \cos e}. \tag{6.11}$$

These functions are discussed more fully in Chapter 8. Note that both are independent of g.

Inserting (6.10) into (6.9) gives

$$I = \int_{\Lambda=-\pi/2}^{\pi/2-g} \int_{L=-\pi/2}^{\pi/2} \frac{J}{\pi} \cos i \cos e \, dA$$
$$= \frac{J}{\pi} a^2 \int_{\Lambda=-\pi/2}^{\pi/2-g} \int_{L=-\pi/2}^{\pi/2} \cos(\Lambda + g) \cos \Lambda \cos^3 L dL d\Lambda.$$

Using the identity $\cos x \cos y = [\cos(x+y) + \cos(x-y)]/2$, this integral becomes

$$I = \frac{J}{2\pi} a^2 \left\{ \int_{\Lambda=-\pi/2}^{\pi/2-g} [\cos(g + 2\Lambda) + \cos g] d\Lambda \right\} \cdot \left\{ \int_{L=-\pi/2}^{\pi/2} \cos^3 L dL \right\},$$

which is readily evaluated and gives

$$I = Ja^2 \frac{1}{\pi} \frac{2}{3} [\sin g + (\pi - g) \cos g]. \tag{6.12}$$

By definition, $I = J\sigma Q_s[\Pi(g)/4\pi]$, where $\sigma = \pi a^2$. However, because we are assuming that radiation does not penetrate into the sphere, there is no absorption, so that $Q_s = 1$. Hence, the phase function of a *Lambert sphere* is

$$\Pi_{\mathrm{L}}(g) = \frac{8}{3} \frac{\sin g + (\pi - g) \cos g}{\pi}. \tag{6.13}$$

This expression was first derived by Schönberg (1929) and is known as the Schönberg function. It may readily be verified that this phase function satisfies the normalization condition, equation (5.13). It is plotted in Figure 6.4.

Similarly, inserting the Lommel–Seeliger law (6.11) into (6.9) gives

$$I = \int_{\Lambda=-\pi/2}^{\pi/2-g} \int_{L=-\pi/2}^{\pi/2} \frac{J}{4\pi} \frac{\cos(\Lambda + g) \cos L \cos \Lambda \cos L}{\cos(\Lambda + g) \cos L + \cos \Lambda \cos L} dA$$
$$= \frac{J}{4\pi} a^2 \int_{\Lambda=-\pi/2}^{\pi/2-g} \int_{L=-\pi/2}^{\pi/2} \frac{\cos(\Lambda + g) \cos \Lambda}{\cos(\Lambda + g) + \cos \Lambda} \cos^2 L dL d\Lambda.$$

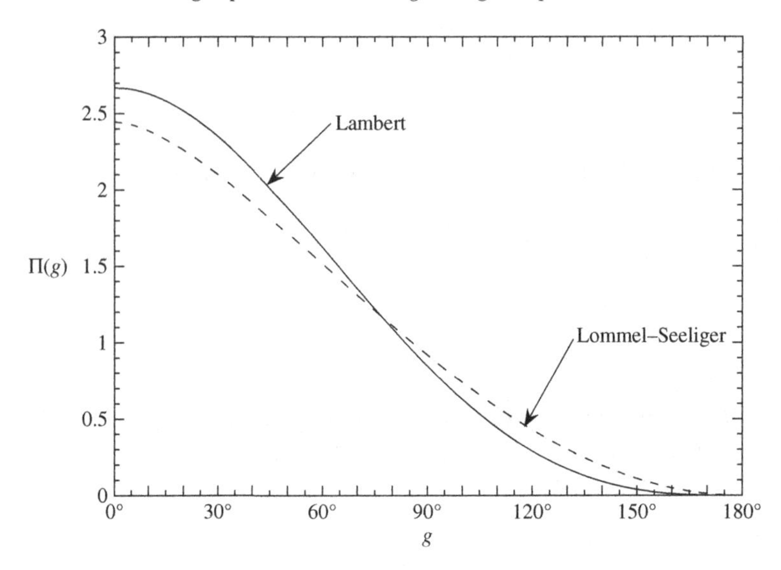

Figure 6.4 Phase functions of Lambert and Lommel–Seeliger spheres.

The integral over L is readily evaluated. The integral over Λ may be evaluated by putting $x = \Lambda + g/2$, so that $\cos(\Lambda + g) = \cos(x + g/2)$ and $\cos\Lambda = \cos(x - g/2)$, giving

$$I = \frac{J}{16}a^2 \int_{x=-\pi/2+g/2}^{\pi/2-g/2} \left[\sec\frac{g}{2}\cos x - \sin\frac{g}{2}\tan\frac{g}{2}\sec x\right] dx$$
$$= \frac{J}{8}a^2\left[1 - \sin\frac{g}{2}\tan\frac{g}{2}\ln\left(\cot\frac{g}{4}\right)\right] = J\pi a^2 \frac{\mathrm{p}(g)}{4\pi}.$$

Hence,

$$\Pi_{LS}(g) = \frac{1}{2}\left[1 - \sin\frac{g}{2}\tan\frac{g}{2}\ln\left(\cot\frac{g}{4}\right)\right].$$

However, this phase function is not normalized. To normalize $\Pi(g)$, set

$$\frac{C}{4\pi}\int_0^{\pi} \frac{1}{2}\left[1 - \sin\frac{g}{2}\tan\frac{g}{2}\ln\left(\cot\frac{g}{4}\right)\right] 2\pi \sin g\, dg = 1,$$

where C is the normalization constant. This integral may be evaluated by letting $y = \cos(g/2)$, noting that

$$\ln[\cot(g/4)] = \ln[(1+y)/(1-y)],$$

and integrating by parts. This gives the phase function of a *Lommel–Seeliger sphere,*

$$\Pi_{LS}(g) = \frac{3}{4(1-\ln 2)}\left[1 - \sin\frac{g}{2}\tan\frac{g}{2}\ln\left(\cot\frac{g}{4}\right)\right]. \tag{6.14}$$

This function is plotted in Figure 6.4 along with the Lambert sphere.

Note that both the Lambert and Lommel–Seeliger functions are strongly backscattering.

6.3.5 Other functions

Several other semi-empirical scattering functions have been introduced in the literature to describe various aspects of scattering by large irregular particles. One of the simplest is that used by Hapke and Nelson (1975) in connection with a study of the clouds of Venus. For the diffracted component of the intensity they used a delta function. They assumed that the nondiffracted part of the phase function is isotropic and that the scattering efficiency could be described by

$$Q_s = S_e + (1 - S_e)e^{-\alpha D}, \tag{6.15}$$

where S_e is the integral of the Fresnel reflection coefficient given by (5.36).

For large particles, Pollack and Cuzzi (1980) described diffraction by an approximation to the circular-hole equation, and external surface scattering by the Fresnel equations. They represented the internally refracted light by a simple expression of the form $ce^{(1-bg)}$, where c and b are empirical constants.

6.4 Theoretical and experimental studies of nonspherical particles

6.4.1 Theory for particles small compared with the wavelength

Although Mie theory is strictly valid only for spherical particles, it still serves as a valuable starting point in discussing scattering by irregular particles. Suppose that the condition $|n|X \ll 1$ is satisfied, where $n = n_r + in_i$ is the complex index of refraction relative to its surroundings, and $X = 2\pi a/\lambda$ is the particle size parameter. Then the change in phase of the applied internal field across the particle is $n_r X$, and the attenuation of the field across the particle is $n_i X$, both of which are small. At any given time the entire particle sees electric and magnetic fields that are essentially uniform, and the instantaneous fields in the vicinity of the particle are essentially the same as if only static fields are applied. Thus, the fields are given, to a good approximation, by solutions of the appropriate electrostatic and magnetostatic equations. That this is indeed the case may be verified for small spherical particles as follows.

The dipole moment induced by a uniform, static, external field E_{e0} on a sphere of radius a and dielectric constant K_e is (see any textbook on advanced electrodynamics, e.g., Stratton (1941)):

$$p_{e0} = 4\pi a^3 \varepsilon_{e0} E_{e0}(K_e - 1)/(K_e + 2). \tag{6.16}$$

If $|n|X = 1$, then the electric field seen by a sphere illuminated by a plane electromagnetic wave of irradiance $J = (\varepsilon_{e0}/\mu_{m0})^{1/2} E_{e0}^2/2$ and frequency ν is $E_{e0}e^{2\pi i\nu t}$, and the induced electric dipole of the sphere is $p_{e0}e^{2\pi i\nu t}$.

It can also be shown (Stratton, 1941) that the power radiated into a vacuum by an oscillating electric dipole of moment $p_e = p_{e0}e^{2\pi i\nu t}$ and frequency ν is given by

$$p_e = 4\pi^3\nu^4\varepsilon_{e0}^{1/2}\mu_{m0}^{3/2}|p_{e0}|^2/3. \tag{6.17}$$

Inserting (6.16) into (6.17) and equating the result to $J\pi a^2 Q_S$, where Q_S is the scattering efficiency, gives

$$Q_S = \tfrac{8}{3}X^4|(K_e - 1)/(K_e + 2)|^2. \tag{6.18}$$

Since $K_e = n^2$, equation (6.18) is identical with (5.11) for the scattering efficiency of a small particle.

Now, a static electric field will induce a dipole moment in any particle, regardless of shape. This suggests that small particles of any shape will scatter light similar to a sphere, resulting in equations similar to (5.16) and (5.17) for a sphere. Figure 6.5

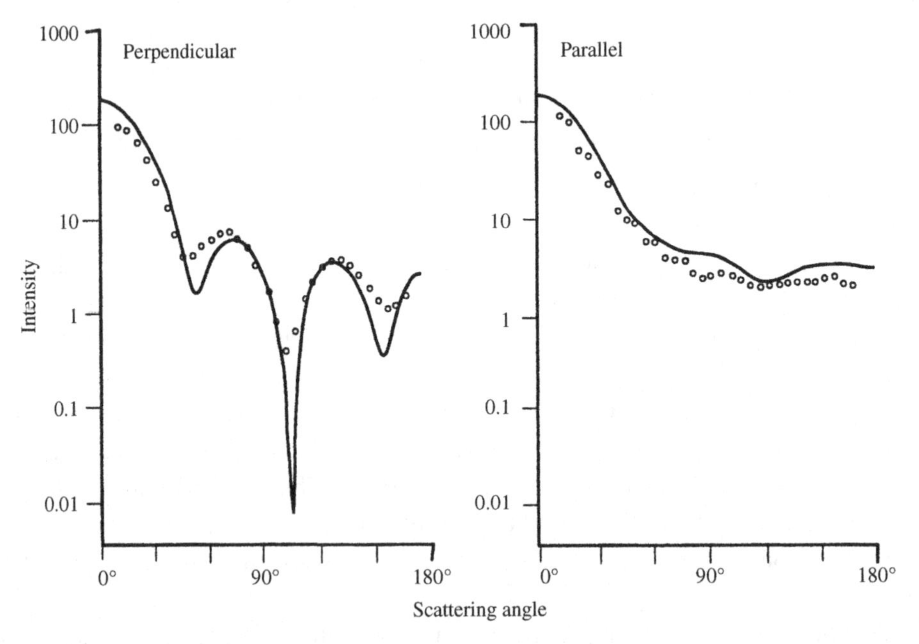

Figure 6.5 Measured angular scattering coefficient of a cube of $X = 3.75$ and $n = 1.57 + i0.006$ (circles), compared with the predictions of Mie theory for a sphere of the same size (lines), for the two polarized components of incident radiation. (Reproduced from Zerull [1976], copyright 1976, with permission of Friedrich Vieweg & Sohn.)

shows the measured phase function of a cube compared with the prediction of Mie theory for a sphere of the same size and refractive index.

Hence, electrostatic theory may be employed to calculate p_{e0}, and the result may be used in (6.17) to calculate Q_S. The advantage of this method is that it allows a wider variety of particle shapes to be treated analytically than does electrodynamic theory. In particular, scattering by oblate and prolate ellipsoids of rotation may be calculated, which includes disks and needles as special cases, and also triaxial ellipsoids.

Scattering by small ellipsoidal particles is discussed in considerable detail by Bohren and Huffman (1983). The net result is that if a particle is *equant*, that is, if its dimensions are roughly the same in all directions, then its efficiencies are not very different from those of a sphere of similar size. In particular, $Q_S \propto X^4$ and $Q_A \propto n_i X$. However, if the particle is not approximately equidimensional, the particle still scatters and absorbs in a dipole-like manner, but its efficiencies differ quantitatively from (6.18).

6.4.2 General theoretical methods

A large number of different types of theoretical approaches to the problem of calculating single-particle scattering have been published, in addition to the direct solution of the wave equation discussed in Chapter 5. Of these, three appear to be the most fruitful and have become dominant in recent years with the advent of high-speed desktop computers. These are the discrete dipole approximation (DDA), the T-matrix method (TMM), and the ray-tracing Monte Carlo approximation (RTMCA). All are capable of handling particles with complex shapes, rough surfaces, and inclusions, as well as systems of several interacting particles. However, they require the detailed specification of the shape and structure of the system. They are highly computer intensive, and the computational requirements increase geometrically with increasing particle size. As the capabilities of computers increase in coming years it is expected that these methods will be able to describe scattering by both larger particles and systems of more particles.

The discrete dipole approximation This method is reviewed by Draine (2000). It was introduced by Purcell and Pennypacker (1973) and further developed by Draine (1988), Draine and Goodman (1993), and Draine and Flatau (1994). Two versions are available for download on the internet.

In the DDA the system under study is replaced by an array of point polarizable electric dipoles that approximate its size and structure as closely as possible. The dipoles interact with the radiation from each other as well as the incident irradiance. For a total of N dipoles this results in $3N$ coupled linear equations. The smaller the distance d between dipoles is, the more accurate the simulation will be, but the resulting increase in N also increases the CPU time and memory required. The maximum spacing is such that the phase change of the field between dipoles is $2\pi |n| d/\lambda < 1$.

T-matrix method (T for transition) This method (also known as the extended boundary condition method) is reviewed by Mishchenko *et al.* (2000b). It was first proposed by Waterman (1965, 1979) and further developed by Mishchenko and his colleagues (Mishchenko *et al.*, 1996, 2002). Scattering and absorption inside a particle due to the incident wave are replaced by charges and currents on the surface of a sphere that completely circumscribes the particle. Mishchenko discovered a way of analytically simplifying scattering from the system if it is averaged over all orientations, thus facilitating the computation of the important problem of scattering by ensembles of randomly oriented particles. He has made his program available for downloading from the internet. In addition to applying the method to a wide range of problems, he has also published a useful list of papers in which the method is applied to scattering problems (Mishchenko *et al.*, 2007).

Ray-tracing Monte Carlo approximation It was shown in Chapter 5 that when $X \gg 1$, solutions for the scattering of light from a sphere obtained using geometric optics give reasonably good approximations to those using Mie theory. This suggests that ray theory may be used on large, nonspherical particles. In the RTMCA rays from a distant source are directed randomly at the system under study, with which they interact according to Fresnel's laws of reflection and refraction. The directions into which they are finally scattered are recorded. In this way the scattering properties are built up after a large number of trials.

This approach has been used by several authors to calculate scattering by particles of regular shape, such as cubes, parallelepipeds, hexagonal cylinders, and other common crystalline forms (e. g., Liou and Coleman, 1980; Liou *et al.*, 1983; Muinonen *et al.*, 1989; Peltoniemi *et al.*, 1989; Vilaplana *et al.*, 2006). It is discussed in detail by Macke (2000) and Muinonen (2000).

6.4.3 Other approaches

Various other schemes for calculating scattering have been proposed. Further references and discussions can be found in the works of Schuerman (1980), Bohren and Huffman (1983), and Mishchenko *et al.* (2000a). Most of the models divide the problem into several parts, typically diffraction, external surface scattering, and internal refraction and scattering, and treat each part separately. Often one part is emphasized, and the rest either ignored or treated in an ad hoc fashion.

Leinert *et al.* (1976) fitted observations of the zodiacal light to models in which the particles are large compared with the wavelength. The diffracted radiation was taken to be identical with that from a sphere (equation [5.28]), and surface-reflected light was described by the Fresnel reflectances (equation [5.34]). Internally refracted light was assumed isotropic and simply adds a constant term to the phase function.

Chylek *et al.* (1976) pointed out that the large resonances displayed by scattering from a perfect sphere, such as peaks in the scattering efficiency and the cloudbows and glories in the phase function, can be attributed to specific terms in the series expansion of the Mie solution. They set these terms equal to zero and assumed that the resulting expression applies to irregular particles. However, though this assumption may be reasonable for particles whose sizes are comparable to the wavelength, it is certainly not true for large, irregular, nonuniform particles.

In general, the scattered wave can be considered as the sum of waves radiating from all the different points within the particle. The amplitudes of these wavelets are proportional to the amplitude of the wave inside the particle and the difference between the scatterer and a vacuum. Several workers have used the so-called *eikonal approximation,* in which the internal wave is replaced by the incident wave phase-shifted along undeviated linear paths. Both Chiappetta (1980) and Perrin and Lamy (1983) used this approach to calculate the intensity forward-scattered into large phase angles, and they used an empirical model to describe the scattering at smaller phase angles. Their models assume that the particles have very rough surfaces. They also assume that the broad backscattered peaks in the particle phase functions are caused by shadows on the surface. Internally refracted light is mainly ignored.

Schiffer and Thielheim (1982a, b) used a geometric-optics approach to calculate the effects of shadows on light reflected from a rough surface. Mukai *et al.* (1982) modeled multiple reflections between facets on the rough surface of a particle using the equation of radiative transfer, although it is not clear that this equation is applicable. Internally refracted light is ignored in the model of Mukai and associates, so that the types of particles to which it may be applicable are extremely limited.

Emslie and Aronson (1973) and Aronson and Emslie (1973, 1975) emphasized the importance of surface roughness to the spectral scattering properties of the particle. In an elaborate numerical model, later extended by Egan and Hilgeman (1978), they assumed a particle with a basic spherical shape. The absorption and scattering of this sphere were calculated using geometric optics. Roughness on the particle surface was assumed to scatter like randomly oriented dipoles distributed over the surface of the sphere. It was also assumed that the rest of the surface scattered light diffusely instead of specularly. This model has had considerable success in explaining features in the reflectance and emittance spectra of powders in the thermal infrared.

6.4.4 Laboratory measurements

Most measurements of particle scattering have been done at visible (Hovenier, 2000) and microwave (Gustafson, 2000) wavelengths. The sizes of particles of greatest interest in remote sensing generally tend to be of the order of a few micrometers. Hence, the particles studied using visible light generally are of the order of 1–10 μm in size. Particles are levitated electrostatically or in a column of gas. The particle is illuminated by a collimated light source and the light scattered into all directions in the plane containing the source and detector measured. Polarized optics may or may not be used. At radio frequencies the targets are several centimeters in size in order to be analogs of particles interacting with visible light. The large size allows better control over the particle characteristics, as well as convenient handling and support. The particle is illuminated by radiation from a transmitting antenna and the scattered radiation detected by a separate movable receiving antenna.

Experimental studies of scattering by large, irregular particles have been reported by several workers, including Richter (1962), Hodkinson (1963), Zerull and Giese (1974), Zerull (1976), Weiss-Wrana (1983), McGuire and Hapke (1995), Volten *et al.* (2001), Barkey *et al.* (2002), and Munoz *et al.* (2006). The study by McGuire and Hapke is particularly illluminating because they systematically varied the shape, surface roughness, absorption coefficient, and density of internal scatterers in their analog particles, and because theirs is one of the few to investigate the effects of internal scatterers. A catalog of scattering measurements on a variety of types of particles at 633 and 442 nm by the Amsterdam group is available for download on the internet.

6.4.5 Summary of theoretical and experimental studies

Based on the studies summarized in the preceding subsections, a number of general inferences concerning the scattering properties of large, irregular particles can be made. The following discussion specifically excludes the Fraunhofer diffraction peak.

(1) A major difference between the scattering properties of irregular particles and perfect spheres is the absence of the large oscillations observed in both the scattering efficiencies of spheres as the size parameter changes and in the resonances in the angular scattering function. The varying dimensions in different directions of the irregularly shaped particles destroys the positive and negative interference effects that cause the resonances. This is illustrated in the study by Hodkinson (1963) of the extinction of visible light by colloidal suspensions of quartz particles as the size changes. Figure 6.6 shows that, in contrast to perfect

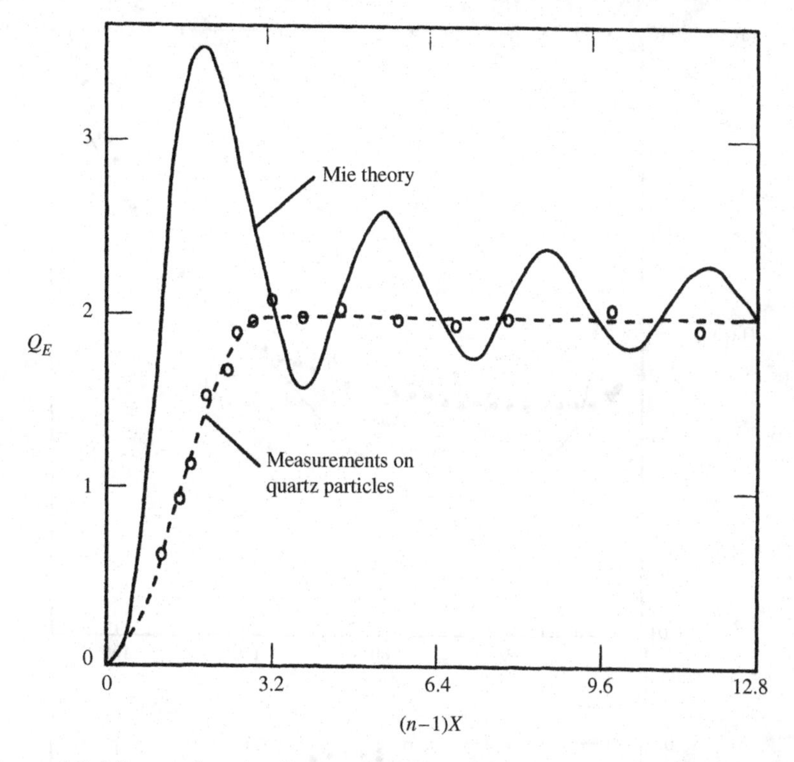

Figure 6.6 Measured extinction coefficients of suspensions of irregular nonabsorbing particles (circles and dashed line) versus size parameter, compared with predictions of Mie theory for perfect spheres (solid line). (Reproduced from Hodkinson [1963], copyright 1963, with permission of Elsevier.)

spheres, the extinction efficiency of the irregular particles increases smoothly and monotonically with X until when $(n_r - 1)X \gtrsim 3$ it levels off at $Q_E \simeq 2$. Figure 6.7 shows the measured phase function and polarization of irregular particles of olivine compared with those calculated for a perfect sphere. Note that there is no sign of a cloudbow or other resonances.

(2) If a particle is translucent, that is, if it has few internal scatterers and is not completely opaque, its phase function is dominated by a strong, broad, forward-scattering lobe caused by refracted, singly transmitted rays (Figures 6.7 and 6.8). This lobe is weaker for irregular particles than for spherical ones, but it occurs for particles of any shape or degree of surface roughness. Its amplitude decreases as the absorption coefficient increases.

(3) At intermediate angles the intensity scattered from a spherical particle is small and independent of absorption coefficient, and the only rays that contribute are those reflected from the surface. In an irregular particle the intensity at intermediate angles is increased markedly, by as much as an order of magnitude,

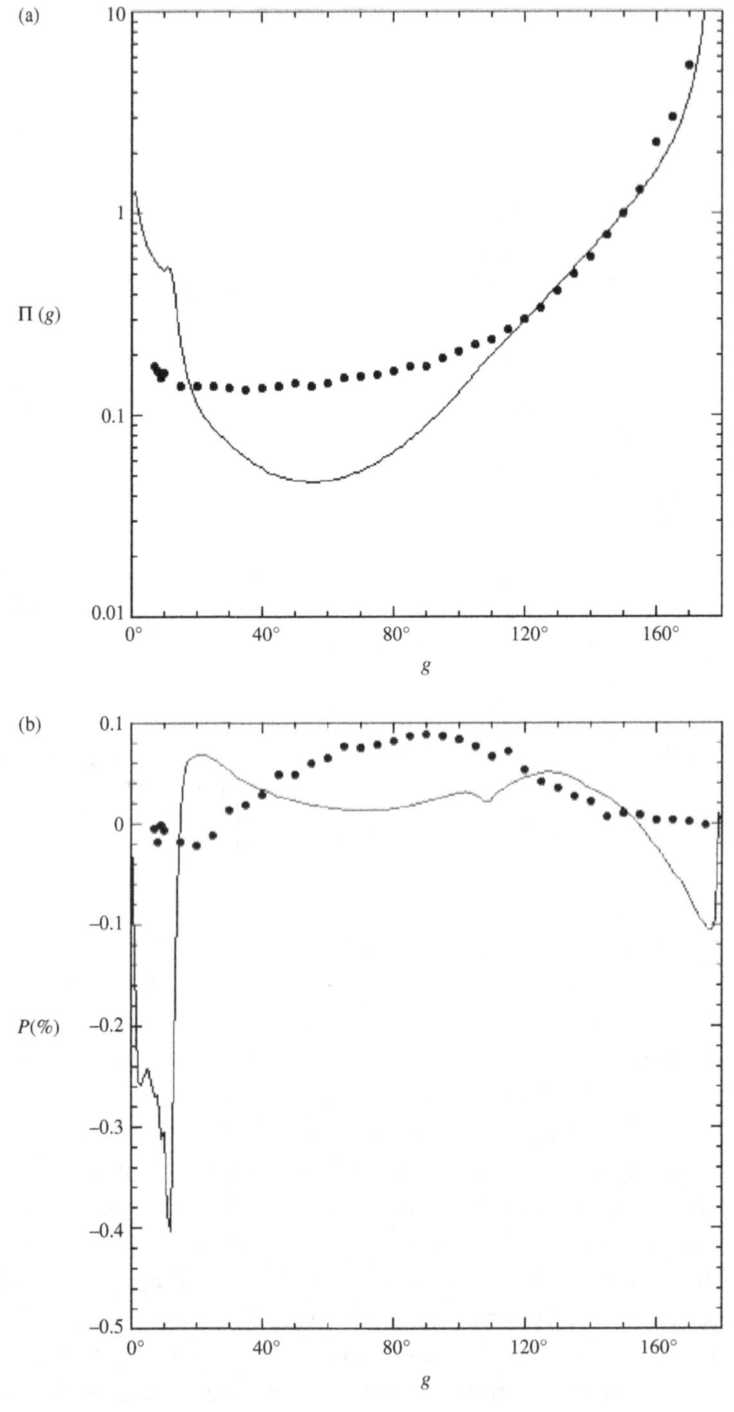

Figure 6.7 (a) Measured phase function of crushed olivine particles of size parameter $X \sim 38$ (dots) compared with predictions of Mie theory for spheres (line). Olivine data from the Amsterdam Data Base (Munoz *et al.*, 2000). (b) Same as Figure 6.7a for the polarization.

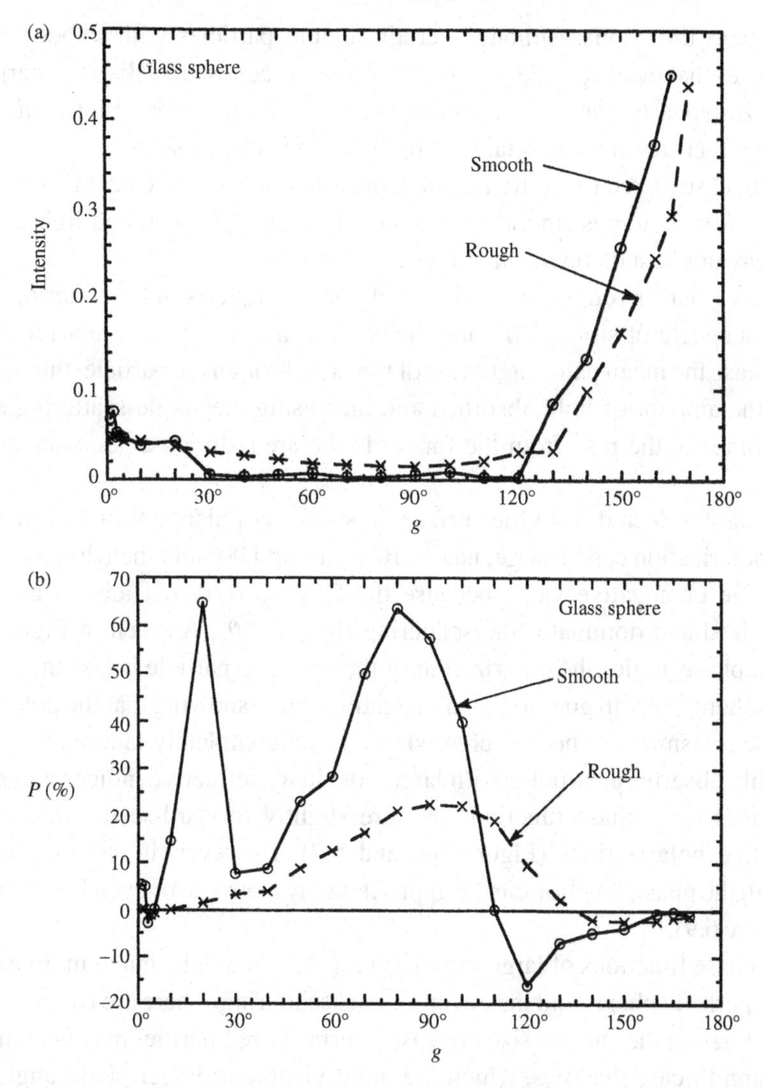

Figure 6.8 (a) Measured phase functions of a large, clear, spherical particle with a smooth surface and the same particle after the surface had been roughened. (b) Measured polarization functions of the particles shown in Figure 6.8a.

over that for a sphere (Figures 6.7 and 6.8). It is sensitive to the absorption coefficient, showing that internally reflected and transmitted light contributes. This is one of the places where Mie theory is grossly incorrect for an irregular particle.

In scattering by an irregular particle, some of the energy that would have gone into the strong, forward-refracted and weaker, backward-internally-reflected peaks if the particle were a sphere is redirected and internally scattered into the sideways directions. Although certain regular particles with smooth external surfaces that meet at angles close to 90°, such as cubes, may display sharp peaks near zero phase because of an internal corner-reflector effect (Liou *et al.*, 1983), such effects are not important for particles of irregular shapes.

(4) The backscattered twice-transmitted, once-internally-reflected lobe in an irregular particle merges smoothly into the intermediately scattered light and may or may not be a distinguishable peak.

(5) An irregular or rough-surfaced particle has a higher single-scattering albedo than a sphere of similar size and dielectric constant. The irregularities tend to decrease the mean path length $\langle D \rangle$ of the rays through the particle, thus decreasing the amount of light absorbed and increasing the single-scattering albedo. In doing so, the rays from the forward lobe are redirected into smaller phase angles.

(6) The light reflected from the surface is positively polarized. In a clear sphere, the polarization can be large, nearly 100% around 90° and then drops to a small positive or negative value because the forward-refracted lobe is negatively polarized and dominates the scattering for $g \gtrsim 2\vartheta_c$. As seen in Figure 6.8 at large phase angles the polarization of the irregular particle is less than that for the sphere, but still goes to a small negative value, showing that the polarization of the transmitted lobe is decreased, but is not completely random.

(7) Highly absorbing particles with large imaginary refractive indices and smooth surfaces have phase functions that are slightly forward-scattering and large positive polarizations (Figures 5.6 and 5.7). However, if their surfaces are rough the phase function can be approximately isotropic or even backscattering (Figure 6.9).

(8) The phase functions of large irregular particles tend to be much more isotropic than perfect spheres and may even be backscattering. There are two reasons for this. First, as the particle size increases, surface irregularities may become large enough to cast shadows, which are more visilble at larger phase angles. The surface elements of large protuberances facing away from the source, with the largest specular reflectances, tend to be less visible than elements with smaller reflectances facing toward the source.

    Second, large real paricles tend to be composite and made up of many smaller particles that can act as internal scatterers and may also be absorbing. The presence of internal scatterers causes major departures from the predictions of Mie theory, as can be seen in Figure 6.10. They are of special importance because internal scatterers are abundant in both laboratory samples and in planetary soils. Most natural particles contain smaller inclusions and bubbles. Usually large particles are polycrystalline, and the internal grain boundaries scatter light. The process of grinding inevitably produces cracks and fractures that radiate

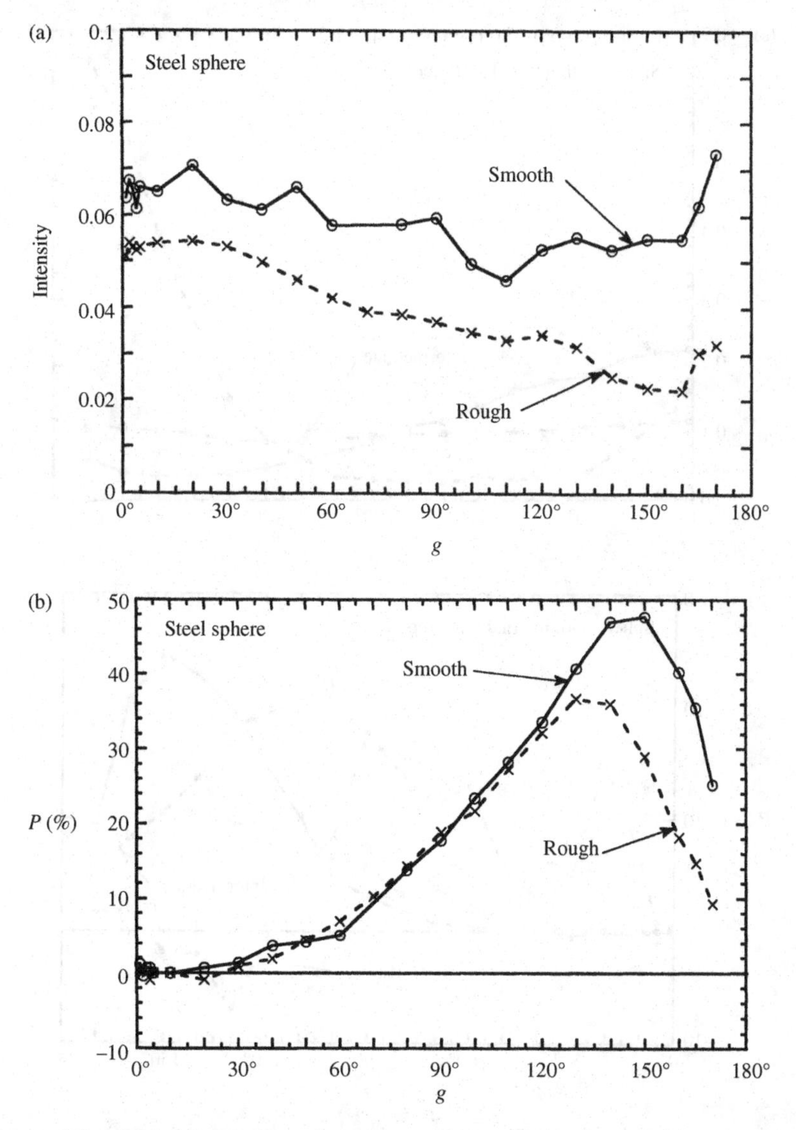

Figure 6.9 (a) Measured phase functions of large steel spheres with smooth and rough surfaces. (b) Polarization functions of the particles shown in Figure 6.9a.

from the surface of a grain into the interior (Tanashchuk and Gilchuk, 1978; Skorobogatov and Usoskin, 1982). The lunar regolith (and presumably the regoliths of other bodies as well) contains large numbers of agglutinate particles, which are welded agglomerates of smaller grains and voids that can act as internal scatterers. A particularly dramatic effect of inclusions may occur if they consist of submicroscopic particles with large imaginary refractive indices. Because such particles are very efficient absorbers, they may increase the

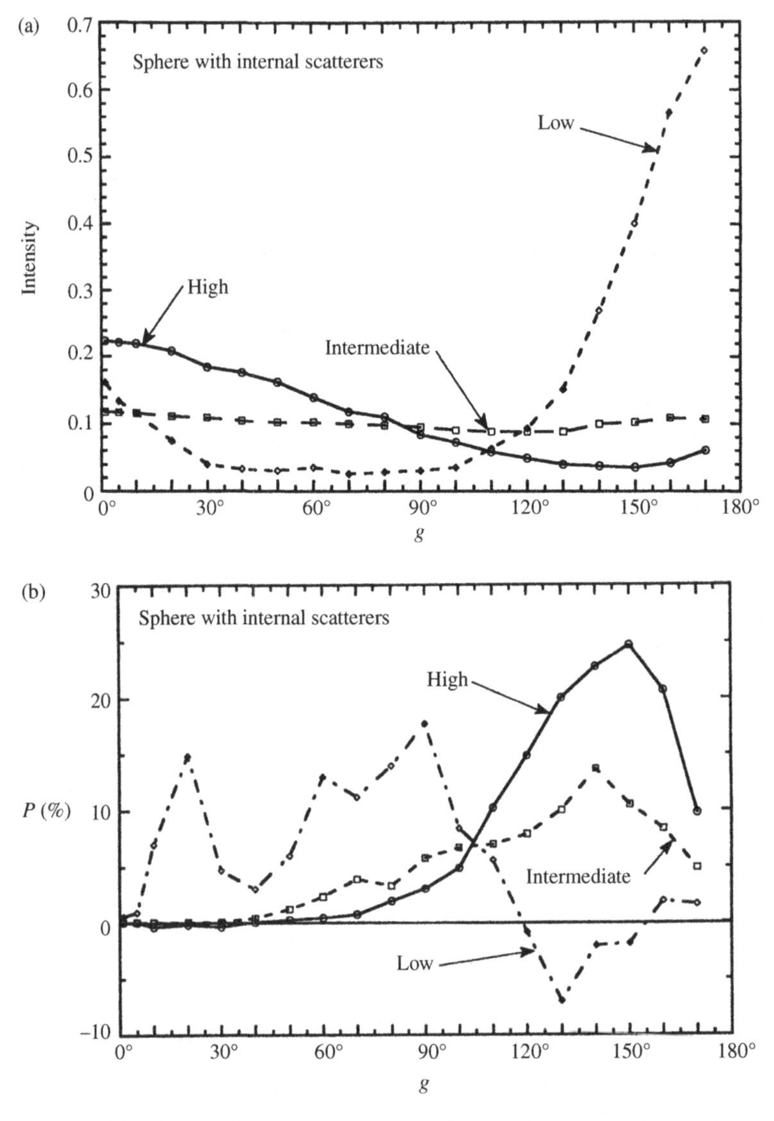

Figure 6.10 (a) Measured phase functions of large spherical particles filled with internal scatterers. The terms "low," "intermediate," and "high" refer to the relative internal scattering coefficients. (b) Measured polarization functions of the particles shown in Figure 6.10a.

effective absorption coefficients of their host particles by orders of magnitude (Hapke, 2001).

If internal scatterers are present, less light is directly transmitted through the particle, and more is scattered out the back and sides after traveling only a short distance through the particle. If the subparticles are nonabsorbing this increases both the single-scattering albedo and the amplitude of the phase function at small and intermediate angles. If the subparticles are absorbing the albedo and amount of light reaching the forward surface are both decreased. As internal scatterers are added, the amplitude of the forward lobe decreases, and the intensity scattered at intermediate phase angles increases, so that the phase function becomes more isotropic. Further increasing the internal scattering causes the phase function to acquire a strong, broad backscattered lobe. The polarization peaks around 150°, implying that the polarization of the internally scattered light is random, so that most of the polarization of the light scattered from the particle will be due to the surface-reflected rays.

The particle phase functions of many planetary regoliths appear to be backscattering. It is of interest to note that the only known types of particles that are backscattering are either filled with internal scatterers or have large imaginary refractive indices and rough surfaces.

(9) McGuire and Hapke (1995) found that the phase functions of their particles could be adequately described by a two-parameter double Henyey–Greenstein function (equation [6.7a]). In this representation of $\Pi(g)$ the parameter c describes the relative amplitudes of the forward-and backscattered lobes and b determines their shapes. A forward-scattering particle has $c < 0$; increasing b increases the amplitudes and decreases the widths of the lobes.

When plotted on a diagram of b versus c all of the particles were found to fall within an L-shaped region of restricted area clustered around the empirical curve

$$c = \left(\frac{0.05}{b - 0.15}\right)^{3/4} - 1, \tag{6.19}$$

as shown in Figure 6.11. Clear, spheroidal particles have $b \sim 0.5 - 0.7$ and $c \sim -0.9$.

Increasing the absorption decreases the single scattering albedo ϖ and decreases b slightly, but does not change c appreciably. Making the particle more irregular in shape, but still clear, increases ϖ, decreases b, and increases c; that is, the particle becomes more isotropic and less strongly forward-scattering. Roughening the surface of the particle or making a composite agglomerate adds internal scatterers and increases c. As more internal scatterers are added, c increases, but b remains in the range 0.15–0.35; that is, the particle becomes more strongly backscattering, while the lobes remain low and wide. Increasing the absorption decreases ϖ and causes a particle to move a short distance in a direction away from either end toward the center of the L-shaped area.

Cord *et al.* (2003) and Shepard and Helfenstein (2007) measured the reflectances of a large number of natural substances in powder form and extracted the values of b and c by fitting a theoretical reflectance model. Most of these parameters were

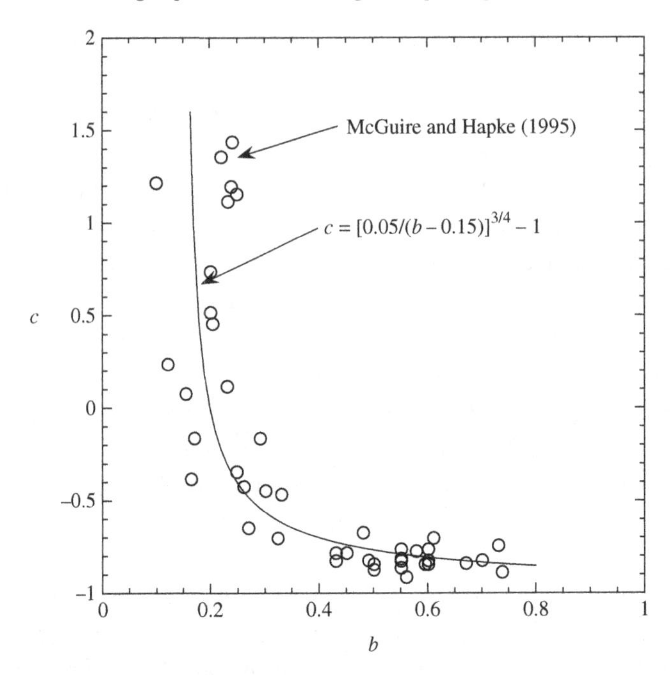

Figure 6.11 Empirical two-parameter double Henyey–Greenstein parameters for large silicate, resin, and metal particles of varied shapes, absorption coefficients, and conditions of surface roughness, and containing differing densities of internal scatterers (data from McGuire and Hapke, 1995).

found to lie close to equation (6.19). Thus b and c are not independent parameters, but appear to be highly correlated.

In summary, almost any change from a perfect, uniform sphere has the effect of increasing ϖ, decreasing b, increasing c, and generally causing the particle to scatter more isotropically than a Mie sphere. For a particle with internal scatterers b is small, and decreasing the absorption increases both ϖ and c, but does not affect b appreciably. Smooth strongly absorbing particles scatter nearly isotropically, with b and c both small; roughening their surfaces decreases ϖ and increases c but b does not change.

6.5 The generalized equivalent-slab model

6.5.1 The scattering efficiency

The exact particle scattering models discussed in this chapter require extensive numerical evaluation, and are inconvenient to use. The empirical models of Section 6.3 require the specification of parameters that are not connected in any obvious way to the size, shape, and complex refractive index of the particle. Hence, a general analytic scattering model would appear to be useful, especially in contrast to spheres which are hardly realistic analogs of natural regolith particles. In this section, the equivalent-slab model derived in Chapter 5 for the nondiffractive portion of the scattering efficiency of large spheres will be extended to ensembles of large irregular particles. The parameters of the model are based on experimental data. Although approximate, the model is given in closed analytic form, and its parameters can be related to the fundamental properties of the particle. An expression for the phase function will be derived, and the model will be further extended to include coated particles.

Following the notation in Chapter 5, write $Q_S = Q_s + Q_d$ where Q_d is the portion of the scattering efficiency associated with diffraction, and Q_s is the nondiffractive portion scattered into smaller phase angles. The diffraction term will be discussed first.

6.5.2 Fraunhofer diffraction

We have seen that the diffraction efficiency of a perfect, isolated, large sphere is $Q_d = 1$. By Babinet's principle (Section 5.4.4.2), the diffraction of a wave around an isolated obstacle is equivalent to that through a hole of identical cross-sectional size and shape in an opaque, infinite screen. Because the total power in the diffraction pattern of the hole is equal to the power passing through the hole, regardless of shape, the diffraction efficiency of the hole is 1. Hence, the average diffraction efficiency of an ensemble of isolated, randomly oriented, irregular particles is also $Q_d = 1$. The diffraction pattern has a phase function sharply peaked in the forward direction with an angular width $\sim 1/X$. For the portion of the phase function associated with diffraction, the Allen function, equation (6.2), may be used.

However, it must be emphasized that Fraunhofer diffraction is only relevant to an isolated particle and does not exist when particles are close together in a regolith or powder.

6.5.3 Nondiffractive scattering

6.5.3.1 The equivalent-slab model for Q_s

The success of the equivalent-slab model in reproducing Q_s for a sphere suggests that we attempt to generalize it to irregular particles. Hence, we again write equation (5.52a) for Q_s,

$$Q_s = S_e + (1 - S_e)\frac{1 - S_i}{1 - S_i\Theta}\Theta. \tag{6.20}$$

The various quantities in this equation have the same meanings as in Section 5.5.6. That is, S_e and S_i are respectively the surface reflection coefficients for light that is externally and internally incident, and Θ is the internal-transmission factor. The first term of (6.20) represents the light externally scattered from the particle surface. The numerator of the second term corrects for the light transmitted once through the particle, and the denominator represents light multiply internally scattered through the particle. Expressions for each of these quantities for irregular particles will be derived in the following section.

6.5.3.2 Exterior-surface reflection

The evaluation of S_e is simple if the particle is *convex*, with smooth, randomly oriented surface facets. Van de Hulst (1957) defined a convex particle as one that "when illuminated from any direction has a light side and a dark side separated by a closed curve on the surface. Any small surface area dS is on the dark side when its outward normal makes an angle $<90°$ with the direction of propagation of the incident light; it is on the light side when this angle is $>90°$." A more intuitive description of a convex particle would be to say that it is one that is without any depressions, dimples, or projections on its surface that can cast shadows on another part of the surface. Convex particles include most simple shapes, such as spheres, ellipsoids, cylinders, and euhedral crystals of most minerals.

The normals to the surface facets of an ensemble of randomly oriented, convex particles are distributed isotropically. This distribution is identical with that of the surface elements of a sphere. Therefore, provided each facet is smooth, the light reflected from the surface facets of the ensemble is the same as for a sphere, so that the phase function associated with external reflection is given by the integral of the Fresnel coefficients, equation (5.36).

The effects of surface roughness on the scattering characteristics of the particle depend on the scale of the roughness. The discussion in Section 4.6 implies that under most circumstances a particle surface may be treated as smooth if the scale of the surface roughness is small compared with the wavelength. More sophisticated theoretical analyses (e.g., Berreman, 1970) indicate that small asperitics and depressions in a surface will scatter light in a manner similar to dipoles suspended just above the surface. Because a dipole scatters light proportionally to $(\text{size}/\lambda)^4$, if the Rayleigh criterion is satisfied then the effects of a few small surface imperfections usually can be ignored. This conclusion is supported by the microwave experiments of Zerull and Giese (1974), which showed that a sphere with surface roughness elements small compared with the wavelength scatters light in a manner that is virtually indistinguishable from the predictions of Mie theory.

A major exception may occur when the complex refractive index is in the region of anomalous dispersion in the vicinity of a strong absorption band. As discussed in Chapter 5, when $K_e = n^2 \simeq -2$, a resonance can occur that may cause the scattering and particularly the absorption efficiencies of small particles to be anomalously large, so that they cannot be ignored. This point has been emphasized by Emslie and Aronson (1973). The effect can be especially important in the thermal infrared, where materials typically have strong restrahlen bands.

A second exception occurs when k is large, as with metals. Figure 6.10 shows that roughening the surface of a metal will markedly decrease the scattering efficiency. The probable explanation is that the scratches and corners on the surface act like Rayleigh absorbers (Section 5.4.2).

The case in which the scale of the roughness is larger than the wavelength is more difficult to treat, because shadowing may be important. Several experimental studies of the effect of well-characterized, large-scale roughness on the reflection from a surface have been carried out, including those by Torrance and Sparrow (1967) and O'Donnell and Mendez (1987). An important conclusion that may be drawn from these studies is that the general character of the scattering remains quasi-specular. That is, the scattering does not change from specular to diffuse, as has been conjectured by many authors. Rather, the reflected light is redirected into a relatively broad peak that is approximately centered in the specular direction. Even at glancing incidence and reflection, the character of the scattering remains quasi-specular. Apparently, a ground or frosted surface scatters light diffusely because of the subsurface fractures created by the grinding process, rather than because of the irregular surface geometry.

Thus, to a first approximation the scattering from the surface of an irregular particle can be treated as quasi-specular, so that the exterior surface reflection coefficient of an irregular particle is S_e, the same as that of a sphere, equation (5.36). A convenient analytic approximation to S_e is given by equation (5.37) and plotted in Figure 5.12. Repeating it here for convenience,

$$S_e = 0.0587 + 0.8543R(0) + 0.0870R(0)^2,$$

where

$$R(0) = \frac{(n_r - 1)^2 + n_i^2}{(n_r + 1)^2 + n_i^2}$$

is the normal specular reflection coefficient.

For particles with very rough surfaces containing large numbers of small scattering elements, it can be assumed that the surface asperities and depressions behave like quasi-independent small scattering particles, so that the effective scattering efficiency can be treated like a mixture of large and small particles. Experimental support for this assumption is given in Section 6.5.4. Scattering by mixtures is treated in Chapter 10.

6.5.3.3 Internal surface reflection

In a perfect sphere, the angle at which a refracted ray is incident on the inside of the sphere is equal to that with which it is refracted into the sphere, so that

$S_i = S_e$. However, as emphasized by Melamed (1963), for an ensemble of irregular particles the two angles are uncorrelated, and a refracted ray will have virtually equal probabilities of encountering interior surfaces at all orientations. Thus, a reasonable expression for the interior reflection coefficient is the integral of the internal Fresnel reflection coefficients over all angles:

$$S_i = \int_0^{\pi/2} [R_\perp(\vartheta') + R_\parallel(\vartheta')] \cos\vartheta' \sin\vartheta' d\vartheta' y = 2\int_0^{\pi/2} R(\vartheta') \cos\vartheta' \sin\vartheta' d\vartheta', \tag{6.21}$$

where ϑ' is the angle of incidence for the interior rays. This expression is identical in form with (5.36) for S_e, except that now $R(\vartheta')$ refers to internally reflected light. The curve for S_i calculated by numerical integration for n real is shown as a function of n in Figure 6.12.

A convenient approximate analytic expression for S_i may be derived as follows. As may be seen in Figure 4.5, when $0 < \theta < \theta_c$ the average of the internal polarized reflectivities is nearly constant and equal to its value $R(0)$ at $\theta = 0$. It then rises abruptly to 1 at $\theta = \theta_c$, and is equal to 1 for $\vartheta'_c < \vartheta' < \pi/2$. Hence, S_i can be

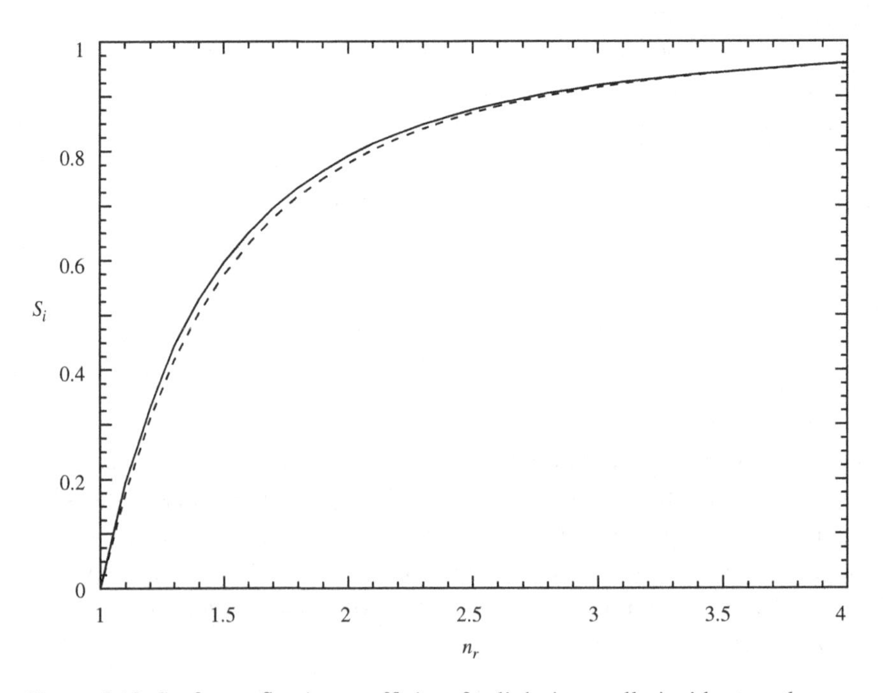

Figure 6.12 Surface reflection coefficient for light internally incident on the surface of the particle versus the real part of the refractive index for $n_i << 1$. The solid line is the exact expression, equation (6.21); the dashed line is the approximation, equation (6.23).

approximated by

$$S_i \simeq 2\left[\int_0^{\vartheta_c'} R(0)\cos\theta\sin\theta d\theta + \int_{\vartheta_c'}^{\pi/2}\cos\theta\sin\theta d\theta\right], \tag{6.22}$$

Carrying out this integration gives

$$S_i \simeq 1 - \sin^2\theta_c[1 - R(0)],$$

where $\sin\theta_c = 1/n_r$, and n_r, is the real part of the refractive index of the particle relative to the surrounding medium. Empirically, it was found that a better fit can be obtained if equation (5.37) for S_e is substituted for $R(0)$, which gives

$$S_i = 1 - \frac{1}{n_r^2}[0.9413 - 0.8543R(0) - 0.0870R(0)^2]. \tag{6.23}$$

This expression for S_i is plotted in Figure 6.12 for n real, where it is compared with the exact expression, equation (6.21). In practice, it is seldom necessary to calculate S_i when n_i is large enough to affect $R(0)$, because then the particle is opaque and the refracted wave is absorbed before it reaches the far surface.

6.5.3.4 *The internal-transmission factor*

One might try to estimate Θ by carrying out ray-tracing calculations on a variety of models of specific shapes, hoping that the results would have some resemblance to the transmission of irregular particles. However, this approach is of questionable value, particularly if it is realized that accounting for effects of particle shape is not the only difficulty: often the particles of interest are not clear, but are full of internal scatterers. Instead of that approach, we will consider several possible general models and use the results of experiments to choose the best one. Four models for Θ will be considered: exponential, Melamed, internal scattering, and double exponential. These are illustrated schematically in Figure 6.13.

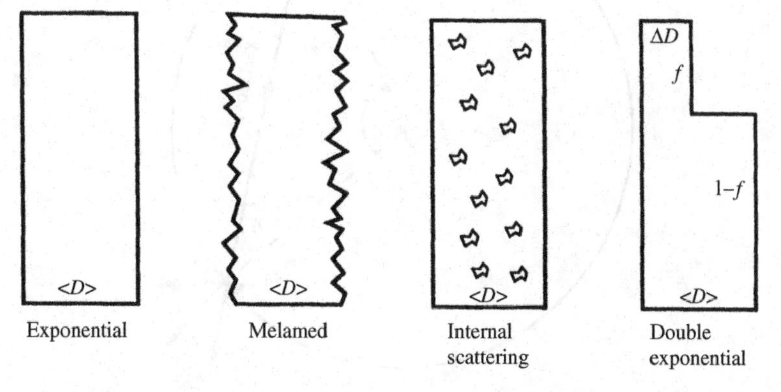

Figure 6.13 Schematic diagram of the various models for the scattering efficiency.

The exponential model It was shown in Chapter 5 that for a sphere the internal-transmission factor can be approximated by an exponential function. Hence, the first and simplest model to be considered is of the form

$$\Theta = e^{-\alpha\langle D\rangle}, \tag{6.24}$$

where $\langle D\rangle$ is the average distance traveled by all transmitted rays during one traverse of the particle. If the particle is spherical, $\langle D\rangle \simeq 0.9D$. However, if the particle is irregular, then $\langle D\rangle$ can be quite different from D and in general will be smaller.

The Melamed model Another possibility for Θ was suggested by Melamed (1963). In his model, the external reflection coefficient is assumed to be S_e, and Melamed was the first to introduce the idea that the internal reflection coefficient should be S_i, which we have adopted here.

To calculate Θ Melamed assumes a clear particle of spherical shape and diameter $D = 2a$. Radiance emerging from any point of an inner surface after being either transmitted or reflected is assumed to have an angular distribution given by Lambert's law, $(1/\pi)\cos\vartheta'$, where ϑ' is the angle between the surface normal and the direction of the radiance (see Figure 6.14). An incremental area dA a

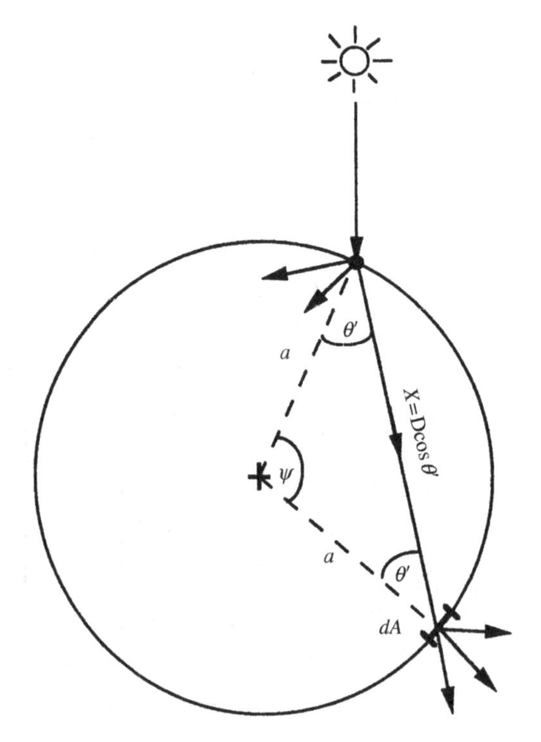

Figure 6.14 Melamed model for the internal-transmission factor.

distance $x = D\cos\vartheta'$ away on the inner surface of the sphere subtends a solid angle $dA\cos\vartheta'/x^2$ from the point. The radiance is attenuated by a factor $e^{-\alpha x}$ in traveling from the point to dA. Thus, the total fraction of light emitted from the point that reaches all areas on the interior surface of the particle is

$$\Theta = \int_{\psi=0}^{\pi} \frac{1}{\pi}\cos\vartheta' e^{-\alpha x}\frac{dA\cos\vartheta'}{x^2},$$

where $dA = 2\pi a^2 \sin\psi d\psi$, and $\psi = \pi - 2\vartheta'$. It is readily seen that $x = a\sin\psi$, so that $dA = 2\pi x dx$. Hence,

$$\Theta = \int_0^D e^{-\alpha x}\frac{2x}{D^2}dx = \frac{2}{(\alpha D)^2}\left[1 - e^{-\alpha D}(1+\alpha D)\right].$$

Generalizing to irregular particles by replacing D by $\langle D\rangle$ gives

$$\Theta = \frac{2}{(\alpha\langle D\rangle)^2}\left[1 - e^{-\alpha D}(1+\alpha\langle D\rangle)\right]. \tag{6.25}$$

The internal-scattering model The importance of internal scatterers in natural particles has already been emphasized. One approach to investigating their effects on Θ would be to attempt to find the radiance inside a spherical absorbing particle containing embedded scatterers. In order to do that it would be necessary to solve the equation of radiative transfer in a spherical geometry. The radiative-transfer equation will be introduced in Chapter 7, with exact and approximate solutions obtained in subsequent chapters. However, the solution of this equation is a formidable problem that requires numerical solution by a computer for reasonable accuracy. Thus, we seek a simpler approximate analytic model.

We have seen that the transmission factor of a perfect sphere can be approximated quite well by $\exp(-a\langle D\rangle)$, which is equivalent to the transmission of a slab of thickness $\langle D\rangle$. This suggests that the transmission of an absorbing slab containing embedded scatterers might provide a suitable approximation. Consider the following model: diffuse radiation is assumed to be incident from above on a plane slab of thickness L, refractive index n, absorption coefficient α, and with distributed scattering coefficient s, which describes the scattering by the internal embedded scattering centers. These are assumed to scatter light isotropically. The physical meaning of s is similar to that of α: the intensity is attenuated by a factor of e^{-1} after traveling a distance $1/s$, but the attenuation is by scattering, rather than by absorption. As with the other models, the slab is assumed to have external reflection coefficient S_e and internal reflection coefficient S_i. It will also be assumed that any radiation transmitted through or reflected from an internal surface has a diffuse angular distribution.

The problem of the radiation field inside such a slab and the intensity emerging from the top and bottom surfaces will be solved in Chapter 10 using a method known as the two-stream approximation to the equation of radiative transfer. It will be shown that Q_{sB}, the efficiency for scattering light into the back direction, and Q_{sF}, the efficiency for scattering into the forward direction, are given by the relations

$$Q_{sB} + Q_{sF} = Q_s,$$

and

$$Q_{sB} - Q_{sF} = \Delta Q_s,$$

where the scattering efficiency Q_s is given by equation (6.20), with

$$\Theta = \frac{r_i + \exp\left(-\sqrt{\alpha(\alpha+s)}2L\right)}{1 + r_i \exp\left(-\sqrt{\alpha(\alpha+s)}2L\right)}, \tag{6.26}$$

and

$$r_i = \frac{1 - \sqrt{\alpha/(\alpha+s)}}{1 + \sqrt{\alpha/(\alpha+s)}}, \tag{6.27}$$

and the scattering efficiency difference ΔQ_s is

$$\Delta Q_s = S_e + (1 - S_e)(1 - S_i)\frac{\Psi}{1 - S_i \Psi}, \tag{6.28}$$

with

$$\Psi = \frac{r_i - \exp\left(-\sqrt{\alpha(\alpha+s)}2L\right)}{1 - r_i \exp\left(-\sqrt{\alpha(\alpha+s)}2L\right)}. \tag{6.29}$$

If the slab is clear, with $s = 0$, then $\Theta = -\Psi = \exp(-a2L)$, which is identical with equations (5.52a), (5.52b), and (5.56) for a sphere if $2L$ is replaced by $\langle D \rangle$. This shows that the mean path length through the slab is $2L = \langle D \rangle$, where $\langle D \rangle$ is to be interpreted as the length of the average ray that traverses the particle once without being scattered. This gives that the internal-transmission factor of the equivalent slab is

$$\Theta = \frac{r_i + \exp\left(-\sqrt{\alpha(\alpha+s)}\langle D \rangle\right)}{1 + r_i \exp\left(-\sqrt{\alpha(\alpha+s)}\langle D \rangle\right)}, \tag{6.30}$$

and the scattering efficiency difference factor is

$$\Psi = \frac{r_i - \exp\left(-\sqrt{\alpha(\alpha+s)}\langle D \rangle\right)}{1 - r_i \exp\left(-\sqrt{\alpha(\alpha+s)}\langle D \rangle\right)}. \tag{6.31}$$

The double-exponential model It was suggested in the discussion of S_e that surface asperities might scatter as quasi-independent particles, so that Q_s can be treated as a mixture of particles of two different sizes. Internal scattering elements located just under the surface might also behave in this manner. Thus, we will consider a model of the form

$$Q_s = S_e + (1 - S_e)(1 - S_i)\left[(1 - f)\frac{e^{-\alpha\langle D\rangle}}{1 - S_i e^{-\alpha<D>}} + f\frac{e^{-\alpha\Delta D}}{1 - S_i e^{-\alpha\Delta D}}\right], \quad (6.32)$$

where $\langle D\rangle$ is the mean absorption path through the main particle, ΔD is the mean path associated with scattering by the surface asperities or subsurface fractures, and f is the fraction of light scattered by these small "particles."

6.5.4 Experimental choice of models for Θ

The scattering efficiencies corresponding to these four models for Θ are illustrated in Figure 6.15. For very small values of α, all models attenuate light exponentially; that is, $\Theta \simeq \exp(-\alpha\langle D\rangle)$. At large values of $\alpha\langle D\rangle$ the scattering efficiencies all approach S_e, but for the last three models they do so more slowly than for the exponential model, so that Q_s is larger at intermediate values of $\alpha\langle D\rangle$.

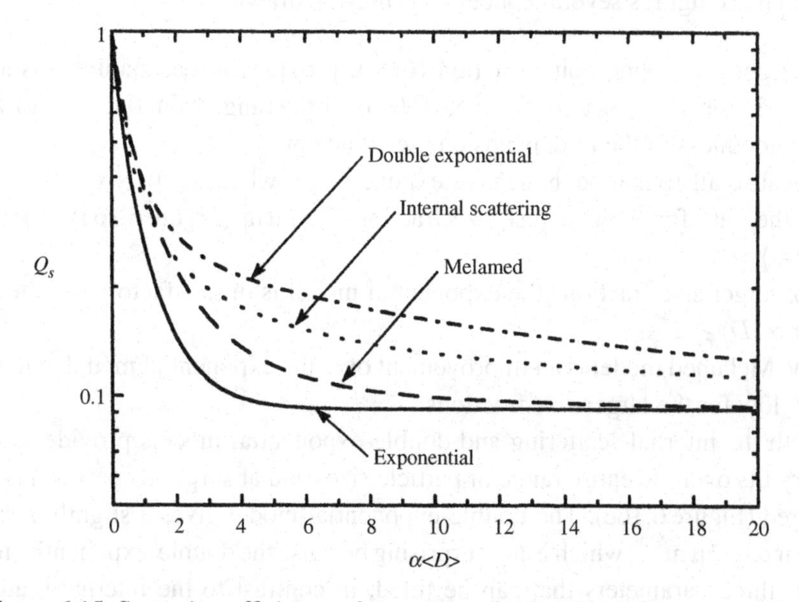

Figure 6.15 Scattering efficiency of a particle of refractive index $n_r = 1.50$ vs. $\alpha\langle D\rangle$ for the four internal-transmission models considered in the text. For the internal-scattering model, $s\langle D\rangle = 5$, and for the double-exponential model, $f = 0.2$ and $\Delta D = 0.05\langle D\rangle$.

The exponential and Melamed models have only one free parameter, $\langle D\rangle$; the internal-scattering model has two, $\langle D\rangle$ and s; and the double-exponential model has three, $\langle D\rangle$, ΔD, and f.

In order to verify equation (6.20) for Q_s and choose the best expression for Θ, the results of measurements on comminuted materials will be used. This is justified because most laboratory powders and planetary regoliths are the products of grinding and crushing.

In a series of measurements on crushed-glass powders (Hapke and Wells, 1981), silicate glass doped with $CoCl_2$ was synthesized, and its absorption coefficient was measured as a function of wavelength in the near ultraviolet, visible, and near infrared by transmission of polished thin sections. The remainder of the batch was ground and separated into several size ranges by wet-sieving. Each size fraction was washed to remove clinging fines. The bidirectional reflectances of the powder fractions were measured over the same wavelength range as the transmission measurements and Q_s as a function of wavelength found using the reflectance theory to be derived in Chapters 8–11. Because $\alpha(\lambda)$ was known, $Q_s(\lambda)$ could be converted to Q_s as a function of α. The four models for Θ were then fitted to the data. The resultant experimental values and fitted curves for each size range are shown in Figures 6.16. (To avoid clutter, curves for all of the models are not shown in all of the figures.)

From these figures several conclusions may be drawn:

(1) The general expression, equation (6.20), provides a reasonably satisfactory description of Q_s as a function of $\alpha\langle D\rangle$ over the range from 0 to 22. However, the goodness of the fit depends on the choice of Θ.
(2) Because all four models behave exponentially when $\alpha\langle D\rangle$ is small, they all fit the data for the smallest size fraction, $<37\,\mu m$ (Figure 6.16a), for which $\alpha\langle D\rangle \lesssim 2$.
(3) For larger size fractions the exponential model is unsatisfactory. It is too low for $\alpha\langle D\rangle \gtrsim 2$.
(4) The Melamed model is an improvement over the exponential model, but is also too low for the larger size fractions.
(5) Both the internal-scattering and double-exponential models provide satisfactory fits over the entire range of particle sizes and absorption coefficients measured (Figure 6.16e). The double-exponential model gives a slightly better fit near $\alpha \sim 2\,\mu m^{-1}$, which is not surprising because the double-exponential model has three parameters that can be fitted, in contrast to the internal-scattering model, which has only two.
(6) The mean ray path length $\langle D\rangle$ is of the same order of magnitude as the particle size, but is smaller than the average particle diameter, as is expected for irregular particles.

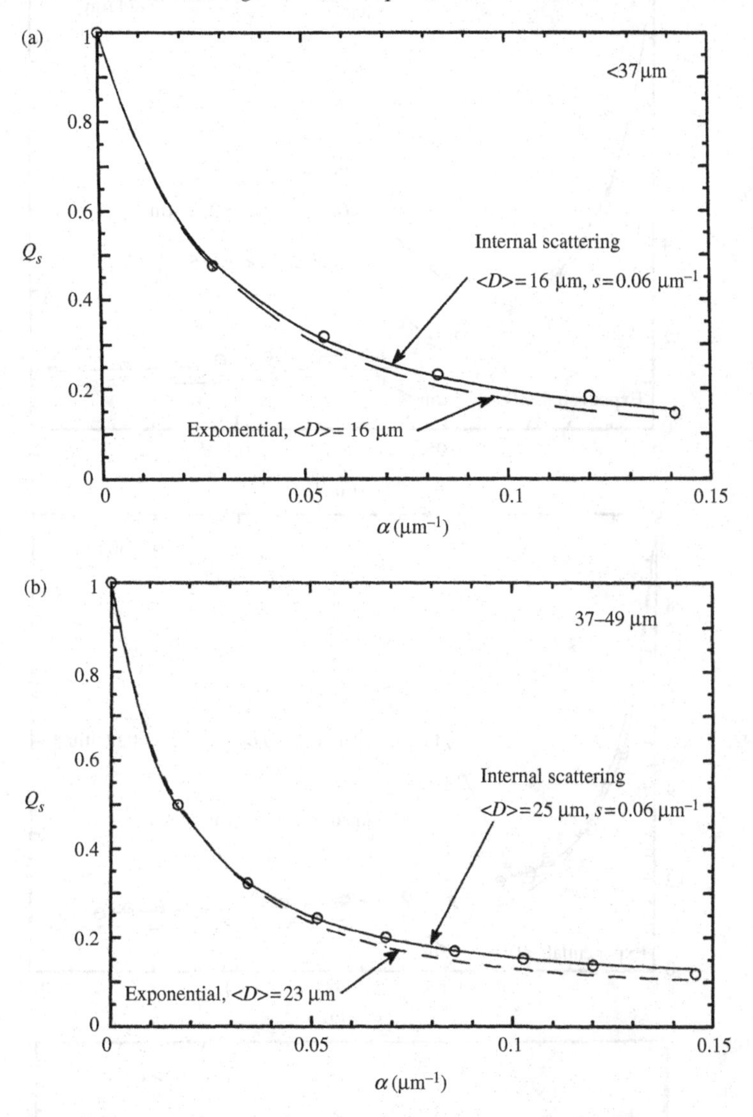

Figure 6.16 (a) Scattering efficiency of crushed cobalt glass of size $<37\,\mu$m as a function of absorption coefficient. The points show the values calculated from reflectance measurements. The lines show the fit of the various models discussed in the text. The Melamed model with $\langle D\rangle = 24\,\mu$m and the double-exponential model with $\langle D\rangle = 18\,\mu$m, $f = 0.2$, and $\Delta D = 60\,\mu$m are indistinguishable from the internal-scattering curve. (b) Same as Figure 6.16a for size 37–49 µm. The internal-scattering curve is indistinguishable from the Melamed model with $\langle D\rangle = 35\,\mu$m and the double-exponential model with $\langle D\rangle = 29\,\mu$m, $f = 0.2$, and $\Delta D = 6.0\,\mu$m. (c) Same as Figure 6.16a for size 37–74 µm. The internal-scattering curve is indistinguishable from the double-exponential model with $\langle D\rangle = 45\,\mu$m, $f = 0.2$, and $\Delta D = 4.5\,\mu$m. (d) Same as Figure 6.16a for size 74–150 µm. The internal-scattering curve is indistinguishable from the double-exponential model with $\langle D\rangle = 80\,\mu$m, $f = 0.2$, and $\Delta D = 5.6\,\mu$m. (e) Same as Figure 6.16a for size $>150\,\mu$m. The exponential and Melamed models fall well below the measured points. This is the only example for which the internal-scattering and double-exponential models are noticeably different.

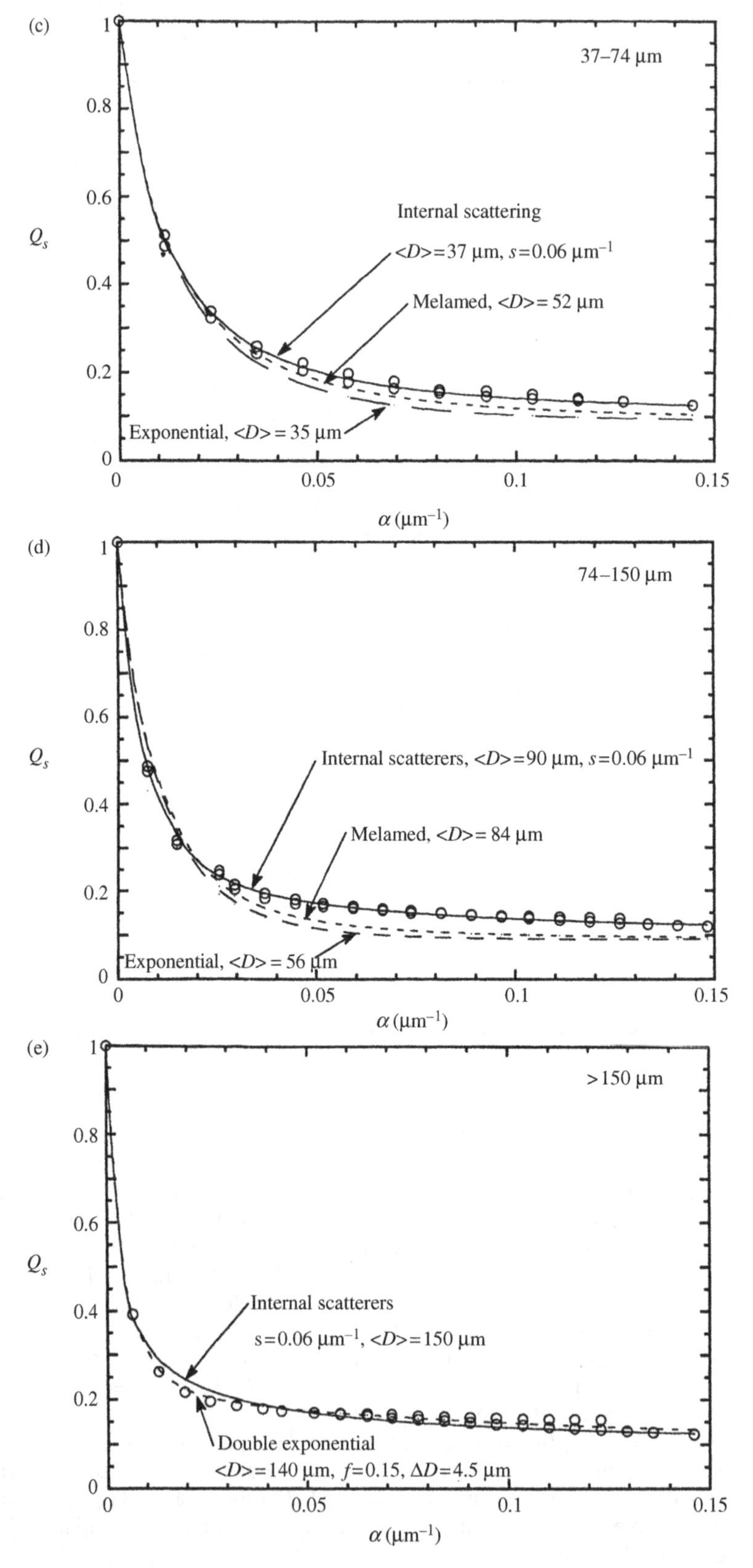

Figure 6.16 (*cont.*)

(7) The excellent fit provided by the double-exponential model over the entire range of α supports the hypothesis that surface asperities and subsurface scatterers can be treated as separate small particles in their effects on Q_s. For the particles of ground silicate glass whose scattering efficiencies are shown in Figure 6.13, the surface imperfections evidently scatter about 15% of the light and behave as particles whose effective sizes are approximately 5 μm.

(8) The best fit value of the internal scattering coefficient is $s = 600\,\mathrm{cm}^{-1}$ for these silicate glass particles. It is likely that both of the successful models, the double-exponential and internal-scattering models, are describing the same phenomenon, the scattering by surface asperities and subsurface fractures, but in mathematically different ways. Hence, the decision as to which one to use is arbitrary. We will adopt the internal-scattering model, equation (6.30), because this model contains the minimum number of free parameters that can still adequately describe Q_s. Note that the exponential model is included as a special case when $s = 0$. If it is desired to use the double-exponential model, this can easily be done using the single-exponential model for Θ and the theory for intimate mixtures that is discussed in Chapter 10.

Some additional properties of the internal-scattering model should be noted. When $\alpha \langle D \rangle \ll 1$, $\Theta \simeq \exp(-\alpha \langle D \rangle)$ independently of s. Thus, to first order, $Q_s \simeq S_e + (1 - S_e)e^{-\alpha \langle D \rangle}$, which is of the same form as expression (6.16), that was used by Hapke and Nelson (1975) to describe Venus cloud particles.

When α is small, the critical path length inside the particle that governs the behavior of Q_s is $\langle D \rangle$, which is a distance somewhat less than the mean diameter of the particle. As α increases, Θ is influenced more and more by s, until, when $\alpha \langle D \rangle \gg 1$, $\Theta \simeq s/4\alpha$, and

$$Q_s \simeq S_e + (1 - S_e)(1 - S_i)\frac{s}{4\alpha}. \tag{6.32}$$

In this case, Q_s is very nearly independent of $\langle D \rangle$, and the critical internal path is the scattering length $1/s$. That is, the portion of the light that has been refracted into the particle and scattered back out has traveled a mean distance of the order of $1/s$. Because $\alpha \langle D \rangle \gg 1$, the particle is nearly opaque, so that most of the refracted light interacts only with a layer of the order of $1/s$ thick on the side of the particle facing the source. Hence, s does not necessarily refer to the scattering coefficient throughout the whole interior of the particle, but characterizes the imperfections close to the surface. For this reason, s will be referred to as the *near-surface scattering coefficient*. This term is intended to be inclusive and not exclude the possibility that the scatterers are distributed throughout the volume of the particle.

A caveat concerning this model for Q_s must be emphasized, in that it is based on measurements on one type of particle, pulverized silicate glass. The extent to

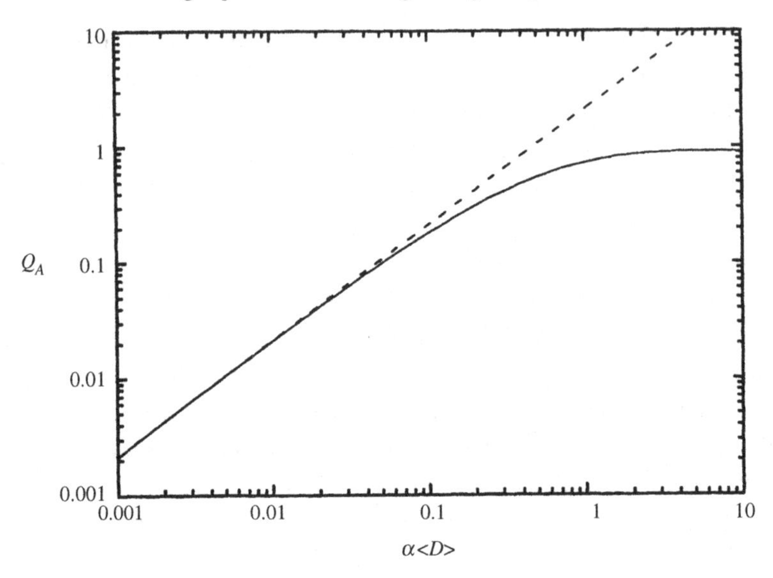

Figure 6.17 Absorption efficiency vs. $\alpha\langle D\rangle$ calculated using (6.20) and (6.26) with $s = 0$. On the log–log plot the dashed line has unit slope, showing that $Q_A \propto \alpha$ for $\alpha\langle D\rangle \lesssim 0.1$.

which these particles are representative has not been experimentally explored. Of particular interest is the appropriate value of s for other types of particles

6.5.5 The espat function

In Chapter 5 a quantity called the *espat function* was introduced. We will now consider its properties in detail. When $n_i = 1$, the amount of light absorbed must be proportional to the product of the absorption coefficient and the volume of the particle, so that Q_A must be proportional to $\alpha\langle D\rangle$. In Figure 6.17 the quantity $Q_A = 1 - Q_s$ calculated from (6.20) and (6.26) is plotted versus $\alpha\langle D\rangle$ for $s = 0$. This figure shows that Q_A is indeed linearly proportional to α, but that the linear region is only in the range $0 < \alpha\langle D\rangle \lesssim 0.1$.

However, consider the quantity

$$W \equiv \frac{Q_A}{Q_S} = \frac{Q_E - Q_S}{Q_S} = \frac{1-\varpi}{\varpi}. \tag{6.33}$$

For a large isolated particle $Q_E \simeq 2$. However, it will be argued in Chapter 7 that when the particles are close enough together to touch, as in a powder or regolith, Fraunhofer diffraction does not exist. In that case $Q_d = 0$, $Q_E \approx 1$, $Q_S = Q_s = \varpi$, and $W = (1 - Q_s)/Q_s$. Inserting expression (6.20) for Q_s into (6.33), W becomes

$$W = \frac{1 - S_e}{1 - S_i} \frac{1/\Theta - 1}{1 + [S_e/(1 - S_i)](1/\Theta - 1)}. \tag{6.34}$$

Using (6.26) for Θ, the quantity $1/\Theta - 1$ may be expanded for small $a\langle D\rangle$ to give

$$1/\Theta - 1 \simeq (\alpha\langle D\rangle) + \frac{1}{2}\left[1 - \frac{1}{6}(s\langle D\rangle)\right](\alpha\langle D\rangle)^2 + L. \tag{6.35}$$

Hence, to first order in $\alpha\langle \mathrm{D}\rangle$,

$$W \simeq \frac{1 - S_e}{1 - S_i}\alpha\langle D\rangle. \tag{6.36}$$

In Figure 6.18 W is plotted versus $[(1 - S_i)/(1 - S_i)]\alpha\langle D\rangle$ for $n = 1.50$ and several values of $s\langle D\rangle$. This figure shows that when $s\langle D\rangle$ is not too large, W is approximately linearly proportional to α over a much larger range of $\alpha\langle D\rangle$ than Q_A. There is no physical reason for this quasi-linear behavior when $\alpha\langle D\rangle$ is larger than 0.1; it is simply a serendipitous mathematical coincidence. However, as $\alpha\langle D\rangle$ becomes large, the slope of W decreases until W saturates at

$$W = (1 - S_e)/S_e. \tag{6.37}$$

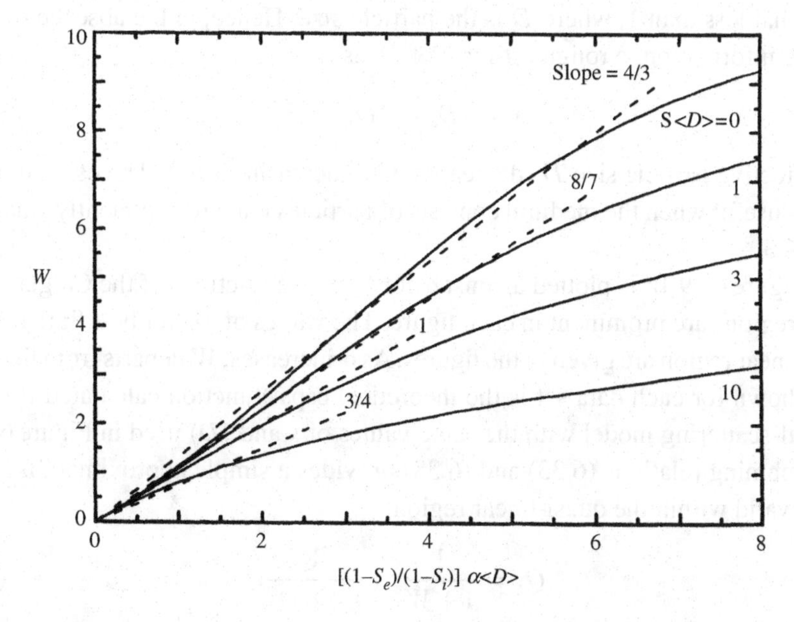

Figure 6.18 Espat function W of a particle calculated using (6.30) for several values of $s < D >$. The solid lines show the calculated values of W. The dashed lines show the straight-line approximations to W. The slopes of the straight lines are as indicated.

When $\alpha\langle D\rangle$ is small, Figure 6.18 shows that W approximates, although it is not exactly equal to, a straight line that is proportional to $[(1-S_e)/(1-S_i)]\alpha\langle D\rangle$. However, the constant of proportionality and the length of the linear region both depend on s. When $s\langle D\rangle = 1$, the constant of proportionality is about 4/3, and the quasi-linear region extends out to $[(1-S_e)/(1-S_i)]\alpha\langle D\rangle \simeq 6$, or $\alpha\langle D\rangle \simeq 3$. As s increases, the constant and the length decrease until when $s\langle D\rangle = 10$, the constant is 3/4 and the length of the quasi-linear region is less than 1.

In the quasi-linear region we may write

$$W \simeq \alpha D_e, \tag{6.38a}$$

where D_e is an effective particle size given by

$$D_e = C\frac{1-S_e}{1-S_i}\langle D\rangle, \tag{6.38b}$$

and C is a constant that depends on s, but $C \sim 1$. Thus, the W function is an absorption optical thickness. Hence, this quantity will be called the effective single-particle absorption thickness function, or *espat function*.

Except in regions of anomalous dispersion, n_r is very nearly independent of wavelength, so that D_e is approximately constant also. For many substances of interest in remote sensing, $D_e/\langle D\rangle$ is somewhat greater than 2, while $\langle D\rangle/D$ is somewhat less than 1, where D is the particle size. Hence, in the absence of more definite information, a rough estimate of D_e is

$$D_e \simeq 2D. \tag{6.39}$$

The effective particle size D_e decreases with increasing $s\langle D\rangle$. The espat function is most useful when the medium consists of particles that are sufficiently small that $s\langle D\rangle \lesssim 3$.

In Figure 6.19 W is plotted against α for four size fractions of the Co glass. The linear regions are prominent in each figure. The values of D_e for best fit to the data in the linear region are given in the figures. As α increases, W departs from linearity. Also shown for each data set is the theoretical espat function calculated from the internal-scattering model with the same values of s and $\langle D\rangle$ used in Figure 6.16.

Combining relations (6.33) and (6.38) provides a simple approximation for Q_s that is valid within the quasi-linear region:

$$Q_s = \frac{1}{1+W} \simeq \frac{1}{1+\alpha D_e}. \tag{6.40}$$

Because W is not independent of $s\langle D\rangle$, the range in $\alpha\langle D\rangle$ over which this expression is valid depends on the particle size. However, even for the largest particles of Figure 6.19, W is linear for $\alpha D_e \lesssim 3$, or $\varpi = Q_s \gtrsim 0.25$. It can be calculated

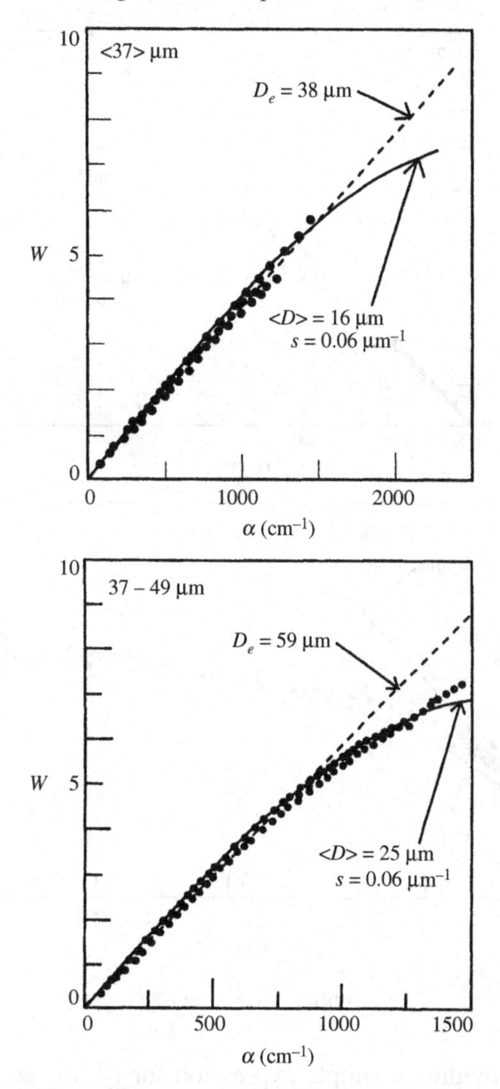

Figure 6.19 Espat function W vs. absorption coefficient for four size fractions of the cobalt silicate glass. W is calculated from the measured bidirectional reflectance of the powdered glass; α is measured from transmission. Also given in each figure is the effective particle size D_e calculated from the slope of the linear part of the curve, and the average path length $\langle D \rangle$ and internal scattering coefficient s calculated from fitting the internal-scattering model to the data. (Reproduced from Hapke and Wells [1981], copyright 1981 by the American Geophysical Union.)

from the reflectance expressions that will be derived in later chapters that $\varpi \simeq 0.25$ corresponds to particulate media with reflectances of the order of 0.07. Thus, expression (6.40) for Q_S may be used if the reflectance is greater than about 7%, provided that the density of internal scatterers is not too high.

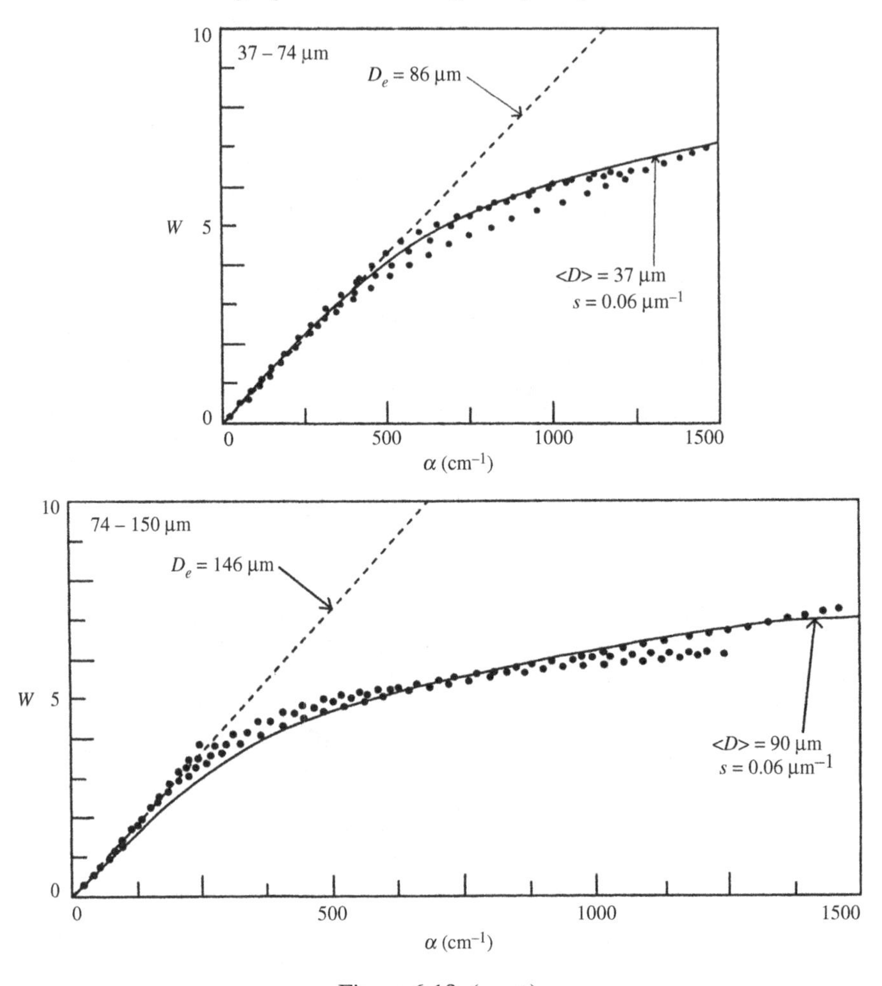

Figure 6.19 (*cont.*)

In addition to providing a simple expression for Q_s the espat function can be very useful for extracting absorption coefficients from reflectance measurements. This will be discussed in more detail in Chapter 14. However, it can be used only if neither $\alpha(D)$ nor $s\langle D\rangle$ is large. It will also be seen in Chapter 14 that the linear relation between W and α does not hold in general for mixtures of different types of particles.

6.5.7 The particle phase function of the equivalent-slab model

From the relations $Q_s = Q_{sB} + Q_{sF}$ and $\Delta Q_s = Q_{sB} - Q_{sF}$, the fractions of Q_s scattered in the back and forward directions are, respectively

$$\frac{Q_{sB}}{Q_s} = \frac{Q_s + \Delta Q_s}{2Q_s}, \tag{6.41a}$$

and

$$\frac{Q_{sF}}{Q_s} = \frac{Q_s - \Delta Q_s}{2Q_s}. \tag{6.41b}$$

Based on the results of McGuire and Hapke (1995), Domingue and Verbiscer (1997), Hartman and Domingue (1998), Cord *et al.* (2003), and Shepard and Helfenstein (2007), the two-parameter double Henyey–Greenstein function, $\Pi_{HG2}(g)$ (equation [6.7a]), appears to be able to describe a wide variety of particles, and so will be used in the general equivalent-slab model. The parameters in the function can be obtained by equating the strength of the backscattering term with Q_{sB}/Q_s and the forward-scattering term with Q_{sF}/Q_s, giving $Q_{sB}/Q_s = (1+c)/2$ and $Q_{sF}/Q_s = (1-c)/2$. Solving these for c gives

$$c = \frac{Q_{sB}}{Q_s} - \frac{Q_{sF}}{Q_s} = \frac{\Delta Q_s}{Q_s}. \tag{6.42}$$

The lobe shape parameter b can then found by using the McGuire–Hapke relation, equation (6.19), between b and c. Solving this equation for b gives

$$b = 0.15 + \frac{0.05}{(1+c)^{4/3}} = 0.15 + \frac{0.05}{(1+\Delta Q_s/Q_s)^{4/3}}. \tag{6.43}$$

6.5.8 Coated particles

Coated particles are common in nature. They are often produced by weathering processes operating on rocks and soil grains on the surfaces of a planet. To account for the effects of coatings the equivalent-slab model can be extended with the following assumptions: (1) the coatings are irregular in thickness so that no interference effects occur; (2) the coatings absorb light but are sufficiently thin that internal scattering within them is negligible.

Notation: let S_{jk} be the Fresnel reflection coefficients averaged over angle for rays incident from medium j reflected from the interface between mediums j and k. If $n_{rj} < n_{rk}$ then S_{jk} is calculated using equation (5.37) for external incidence; if $n_{rj} > n_{rk}$ then S_{jk} is calculated using equation (6.23) for internal incidence. Let subscript 0 denote values of quantities in the medium in which the coated particle is imbedded (usually air or vacuum); let subscript 1 denote the coating; and subscript 2 denote the host coated particles. Let the average ray path length through the coating be $< l >$ and the average path length through the coated particle be $< D >$.

Using the same formalism as was used in Chapter 5 to derive Q_s (Figure 5.15), but keeping the reflected and transmitted parts separate, it may readily be shown that the coefficient for the reflection of light from the surface of the coating on the back side of the particle for light incident from outside is

$$R_B = S_{01} + (1 - S_{01})(1 - S_{12}) \frac{S_{12} \exp(-2\alpha_1 < l >)}{1 - S_{10} S_{12} \exp(-2\alpha_1 < l >)} \tag{6.44a}$$

and the transmission coefficient is

$$T_B = (1 - S_{21})(1 - S_{10}) \frac{\exp(-2\alpha_1 < l >)}{1 - S_{10} S_{12} \exp(-2\alpha_1 < l >)}. \tag{6.44b}$$

Similarly, the reflection coefficient of the coating on the forward side of the particle for light incident from inside the particle is

$$R_F = S_{21} + (1 - S_{21})(1 - S_{12}) \frac{S_{10} \exp(-2\alpha_1 < l >)}{1 - S_{12} S_{10} \exp(-2\alpha_1 < l >)}, \tag{6.45a}$$

and the transmission coefficient is

$$T_F = (1 - S_{21})(1 - S_{10}) \frac{\exp(-2\alpha_1 < l >)}{1 - S_{12} S_{10} \exp(-2\alpha_1 < l >)}. \tag{6.45b}$$

Then the equations for the scattering efficiency and phase function of the equivalent-slab approximation of an uncoated particle can be used for a coated particle by making the following substitutions: $S_e \to R_B$, $(1–S_e) \to T_B$, $S_i \to R_F$, $(1–S_i) \to T_F$, giving

$$Q_s = R_B + T_B T_F \frac{\Theta}{1 - R_F \Theta}, \tag{6.46a}$$

$$\Delta Q_s = R_B + T_B T_F \frac{\Psi}{1 - R_F \Psi}, \tag{6.46b}$$

where Θ and Ψ refer to the interior of the coated particle and are the same as equations (6.30) and (6.31), respectively. The phase function is the two-parameter double Henyey–Greenstein function, equation (6.7a) with $c = \Delta Q_s / Q_s$ and b given by equation (6.43).

This completes the derivation of the equivalent-slab model.

6.5.9 Summary of the equivalent-slab model for irregular particles

For convenience, the equations of the equivalent-slab model for $X <\sim 1$ are collected here. The radiance scattered by a single particle is given by

$$I(g) = J \sigma Q_s \frac{\Pi(g)}{4\pi}. \tag{6.46}$$

When $X \ll 1$, the dipole expressions for spheres (Chapter 5.4.2) may be used to calculate Q_s and Π(g). In the region where $X \gtrsim 1$, the irregular particle shapes and the fact that most natural assemblages of particles have wide size distributions cause the optical parameters to approach their large X values smoothly and monotonically, without the large oscillations seen in spheres. Similarly, resonances in $\Pi(g)$ such as cloudbows and glories do not occur.

When the particle is large the quantity $Q_s\Pi(g)$ can be divided into diffractive and nondiffractive components $Q_s\Pi(g) = Q_s\Pi_s(g) + Q_d\Pi_d(g)$. The scattering coefficient is $Q_S = Q_d + Q_s$, where Q_d is the part of Q_S associated with diffraction, and $Q_d = 1$ for an isolated particle. The diffractive phase function associated with Q_d can be approximated by the Allen function, equation (6.2). However, if the particle is not isolated the diffraction is changed both qualitatively and quantitatively. This will be discussed in Chapter 7.

The portion of the scattering not associated with diffraction is given by three quantities: the scattering efficiency Q_s , the scattering efficiency difference ΔQ_s , and the particle phase function $\Pi(g)$. The scattering efficiency is given by

$$Q_s = S_e + (1 - S_e)\frac{1 - S_i}{1 - S_i\Theta}\Theta, \tag{6.47}$$

where the transmission function of the particle is

$$\Theta = \frac{r_i + \exp\left(-\sqrt{\alpha(\alpha+s)}\langle D\rangle\right)}{1 + r_i\exp\left(-\sqrt{\alpha(\alpha+s)}\langle D\rangle\right)}, \tag{6.48a}$$

$$r_i = \frac{1 - \sqrt{\alpha/(\alpha+s)}}{1 + \sqrt{\alpha/(\alpha+s)}}, \tag{6.48b}$$

$$S_e = 0.0587 + 0.8543R(0) + 0.0870R(0)^2, \tag{6.49a}$$

$$R(0) = \frac{(n_r - 1)^2 + n_i^2}{(n_r + 1)^2 + n_i^2}, \tag{6.49b}$$

and

$$S_i \simeq 1 - \frac{1}{n_r}[0.9413 - 0.8543R(0) - 0.0870R(0)^2]. \tag{6.50}$$

The scattering efficiency difference is given by

$$\Delta Q_s = S_e + (1 - S_e)(1 - S_i)\frac{\Psi}{1 - S_i\Psi}, \tag{6.51}$$

where

$$\Psi = \frac{r_i - \exp\left(-\sqrt{\alpha(\alpha+s)}\,\langle D\rangle\right)}{1 - r_i\exp\left(-\sqrt{\alpha(\alpha+s)}\,\langle D\rangle\right)}. \tag{6.52}$$

The particle phase function is described by the two-parameter double Henyey–Greenstein function

$$\Pi(g) = \Pi_{HG2}(g) = \frac{1+c}{2}\frac{1-b^2}{(1-2b\cos g + b^2)^{3/2}} + \frac{1-c}{2}\frac{1-b^2}{(1+2b\cos g + b^2)^{3/2}}, \tag{6.53}$$

where

$$c = \frac{\Delta Q_s}{Q_s}, \tag{6.54}$$

$$b = 0.15 + \frac{0.05}{(1+\Delta Q_s/Q_s)^{4/3}}. \tag{6.55}$$

The scattering effifciency and phase function of coated particles can be calculated using the formalism in section 6.5.8.

A useful approximation for the scattering efficiency, which may be used if the particle is not too absorbing and if the near-surface internal scattering density s is not too large, is

$$Q_s = \frac{1}{1+W} \simeq \frac{1}{1+\alpha D_e}. \tag{6.56}$$

where D_e is a size of the order of twice the particle diameter.

6.6 Computer programs and databases

Useful computer programs are available at the following internet sites:

the Dauger program for calculating the diffraction pattern of an object of arbitrary cross-section: daugerresearch.com/fresnel/index.shtml;

the DDSCAT program for the DDA method: www.astro.princeton.edu/~draine/DDSCAT;

the Amsterdam program for the DDA method: www.science.uva.nl/reserch/scs/software/adda/index.html;

the Mishchenko *et al.* program for the T-matrix method: www.gis.nasa.gov/~crimm.

A catalog of particle scattering measurements by the Amsterdam group (Munoz *et al.*, 2000) may be downloaded at www.laa.es/scattering/.

7

Propagation in a nonuniform medium: the equation of radiative transfer

7.1 Introduction

Virtually every natural and artificial material encountered in our environment is optically nonuniform on scales appreciably larger than molecular. The atmosphere is a mixture of several gases, submicroscopic aerosol particles of varying composition, and larger cloud particles. Sands and soils typically consist of many different kinds and sizes of mineral particles separated by air, water, or vacuum. Living things are made of cells, which themselves are internally inhomogeneous and are organized into larger structures, such as leaves, skin, or hair. Paint consists of white scatterers, typically TiO_2 particles, held together by a binder containing the dye that gives the material its color. These examples show that if we wish to interpret the electromagnetic radiation that reaches us from our surroundings quantitatively, it is necessary to consider the propagation of light through nonuniform media.

Although the equations for this problem can be formally written down (Ishimaru, 1978), their solution to produce useful, practical answers is another matter. The exact solution of Maxwell's equations for this class of problems is possible today only for ensembles of a small number of particles of relatively simple shapes, even with the help of modern high-speed computers. Several persons have obtained direct solutions of Maxwell's equations for a few interacting spheres (e.g., Liang and Lo, 1967; Bruning and Lo, 1971a, b; Fuller and Kattawar, 1988a, b; Xu, 1995). Lumme *et al.* (1997) used the discrete dipole approximation to synthesize scattering by several particles. Mishchenko and his colleagues have calculated scatterings by rotationally averaged systems of multiple spheres using the T-matrix method (e.g., Mishchenko *et al.*, 1995, 2007; Mishchenko and Mackowski, 1996; Mishchenko and Liu, 2007). While these studies have been invaluable in illuminating the electromagnetic interactions between particles, they are hardly realistic models of regoliths and powders. Hence, we must resort to approximate methods whose validity must be judged by the accuracy with which they describe and predict observations. Two of these methods are effective-medium theories and the equation of radiative transfer.

7.2 Effective-medium theories

One such approximate method is known as *effective-medium theory*, which attempts to describe the electromagnetic behavior of a geometrically complex medium by a uniform dielectric constant that is a weighted average of the dielectric constants of all the constituents. The various approaches differ chiefly in their specifications of the weighting factors. Examples are the models of Maxwell-Garnett (1904), Bruggeman (1935), Stroud and Pan (1978), and Niklasson *et al.* (1981); see also Bohren and Huffman (1983) and Bohren (1986).

One of the most widely used is the Maxwell-Garnett model, which will be summarized briefly. It is assumed that the medium consists of a vacuum of dielectric constant $K_{e1} = 1$, in which approximately spherical particles of radius a and complex dielectric constant K_{e2} are suspended. Both the particles and their separations are assumed to be small compared with the wavelength. Radiation having an electric field of strength $\mathbf{E}_e$ is incident on the medium and induces a dipole moment $\mathbf{p}_e$ in each particle that radiates coherently with the incident light. Consider an element of volume in the medium a small fraction of a wavelength in size. From the equations in Section 2.3.1 the electric displacement in that volume element is given by

$$\mathbf{D}_e = \varepsilon_{e0}\mathbf{K}_e\mathbf{E}_e = \varepsilon_{e0}\mathbf{E}_e + \mathbf{P}_e, \tag{7.1}$$

where $\mathbf{E}_e$ is the external field, K_e is the average dielectric constant to be determined, $\mathbf{P}_e$ is the electric polarization or dipole moment per unit volume, given by $\mathbf{P}_e = N\mathbf{p}_e$, N is the number of particles per unit volume, $\mathbf{p}_e$ is their induced dipole moment, $\mathbf{p}_c = \alpha_e\varepsilon_{e0}\mathbf{E}_{\text{loc}}$, where α_e is the electric polarizability, and $\mathbf{E}_{\text{loc}}$ is the local electric field. At any particle, $\mathbf{E}_{\text{loc}}$ is the sum of the applied field plus the fields due to the surrounding particles.

To calculate $\mathbf{P}_e$, the Clausius–Mossotti/Lorentz–Lorenz model (Section 2.3.2), which was originally derived to describe intermolecular fields, is also assumed to be a valid description of the fields between the larger particles. Then,

$$\mathbf{P}_e = N\alpha_e\varepsilon_{e0}\mathbf{E}_e/(1 - N\alpha_e/3). \tag{7.2}$$

The electric dipole moment of a sphere is given by equation (6.16),

$$\mathbf{p}_e = 4\pi a^3\varepsilon_{e0}\mathbf{E}_{\text{loc}}(K_{e2} - 1)/(K_{e2} + 2).$$

(As discussed in Chapter 6, this expression also results from the first term in the Mie solution for scattering of a plane wave by a sphere.) Hence,

$$\alpha_e = 4\pi a^3(K_{e2} - 1)/(K_{e2} + 2). \tag{7.3}$$

Combining equations (7.2) and (7.3) gives

$$\mathbf{P}_e = \varepsilon_{e0}\mathbf{E}_e \frac{N4\pi a^3[(K_{e2}-1)/(K_{e2}+2)]}{1-\frac{1}{3}N4\pi a^3[(K_{e2}-1)/(K_{e2}+2)]},$$

and inserting this into (7.1) we obtain

$$\overline{K_e} = 1 + \frac{3\phi[(K_{e2}-1)/(K_{e2}+2)]}{1-\phi[(K_{e2}-1)/(K_{e2}+2)]}, \tag{7.4}$$

where $\phi = N4\pi a^3/3$ is the total fraction of the volume occupied by the particles and is called the *filling factor.* The quantity $1-\phi$ is the *porosity*. If, instead of a vacuum, the particles are embedded in a matrix of dielectric constant $K_{e1} \neq 1$ then K_{e2} and K_e are the dielectric constants relative toK_{e1}, so that

$$K_\varepsilon = K_{e1} + \frac{3\phi K_{e1}[(K_{e2}-K_{e1})/(K_{e2}+2K_{e1})]}{1-\phi[(K_{e2}-K_{e1})/(K_{e2}+2K_{e1})]}. \tag{7.5}$$

Equation (7.5) is the Maxwell-Garnett effective-medium expression.

Unfortunately, effective-medium theories have two major difficulties. The first is that they deal only with media of particles much smaller than the wavelength. The second is that they are not capable of explaining an observation familiar to every child the world over: the color and brightness of the clear sky. The reason is that scattering by the individual particles is neglected.

This neglect is often justified by the following argument. Suppose the radiation is observed exactly from the direction into which it is propagating. Then the scattered radiation combines coherently with the incident light to produce a wave characterized by a modified dielectric constant. If the medium is observed at some other angle, all the particles in any increment of volume small compared with the wavelength radiate toward the observer. However, a second volume along the line of sight can always be found whose distance differs from that of the first volume by exactly one-half wavelength. The particles in the second volume also radiate toward the observer, but the contributions of the two volume elements are exactly out of phase and cancel.

The fallacy in this argument is that it is valid only if the density of particles is perfectly uniform. However, if the various volume elements do not contain equal numbers of particles, they will not radiate equally, and the coherent cancellation will be incomplete. Let the average number of particles in a volume element be $\langle N \rangle$, and the actual number of particles in the jth volume element be $N_j = \langle N \rangle + \delta N_j$. Let the phase of the wave at the observer radiated by the jth element be ϕ_j. Then, if the density deviations are uncorrelated from place to place, the part of the scattered intensity that is incompletely canceled is proportional to

$$\left|\sum_j \delta N_j e^{i\phi_j}\right|^2 = \sum_j \left((\delta N_j)^2 + \left|\sum_{k \neq j} \delta N_k \delta N_j e^{(i(\phi_k - \phi_j)}\right| \right)$$

$$= \sum_j \left((\delta N_j)^2 + \delta N_j \sum_{k \neq j} \delta N_k \right) = \sum_j (\delta N_j)^2, \tag{7.6}$$

where the sum is taken over all the volume elements along the line of sight. The sum over δN_k is zero, because, by definition, the average value of δN_k is zero.

Although the summation in equation (7.6) is over the entire line of sight, the critical distance is the wavelength of light. The medium may be conceptually divided into compartments whose dimensions are $\lambda/2$. Then, if the medium is perfectly uniform, the light scattered by the particles in one compartment will be exactly canceled by light scattered in a close neighbor along the line of sight. However, if the medium is inhomogeneous, the cancellation will be incomplete. The variations in particle density are equivalent to variations in the local dielectric constant or refractive index. Thus, the properties of the medium are effectively controlled by the root-mean-square (rms) average of the fluctuations in the local index of refraction over a volume element whose dimensions are of the order of the wavelength.

In gaseous media the density fluctuations can be described by a Poisson distribution in which the mean-square deviation is

$$\left\langle (\delta N_j)^2 \right\rangle = N_j, \tag{7.7}$$

so that, in this case, the scattered intensity is proportional to the number of particles along the line of sight. Because the scattering efficiency of a small particle is inversely proportional to the fourth power of the wavelength, blue light is scattered more efficiently than red, thus accounting for the color of the sky.

These concepts also have applications to the study of gases near their critical points, where they are known as *theories of critical opalescence* (e.g., Kocinski and Wojtczak, 1978). However, they are deficient on two counts: they fail to describe how the scattered intensity is modified by further scattering as it propagates through the medium, and they break down completely when the particles are larger than the wavelength.

7.3 The transport of radiation in a particulate medium

7.3.1 Concepts and definitions

In most applications of reflectance spectroscopy we are interested in the quantitative amount of light scattered by an infinitely thick layer of large particles. One way of calculating this is to consider the microscopic processes that occur when a typical particle interacts with the radiation field inside the medium. Collimated light incident on the medium is partly absorbed and partly scattered by direct encounter with particles in the upper layers. The light that passes between these particles and the light scattered by them illuminates the given particle where it is partly absorbed and scattered into all directions. The particle is heated by the light it absorbs and emits thermal radiation. Some of the radiation scattered or emitted by the particle into an upward-going direction, passes between particles and escapes from the medium. It is the combined light from all the particles that escapes from the upper surface of the medium that we wish to calculate.

In order to determine the amount of radiation escaping into a particular direction several questions must be answered. (1) What is the transmissivity of the medium; i.e., what fraction of radiation survives after traveling a certain distance through the medium? (2) What is the appropriate particle phase function? (3) How can the component of light scattered several times within the medium be calculated?

We begin by defining a number of quantities commonly used in radiative transport problems. Consider a medium of identical particles much larger than the wavelength separated by distances that are random, but that are, on the average, large compared with the particle sizes. An example might be a cloud. Consider a slab of the medium of area ΔA and thickness Δs, through which radiance I propagates (Figure 7.1). The volume of the slab $\Delta s \Delta A$ is assumed to be large compared with the volume of an individual particle in such a way that ΔA is much larger than the geometric cross-sectional area of an individual particle, but Δs is so small that no particle

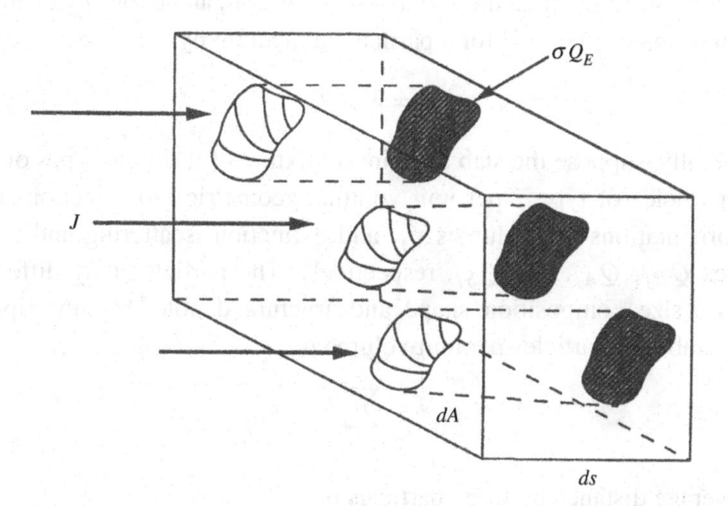

Figure 7.1 Extinction by a sparsely packed distribution of particles in an optically thin slab.

shields another within the incremental volume element. Assume that the particles are randomly oriented and positioned. Let N be the number of particles per unit volume, σ be the particle geometric cross-sectional area averaged over all orientations, and Q_E be their extinction efficiencies, as defined in Chapters 5 and 6. Because the spacing between the particles is large their properties may be considered to be the same as if they were isolated.

There is a total of $N\Delta s\Delta A$ particles in the slab, so that as the light travels through it the radiant power intercepted by the particles in the volume decreases by an amount $dP_E = \Delta I(s)\Delta A = -I(s)N\Delta s\Delta A\sigma Q_E$, where $I(s)$ is the radiance. Assuming that ΔI and Δs can be made sufficiently small, we can pass to the limit, and this equation becomes

$$\frac{dI(s)}{ds} = -I(s)N\sigma Q_E, \tag{7.8}$$

which can be integrated

$$I(s) = I(0)\exp(-N\sigma Q_E s), \tag{7.9}$$

or if N varies with position,

$$I(s) = \exp(-\int_0^s N(s')\sigma Q_S ds'). \tag{7.10}$$

The last two equations are forms of Beer's law similar to equation (2.95a) which describes the decrease of intensity in a continuous absorbing medium. This suggests that a quantity E called the *extinction coefficient*, analogous to the absorption coefficient α, may be defined for a particulate medium by

$$E = N\sigma Q_E. \tag{7.11}$$

More generally, suppose the slab contains a mixture of different types of particles with N_j particles of type j per unit volume, geometric cross sections averaged over all orieintations σ_j, volumes v_j, and extinction, scattering and absorption efficiencies Q_{Ej}, Q_{Aj}, and Q_{Sj}, respectively. The particles may differ in such properties as size, composition, shape, and structure, denoted by subscript j. Then the total number of particles per unit volume is

$$N = \sum_j N_j, \tag{7.12}$$

and the average distance between particles is

$$L = N^{-1/3}. \tag{7.13}$$

The fraction of space within the medium occupied by particles is the *filling factor*

$$\phi = \sum_j N_j v_j = N v \tag{7.14}$$

where $v = \phi/N$ is the average volume of a particle. The quantity $1\text{–}\phi$ is the *porosity*.

The *volume-average particle cross-sectional area* is

$$\sigma = \left(\sum_j N_j \sigma_j \right) / N. \tag{7.15}$$

Define the *average particle size* as the diameter of an equivalent sphere with the average cross-sectional area σ,

$$D = \sqrt{4\sigma/\pi}. \tag{7.16}$$

Putting these definitions together, the *volume extinction coefficient* of a particulate medium is

$$E = \sum_j N_j \sigma_j Q_{Ej} = N\sigma Q_E, \tag{7.17a}$$

where

$$Q_E = E/n\sigma \tag{7.17b}$$

is the *volume-average extinction efficiency*. Since extinction is the sum of absorption and scattering, the *volume-average absorption and scattering coefficients* and *efficiencies* can be defined similarly as, respectively,

$$A = \sum_j N_j \sigma_j Q_{Aj} = N\sigma Q_A, \tag{7.18a}$$

where

$$Q_A = A/N\sigma, \tag{7.18b}$$

and

$$S = \sum_j N_j \sigma_j Q_{Sj} = N\sigma Q_S, \tag{7.19a}$$

where

$$Q_S = S/N\sigma. \tag{7.19b}$$

Since $Q_{Ej} = Q_{Aj} + Q_{Sj}$, $Q_E = Q_A + Q_S$, and $E = A + S$; Q_A and Q_S are the *volume-average absorption and scattering efficiencies* respectively.

The *volume angular scattering coefficient* is defined as

$$G(g) = \sum_j N_j \sigma_j Q_{sj} \Pi_j(g) = N\sigma Q_S p(g) = Sp(g), \tag{7.20a}$$

where $\Pi_j(g)$ is the phase function of the jth type of particle and

$$p(g) = G(g)/S \tag{7.20b}$$

is the *volume-average single-particle phase function.* Since each $p_j(g)$ must satisfy the normalization condition, equation (5.13), S is the integral of $G(g)$: $S = (1/2)\int_0^\pi G(g)\sin g dg$, and $(1/2)\int_0^\pi p(g)\sin g dg = 1$.

The *volume-average single scattering albedo* is

$$w = S/E. \tag{7.21a}$$

It is convenient to define a related quantity called the *albedo factor*

$$\gamma = \sqrt{1-w}. \tag{7.21b}$$

The *volume-average asymmetry factor is*

$$\xi = <\cos\theta> = -\sum_j \frac{1}{2}\int_0^\pi p(g)\cos g \sin g dg. \tag{7.22}$$

The *extinction mean free path* Λ_E is the average distance a photon travels through the medium before being extinguished by either absorption or scattering,

$$\Lambda_E = \left(\int_0^\infty s'\, e^{-Es'} ds'\right) / \left(\int_0^\infty e^{-Es'} ds'\right) = 1/E; \tag{7.23}$$

similarly the *absorption and scattering mean free paths* Λ_A and Λ_S are the reciprocals of their respective coefficients. Finally, two other quantitties frequently encountered are the *transport coefficient*

$$S_T = S(1-\xi), \tag{7.24a}$$

and, its reciprocal, the *transport mean free path*,

$$\Lambda_T = 1/S(1-\xi). \tag{7.24b}$$

If the distribution of properties in these definitions is continuous the summations are replaced by integrations. In general, each of these quantities may depend on both location and direction of the incident radiance, although this has not been indicated explicitly.

Thus, with these definitions a discontinuous medium of randomly positioned and oriented particles may be described as if it were a quasi-continuous medium. However, defining the various coefficients in this way results in a subtle, but major, change in the physical interpretation of the propagation of radiation through a medium. In a continuous medium the absorbers and scatterers are smoothly distributed. That is, the radiation encounters them everywhere, so that the intensity is reduced equally across the wave front. However, in the nonuniform medium, the absorbers and scatterers are localized, so that the local intensity is drastically perturbed. The part of the wave front traversing a layer that passes between the particles is relatively unchanged, but the parts that encounter particles are extinguished. Even if the particles are smaller than the wavelength, the energy is no longer distributed uniformly across the wave front of the transmitted light. It is the intensity averaged over an area of the wave front whose dimensions are much larger than both the particle separation and the wavelength that must be considered to be exponentially attenuated.

If the medium consists of several different types of particles the various quantities defined above must be averaged over distances that not only are greater than the particle sizes and separations but are large enough to contain a representative sample of the medium. In a gas the fluctuations in the molecular density are proportional to the number of molecules per unit volume and are uncorrelated; hence, these definitions may be used to calculate radiative transport in atmospheres, as well as in clouds of well-separated larger particles.

7.3.2 The equation of radiative transfer in a quasi-continuous medium

The formalism that is commonly used to calculate how the processes of emission, absorption, and scattering control the propagation of electromagnetic waves within a complex medium is a form of the Boltzmann transport equation known as the *equation of radiative transfer*. The fundamental assumption of this formalism is that the medium can be treated as if it were a continuous fluid, each incremental portion of which interacts with other portions independently and incoherently through the processes of absorption, scattering, and emission. The photons are treated as if they were particles diffusing through the medium in a manner similar to gas diffusing through a complicated structure or neutrons diffusing through a nuclear reactor. The theory is not applicable to a medium consisting of particles that are uniformly spaced and regular in shape. However, Mishchenko (2002) has shown that for media of widely spaced discrete particles the radiative transfer equation can be derived from statistical electromagnetic theory. Hence, this equation can be directly applied to one of the two media that are of greatest interest in remote sensing: planetary atmospheres. The theory's applicability to the other type of medium, powders and planetary regoliths where the particles are close together, has not been rigorously established. Nevertheless, as will be seen, models based on the assumption that the radiative transfer equation is valid for these media as well enjoy considerable experimental support, and, therefore, will be the principal type of model used in this book.

In the first part of this chapter the discussion will be confined to continuous media and media of particles sufficiently far apart that equations (7.9) or (7.10) are valid. The various changes that must be made in order to describe media in which the particles are close together will be discussed later. Let $I(s, \Omega)$ be the radiance at position s, propagating into direction Ω within a continuous or quasi-continuous medium that both scatters and absorbs. The units of $I(s, \Omega)$ are power per unit area per unit solid angle. In general, unless stated otherwise, $I(s, \Omega)$ will also be a function of wavelength or frequency; however, in the interest of economy of notation, this dependence usually will not be denoted explicitly. Suppose s lies on the base of a right cylinder of area dA, length ds, and volume $dsdA$, where ds points in the direction of Ω (Figure 7.2). Then the radiant power at s passing through the base contained in a cone of solid angle $d\Omega$ about Ω is $I(s, \Omega)dAd\Omega$. Similarly, the power emerging from the top of the cylindrical volume into $d\Omega$ is $I(s + ds, \Omega)dAd\Omega = \{I(s, \Omega) + [\partial I(s, \Omega)/\partial s]ds\}dAd\Omega$. The difference between the power emerging from the top and that entering at the

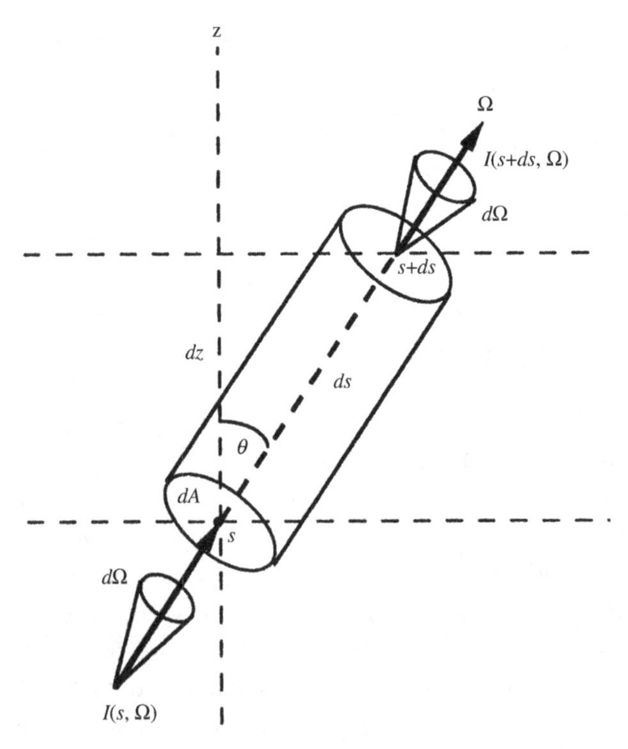

Figure 7.2 Changes in the radiance as it travels a distance *ds* through an absorbing and scattering medium.

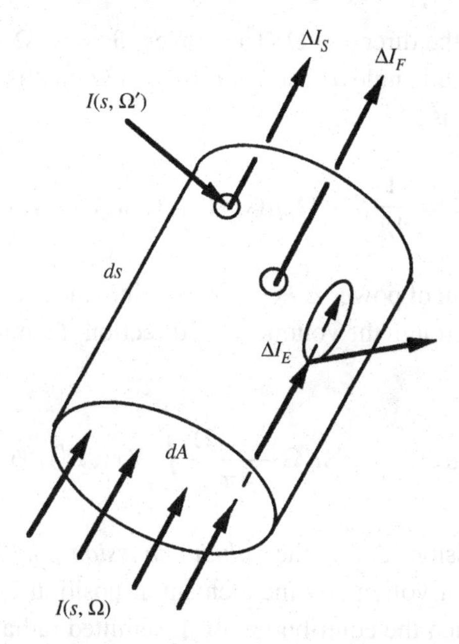

Figure 7.3 Schematic diagram of the changes in the radiance as it traverses the cylindrical volume *dsdA*.

bottom is $\Delta P = (\partial I/\partial s)dsdAd\Omega$. More generally, $\partial I/\partial s$ is the divergence of I (see Appendix A) in the direction of Ω: $\frac{\partial I}{\partial s} = \Omega \cdot \mathrm{div} I$.

This change in radiant power is due to various processes occurring in the cylinder that add or subtract energy from the beam. In this book we will be concerned with three such processes: absorption, scattering, and thermal emission, as illustrated schematically in Figure 7.3. Fluorescent emission will not be considered.

Since the medium is continuous, it is assumed that both the absorption and scattering can be described by a relation of the form of Beer's law, $I(s_2) = I(s_1)T(s_1, s_2)$, where T is the *transmissivity function* in the medium,

$$T(s_1, s_2) = \exp\left[-\int_{s_1}^{s_2} E(s, \Omega)ds\right]. \tag{7.25}$$

Later in the chapter it will be shown that this equation must be modified for media of closely packed particles; however, for now it will be assumed to be valid. Then the decrease in power due to extinction as the radiance propagates through the volume element is

$$\Delta P_E = -E(s, \Omega)I(s, \Omega)dsdAd\Omega. \tag{7.26}$$

Scattering increases as well as decreases the power in the beam, because light of intensity $I(s, \Omega')$ propagating through the volume element *dsdA* in a direction Ω'

can be scattered into the direction Ω. The power $(\partial P_{SC}/\partial\Omega)d\Omega'$ passing through $dsdA$ into a cone of solid angle $d\Omega'$ about direction Ω' that is scattered into a cone $d\Omega$ about direction Ω is

$$\frac{\partial P_{SC}}{\partial\Omega}d\Omega' = \frac{1}{4\pi}S(s,\Omega)p(s,\Omega',\Omega)I(s,\Omega')dsdAd\Omega'd\Omega. \tag{7.27}$$

To find the total amount of power ΔP_{SC} scattered into the beam, the contribution of intensities traveling through the volume in all directions Ω' must be added together, so that

$$\Delta P_{SC} = \int_{4\pi}\frac{\partial P_{SC}}{\partial\Omega}d\Omega' = dsdAd\Omega\frac{S(s,\Omega)}{4\pi}\int_{4\pi}I(s,\Omega')p(s,\Omega',\Omega)d\Omega'. \tag{7.28}$$

Turning next to emission, define the *volume emission coefficient* $F(s,\Omega)$ as the power emitted per unit volume by the element at position s into unit solid angle about direction Ω. Then the contribution of the emitted radiation to the change in power is

$$\Delta P_F = F(s,\Omega)dsdAd\Omega. \tag{7.29}$$

In general, there are at least four processes that contribute to the volume emission: singly scattered incident irradiance from a source, thermal emission, fluorescence and luminescence, and stimulated emission. Fluorescence and luminescence are outside of the scope of this book. Stimulated emission is important in gaseous media that are not in thermodynamic equilibrium, but usually not in particulate media. It can be effectively included by adding a negative component to the volume absorption coefficient. Hence, only single scattering and thermal emission will be considered in detail; their volume coefficients will be denoted by $F_S(s,\Omega)$ and $F_T(s,\Omega)$ respectively. Then

$$F(s,\Omega) = F_s(s,\Omega) + F_T(s,\Omega). \tag{7.30}$$

Single scattering will be discussed first. In a large number of problems the medium is illuminated by highly collimated radiation $J(\Omega_\iota)$ incident on the top surface from a point source that may be considered to be located effectively at an infinite distance above the medium in a direction Ω_0 from the medium making an angle i from the zenith. The vector Ω_i points in the exact opposite direction to Ω_0. At any point s below the surface of the medium the irradiance that has not been extinguished is $J\delta(\Omega-\Omega_i)\exp[-\int_s^\infty E(s',\Omega)ds']$, where δ is the delta function. The irradiance that is scattered once by the medium is a source of diffuse radiance that contributes an amount

$$\mathcal{F}_s(s,\Omega) = \frac{1}{4\pi}\int_{4\pi} J\delta(\Omega'-\Omega_i)\exp\left[-\int_s^\infty E(s',\Omega)ds'\right]G(s,\Omega',\Omega)d\Omega'$$
$$= \frac{J}{4\pi}S(s,\Omega)p(s,\Omega_i,\Omega)\exp\left[-\int_s^\infty E(s',\Omega_i)ds'\right] \quad (7.31)$$

to the volume emission coefficient. The second emission process, thermal radiation $\mathcal{F}_T(s,\Omega)$, will be considered in detail in Chapter 13. Equating the sum of all the contributions $(\Delta P_{SC}+\Delta P_E+\Delta P_F)$ to $(\delta I/\partial s)dsdAd\Omega$ and dividing by $dsdAd\Omega$ gives

$$\frac{\partial I(s,\Omega)}{\partial s} = -E(s,\Omega)I(s,\Omega) + \frac{S(s,\Omega)}{4\pi}\int_{4\pi} I(s,\Omega')p(s,\Omega',\Omega)d\Omega' + \mathcal{F}(s,\Omega). \quad (7.32)$$

Equation (7.32) is the general form of the equation of radiative transfer.

In most applications of interest the medium is horizontally stratified. Let the positive z-axis point in the vertical direction, and let ds make an angle ϑ with dz, so that $ds = dz/\cos\vartheta$ (Figure 7.1). Making this substitution in (7.34) and dividing through by $E(z,\Omega)$ gives

$$\frac{\cos\vartheta}{E(z,\Omega)}\frac{\partial I(z,\Omega)}{\partial z} = -I(z,\Omega) + \frac{1}{E(z,\Omega)}\frac{1}{4\pi}\int_{4\pi} I(z,\Omega')G(z,\Omega',\Omega)d\Omega' + \frac{\mathcal{F}(z,\Omega)}{E(z,\Omega)}. \quad (7.33)$$

Define the *source function*,

$$F(z,\Omega) = \mathcal{F}(z,\Omega)/E(z) = \frac{J}{4\pi}\exp\left[-\int_s^\infty E(s')ds'\right]\frac{S(s,\Omega)p(z,\Omega_I,\Omega)}{E(z)} + F_T(z,\Omega) \quad (7.34)$$

where

$$F_T = \mathcal{F}_T/E. \quad (7.35)$$

In many applications, including the important case of an ensemble of randomly oriented particles, S, A, and E are independent of Ω and p depends only on the angle between Ω' and Ω, rather than on the two directions separately. Let θ' be the scattering angle between the directions into which Ω' and Ω are pointing, and g' be the phase angle between the directions from which the radiance is coming and into which it is scattered. Then the angular scattering phase function can be equivalently described by $p(s,\theta')$ or by $p(s,g')$, as convenient. In what follows it will be assumed that the medium is horizontally stratified and isotropic.

Define the *optical depth* τ,

$$\tau = \int_z^\infty E(z')dz' = \int_s^\infty E(s')ds'/\cos\vartheta, \quad (7.36)$$

so that

$$d\tau = -E(z)dz = -E(s)ds/\cos\vartheta. \tag{7.37}$$

The optical depth τ is a dimensionless vertical distance expressed in units of the extinction length $1/E$. The altitude z can be expressed equivalently in terms of τ. Radiance emitted vertically upward at altitude z within the scattering medium is reduced by a factor $e^{-\tau}$ by extinction as it propagates to the top of the medium. Conversely, light incident vertically on the top of the medium will be reduced in intensity by the same factor $e^{-\tau}$ as it penetrates to the altitude corresponding to τ.

Making these substitutions and remembering that the volume single-scattering albedo is $w(z) = S(z)/E(s)$, the equation of radiative transfer for a horizontally stratified medium may be written

$$-\cos\vartheta \frac{\partial I(\tau,\Omega)}{\partial\tau} = -I(\tau,\Omega) + \frac{w(\tau)}{4\pi}\int_{4\pi} I(\tau,\Omega')p(\tau,g')d\Omega'$$
$$+ J\frac{w(\tau)}{4\pi}p(\tau,g)e^{-\tau/\cos i} + F_T(\tau,\Omega), \tag{7.38}$$

where g is the phase angle between Ω_0 and Ω. Note that none of the quantities *E, A, S, G*, or ϕ appear explicitly in the radiative-transfer equation when it is written in this form, only ratios in the form of w and p. The advantage of writing the equation in the form of (7.38) is that in many applications the parameters A, S, G, and E all have the same dependence on altitude through being proportional to a common function of z, such as the density of the medium. In these cases the ratios w and p are independent of z or τ, which simplifies the problem considerably.

7.4 Radiative transfer in a medium of arbitrary particle separation

7.4.1 Introduction

If the medium consists of well-separated particles the transmissivity function is given by equation (7.25) and, to a good approximation, $p(\tau, g)$ can be taken to be the same as that of an isolated particle. It is often assumed that this is also true when the particles are touching. However, it will be seen that several major changes in the nature of the scattering phenomena occur as the particles move close together. These include the following: the Fraunhofer diffraction pattern in the single-particle phase function no longer exists; the transmissivity function is altered; and coherent and collective effects between particles become important. These changes will be discussed in this section.

7.4.2 Fresnel diffraction in particulate media

The diffraction pattern of a particle is one of the most misunderstood concepts in the theory of the reflectance of a particulate medium. In order to calculate the multiply scattered component of the reflectance it is necessary to compute the radiance scattered by one particle that falls on another located a *short* distance away, not the radiance at infinity. A common assumption is that the Fraunhofer diffraction pattern (Figure 5.8) is unchanged. However, this assumption is incorrect on two counts. First, the radiant power in the Fraunhofer pattern comes from the portion of the wave that passes by the particle. In a close-packed medium much of the space around the particle is blocked by other particles, which decreases the intensity available for diffraction. To assume that the Fraunhofer pattern is unchanged violates the law of conservation of energy.

Second, the diffraction pattern is caused by interference between the Huygens wavelets generated by the part of the wavefront passing by the particle. In order to evaluate the integral over the wavelets the Fraunhofer formalism depends on a number of mathematical approximations that are valid only if both the source of radiation and the detector are located an infinite distance from the particle. If the particles are sufficiently far apart in a sparse medium these approximations are valid, but they break down completely when the particles are close together. The most obvious manifestation of this breakdown is that Fraunhofer diffraction predicts that shadows do not exist!

Diffraction when the source and detector are at arbitrary distances from an obstacle was investigated in the nineteenth century by Augustin-Jean Fresnel, and is known as *Fresnel diffraction*. Fraunhofer diffraction is a special case of Fresnel diffraction that is valid only if the *Fresnel condition,*

$$\frac{D^2}{4\lambda}\left(\frac{1}{l_S}+\frac{1}{l_D}\right) << 1, \tag{7.39}$$

is met, where D is the average particle size, and l_S and l_D are the distances from the particle to the source and detector, respectively. For multiply scattered light in a particulate medium the "source" is the light scattered by one particle that illuminates a second particle, and the "detectors" are the layers of other particles behind the second particle on which the diffracted light falls. Thus $l_S \sim l_D \sim L$, so the Fresnel condition becomes

$$\frac{L}{D} >> \frac{D}{4\lambda} = \frac{X}{4\pi}, \tag{7.40}$$

where X is the average size parameter. If $X = 100$ the particles must be separated by more than 8 particle diameters. Fresnel developed graphical methods of carrying out the summation over all the Huygens wavelets, but today with modern computers it is more convenient to evaluate the integrals numerically. Several programs are available for this on the internet.

Figure 7.4 illustrates how the diffraction pattern of an opaque disk changes with distance and with viewing conditions. These figures were calculated using

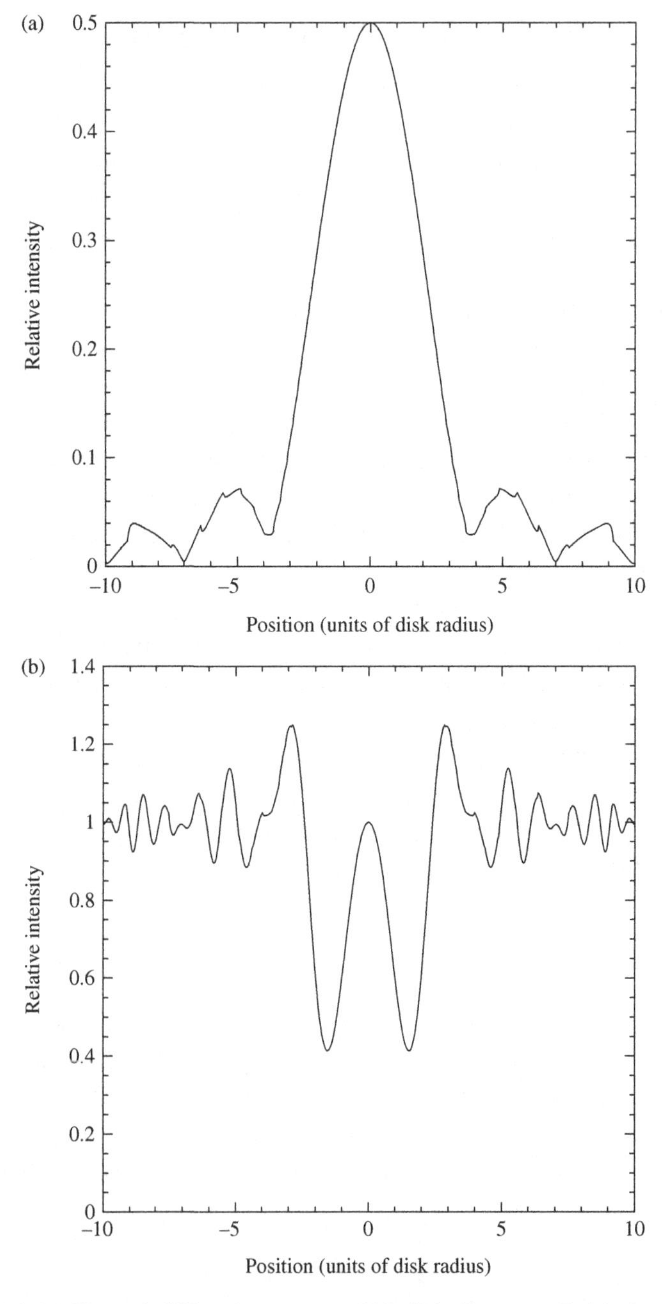

Figure 7.4 (a) Fresnel diffraction pattern 100 disk diameters behind an opaque disk viewed by a detector that accepts only light directly scattered by the disk. Compare with the Fraunhofer diffraction pattern, Figure 5.8. (b) Fresnel diffraction pattern of light 100 diameters behind an opaque disk viewed by a detector that accepts both the incident and the diffracted light. This is the diffraction pattern of a particle that would fall on a screen or on another particle 100 diameters away. (c) Fresnel diffraction one disk diameter directly behind an opaque disk illuminated by a source one disk diameter in front of the disk. Note the shadow. This is the diffraction pattern of a particle that would fall on a screen or on another particle located one diameter behind the first.

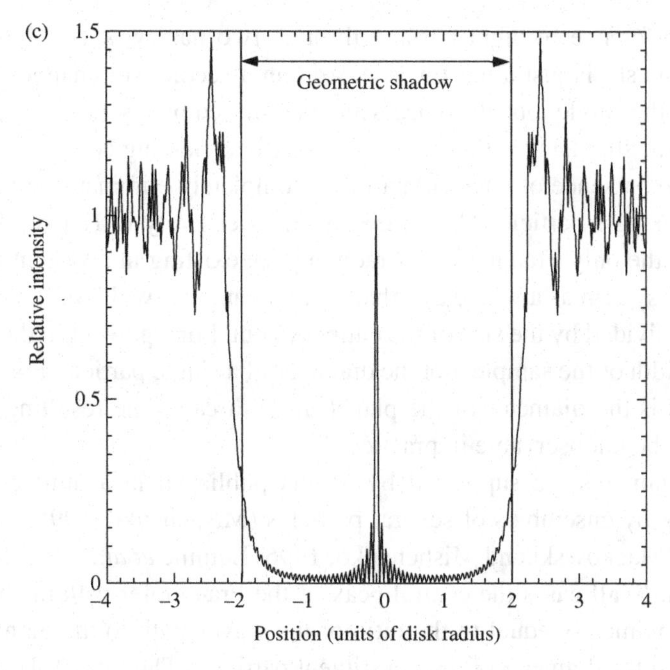

Figure 7.4 (*cont.*)

the Fresnel Diffraction Explorer program created by D. Dauger (see Section 7.7). Figure 7.4a is the diffraction pattern of an opaque disk illuminated by light from a point source at a distance of 100 particle diameters viewed by a detector at a distance of 100 diameters. In this figure the detector is assumed to be well collimated so that it views only the light diffracted by the particle and excludes the incident irradiance. Compare with Figure 5.8. However, if the incident irradiance is included, as would be the case for multiply scattered light falling on layers of other particles in a sparsely packed medium, the total intensity pattern is shown in Figure 7.4b. The reason for the differences between the two figures is that the peak is superposed coherently onto the incident irradiance. By contrast Figure 7.4c shows the diffraction pattern when the source and screen are both at distances of one diameter from the plane of the disk, which would be a typical interparticle separation in a closely spaced powder. The main feature is a shadow behind the particle. The edges of the shadow are not sharp but have a pronounced ripple structure. There is also a very thin central spike. As the distance from the disk increases, the ripples and central peak broaden and fill in the shadow.

The change in the nature of the diffraction with distance is not the only effect of decreasing the filling factor. Even the diffraction pattern of light from a distant source scattered once by particles in a regolith and viewed by a distant detector is drastically altered. To understand this, recall the discussion in freshman physics class of interference by light passing through two narrow slits. The diffraction pattern of one slit is just a narrow line. Adding a second slit changes the pattern completely: the single lobe disappears and is replaced by a series of parallel lines. Adding a third slit changes the pattern again into something completely different. Similarly the presence of adjacent particles completely eliminates the Fraunhofer pattern of a single particle. This is discussed extensively in Hapke (1999), who showed that the only Fraunhofer diffraction peak existing in a system of particles is that of the system as a whole and that it has an angular width of the order of the wavelength divided by the size of the entire system. For a powder in the laboratory this is the width of the sample, not the diameter of a single particle. For a planetary regolith this is the diameter of the planet. In both cases the resulting peak is so narrow as to be unobservable in practice.

These arguments are supported by results published in a number of papers on scattering by ensembles of several particles (Mishchenko, 1995; Mishchenko *et al.*, 1995; Mackowski and Mishchenko, 1996; Lumme *et al.*, 1997; Kolokolova *et al.*, 2006). In all cases the central peak of the Fraunhofer patterns had angular widths approximately equal to the ratio of the wavelength to the diameter of the ensemble, not the diameter of the constituent particles. The loss of the Fraunhofer peak of the individual particles is also demonstrated by observations of the Moon. If the diffraction peaks of the individual particles of the lunar regolith existed the crescent Moon should be brighter per unit area than the full Moon. In fact, however, the Moon is difficult to observe when less than a day from new and has never been reported at less than 7° from new.

It would appear to be a hopeless task to calculate the exact Fresnel diffraction pattern behind all of the irregularly shaped particles in a layer of close-packed powder. However, it may be assumed that particles randomly located behind the layer are statistically equally illuminated by all the different parts of the diffraction pattern. Thus, a reasonable approximation is to average the transmitted light over the whole pattern of shadows and ripples and replace it by a uniform wave front that is reduced from the incident intensity in proportion to the fraction of area occupied by the open spaces between the particles. This is equivalent to removing the diffraction pattern from $\Pi(g)$ and renormalizing the undiffracted remainder according to equation (5.13). Similarly, $Q_d = 1$ must be subtracted from Q_S and Q_E, and w recalculated, so that now

$$w = \frac{Q_s}{Q_s + Q_A}. \tag{7.41}$$

For particles much larger than the wavelength $Q_E = Q_s + Q_A \approx 1$, so $w \approx Q_s$. In the remainder of this book it will be assumed that $Q_d = 0$ whenever we are dealing with a close-packed powder.

7.4.3 Coherent effects in a close-packed medium

As discussed in Section 5.3, when light is incident on a particle the space outside the surface within roughly a wavelength contains near and evanescent fields. If the filling factor of the medium is low these fields do not carry energy away from a particle. However, if two particles are so close that portions of their surfaces are nearly touching then, as discussed in Chapter 4, the near fields of one particle can induce charges and currents on the adjacent surface of the other particle and transfer energy from one particle to the other, which constitutes a form of scattering. As the porosity decreases chains and clumps of particles begin to act collectively like a single, larger particle. These effects are illustrated in several papers giving exact solutions of scattering by many-particle systems using the T-matrix method (Mishchenko, 1995; Mackowski and Mishchenko, 1996; Tishkovets *et al.*, 1997, 2004; Mishchenko *et al.*, 2007).

From the definitions in Section 7.3.1 the average center-to-center spacing between particles is L. Thus the statistical condition for particles in a close-packed medium to be within about a wavelength of each other is $L - D < \lambda$, where D is the average particle size. Dividing by D and cubing gives

$$\left(\frac{L}{D}\right)^3 = \frac{1}{ND^3} = \frac{\pi/6}{\phi} < \left(1 + \frac{\lambda}{D}\right)^3,$$

or

$$\phi > \frac{\pi/6}{(1+\lambda/D)^3} = \frac{0.524}{(1+\lambda/D)^3}. \tag{7.42}$$

Equation (7.42) is plotted in Figure 7.5. It shows that when $D >> \lambda$ collective effects might be expected to occur when the filling factor is greater than about 50%, but should be important in all close-packed media in which particles of the order of the wavelength and smaller are abundant. Below the line the medium may be considered to be a nearly continuous void containing some solid particles. Above the line the optical properties begin to change over to those of a nearly continuous solid containing some void spaces.

Another coherent phenomenon called weak localization or coherent backscattering also occurs in particulate media. This will be discussed in detail in Chapter 9.

7.4.4 The transmissivity of a particulate medium

It was seen in Section 7.3 that the transmissivity function in a sparsely packed particulate medium is (equation 7.25), $T(s) = I(s)/I(0) = \exp(-Es)$, where $E =$

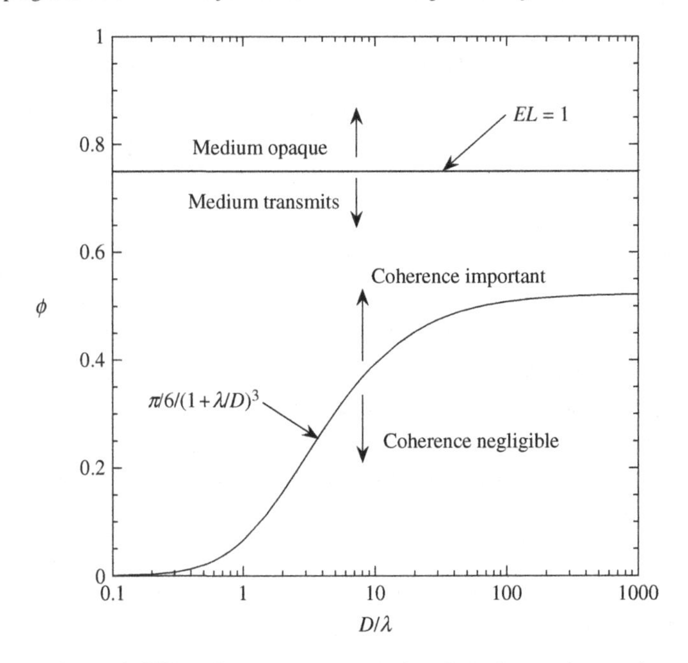

Figure 7.5 Critical filling factors vs. D/λ, showing the regions where collective and coherent interference effects are important and where the medium becomes opaque to directly transmitted radiance. (Reproduced from Hapke [2008], copyright 2008, with permission of Elsevier.)

$N\sigma Q_E$ is the extinction coefficient. However, in a medium where the particles are close together this expression is incorrect.

Ishimaru and Kuga (1982) measured the extinction coefficient as a function of filling factor for colloidal suspensions of latex spheres of sizes ranging from about $\lambda/6$ to 26λ. The results are shown in Figure 7.6, which plots the ratio of the extinction coefficients measured at various filling factors to the coefficient when the suspension is dilute. As expected from the discussion of the previous section, the coefficient for particles comparable and smaller than the wavelength decreases with increasing filling factor ϕ as collective effects become more and more important. However, for particles larger than the wavelength the extinction coefficient increases as the particles come closer together.

The reason that equation (7.25) is incorrect is that it assumes that the fraction of light blocked by particles in a thin slab of thickness ds is negligible compared with the fraction of open space. While this is a valid assumption for a sparsely packed medium it is not true when the particles are close together. In Hapke (1986, 1993) it was proposed that E be replaced by $-E\ln(1-\phi)/\phi$. However, this expression is only approximate. A more rigorous expression can be derived as follows.

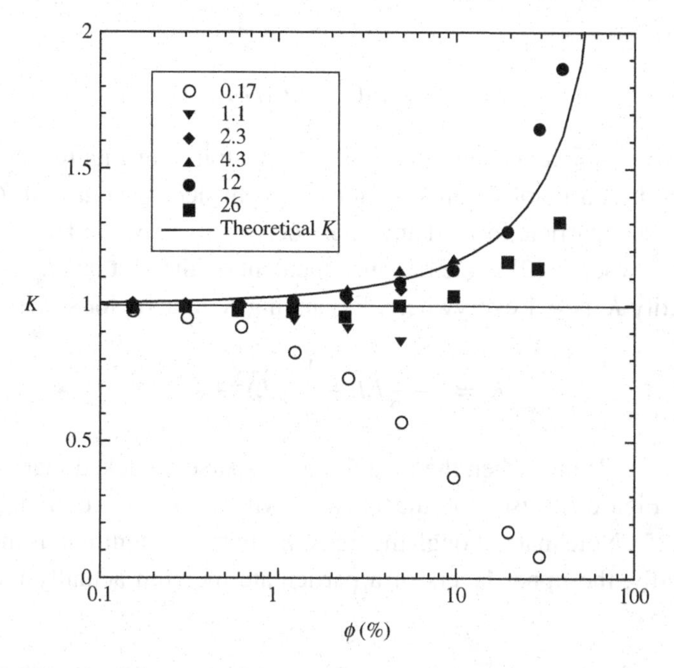

Figure 7.6 Ratio of the extinction coefficients of colloidal suspensions of latex spheres with varying filling factors to that in a low-density medium vs. filling factor ϕ. The numbers give the ratio of particle diameter to wavelength. The solid line is the theoretical expression (7.45b). (Data from Ishimaru and Kuga [1982].)

According to the discussion in Section 7.3.1 a particulate medium may be thought of as being made up of a lattice of imaginary cubes with edges of length L, with the center of each particle located somewhere inside each cube. In order that the particles do not overlap each cube can contain only one particle. Consider a wave of radiance $I(s)$ propagating through the medium as it passes through a slab of area A and thickness L containing a representative distribution of particles and oriented perpendicularly to the direction of propagation. The slab contains NAL particles with extinction cross sections σQ_E, where N is the number of particles per unit volume. Then the amount of light that encounters particles in the volume and is extinguished is $A\Delta I_E = I(NAL)\sigma Q_E = AIEL$. The amount that is transmitted between the particles is $A\Delta I_T = A1 - A\Delta I_E = AI(1 - EL)$. Hence, the fraction of the radiance incident on the slab that is unobstructed and transmitted is $\Delta I_T/I = 1 - EL$.

Assuming that the particles are randomly positioned within each box, the probability of transmission through several layers is the product of the individual probabilities. Hence, the fraction of light remaining after traversing a distance s consisting of $N = s/L$ layers is

$$T(s) = (1 - EL)^N = \exp[N \ln(1 - EL)] = \exp(-KEs), \tag{7.43}$$

where

$$K = -\ln(1 - EL)/EL. \tag{7.44}$$

is the *porosity coefficient*. Equation (7.43) is illustrated in Figure 7.7, where it is seen to be a discontinuous stair function. The physical meaning of $T(s)$ is that the radiance on a particle located anywhere between $s = NL$ and $s = (N+1)L$ is $I(0)(1–EL)^N$, where $I(0)$ is the radiance incident on the first layer.

The quantity K is ≥ 1 everywhere. Expanding K in a Taylor series,

$$K = 1 + \frac{1}{2}EL + \frac{1}{3}(EL)^2 + \cdots$$

shows that $K \to 1$ only when the medium is so sparsely packed that $EL \ll 1$. For comparison, Figure 7.7 also plots the transmissivity of a sparse continuous medium, equation (7.25). Note that although the transmissivity of a continuous medium falls off less rapidly, the upper layers of a particulate medium actually receive more

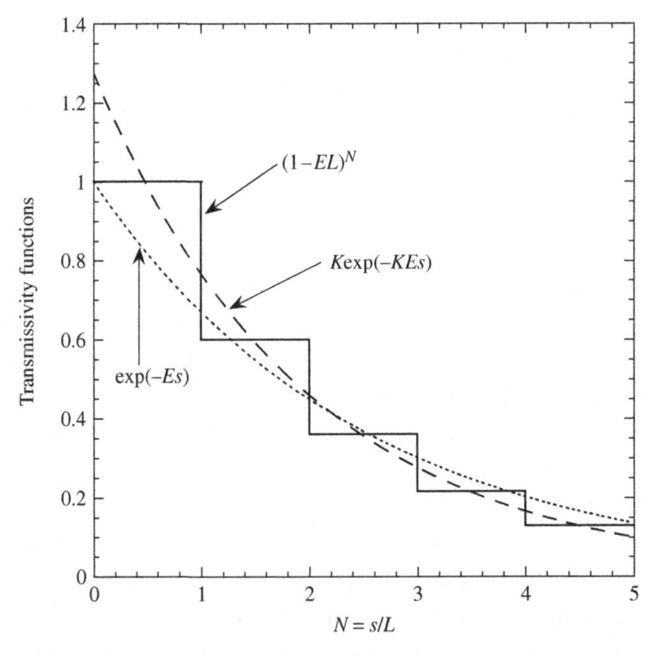

Figure 7.7 Transmissivities vs. number of the layer $N = s/L$ for the case $EL = 0.4$. The discontinuous stair function, equation (7.43), is plotted as the solid line, and its continuous approximation, equation (7.46), as the dashed line. For comparision the transmissivity of a continuous medium, equation (7.25), is shown as the dotted line. (Reproduced from Hapke [2008], copyright 2008, with permission of Elsevier Publishing Co.)

illumination than the same thickness of a continuous medium. Now,

$$EL = N\sigma Q_E L = N^{2/3}\sigma Q_E = \left[N(\sigma Q_E)^{3/2}\right]^{2/3} = \left[\sum_j N_j v_j \frac{(\sigma_j Q_{Ej})^{3/2}}{v_j}\right]^{2/3}.$$

If the particles are equant $[\sigma_j^{3/2}/v_j]^{2/3} \approx [\pi^{3/2}/(4\pi/3)]^{2/3} = 1.209$, and for particles large compared to the wavelength, $Q_{Ej} \approx 1$. Thus

$$EL \approx 1.209\left(\sum_j N_j v_j\right)^{2/3} = 1.209\phi^{2/3}, \tag{7.45a}$$

and

$$K = \frac{-\ln(1 - 1.209\phi^{2/3})}{1.209\phi^{2/3}}. \tag{7.45b}$$

The medium bccomes opaque when $EL = 1$ or $\phi = 1.209^{-3/2} = 0.752$, even though the porosity is >0. This limit is shown in Figure 7.5. In the opposite limit of small EL equation (7.43) becomes the continuous exponential function $\exp(-Es)$. Equation (7.45b) for K is plotted as the solid line in Figure 7.6.

A convenient approximation to K if the filling factor is not too large is given by

$$\begin{aligned} K &= 1 + EL/2 + (EL)^2/3 + \cdots \\ &= \{1 - [EL/2 + (EL)^2/3 + \cdots] + [EL/2 + (EL)^2/3 + \cdots]^2 + \cdots\}^{-1} \\ &= \{1 - EL/2 + (EL)^2/12 + \cdots\}^{-1} \end{aligned}$$

At the upper limit of $\phi \approx 0.5$ for coherent limits to be negligible, $(\mathrm{EL})^2/12 = 3.7\%$. Thus to within an error of $<4\%$,

$$K \approx 1/(1 - EL/2) = 1/(1 - 0.605\phi^{2/3}). \tag{7.45c}$$

7.4.5 Radiative transfer in a particulate medium of arbitrary filling factor

If the medium is continuous the equation of radiative transfer takes the form of equation (7.38), in which the various parameters are simply constants of the medium that must be specified. When the medium consists of widely spaced particulates the same form of the equation of radiative transfer may be used, in which the parameters are determined by the properties of the particles making up the medium as defined in Section 7.3.2. However, as the filling factor increases, the radiative transfer equation in its usual form, equation (7.38) is no longer correct.

Consider radiance $I(z,\Omega)$ propagating through a horizontally stratified, isotropic particulate medium at a depth z in a direction Ω making an angle ϑ with the vertical z-axis. Thermal radiation will be neglected. Then the changes $\Delta I(z,\Omega)$ in the radiance as it passes through a monolayer of particles of thickness $\Delta s = L$ perpendicular to Ω are

$$\Delta I(z,\Omega) = -ELI(z,\Omega) + \frac{wEL}{4\pi}\int_{4\pi} I(z,\Omega')p(g')d\Omega' + J\frac{SL}{4\pi}T(z/\cos i),$$

where $T(s)$ is given by equation (7.43). This expression is a difference equation. In order to make it more analytically tractable it is desirable to approximate it by a quasi-continous medium model. In order to do this we will make two assumptions. This first is that to a sufficient approximation $I(s)$ statistically varies linearly through a layer of thickness L. Then ΔI may be written

$$\Delta I(z,\Omega) = \frac{\Delta I(z,\Omega)}{\Delta s}L \approx \frac{\partial I(z,\Omega)}{\partial s}L = \cos\vartheta\frac{\partial I(z,\Omega)}{\partial z}L = -\cos\vartheta\frac{\partial I(z,\Omega)}{\partial \tau}EL.$$

The second assumption is that the discontinuous transmissivity function can be approximated by a continuous one. If s is allowed to be a continuous variable the function $\exp(-KEs)$ falls off correctly with s; however, this function passes through the left edges of each step, so that the average illumination on any given layer is too low. In order to correct this we replace the simple exponential by $C\exp(-KEs)$, where C is a constant determined by requiring that the radiance, $I(0)(1-EL)^N$, incident on a particle located between $s = NL$ and $s = (N+1)L$ in the discontinuous function, be equal to the average radiance, $(I(0)/L)\int_{NL}^{(N+1)L} C\exp(-KEs)ds$, between these two distances in the continuous function. This gives $C = K$. Hence, the transmissivity function becomes

$$T(s) = K\exp(-KEs), \tag{7.46}$$

where K is given by (7.44) and $T(s)$ is now assumed to be continuous; if the medium consists of equant particles EL is given by (7.45a). Equation (7.46) is plotted in Figure 7.7, where it is seen to pass through the central points of each step.

If the medium consists of spherical particles all of one size K can also be derived numerically from a quantity called the structure factor using the Percus–Yevic pair correlation function (Mishchenko and Macke, 1997). However, equation (7.43) is much more general and applies to particles of any shape and arbitrary size distribution. It also has the advantage of being analytic.

Making these assumptions the quasi-continuous approximation to the radiative transfer equation (7.38) becomes

$$-\cos\vartheta\frac{\partial I(\tau,\Omega)}{\partial\tau} = -I(\tau,\Omega) + \frac{w}{4\pi}\int_{4\pi} I(\tau,\Omega')p(g')d\Omega' + J\frac{w}{4\pi}p(g)Ke^{-K\tau/\mu_0}, \tag{7.47}$$

where g is the phase angle between Ω_0 and Ω, and g' is the phase angle between Ω' and Ω. This equation has the same form as the usual equation of radiative transfer except for the factors of K in the last term and the implicit lack of a Fraunhofer contribution to w and $p(g)$.

7.4.6 Mean free paths in a particulate medium

The extinction mean free path in a particulate medium can be readily computed.

$$\Lambda_E = \frac{\int_0^\infty sT(s)ds}{\int_0^\infty T(s)ds} = \frac{\sum_{N=0}^{\infty} \int_N^{N+1} s(1-EL)^N ds}{\sum_{N=0}^{\infty} \int_N^{N+1} (1-EL)^N ds} = \frac{\sum_{n=0}^{\infty} (N+1/2)L(1-EL)^N}{\sum_{N=0}^{\infty} (1-EL)^N}$$

$$= \frac{L(1-EL)[1+2(1-EL)+(1-EL)^2+\cdots]+(1/2)[1+(1-EL)+(1-EL)^2+\cdots]}{1+(1-EL)+(1-EL)^2+\cdots}$$

$$= \frac{L(1-EL)[1-(1-EL)^2]^{-1}}{[1-(1-EL)]^{-1}} + \frac{1}{2} = \frac{1-EL/2}{E}. \quad (7.48a)$$

Thus, in a particulate medium the extinction mean free path is reduced by a factor of $(1-EL/2)$ over that of a continuous medium. If the quasi-continuous approximation to the transmissivity, equation (7.46), is used to compute Λ_E, the expression

$$\Lambda_E = \frac{\int_0^\infty sK\exp(-KEs)ds}{\int_0^\infty K\exp(-KEs)ds} = \frac{1}{KE} \quad (7.48b)$$

is obtained. Equations (7.45) shows that (7.48a) and (7.48b) are equal to within an error of $< 4\%$. For internal consistency, equation (7.48b) will be used whenever the quasi-continuous approximation is employed.

The absorption, scattering, and transport mean free paths are given by relations identical to (7.48b). In particular the transport mean free path is

$$\Lambda_T = 1/KS(1-\xi). \quad (7.48c)$$

7.5 Methods of solution of radiative-transfer problems

7.5.1 Introduction

The equation of radiative transfer is a linear integrodifferential equation. In spite of the fact that it is one of the most important equations in astrophysics and remote sensing, it has proved to be remarkably intractable, and no exact analytic solution in closed form has been obtained. Therefore, numerical computer methods must be used if a high degree of accuracy is desired; otherwise one must be satisfied with approximate analytic solutions. It must be emphasized again that when equation

(7.38) is used to find the radiance in particulate media the solutions implicitly apply to averages over dimensions in the media that are large compared with both the wavelength and the particle separation and assume that the particles are well separated.

A large number of different methods have been developed to obtain solutions of the radiative transfer equation of varying degrees of accuracy. These are reviewed in Lenoble (1985) and utilized in many references, including Ambartsumian (1958), Chandrasekhar (1960), Kourganoff (1963), Hansen and Travis (1974), Sobolev (1975), Ishimaru (1978), Van de Hulst (1980), Gerstl and Zardecki (1985a), and Mishchenko *et al.* (1999). The reader is referred to those works for details of a particular method. In this section I will outline a few that are widely used.

7.5.2 The Monte Carlo method

This numerical method is potentially the most accurate, especially for complicated geometries (including spherical atmospheres) and highly anisotropic particle phase functions, but requires large amounts of computer time. Photons are injected into the medium, and at each computational step they are presented with a probability of being scattered through a given angle or absorbed. Each photon is followed until it either is absorbed or leaves the medium. The process is continued until adequate statistics are built up for all directions of scattering. Photons that contribute to the observed brightness can be calculated as if they travel either from the source to the detector or in the opposite direction, as convenient.

Although the accuracy of this method is high, in order to use it one must specify the detailed position, orientation, and shape of each object in addition to its scattering properties. This is seldom known in either remote-sensing or laboratory applications, where a statistical description of the properties of the scatterers is usually all that is available.

7.5.3 The radiosity method

The radiosity (Borel *et al.*, 1991) of a differential area is defined as the sum of the radiance emitted and reflected from that area, plus the total radiance scattered and transmitted onto the area from other differential areas. An energy-balance equation is set up for each object in the medium. This gives N coupled differential equations, where $\mathcal{N}$ is the number of objects. The same advantages and disadvantages as in the Monte Carlo method also applies to this technique.

7.5.4 The doubling method

This numerical method was pioneered by Van de Hulst (1980) and is widely used in calculations of radiative transfer in planetary atmospheres (e.g., Hansen and

Travis, 1974) and other horizontally stratified media. It is efficient in computer time and is especially useful when the particle phase function is anisotropic. The polarization as well as the intensity can be readily calculated. The concept is simple, although its practical realization may be complicated. Suppose the amount of light scattered into all angles by a layer of finite optical thickness illuminated from all directions is known. The method uses a general algorithm that calculates the amount of light scattered into all angles by two identical layers of known reflectance and transmittance, when the top layer is illuminated from a single direction above it.

The calculation starts with a layer that is so thin that its reflectance and transmittance into all angles from sources incident on it from all directions can be written down by inspection. A typical optical thickness might be 2^{-25}. A second, identical layer is added, and the reflectance and transmittance of the combined layers are calculated using the algorithm. This layer is again doubled and the process continues until the desired thickness is obtained. For instance, if the properties of a layer of optical thickness 2^{+25} are desired, only 50 repetitions of the algorithm are required. The technique is readily modified to allow calculations of layers of differing properties.

7.5.5 The Eddington approximation

In this method the solution of the radiative-transfer equation is assumed to be of the form $I(\tau,\Omega) = I_0(\tau) + I_1(\tau)\cos\vartheta$. This solution is substituted into (7.38), and the resulting equation is integrated over solid angle, giving one equation in $I_0(\tau)$ and $I_1(\tau)$. Next this solution is multiplied by $\cos\vartheta$, substituted into (7.38), and integrated over Ω, giving a second equation for $I_0(\tau)$ and $I_1(\tau)$. The resulting equations are

$$-\frac{1}{3}\frac{dI_1(\tau)}{d\tau} = -I_0(\tau) + w[I_0(\tau) + \xi I_1(\tau)] + Jwe^{-\tau/\cos i} + 4\pi F_T(\tau)$$

and

$$-\frac{1}{3}\frac{dI_0(\tau)}{d\tau} = -\frac{1}{3}I_1(\tau) + Jw\xi e^{-\tau/\cos i},$$

where ξ is the cosine asymmetry factor. The two equations may then be solved simultaneously.

7.5.6 Integral equation formulation

The partial differential equation (7.38) can be converted into an integral equation as follows. The term on the left plus the first term on the right-hand side of the equation can be put into the form:

$$-\cos\vartheta\frac{\partial I(\tau,\Omega)}{\partial\tau} + I(\tau,\Omega) = -\cos\vartheta\, e^{\tau/\cos\vartheta}\frac{\partial}{\partial\tau}[e^{-\tau/\cos\vartheta}I(\tau,\Omega)].$$

Hence, the radiative-transfer equation can be written

$$\frac{\partial}{\partial \tau}\left[e^{-\tau/\cos\vartheta} I(\tau,\Omega)\right] = -\frac{e^{-\tau/\cos\vartheta}}{\cos\vartheta}\left[\frac{w(\tau)}{4\pi}\int_{4\pi} I(\tau,\Omega')p(\tau,\Omega',\Omega)d\Omega' + F(\tau,\Omega)\right].$$

To find the radiance at optical depth τ both sides must be integrated between 0 and ∞.

For $\cos\theta < 0$ the radiance is traveling in the downward direction so that it all comes from levels between 0 and τ, and the boundary condition is that there are no sources of multiply scattered radiance above $\tau = 0$. For $\cos\theta > 0$ the radiance is traveling in the upward direction and it all comes from levels between τ and ∞; the boundary condition is that the radiance remains finite as $\tau \to \infty$. This gives

$$I(\tau,\Omega) = -\left\{\frac{e^{\tau/\cos\vartheta}}{\cos\vartheta}\int_0^{\tau}\left[\frac{w(\tau')}{4\pi}\int_{4\pi} I(\tau',\Omega')p(\tau',\Omega',\Omega)d\Omega' + F(\tau',\Omega)\right]e^{-\tau'/\cos\vartheta}d\tau'\right\}_{\cos\theta<0}$$
$$+\left\{\frac{e^{\tau/\cos\vartheta}}{\cos\vartheta}\int_{\tau}^{\infty}\left[\frac{w(\tau')}{4\pi}\int_{4\pi} I(\tau',\Omega')p(\tau',\Omega',\Omega)d\Omega' + F(\tau',\Omega)\right]e^{-\tau'/\cos\vartheta}d\tau'\right\}_{\cos\theta>0}.$$

This may be solved using the method of successive approximations. A trial solution for $I(\tau',\Omega')$ is inserted into the double integral on the right-hand side of (7.38), and a new solution is calculated, either analytically or numerically. This procedure is repeated until the desired accuracy is obtained.

7.5.7 The multistream method

The *multistream method*, also called the *discrete ordinates method*, can be as accurate as computer time allows. An example is the DISORT program (Stamnes *et al.*, 1988). However, it is also useful for obtaining approximate solutions of the radiative transfer equation. The sphere of all propagation directions Ω is broken up into $\mathcal{N}$ regions of solid angle $\Delta\Omega_j$, which need not be equal. The equation of radiative transfer, equation (7.38), is integrated in solid angle over each of the regions $\Delta\Omega_j$ and each resulting equation divided by $\Delta\Omega_j$, giving N equations of the form

$$-\frac{1}{\Delta\Omega_j}\int_{\Delta\Omega j}\frac{\partial I(\tau,\Omega)}{\partial\tau}\cos\vartheta d\Omega = -\frac{1}{\Delta\Omega_j}\frac{\partial}{\partial\tau}\int_{\Delta\Omega j} I(\tau,\Omega)\cos\vartheta d\Omega$$
$$= -\frac{1}{\Delta\Omega_j}\int_{\Delta\Omega j} I(\tau,\Omega)d\Omega + \frac{w}{4\pi}\frac{1}{\Delta\Omega_j}\int_{\Delta\Omega_j}\int_{4\pi} p(g')I(\tau,\Omega')d\Omega' d\Omega \quad (7.49)$$
$$+J\frac{w}{4\pi}e^{-\tau/\cos i}\frac{1}{\Delta\Omega_j}\int_{\Delta\Omega j} p(g)d\Omega + \frac{1}{\Delta\Omega_j}\int_{\Delta\Omega j} F_T(\tau)d\Omega.$$

Define the following average quantities for the jth zone:

$$I_j(\tau) = \frac{1}{\Delta\Omega_j}\int_{\Delta\Omega j} I(\tau,\Omega)d\Omega, \tag{7.50}$$

$$\mu_j = \frac{1}{\Delta\Omega_j}\int_{\Delta\Omega_j} \cos\vartheta\, d\Omega, \tag{7.51}$$

$$p_{kj} = \frac{1}{\Delta\Omega_k}\frac{1}{\Delta\Omega_j}\int_{\Delta\Omega k}\int_{\Delta\Omega j} p(g')\, d\Omega' d\Omega, \tag{7.52}$$

$$F_{Ij}(\tau) = \frac{1}{\Delta\Omega_j}\int_{\Delta\Omega_j} F_T(\tau,\Omega)d\Omega. \tag{7.53}$$

Note that p_{kj} is the fraction of the radiance traveling in the direction of the center of region k that is scattered into region j.

In (7.49), $I(\tau,\Omega)$ is replaced by its average value $I_j(\tau)$, which may then be taken out of the integrals over angle. Then the equation for the intensity in the jth directional region becomes

$$\mu_j dI_j(\tau)/d\tau = -I_j(\tau) + (w/4\pi)\sum_{k=1}^{\mathcal{N}} \Delta\Omega_k p_{kj} I_k(\tau)$$
$$+ J\frac{w}{4\pi} p_{mk} e^{-\tau/\cos i} + F_{Tj}(\tau), \tag{7.54}$$

where the region Ω_m contains the direction to the collimated source, and j takes all integer values between 1 and N. Thus, the partial integrodifferential equation (7.38) is replaced by a series of N linear, first-order, coupled differential equations that are amenable to well-established numerical solution. In principle, the mesh can be made as fine as computer time allows, and the calculations can approximate the exact solution as closely as desired.

If N is small enough, the equations can be solved analytically. In particular, if $N = 2$, the method is known as the two-stream or Schuster–Schwarzschild method (Schuster, 1905). It will be used extensively in this book.

The multistream method was used by Chandrasekhar (1960) in his classic treatise on radiative transfer. The directional sphere is divided into N regions in polar angle ϑ, where N is an even integer. The boundaries are chosen to be the zeros of $P_{\mathcal{N}}(\cos\vartheta)$, the Legendre polynomial of order N. It was shown by Gauss that this choice of fixing the boundaries results in an approximation that is accurate to order $2N$, and the technique is known as the method of Gaussian quadratures.

For example, for isotropic scatterers $[p(g) = 1]$ and $F_T = 0$, in each region equation (7.49) has a solution of the form $I(\tau,\Omega) = \mathrm{A}e^{-a\tau}/[1 + a\cos\vartheta] + \mathrm{B}e^{-\tau/\cos i}/(1+\cos\vartheta/\cos i)$. However, this solution cannot be made to satisfy the boundary conditions if the quantities A, B, and a are the same for all directions.

In Chandrasekhar's method the solution is forced to be of this form within each Gaussian region, but the "constants" are different in different regions.

7.5.8 The method of invariance

The method of invariance was pioneered by Ambartsumian (1958) in his classic treatise on radiative transfer. Although it is not derived directly from the equation of radiative transfer, Chandrasekhar (1960) showed that it gives the same answer. It allows one to obtain exact expressions for the reflectance and emittance of a horizontally stratified, semi-infinite particulate medium without explicitly solving the radiative-transfer equation. It relies on the fact that the reflectance of, or thermal emission from, an infinitely thick medium is not changed by adding a thin identical layer to the top of the original surface.

Denote the direction from the surface of the medium to a distant source by $\Omega_0 = \Omega(i, \psi_0)$, and the direction to a distant detector by $\Omega = \Omega(e, \psi_e)$. The direction of the incident ray is exactly opposite the direction to the source and is denoted by $\Omega_i = \Omega\iota(i_i = \pi - i, \psi\iota = \pi + \psi_0)$. Then the bidirectional reflectance of the medium can be denoted by $r(\Omega_0, \Omega)$ or $r(\Omega_i, \Omega)$, and the volume-average phase function by $p(\Omega_0, \Omega)$ or $p(\Omega_i, \Omega)$, as convenient. Let a thin layer of identical material of thickness Δz be added to the top of the medium. Assume that its optical thickness $\Delta\tau = -E\Delta z$ is so small that interactions of light with the layer involving powers of $\Delta\tau$ greater than 1 can be ignored. Then the layer will cause five distinct changes that are each proportional to $\Delta\tau$ in the scattered light. These changes are shown schematically in Figure 7.8.

(1) Light passing through the added layer is reduced by extinction (Figure 7.8a), once by a factor $\exp(-\Delta\tau/\cos i)$ on the way in, and once by a factor $\exp(-\Delta\tau/\cos e)$ on the way out. Thus, if this effect were the only one operating the emergent radiance would be

$$I = J\exp\left[-\Delta\tau\left(\frac{1}{\cos i}+\frac{1}{\cos e}\right)\right]r(\Omega_i, \Omega)$$
$$\simeq J\left[1-\Delta\tau\left(\frac{1}{\cos i}+\frac{1}{\cos e}\right)\right]r(\Omega_i, \Omega).$$

Let $\mu = \cos e$ and $\mu_0 = \cos i$. Then the first-order change in the emergent radiance is

$$\Delta I_1 = -J\Delta\tau\left(\frac{1}{\mu_0}+\frac{1}{\mu}\right)r(\Omega_i, \Omega) \tag{7.55a}$$

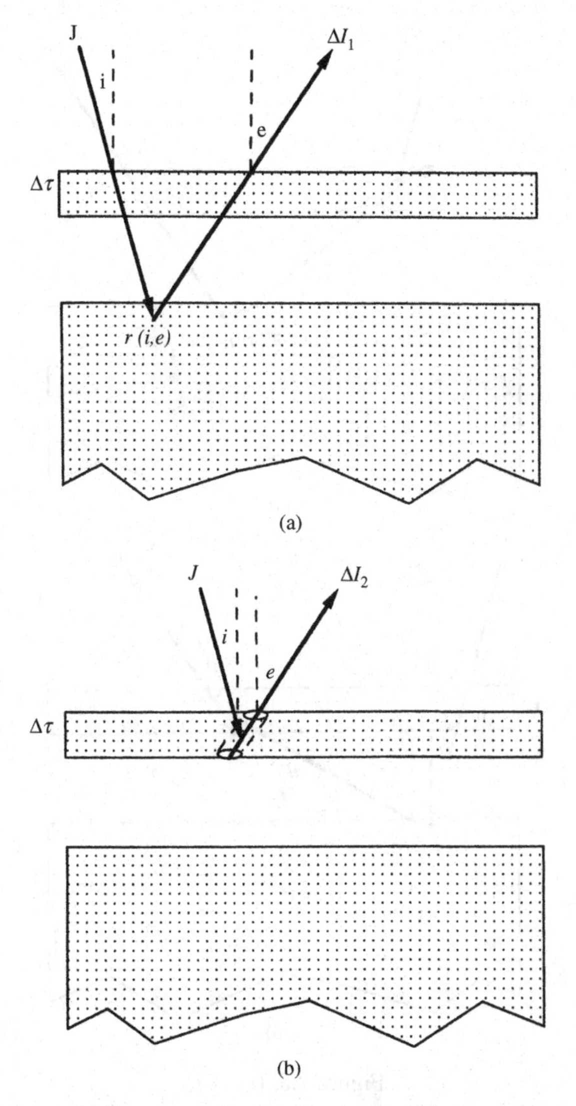

Figure 7.8 Schematic diagram of the five first-order changes in the scattered radiance caused by adding a thin layer of optical thickness $\Delta\tau$ to the top of an infinitely thick medium of bidirectional reflectance r.

(2) The added layer scatters an additional amount of light toward the detector (Figure 7.8b). Consider a cylindrical volume coaxial with a scattered ray emerging from the layer in the direction toward the detector with unit cross-sectional area and length $\Delta z/\mu$. Then there are $N\Delta z/\mu$ particles in the volume. Incident light J will be scattered by any particle with its center in this volume, and this scattered radiance will be added to the emergent radiance scattered by the lower medium. The increase in radiance due to this effect for particles having

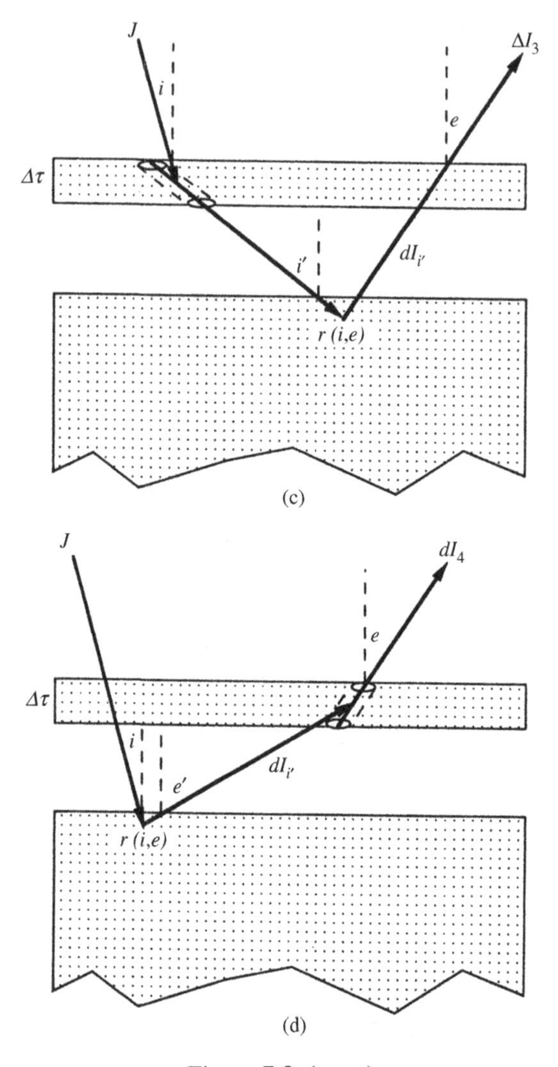

Figure 7.8 (*cont.*)

a general phase function is

$$\Delta I_2 = JN\sigma Q_S \frac{\Delta z}{\mu}\frac{p(\Omega_i,\Omega)}{4\pi} = J\frac{wp(\Omega_i,\Omega)}{4\pi}\frac{\Delta\tau}{\mu}. \tag{7.55b}$$

(3) Light scattered by the added layer in the downward direction is an additional source of illumination of the lower medium (Figure 7.8c). Consider a volume within the layer of unit cross-sectional area and length $\Delta z/|\mu_i'|$ about a direction $\Omega_i'(\mu_i', \psi_i)$, where $\mu_i' = \cos i_i'$ containing $N\Delta z/|\mu_i'|$ particles. The radiance scattered by these particles into a downward direction incident on the lower

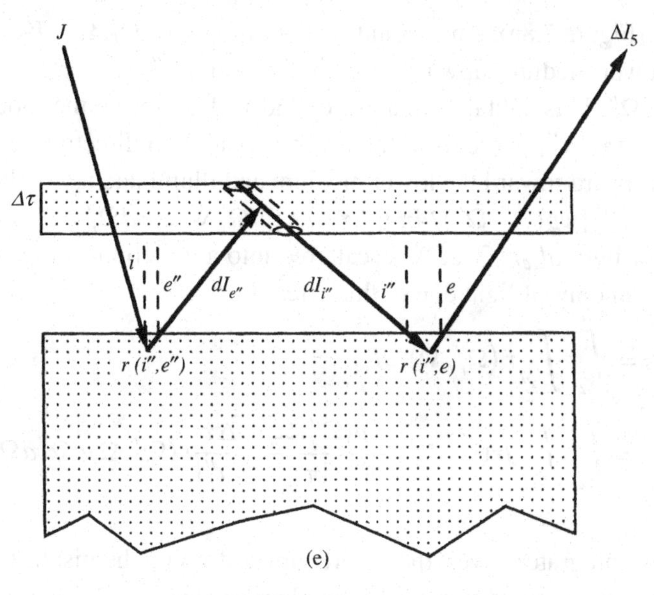

Figure 7.8 (*cont.*)

medium within a small increment of solid angle $d\Omega_i' = \sin i_i' di_i' d\psi_i'$ about Ω_i' is $dI_i' = J[wp(\Omega_i, \Omega_i')/4\pi][\Delta\tau/|\mu_i'|]d\Omega_i'$. An amount $dI_i' r(\Omega_i', \Omega)$ is scattered by the lower medium into a direction toward the detector. To find the total additional light reflected, this quantity must be integrated over all Ω_i' in the downward-going hemisphere. Hence,

$$\Delta I_3 = \int_{\Omega_i'} r\left(\Omega_i', \Omega\right) dI_i' = \int_{\Omega_i'} r(\Omega_i', \Omega) J \frac{wp(\Omega_i, \Omega_i')}{4\pi} \frac{\Delta\tau}{|\mu_i'|} d\Omega_i'. \quad (7.55c)$$

(4) Light scattered upward by the lower medium illuminates the added layer (Figure 7.8d), which scatters some of this light toward the detector. The light scattered by the medium into an increment of solid angle $d\Omega' = \sin e' de' d\psi'$ around a direction Ω' is $dI_e' = Jr(\Omega_i, \Omega')d\Omega'$. This light illuminates a cylindrical volume of unit cross-sectional area containing $N\Delta z/\mu$ particles in the layer coaxial with direction Ω. The light scattered within the volume toward the detector is $dI_e'[wp(\Omega', \Omega)/4\pi][(\Delta\tau/\mu]$. Integrating this over all Ω' in the upward-going hemisphere gives

$$\Delta I_4 = \int_{\Omega'} \frac{wp(\Omega', \Omega)}{4\pi} \frac{\Delta\tau}{\mu} dI_e' = \int_{\Omega'} Jr(\Omega_i, \Omega') \frac{wp(\Omega', \Omega)}{4\pi} \frac{\Delta\tau}{\mu} d\Omega'. \quad (7.55d)$$

(5) Light scattered upward by the lower medium is scattered back down by the added layer and provides an additional source of illumination of the lower

medium (Figure 7.8e). An amount of light $dI'' = Jr(\Omega_i, \Omega'')d\Omega''$ is scattered by the lower medium upward into solid angle $d\Omega'' = \sin e'' de'' d\psi''$ about a direction Ω''. This light illuminates a cylindrical volume in the added layer containing $N\Delta z/|\mu_i''|$ particles in the layer coaxial with direction Ω_i'' that scatter light downward toward the lower medium and illuminate it with light of intensity $dI_{ie} = dI_e''[wp(\Omega'', \Omega_i'')/4\pi][\Delta\tau/|\mu_i''|]d\Omega_i''$ where $d\Omega_i'' = \sin i_i'' di_i'' d\psi_i''$. An amount of light $dI_{ie} r(\Omega_i'', \Omega)$ is scattered into a direction toward the detector. The total amount of light due to this effect is

$$\Delta I_5 = \int_{\Omega''}\int_{\Omega_i''} r\left(\Omega_i'', \Omega\right)\, dI_{ie}$$

$$= \int_{\Omega_i''}\int_{\Omega''} Jr(\Omega_i, \Omega'')\frac{wp(\Omega'', \Omega_i'')}{4\pi}\frac{\Delta\tau}{|\mu_i''|} r(\Omega_i'', \Omega)d\Omega'' d\Omega_i'', \qquad (7.55e)$$

where Ω'' is integrated over the entire upward-going hemisphere and Ω_i'' is integrated over the entire downward-going hemisphere.

The sum of the changes ΔI_1 through ΔI_5 must be zero. Hence,

$$\frac{\mu_0 + \mu}{\mu_0\mu} r(\Omega_i, \Omega) = \frac{wp(\Omega_i, \Omega)}{4\pi\mu}$$
$$+ \int_{\Omega_i'} r(\Omega_i', \Omega)\frac{wp(\Omega_i, \Omega_i')}{4\pi|\mu_i'|}d\Omega_i' + \int_{\Omega'} r(\Omega_i, \Omega')\frac{wp(\Omega', \Omega)}{4\pi\mu}d\Omega'$$
$$+ \int_{\Omega_i}^{''}\int_{\Omega''} r(\Omega_i, \Omega'')\frac{wp(\Omega'', \Omega_i'')}{4\pi|\mu_i''|} r(\Omega_i'', \Omega)d\Omega'' d\Omega_i''. \qquad (7.56)$$

Let Ω_0' and Ω_0'' be the directions opposite to Ω_i' and Ω_I'', respectively. Then $|\mu_i'| = \mu_0'$ and $|\mu_i''| = \mu_0''$. Define the function $L(\Omega_i, \Omega)$ by

$$r(\Omega_i, \Omega) = \frac{w}{4\pi}\frac{\mu_0}{\mu_0 + \mu}L(\Omega_i, \Omega). \qquad (7.57a)$$

Making these substitutions, equation (7.56) can be put into the form

$$L(\Omega_i, \Omega) = p(\Omega_i, \Omega) + \frac{w}{4\pi}\mu\int_{\Omega_i'} p(\Omega_i, \Omega_i')\frac{L(\Omega_i', \Omega)}{\mu_0' + \mu}d\Omega_i'$$
$$+ \frac{w}{4\pi}\mu_0\int_{\Omega'}\frac{L(\Omega_i, \Omega')}{\mu_0 + \mu'}p(\Omega', \Omega)d\Omega'$$
$$+ \left(\frac{w}{4\pi}\right)^2\mu_0\mu\int_{\Omega_i''}\int_{\Omega''}\frac{L(\Omega_i, \Omega'')}{\mu_0 + \mu''}p(\Omega'', \Omega_i'')\frac{L(\Omega_i'', \Omega)}{\mu_0'' + \mu}d\Omega'' d\Omega_i''. \qquad (7.57b)$$

Equations (7.57a) are the general form of the Ambartsumian invariance relation. They can be solved numerically for L and r.

7.6 Computer programs

The computer program Fresnel Diffraction Explorer created by Dean Dauger computes both Fresnel and Fraunhofer diffraction patterns for a variety of disk shapes. It can be downloaded at http://daugerresearch.com/fresnel/index.shtml.

The DISORT program uses the discrete ordinate method to compute the radiance within and emerging from a horizontally stratified medium. The properties of the layers can be different. It can be downloaded at ftp://climate1.gsfc.nasa.gov/wiscombe/.

Mishchenko *et al.* (1999) have developed a code based on the invariance principle that calculates the reflectance of a medium consisting of spherical particles of arbitrary size and refractive index. It can be downloaded at www.giss.nasa.gov/~crimm/publications/.

8
The bidirectional reflectance of a semi-infinite medium

8.1 Introduction

In the following chapters a variety of expressions for several different types of reflectances and related quantities frequently encountered in remote sensing and diffuse reflectance spectroscopy will be given, including empirical formulae, and solutions to the equation of radiative transfer. Approximate analytic solutions to the radiative-transfer equation will be developed. As was discussed in Chapter 1, even though such analytic solutions are not exact, they are useful because there is little point in doing a detailed, exact calculation of the reflectance from a medium when the scattering properties of the particles that make up the medium are unknown and the absolute accuracy of the measurement is not high. In most of the cases encountered in remote sensing an approximate analytic solution is much more convenient and not necessarily less accurate than a numerical computer calculation.

In keeping with this discussion, polarization will be ignored in the derivations. This neglect is justified because most of the applications of interest involve the interpretation of remote-sensing or laboratory measurements in which the polarization of the incident irradiance is usually small. Although certain particles, such as Rayleigh scatterers or perfect spheres, may polarize the light strongly at some angles, the particles encountered in most applications are large, rough, and irregular, and the polarization of the light scattered by them is relatively small (Liou and Schotland, 1971; see also Chapter 6). Hence, to first order, both the incident radiation and scattered radiation may be assumed to be unpolarized.

8.2 Reflectances

In this book we wish to develop formalisms that will allow the estimation of properties of a medium from the way it scatters or emits electromagnetic radiation from its upper surface. The terms *reflectance* and *reflectivity* both refer to the fraction of incident light scattered or reflected by a material. Although they are sometimes

Table 8.1. *Types of reflectances and their symbols*

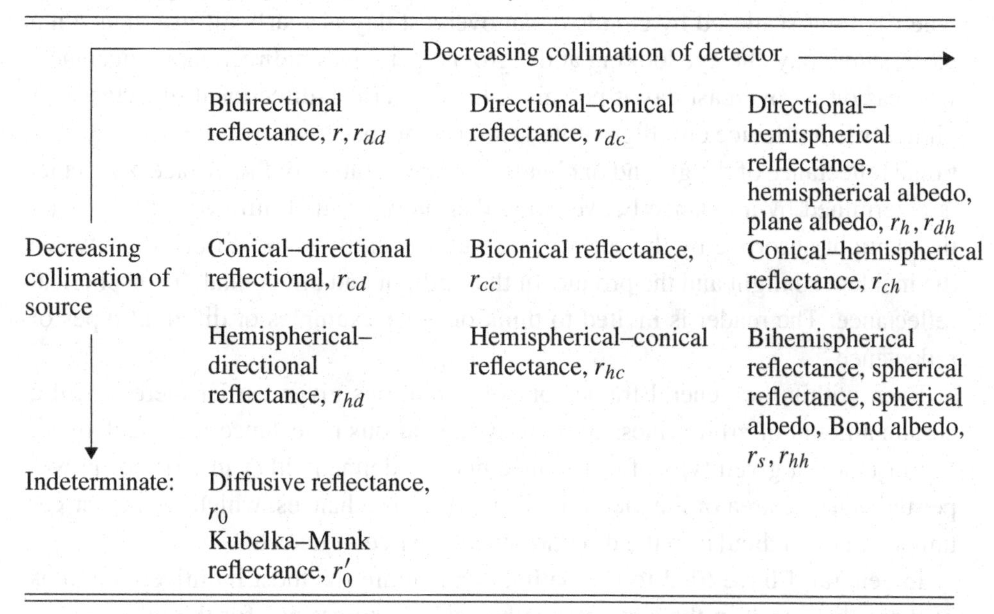

	Decreasing collimation of detector →		
Decreasing collimation of source ↓	Bidirectional reflectance, r, r_{dd}	Directional–conical reflectance, r_{dc}	Directional–hemispherical relflectance, hemispherical albedo, plane albedo, r_h, r_{dh}
	Conical–directional reflectional, r_{cd}	Biconical reflectance, r_{cc}	Conical–hemispherical reflectance, r_{ch}
	Hemispherical–directional reflectance, r_{hd}	Hemispherical–conical reflectance, r_{hc}	Bihemispherical reflectance, spherical reflectance, spherical albedo, Bond albedo, r_s, r_{hh}
Indeterminate:	Diffusive reflectance, r_0 Kubelka–Munk reflectance, r_0'		

used interchangeably, *reflectance* has the connotation of the diffuse scattering of light into many directions by a geometrically complex medium, whereas *reflectivity* refers to the specular reflection of radiation by a smooth surface. Reflectivity was discussed in detail in Chapter 4, and reflectance will be the topic of the next several chapters.

There are many kinds reflectance, depending on the geometry, so that the term must be appropriately qualified to be unambiguous. In modern usage (Nicodemus, 1970; Nicodemus *et al.*, 1977) the word is preceded by two adjectives, the first describing the degree of collimation of the source, and the second that of the detector. The usual adjectives are *directional*, *conical*, or *hemispherical*. For example, the directional–hemispherical reflectance is the total fraction of light scattered into all directions in the upward-going hemisphere by a surface illuminated from above by a highly collimated source. If the two adjectives are identical, the prefix *bi-* is used. Thus, the bidirectional reflectance is the same as the directional–directional reflectance. The various reflectances and the symbols that will be used to represent them in this book are summarized in Table 8.1.

In reality, all measured reflectances are biconical, because neither perfect collimation nor perfect diffuseness can be achieved in practice. However, many situations in nature approach the ideal sufficiently that the other quantities are useful approximations. To give several examples: because the Sun subtends only

0.5° as seen from Earth, sunlight can be treated as collimated in most applications, whereas light scattered by clouds on an overcast day is nearly diffuse; thus, on a clear, sunny day, the eye looking at the ground perceives bidirectional reflectance, whereas on an overcast day it perceives hemispherical–directional reflectance. A photosensitive device on a high-altitude aircraft or spacecraft measures the bidirectional reflectance of the ground or clouds. The temperature of the surface of a planet is determined by a balance between the thermally emitted infrared radiation and the sunlight absorbed by the surface; the latter quantity is the difference between the incident sunlight and the product of the sunlight and directional–hemispherical reflectance. The reader is invited to think of other examples of different types of reflectance.

Even within the general framework of definitions given earlier there is still a certain amount of arbitrariness in the way the various reflectances can be defined. For instance, a given type of reflectance may be defined either in terms of power per unit surface area of the medium or in terms of radiances, which are power per unit area perpendicular to the direction to the source or detector.

In general, I have tried to use definitions that are the most intuitively obvious or those that result in the simplest mathematical expressions for the reflectances. Nicodemus *et al.* (1977) have suggested a system of definitions and notation that is in wide use. Although I have tried to follow that system where convenient, some of the definitions used here differ from those of Nicodemus and associates. Sometimes a symbol proposed by those authors is already being widely used to represent a different quantity. For instance, Nicodemus and associates propose using ρ as the symbol for reflectance. However, ρ is already commonly use to denote mass density; thus, I use r for reflectance.

In general, two subscripts are added to r to indicate the degrees of collimation of the source and detector (e.g., r_{dh} = directional–hemispherical reflectance). However, although this terminology is precise, it is unwieldy. Hence, whenever the meaning is unambiguous, the following contractions will be used. (1) The bidirectional reflectance, denoted by r_{dd}, occurs so often in this book that the simple symbol r with no subscript will be used for it most of the time. (2) Most commercial reflectance spectrometers measure the directional–hemispherical reflectance, denoted by r_{dh}, but this quantity is also widely used in astrophysics, where it is called the hemispherical albedo and the plane albedo. Thus, for convenience, it will frequently be referred to simply as the *hemispherical reflectance* and denoted by r_h. (3) In Chapter 10 it will be shown that the bihemispherical reflectance, denoted by r_{hh}, is equivalent to a quantity known in astrophysics as the spherical reflectance, spherical albedo, and Bond albedo. Hence, the bihemispherical reflectance will be called the *spherical reflectance* and denoted by r_s.

An approximate analytic expression for the bidirectional reflectance of a particulate medium of infinite thickness will be derived and discussed in detail in this chapter. The opposition effect will be deferred until Chapter 9. The effect of large-scale surface roughness is treated in Chapter 10. The reflectances that involve integration over hemispheres are discussed in Chapter 11. In the remote sensing of bodies of the solar system, several additional types of reflectances are in use and are referred to as albedos and phase functions. These will be defined and considered in detail in Chapter 11 also.

8.3 Geometry and notation

The geometry and nomenclature that will be used in the remainder of the book are defined in this section. The geometry is illustrated schematically in Figure 8.1. Collimated light (irradiance) J from a source of radiation is incident on the upper surface of a scattering medium. The normal **N** to the surface is parallel to the z-axis, and the direction to the source makes an angle i with **N**. The light interacts with the medium, and some of the rays emerge from an element ΔA of the surface traveling toward a detector in a direction that makes an angle e with **N**. The plane containing the incident ray and **N** is the *plane of incidence*, and that containing the emerging ray and **N** is the *plane of emergence.* The azimuthal angle between the planes of incidence and emergence is ψ. The angle between the directions to the source and detector as seen from the surface is the phase angle g. The plane containing the incident and emergent rays is the *scattering plane*. If the planes of emergence and

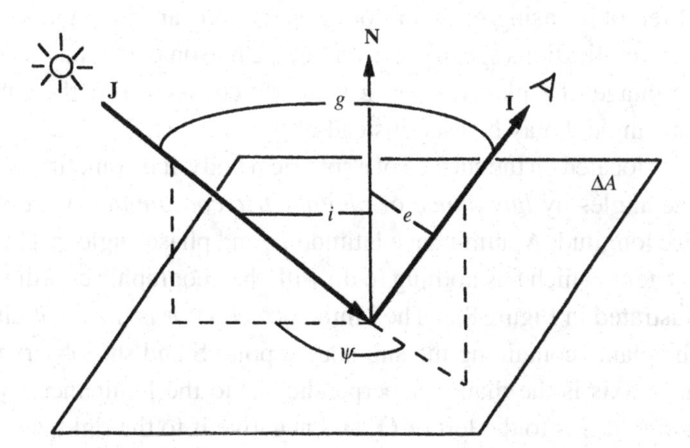

Figure 8.1 Schematic diagram of bidirectional reflectance from a surface element ΔA, showing the various angles. The plane containing **J** and **I** is the scattering plane. If the scattering plane also contains **N**, it is called the principal plane.

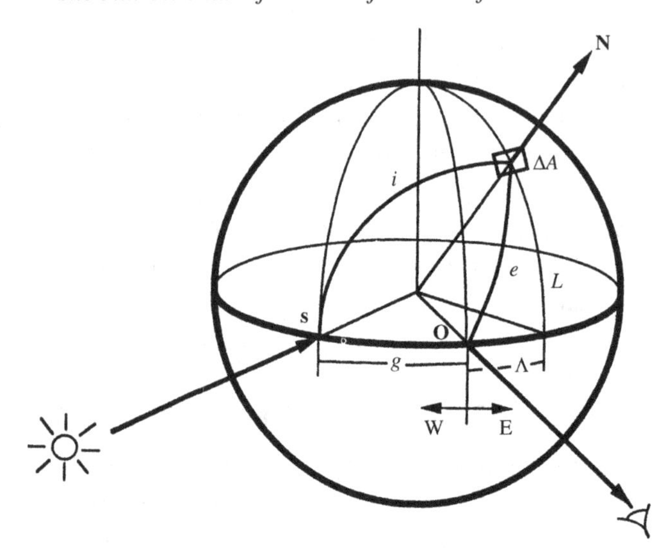

Figure 8.2 Schematic diagram of luminance coordinates on a spherical planet.

incidence coincide ($\psi = 0$ or $180°$), their common plane is called the *principal plane*.

As in Chapter 7, a commonly used notation, which will be followed in this book, is to let

$$\mu = \cos e, \ \ \mu_0 = \cos i. \tag{8.1}$$

In general, three angles are needed to specify the geometry. The angles usually used in terrestrial remote-sensing or laboratory applications are i, e, and ψ. However, most planetary applications specify i, e, and g, the reason being that in a spacecraft or telescopic image of a planet, often g is nearly constant over the entire image. The scattering angle θ may be used instead of g.

When ΔA is located on the surface of a spherical body, it is sometimes convenient to specify the angles by *luminance* or *photometric coordinates*, which consist of the luminance longitude Λ, luminance latitude L, and phase angle g. This spherical coordinate system, which has nothing to do with the geographic coordinates on the planet, is illustrated in Figure 8.2. The luminance equator is the great circle on the surface of the planet containing the sub-source point **S** and sub-observer point **O**. The luminance axis is the diameter perpendicular to the luminance equator. The phase is positive if **S** is to the left of **O**, and negative if to the right, as seen by the observer. The prime luminance meridian passes through **O**. Luminance east and positive longitudes are to the right of **O**; luminance west and negative longitudes are to the left of **O**.

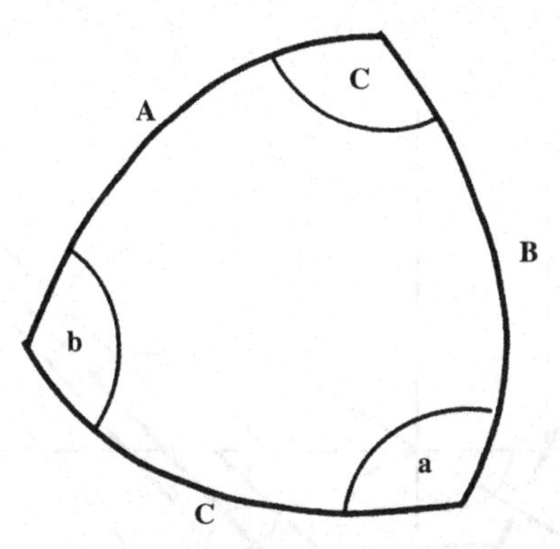

Figure 8.3 Spherical triangle.

The relationships among the various coordinate systems may be found using the law of cosines for spherical triangles. Suppose a triangle whose sides are great circles on the surface of a sphere has sides **A**, **B**, and **C**, which are separated by interior angles **a**, **b**, and **c** (Figure 8.3).

The law of cosines states that

$$\cos \mathbf{C} = \cos \mathbf{A} \cos \mathbf{B} + \sin \mathbf{A} \sin \mathbf{B} \cos \mathbf{c}. \tag{8.3}$$

Identical relations hold between the other sides and angles. Applying the law of cosines to the triangle $\mathbf{SO}\Delta A$ in Figure 8.2 gives

$$\cos g = \cos i \cos e + \sin i \sin e \cos \psi. \tag{8.4}$$

Applying the law to the triangle $\boldsymbol{SE}\Delta A$ gives

$$\cos i = \cos(\Lambda + g) \cos L, \tag{8.5}$$

and to the triangle $\mathbf{OE}\Delta A$,

$$\cos e = \cos \Lambda \cos L. \tag{8.6}$$

When $g \geq 0$, the *terminator*, the boundary between day and night, is located along the meridian of luminance longitude where $i = \pi/2$, corresponding to $\Lambda = \pi/2 - g$, and the bright limb occurs at $e = \Lambda = -\pi/2$.

8.4 The radiance at a detector viewing a horizontally stratified medium

In most laboratory and remote-sensing applications the quantity of interest is the radiance received by a detector viewing a horizontally stratified, optically thick medium of particles that may scatter, absorb, and emit. The geometry is indicated schematically in Figure 8.4. Let z be the vertical distance on the axis perpendicular

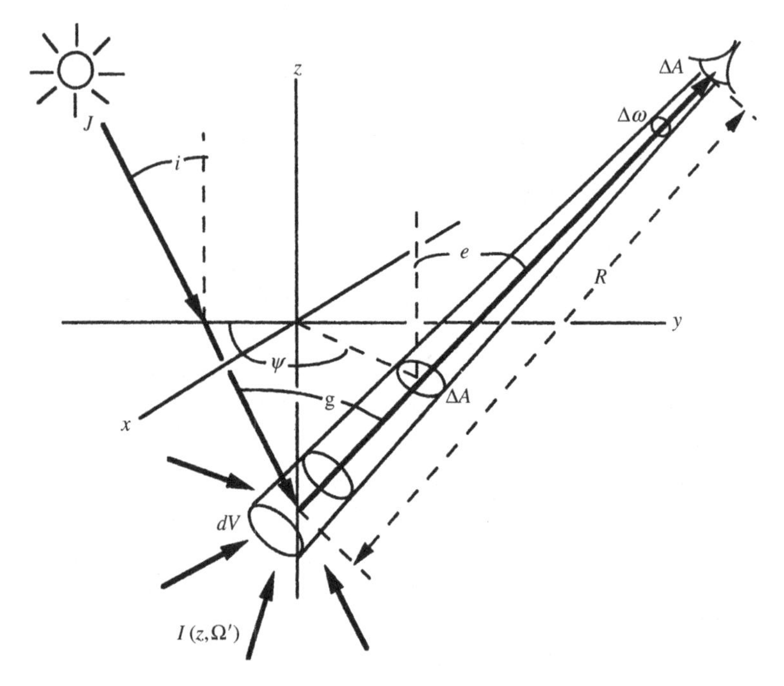

Figure 8.4 Geometry of scattering from within a particulate medium. The nominal surface of the medium is the x–y plane. Add distance Q.

to the planes of stratification. The distribution of particles with altitude z is arbitrary, except that the particle density $N \to 0$ as $z \to \infty$, corresponding to $\tau \to 0$. The space above the $\tau = 0$ level is empty except for a distant source of collimated irradiance J that illuminates the medium and a detector that views the medium. The sensitive area of the detector is Δa, and it responds to light that is incident only within a small solid angle $\Delta\omega$ from a direction making an angle e with the vertical. The detector views an area ΔA on the $\tau = 0$ level of the medium a distance R_0 away. (To avoid clutter R_0 is not shown on the figure.)

Let the power emerging from ΔA in a direction Ω toward the detector be ΔP. The projected area perpendicular to Ω is $\Delta A\mu$, and the solid angle subtended at ΔA by the detector is $\Delta\Omega = \Delta a/R_0^2$. Then the radiance emerging from the surface is

$$I(0, \Omega) = \frac{\Delta P}{(\Delta A\mu)\Delta\Omega} = \frac{\Delta P}{(\Delta A\mu)(\Delta a/R_0^2)}. \tag{8.7}$$

Now, the solid angle subtended by ΔA at the detector is $\Delta\omega = \Delta A\mu/R_0^2$, so that the radiance at the detector is

$$I_D = \frac{\Delta P}{\Delta a\Delta\omega} = \frac{\Delta P}{\Delta a(\Delta A\mu/R_0^2)} = I(0, \Omega). \tag{8.8}$$

Thus the radiance at the detector is equal to the radiance emerging from the surface of the medium in the direction of the detector. Now, $I_D = I(0, \Omega)$ is the power emitted by the medium into unit solid angle about a direction making an angle e with the vertical per unit area perpendicular to this direction. The projection of this unit area onto the apparent surface has area sec e. Hence, the power emitted into unit solid angle from unit area of the apparent surface of the medium is

$$Y(0, \Omega) = I(0, \Omega)/\sec e = I(0, \Omega)\mu. \tag{8.9}$$

8.5 Empirical reflectance expressions

8.5.1 Lambert's law (the diffuse-reflectance function)

Before deriving more rigorous solutions for the reflectances several empirical expressions that are widely used because of their mathematical simplicity will be presented. The first of these is known as Lambert's law and is the simplest and one of the most useful analytic bidirectional reflectance functions. Although no natural surface obeys Lambert's law exactly, many surfaces, especially those covered by bright, flat paints, approximate the scattering behavior described by this expression.

Lambert's law is based on the empirical observation that the brightnesses of many surfaces are nearly independent of the direction from which the surface is viewed, that is, they are independent of e and ψ, and also on the fact that the brightness of any surface must be proportional to the amount of light incident on each unit area and, hence, proportional to cos i. Because light scattering is a linear process, the scattered radiance $I(i, e, \psi)$ must be proportional to the incident irradiance J. Thus, the bidirectional reflectance of a surface that obeys Lambert's law is defined to be

$$r_L(i, e, \psi) = I(i, e, \psi)/J = K_L \cos i, \tag{8.10}$$

where K_L is a constant.

From equation (8.9) the total power scattered by unit area of a Lambert surface into all directions of the upper hemisphere is

$$P_L = \int_{2\pi} I(i, e, \psi)\mu d\Omega = \int_{e=0}^{\pi/2} \int_{\psi=0}^{2\pi} JK_L \cos i \cos e \sin e \, de \, d\psi = \pi JK_L \mu_0. \tag{8.11}$$

Since the incident power per unit surface area is $J\mu_0$, the fraction of the incident irradiance scattered by unit area of the surface back into the entire upper hemisphere is the *Lambert albedo* $A_L = P_L/J\mu_0 = K_L\pi$, so that $K_L = A_L/\pi$. Thus, Lambert's law is $I(i, e, \psi) = (J/\pi)A_L\mu_0$,, and the Lambert reflectance is

$$r_L(i, e, \psi) = \frac{1}{\pi} A_L \mu_0, \tag{8.12}$$

where A_L is the directional–hemispherical reflectance of a Lambert surface. A surface whose scattering properties can be described by Lambert's law is called a *diffuse surface* or *Lambert surface*. If $A_L = 1$, the surface is a *perfectly diffuse surface*.

Lambert's law is widely used as a mathematically convenient expression for the bidirectional reflectance when modeling diffuse scattering. In fact, it does provide a reasonably good description of the reflectances of high-albedo surfaces, like snow or flat, light-colored paint, but the approximation is poor for dark surfaces, such as soils or vegetation.

8.5.2 Minnaert's law

The power emitted per unit surface area per unit solid angle by a medium obeying Lambert's law is $Y = J r_L \mu_0 = (J/\pi) A_L \mu_0 \mu$. In this expression the reflected power per unit area is proportional to $(\mu_0 \mu)^1$. Minnaert (1941) suggested generalizing Lambert's law so that the power emitted per unit solid angle per unit area of the surface be proportional to $(\mu_0 \mu)^\upsilon$. This leads to a bidirectional reflectance function of the form

$$r_M(i, e, \psi) = A_M \mu_0^\upsilon \mu^{\upsilon - 1}, \tag{8.13}$$

where A_M and υ are empirical constants. Equation (8.13) is known as Minnaert's law, υ is the Minnaert index, and A_M is the *Minnaert albedo*. If $\upsilon = 1$, Minnaert's law reduces to Lambert's law, and $A_M = A_L/\pi$.

It is found empirically that Minnaert's law approximately describes the variation of brightness of many surfaces over a limited range of angles (cf. Thorpe, 1973; Veverka *et al.*, 1978c; McEwan, 1991), provided that A_M and υ are both allowed to be functions of phase and azimuth angles. However, the general law breaks down completely at the limb of a planet where $e = 90°$: If $\upsilon < 1$ the calculated brightness becomes infinite, and if $\upsilon > 1$ the limb brightness is zero, neither of which agrees with observations of any real body of the solar system.

8.5.3 The general Lommel–Seeliger law

We have already encountered the Lommel–Seeliger law in Chapter 6. It will be derived later in this chapter, but is included here for completeness and because it approximately describes dark particulate surfaces, like the Moon (Hapke, 1963). The general Lommel–Seeliger law is

$$r_{LS} = K_{LS} \frac{\mu_0}{\mu_0 + \mu} f(g), \tag{8.14}$$

where K_{LS} is a constant and $f(g)$ is a function of phase angle.

8.5.4 Combined Lambert–Lommel–Seeliger law

Since Lambert's law provides a rough description of surfaces of high albedo and the Lommel–Seeliger law of low-albedo surfaces, several persons (e.g., Buratti and Veverka, 1985; McEwan, 1991) have suggested combining them into a relation of the form

$$r_{LLS} = K_{LLS}\frac{\mu_0}{\mu_0+\mu}f(g) + (1-K_{LLS})\mu_0. \tag{8.15}$$

8.6 The diffusive reflectance

8.6.1 Equations

To introduce the two-stream method of solution of the equation of radiative transfer, we will now derive one of the simplest and most useful expressions in reflectance theory, the *diffusive reflectance*. In spite of the fact that this expression is an approximation, its mathematical simplicity allows analytic expressions to be obtained for a variety of scattering problems. These expressions provide surprisingly good first-order estimates in many applications.

In a very large number of papers in the literature involving the scattering of light within and from planetary surfaces and atmospheres, high-precision numerical techniques have been used to obtain answers that could have been calculated to a satisfactory degree of accuracy much more conveniently using the diffusive reflectance. In addition, it will be seen that the diffusive reflectance appears in the mathematical expressions for several other types of reflectances, or that other reflectances reduce to it at special angles. Thus, the diffusive reflectance is representative of solutions of the radiative-transfer equation. An expression known as the Kubelka–Munk equation, which is widely used to interpret reflectance data, will be seen in Chapter 11 to be a form of the diffusive reflectance.

The problem to be solved is the following. Radiant powder $P_{inc} = \pi I_0$ is incident on each unit area of a plane corresponding to $\tau = 0$ that separates an empty upper half-space from an infinitely thick lower half-space filled with a medium that scatters and absorbs light. The collimation of the incident light is unspecified. The simplifying assumption that characterizes the diffusive-reflectance formalism is that the incident radiance does not penetrate directly into the medium, but rather that as soon as the incident radiance crosses the $\tau = 0$ level it is converted into one with an angular distribution that is uniform in all directions in the downward-going hemisphere. Because it is assumed that there are no incident or thermal sources of radiance within the medium, the source function in the radiative-transfer equation $F(\tau, \Omega) = 0$. The problem is to find the radiance inside the medium and the scattered radiance emerging from the $\tau = 0$ plane in the upward direction into the upper half-space.

With these conditions the multistream approximation for the jth region, equation (7.49), becomes

$$-\mu_j dI_j(\tau)/d\tau = -I_j(\tau) + (w/4\pi)\sum_{k=1}^{N} \Delta\Omega_k P_{kj} I_k(\tau). \tag{8.16}$$

We will use the two-stream approximation, so that $N = 2$. The two regions in the directional solid angle are chosen to be the upward-going hemisphere, denoted by $j = 1$, and the downward-going hemisphere, denoted by $j = 2$. Then $\Delta\Omega_1 = \Delta\Omega_2 = 2\pi$, and equation (7.51) becomes

$$\mu_1 = \frac{1}{2\pi}\int_0^{\pi/2} \cos\vartheta\, 2\pi \sin\vartheta\, d\vartheta = \frac{1}{2},$$

$$\mu_2 = \frac{1}{2\pi}\int_{\pi/2}^{\pi} \cos\vartheta\, 2\pi \sin\vartheta\, d\vartheta = -\frac{1}{2}.$$

The angular-scattering properties of the medium will be characterized by the hemispherical asymmetry factor, β, defined such that the amount of light forward-scattered by a scattering center of the medium is $p_{11} = p_{22} = 1 + \beta$, and that backscattered is $p_{12} = p_{21} = 1 - \beta$. Then equations (8.16) become

$$-\frac{1}{2}\frac{dI_1}{d\tau} = -I_1 + \frac{w}{2}[(1+\beta)I_1 + (1-\beta)I_2], \tag{8.17a}$$

$$\frac{1}{2}\frac{dI_2}{d\tau} = -I_2 + \frac{w}{2}[(1-\beta)I_1 + (1+\beta)I_2], \tag{8.17b}$$

The parameters w and β are assumed to be independent of τ. The total downward-going power per unit area emerging from the interface at the $\tau = 0$ level is

$$P_{\text{in}} = \pi I_0 = \int_0^{\pi/2} I_2(0)\cos\vartheta\, 2\pi \sin\vartheta\, d\vartheta = \pi I_2(0)$$

Thus the boundary condition at $\tau = 0$ is $I_2(0) = I_0$. The other boundary condition is that $I_1(\tau)$ and $I_2(\tau)$ remain finite as $\tau \to \infty$.

8.6.2 Solution for isotropic scatterers

If the scattering centers of the medium scatter light isotropically, then $\beta = 0$, and equations (8.17) are

$$-\frac{1}{2}\frac{dI_1}{d\tau} = -I_1 + \frac{w}{2}(I_1 + I_2), \tag{8.18a}$$

$$\frac{1}{2}\frac{dI_2}{d\tau} = -I_2 + \frac{w}{2}(I_2 + I_1). \tag{8.18b}$$

These are two simultaneous equations that can be solved for the upward and downward radiances. The easiest way of solving them is to let

$$\varphi = \frac{1}{2}(I_1 + I_2), \tag{8.19a}$$

and

$$\Delta\varphi = \frac{1}{2}(I_1 - I_2), \tag{8.19b}$$

so that

$$I_1 = \varphi + \Delta\varphi \tag{8.19c}$$

and

$$I_2 = \varphi - \Delta\varphi. \tag{8.19d}$$

Physically, $\varphi(\tau)$ is the radiance averaged over all directions within the medium.

By alternately adding and subtracting (8.18a) and (8.18b) and making these substitutions, the two-stream equations can be put into the form

$$\frac{1}{2}\frac{d\Delta\varphi}{d\tau} = (1 - w)\varphi, \tag{8.20a}$$

$$\frac{1}{2}\frac{d\varphi}{d\tau} = \Delta\varphi. \tag{8.20b}$$

Differentiating (8.20b) and substituting the result into (8.20a) gives

$$\frac{d^2\varphi}{d\tau^2} = 4(1 - w)\varphi. \tag{8.21}$$

It may be readily verified by direct substitution that the general solution of (8.21) is

$$\varphi(\tau) = Ae^{-2\gamma\tau} + Be^{2\gamma\tau}, \tag{8.22a}$$

where

$$\gamma = (1 - w)^{1/2} \tag{8.22b}$$

is the *albedo factor*, and A and B are constants to be determined by the boundary conditions. Then, from (8.20b),

$$\Delta\varphi = \frac{1}{2}\frac{d\varphi}{d\tau} = -\gamma Ae^{-2\gamma\tau} + \gamma Be^{2\gamma\tau}. \tag{8.22c}$$

Converting back to I_1 and I_2,

$$I_1(\tau) = \varphi + \Delta\varphi = A(1 - \gamma)e^{-2\gamma\tau} + B(1 + \gamma)e^{2\gamma\tau}, \tag{8.23a}$$

$$I_2(\tau) = \varphi - \Delta\vartheta = A(1 + \gamma)e^{-2\gamma\tau} + B(1 - \gamma)e^{2\gamma\tau}, \tag{8.23b}$$

In an infinitely thick medium, φ must remain finite as $\tau \to \infty$; hence $B = 0$. Applying the boundary condition at $\tau = 0$ gives

$$I_2(0) = A(1+\gamma) = I_0.$$

Hence

$$A = I_0/(1+\gamma),$$

and the radiance in the medium is

$$I_1(\tau) = I_0 \frac{1-\gamma}{1+\gamma} e^{-2\gamma\tau}, \tag{8.24a}$$

$$I_2(\tau) = I_0 e^{-2\gamma\tau}. \tag{8.24b}$$

Now, the total scattered power emerging in all upward directions from the unit area of the surface at $\tau = 0$ is

$$P_{em} = \int_0^{\pi/2} I_1(0) \cos\vartheta 2\pi \sin\vartheta d\vartheta = \pi I_1(0) = \pi I_0 \frac{1-\gamma}{1+\gamma}.$$

The *diffusive reflectance*, which will be denoted by r_0, is P_{em}/P_{in}. Thus,

$$r_0(w) = \frac{1-\gamma}{1+\gamma}. \tag{8.25}$$

Let us explore some of the properties of $r_0(w)$. The diffusive reflectance is plotted against w in Figure 8.5. Note that $r_0(0) = 0$ and $r_0(1) = 1$. Using Taylor's theorem, equation (8.25) may be expanded into a power series in w,

$$r_0(w) = \frac{1}{4}w + \frac{1}{8}w^2 + \frac{5}{64}w^3 + \cdots. \tag{8.26}$$

In this series the term proportional to w^j is the contribution of the jth order of multiple scattering to r_0. Thus, the contribution of single scattering is $w/4$, and that of double-scattered light is $w^2/8$, and so on. The total contribution of multiple scattering is $r_0 - w/4 = r_0[1 - (1 + r_0)^{-2}]$. For surfaces of very high albedo, multiple scattering is seen to contribute about 75% of the scattered light.

For small values of w, only single scattering is important, and the curve is the straight line $r_0(w) \simeq w/4$. As w increases, the contribution of higher-order scattering becomes significant, and r_0 increases nonlinearly, until at the point $r_0(1) = 1$ the slope is infinite.

Equation (8.25) is readily solved for γ and w in terms of r_0:

$$\gamma = (1 - r_0)/(1 + r_0), \tag{8.27}$$

$$w = 4r_0/(1 + r_0)^2. \tag{8.28}$$

These expressions are useful for rapid estimates of w or γ from measured values of reflectance.

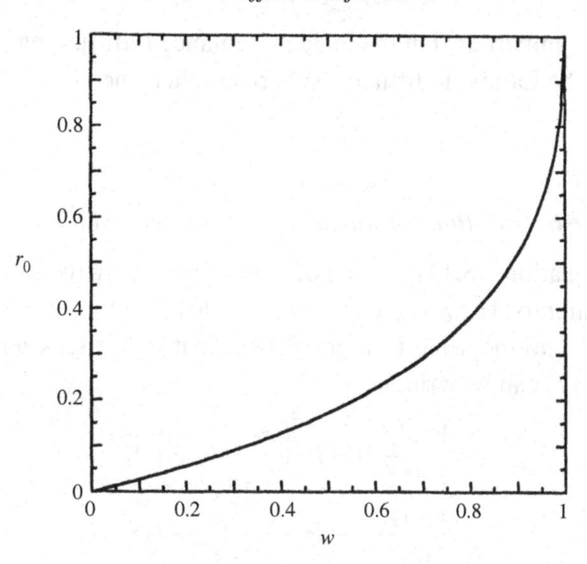

Figure 8.5 Diffusive reflectance as a function of single-scattering albedo.

The answers to a number of interesting questions may be estimated using the diffusive-reflectance model. For example, how many scatterings, on the average, does a photon undergo before being scattered out of a medium? This quantity can be calculated by noting that the total fraction of light absorbed by the medium is $1 - r_0$ and that the fraction absorbed at each scattering is$1 - w$. Therefore, the average number of scatterings, $\mathcal{N}$, is

$$\mathcal{N} = (1 - r_0)/(1 - w) = (1 + r_0)^2/(1 - r_0).$$

Thus, in a powder with particles having, $w = 0.5$, $r_0 = 0.18$, and the average photon emerging from the surface has undergone $\mathcal{N} = 1.7$ scatterings. If $w = 0.9$, $r_0 = 0.5$ and approximately $\mathcal{N} \approx 5$; if $w = 0.99$, $r_0 = 0.8$ and $\mathcal{N} \approx 18$.

8.6.3 The Lambert–diffusive-scattering law

The diffusive reflectance may be combined with Lambert's law by setting the total incident power per unit area $\pi I_0 = J\mu_0$ and assuming that the emergent radiance is independent of e. Then the Lambert–diffusive expressions for the bidirectional and hemispherical reflectances are, respectively,

$$r_L = \frac{1}{\pi} r_0 \mu_0$$

and

$$r_{hL} = r_0.$$

If the surface is Lambertian, but the incident radiance is diffuse and the same in all directions, then the Lambert–diffusive spherical reflectance is

$$r_{sL} = r_0.$$

8.6.4 Diffusive solution for anisotropic scatterers: similarity relations

If $\beta \neq 0$, then equations (8.17) may readily be solved using the same procedure as for isotropic scatterers. However, it is instructive to ask if it is possible to transform equations (8.17) into the same form as (8.18). That is, we seek quantities τ^* and w^* such that (8.17) can be written

$$-\frac{1}{2}\frac{dI_1}{d\tau^*} = -I_1 + \frac{w^*}{2}(I_1 + I_2), \tag{8.29a}$$

$$\frac{1}{2}\frac{dI_2}{d\tau^*} = -I_2 + \frac{w^*}{2}(I_2 + I_1) \tag{8.29b}$$

Equating the coefficients of I_1 and I_2 in (8.29) to those in (8.17) gives the following simultaneous equations:

$$w^*\tau^* = (1-\beta)w\tau,$$

$$\tau^* - \frac{1}{2}w^*\tau^* = \tau - \frac{1}{2}(1+\beta)w\tau.$$

Solving these yields

$$\tau^* = (1-\beta w)\tau, \tag{8.30a}$$

$$w^* = \frac{1-\beta}{1-\beta w}w. \tag{8.30b}$$

Equations (8.30) are known as *similarity relations*.

Because equations (8.29) are of the same form as (8.18), solutions for media of anisotropic scatterers are the same as for isotropic scatterers, except that the quantities τ and w are replaced by τ^* and w^*, respectively. Thus,

$$I_1 = \frac{1}{2}[A(1-\gamma^*)e^{-2\gamma*\tau*} + B(1+\gamma^*)e^{2\gamma*\tau*}], \tag{8.31a}$$

$$I_2 = \frac{1}{2}[A(1+\gamma^*)e^{2\gamma*\tau*} + B(1-\gamma^*)e^{2\gamma*\tau*}], \tag{8.31b}$$

where, for a semi-infinite medium, $B = 0$, $A = 2I_0/(1+\gamma^*)$, and

$$\gamma^* = [1-w^*]^{1/2} = [(1-w)/(1-\beta w)]^{1/2}, \tag{8.32a}$$

so that

$$\gamma^*\tau^* = [(1-w)(1-\beta w)]^{1/2}\tau. \tag{8.32b}$$

The diffusive reflectance becomes

$$r_0(w) = \frac{1-\gamma^*}{1+\gamma^*}. \tag{8.33}$$

When the particles are fully backscattering, $\beta = -1$ and $r_0 = \left(\sqrt{1+w} - \sqrt{1-w}\right) / \left(\sqrt{1+w} + \sqrt{1-w}\right)$. As in the case in which $\beta = 0$, $r_0(0) = 0$ and $r_0(1) = 1$, but when $w \ll 1, r_0 \simeq w/2$, rather than $w/4$. When the particles are fully forward-scattering, $\beta \to 1$, then $\gamma^* \to 1$, and $r_0 \to 0$, except when $w = 1$, in which case $r_0 = 1$. That is, theoretically, if $\beta = 1$, then all of the radiance is scattered deeper into the medium and escapes only if the particles are perfectly nonabsorbing. This, of course, is not a situation that is realizable in practice, and it shows that this formalism breaks down if $|\beta|$ is too close to 1. In general, when $\beta > 0$, r_0 is smaller than its value for $\beta = 0$, and larger when $\beta < 0$.

8.7 The bidirectional reflectance

8.7.1 Introduction

The bidirectional reflectance of a medium is defined as the ratio of the scattered radiance at the detector to the collimated incident irradiance. We will begin by considering the simple case of scattering by a particulate medium in which the single-scattering albedos of the particles are so small that multiply scattered light can be neglected. This will give a well-known expression, the Lommel–Seeliger law, which has already been mentioned. Following that derivation, more general relations that include the effects of multiple scattering will be discussed in detail, and useful approximations given. Initially, it will be assumed that the particles are sufficiently far apart that equation (7.38) is valid. The case when the particles are so close together that this equation is no longer correct will be considered later. Only the bidirectional reflectance of a medium of infinite optical thickness will be derived in this chapter. Layered media will be discussed in Chapter 10.

In most laboratory and remote-sensing applications the quantity of interest is the radiance received by a detector viewing a horizontally stratified, optically thick medium of particles that may scatter, absorb, and emit, illuminated by a collimated irradiance. From the result of Section 8.4, this is equivalent to the radiance emerging in a given direction from a surface. The geometry is indicated schematically in Figure 8.4. Let z be the vertical distance on the axis perpendicular to the planes of stratification. The distribution of particles with altitude z is arbitrary, except that the particle density $N \to 0$ at a finite value of z corresponding to $\tau = 0$. The particles in the space below the $\tau = 0$ level are characterized by the volume-average radiative-transfer parameters $E(z)$, $S(z)$, $A(z)$, $G(z, \Omega', \Omega)$, and $F(z, \Omega)$, as defined in Section 7.3.1. The space above the $\tau = 0$ level is empty, except for a distant point source of collimated irradiance J that illuminates a large area on the surface of the medium and a detector that views a smaller region within the illuminated area. The sensitive area of the detector is Δa, and it responds to light that is incident only within a small solid angle $\Delta\omega$.

The light incident on the detector from the medium emerges from an area ΔA formed by the intersection of $\Delta\omega$ with some surface, which is interpreted by a distant observer as the apparent surface of the medium. However, the radiance at the detector actually comes from light scattered or emitted by all the particles in the medium within the detector field of view $\Delta\omega$. If the distribution of scatterers with altitude is a step function, the apparent surface is the actual upper surface of the medium. If the altitude distribution is nonuniform, the apparent surface is often taken to be at the level corresponding to $\tau = 1$.

Consider a volume element $dV = R^2\Delta\omega dR$ located within $\Delta\omega$ at an altitude z in the medium and a distance R from the detector. This volume element is bathed in radiance $I(z, \Omega')d\Omega'$ traveling within solid angle $d\Omega'$ about direction Ω'. Thus, an amount of power $(dV/4\pi)\int_{4\pi} G(z, \Omega', \Omega)I(z, \Omega')d\Omega'$ is scattered by the particles in dV into unit solid angle about the direction Ω between dV and the detector. In addition, an amount of power $F(z, \Omega)dV$ is emitted from dV per unit solid angle toward the detector.

Now, the solid angle of the detector as seen from dV is $\Delta a/R^2$. The radiance scattered and emitted from dV toward the detector is attenuated by extinction by the particles between dV and the detector by a factor $e^{-\tau/\mu}$ before emerging from the medium. Hence, the power from dV reaching the detector is

$$dP_D = \left[\frac{1}{4\pi}\int_{4\pi} G(z, \Omega', \Omega)I(z, \Omega')d\Omega' + F(z, \Omega)\right] dV\frac{\Delta a}{R^2}e^{-\tau/\mu}$$

$$= \left[\frac{1}{4\pi}\frac{S(z)}{E(z)}\int_{4\pi}\frac{G(z, \Omega', \Omega)}{S(z)}I(z, \Omega',)d\Omega' + \frac{F(z, \Omega)}{E(z)}\right] R^2\Delta\omega\frac{E(z)dz}{\mu}\frac{\Delta a}{R^2}e^{-\tau/\mu}$$

$$= -\Delta\omega\Delta a\left[\frac{w(\tau)}{4\pi}\int_{4\pi} p(\tau, \Omega', \Omega)I(\tau, \Omega')d\Omega' + F(\tau, \Omega)\right] e^{-\tau/\mu}\frac{d\tau}{\mu},$$

where we have put $dz = \mu dR = -d\tau/E$.

The total power reaching the detector is the integral of dP_D over all volume elements within $\Delta\omega$ between $z = -\infty$ and $+\infty$, or, equivalently, between $\tau = \infty$ and 0. The radiance I_D at the detector is the power per unit area per unit solid angle. Thus,

$$I_D = \frac{1}{\Delta\omega\Delta a}\int_{z=-\infty}^{\infty} dP_D$$

$$= \int_0^{\infty}\left[\frac{w(\tau)}{4\pi}\int_{4\pi} p(\tau, \Omega', \Omega)I(\tau, \Omega')d\Omega' + F(\tau, \Omega)\right] e^{-\tau/\mu}\frac{d\tau}{\mu}. \quad (8.34)$$

8.7.2 Single scattering: the Lommel–Seeliger law

The contribution to the bidirectional reflectance of a semi-infinite, particulate medium by light that has been scattered only once can be calculated exactly from equation (8.34). It is assumed that there are no thermal sources. Then the source function is $F(\tau, \Omega) = Je^{-\tau/\mu_0} w(\tau) p(\tau, g)$. Because we are ignoring multiple scattering, the integral in equation (8.34) $\int_{4\pi} p(\tau, \Omega', \Omega) I(\tau, \Omega') d\Omega = 0$. The total radiance I_{SS} reaching the detector due to single scattering is thus

$$I_{SS} = J \frac{1}{4\pi} \frac{1}{\mu} \int_0^\infty w(\tau) p(\tau, g) e^{-(1/\mu_0 + 1/\mu)\tau} d\tau.$$

If w and p are independent of z or τ, as is often the case, then the evaluation of this integral is trivial, and gives

$$I_{SS} = J \frac{w}{4\pi} \frac{\mu_0}{\mu_0 + \mu} p(g). \tag{8.35a}$$

When the scatterers are isotropic, $p(g) = 1$, and equation (8.35a) is the *Lommel–Seeliger law*. This law has been generalized in (8.35a) to include nonisotropic scatterers. Except close to zero phase, this expression is a fair description of the light scattered by low-albedo bodies of the solar system, such as the Moon and Mercury (Hapke, 1963, 1971), for which only light that has been scattered once contributes significantly to the brightness.

Using equations (8.5) and (8.6) to transform to luminance coordinates (8.35a) becomes

$$I_{SS} = J \frac{w}{4\pi} \frac{\cos(\Lambda + g)}{\cos(\Lambda + g) + \cos\Lambda} p(g). \tag{8.35b}$$

Note that this expression is independent of luminance latitude L. To a fair approximation, the brightness of the Moon at small phase angles is, in fact, independent of latitude (Minnaert, 1961; Hapke, 1971). The Lommel–Seeliger function $\cos(\Lambda + g)/[\cos(\Lambda + g) + \cos\Lambda]$ is plotted as a function of longitude for three phase angles in Figure 8.6. The surge in brightness near the limb is not observed on the Moon because of the roughness of the lunar surface; effects of macroscopic surface roughness are discussed in Chapter 12.

8.7.3 The bidirectional reflectance of a sparse medium of isotropic scatterers

8.7.3.1 The two-stream solution with collimated source

In this section we will show how multiple scattering may be included in the calculation of the bidirectional reflectance. In order to illustrate both the power and the limitations of the two-stream method, this technique will be used to obtain an approximate solution to the radiative transfer equation for a horizontally stratified,

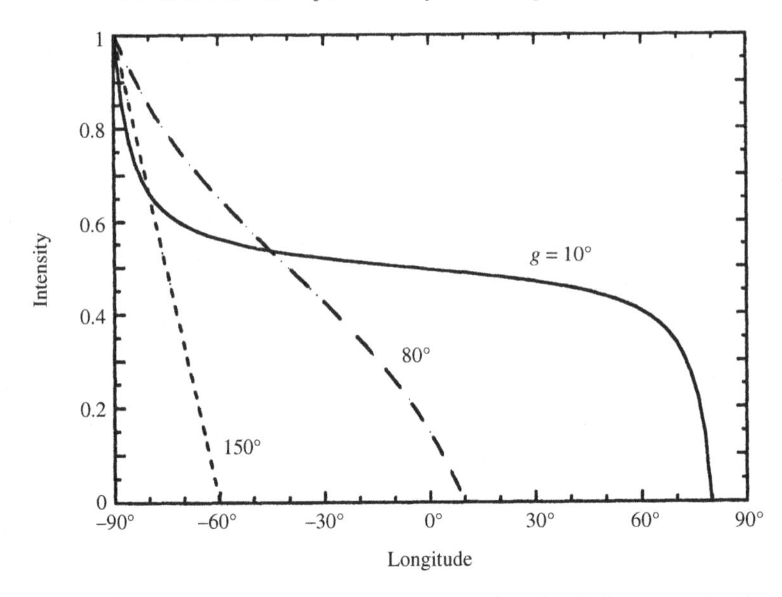

Figure 8.6 Lommel–Seeliger law vs. luminance longitude for several values of the phase angle.

semi-infinite medium of widely spaced isotropic scatterers. The exact solution for the bidirectional reflectance of such a medium will be found in the next section using the method of invariance, and the two solutions will be compared. The reflectance of a medium of closely spaced scatterers will be found later.

The geometry is the same as in Figure 8.1. Collimated irradiance J is incident on a particulate medium. The space above the plane corresponding to $\tau = 0$ is empty, except for the source and detector, and the volume below this plane contains particles that both scatter and absorb. Thermal emission is assumed to be negligible. The properties of the medium are described by the nomenclature of Chapter 7, and w and $p(g)$ are assumed to be independent of τ.

The two-stream method will be used to find an approximate solution to the radiative-transfer equation (7.38) with $p(g)=1$, $F_T=0$, and $F(\tau, g)=(J/4\pi) we^{-\tau/\mu_0}$. This will give the total radiant flux at any optical depth τ of photons that have been scattered one or more times. This flux, together with the incident irradiance, illuminates a layer at some depth within the medium. Finally, the amout of light scattered by the particles in that layer in a direction toward the detector that escapes from the surface will be found.

As in the derivation of the diffusive reflectance, the two-stream form of the equation of radiative transfer is obtained by putting $N=2$ in equations (7.49)–(7.53). The upward-going hemisphere is denoted by subscript $j = 1$, and the downward-going hemisphere by $j=2$. Then $\Delta\Omega_1=\Delta\Omega_2=2\pi$, $\mu_1=\frac{1}{2}$,

$\mu_2 = -\frac{1}{2}$, $P_{kj} = 1$, and equation (7.49) becomes the set of two equations

$$-\frac{1}{2}\frac{dI_1}{d\tau} = -I_1 + \frac{w}{2}(I_1 + I_2) + J\frac{w}{4\pi}e^{-\tau/\mu_0}, \tag{8.36a}$$

$$\frac{1}{2}\frac{dI_2}{d\tau} = -I_2 + \frac{w}{2}(I_1 + I_2) + J\frac{w}{4\pi}e^{-\tau/\mu_0}. \tag{8.36b}$$

The boundary conditions are that the radiance must remain finite everywhere and that there be no sources of diffuse radiation above the upper surface, so that at $\tau = 0$ the downward-going diffuse radiance $I_2(0) = 0$.

As in Section 8.6, put $\varphi = (I_1 + I_2)/2$ and $\Delta\varphi = (I_1 - I_2)/2$, and alternately add and subtract equations (8.36) to obtain

$$-\frac{1}{2}\frac{d\Delta\varphi}{d\tau} = -\gamma^2\varphi + J\frac{w}{4\pi}e^{-\tau/\mu_0}, \tag{8.37a}$$

$$\frac{1}{2}\frac{d\varphi}{d\tau} = \Delta\varphi, \tag{8.37b}$$

where $\gamma = \sqrt{1-w}$ is the albedo factor. Differentiating (8.37b) and inserting into (8.37a) gives

$$-\frac{1}{4}\frac{d^2\varphi}{d\tau^2} = -\gamma^2\varphi + J\frac{w}{4\pi}e^{-\tau/\mu_0}. \tag{8.38}$$

This equation has the solution

$$\varphi(\tau) = Ae^{-2\gamma\tau} + Be^{2\gamma\tau} + Ce^{-\tau/\mu_0}, \tag{8.39}$$

where A, B, and C are constants to be determined from the boundary conditions. Then, from (8.37b),

$$\Delta\varphi = \frac{1}{2}\frac{d\varphi}{d\tau} = -\gamma Ae^{-2\gamma\tau} + \gamma Be^{2\gamma\tau} - \frac{C}{2\mu_0}e^{-\tau/\mu_0}. \tag{8.40}$$

Because I_1 and I_2 must remain finite as $\tau \to \infty$, B=0. Substituting (8.39) with $B = 0$ into (8.38) gives

$$-\frac{1}{4}\left(4\gamma^2 Ae^{-2\gamma\tau} + \frac{C}{\mu_0^2}e^{-\tau/\mu_0}\right) = -\gamma^2(Ae^{-2\gamma\tau} + Ce^{-\tau/\mu_0}) + J\frac{w}{4\pi}e^{-\tau/\mu_0}. \tag{8.41}$$

Now, $e^{-2\gamma\tau}$ and $e^{-\tau/\mu_0}$ are independent functions of τ. Hence, the only way (8.41) can be true for all values of τ is if the coefficients of these functions are separately equal. Equating the coefficients of $e^{-\tau/\mu_0}$ on the left and right sides of (8.41) gives

$$C = \frac{J}{4\pi}\frac{4w\mu_0^2}{4\gamma^2\mu_0^2 - 1}. \tag{8.42}$$

Equating the coefficients of $e^{-2\gamma\tau}$ gives an identity, which confirms that (8.39) is the solution of (8.38). Converting back to I_1 and I_2 gives

$$I_1 = \left[A(1-\gamma)e^{-2\gamma\tau} + C\left(1 - \frac{1}{2\mu_0}\right) e^{-\tau/\mu_0} \right], \tag{8.43a}$$

$$I_2 = \left[A(1+\gamma)e^{-2\gamma\tau} + C\left(1 + \frac{1}{2\mu_0}\right) e^{-\tau/\mu_0} \right]. \tag{8.43b}$$

From equation (8.37b), the boundary condition that $I_2(0) = 0$ is equivalent to

$$\varphi(0) = \frac{1}{2}\frac{d\varphi(0)}{d\tau}. \tag{8.44}$$

Using either form of the boundary condition at $\tau = 0$ gives

$$A = -\frac{1+2\mu_0}{2\mu_0(1+\gamma)} C = -\frac{J}{4\pi}\frac{w2\mu_0(1+2\mu_0)}{(1+\gamma)(4\gamma^2\mu_0^2-1)} = -\frac{J}{4\pi}\frac{(1-\gamma)2\mu_0(1+2\mu_0)}{4\gamma^2\mu_0^2-1}. \tag{8.45}$$

Now $\varphi(\tau) = I_1 + I_2$ is the directionally averaged radiance of photons that have been scattered one or more times. Both I_1 and I_2 contain two terms. The second term is proportional to the source term and is important only within a distance from the surface of a few times the extinction length $1/E$, which is of the order of the particle separation. The first term depends on γ/E, which can be much longer than $1/E$ if the particles have high albedos. Also, note that the radiance is independent of azimuth.

The total amount of light incident on a volume element at an optical depth τ *is* $I(\tau) = 4\pi\varphi(\tau) + J\exp(-\tau/\mu_0)$. Since the particles scatter isotropically a fraction $1/4\pi$ of this is scattered per unit solid angle toward the detector, and a fraction $\exp(-\tau/\mu)$ of this escapes from the surface. Hence, adding up the contribution from all layers, the radiance at the detector is

$$I_D = \int_0^\infty \left[w\varphi(\tau) + J\frac{w}{4\pi}e^{-\tau/\mu_0} \right] e^{-\tau/\mu}\frac{d\tau}{\mu}.$$

Substituting for $\varphi(\tau)$ The integration is straightforward and after a little algebra gives

$$I_D(i,e,g) = J\frac{w}{4\pi}\frac{\mu_0}{\mu_0+\mu}\frac{1+2\mu_0}{1+2\gamma\mu_0}\frac{1+2\mu}{1+2\gamma\mu}. \tag{8.46}$$

Dividing the result by J gives the bidirectional reflectance

$$r(i,e,g) = \frac{w}{4\pi}\frac{\mu_0}{\mu_0+\mu}\frac{1+2\mu_0}{1+2\gamma\mu_0}\frac{1+2\mu}{1+2\gamma\mu}. \tag{8.47}$$

8.7.3.2 Solution using the method of invariance

In this section we will find the exact solution to the reflectance of a horizontally stratified medium of isotropically scattering particles using the method of invariance, equations (7.57), with $p(\Omega_0, \Omega) = I$. This equation can be simplified by realizing that all of the radiance scattered within the medium is independent of azimuth. Thus the integration over azimuth in the last three terms on the right can be performed directly, giving factors of 2π. Then r and L are functions only of i and e, or equivalently, μ_0 and μ, and equation (7.57b) can be put into the form

$$L(\mu_0, \mu) = 1 + \frac{w}{2}\mu \int_0^1 \frac{L(\mu_0', \mu)}{\mu_0' + \mu} d\mu_0' + \frac{w}{2}\mu_0 \int_0^1 \frac{L(\mu_0, \mu')}{\mu_0 + \mu'} d\mu'$$

$$+ \frac{w^2}{4}\mu_0\mu \int_{\mu_0''=0}^1 \int_{\mu''=0}^1 \frac{L(\mu_0'', \mu)}{\mu_0'' + \mu} \frac{L(\mu_0, \mu'')}{\mu_0 + \mu''} d\mu_0'' d\mu''.$$

The last term on the right can be factored into two independent integrals, one over μ_0'' and one over μ''. Furthermore, in this term, μ_0'' and μ'' are simply dummy variables of integration and may be replaced by μ_0' and μ', respectively. If this is done the equation can be factored into

$$L(\mu_0, \mu) = \left[1 + \frac{w}{2}\mu \int_0^1 \frac{L(\mu_0', \mu)}{\mu_0' + \mu} d\mu_0'\right]\left[1 + \frac{w}{2}\mu_0 \int_0^1 \frac{L(\mu_0, \mu')}{\mu_0 + \mu'} d\mu'\right].$$

Written in this form it is seen that the function L is symmetric with respect to μ_0 and μ. Furthermore, the first term in brackets is a function only of μ, and the second term is the identical function of μ_0 only. Denote this function by $H(x)$, where x represents either μ or μ_0. Then L may be written $L(\mu_0, \mu) = H(\mu_0)H(\mu)$, where $H(x)$ is the *Ambartsumian–Chandrasekhar H function*, and is the solution of the integral equation

$$H(x) = 1 + \frac{w}{2} x H(x) \int_0^1 \frac{H(x')}{x + x'} dx', \tag{8.48}$$

and x' is a dummy variable of integration.

Using these results in (7.57a) gives

$$r(i, e, g) = \frac{w}{4\pi} \frac{\mu_0}{\mu_0 + \mu} H(\mu_0) H(\mu), \tag{8.49}$$

where $H(x)$ is a function that satisfies (8.48). Comparing (8.49) with (8.47), we see that the two solutions are of identical form, except that in (8.47) $H(x)$ is approximated by $(1 + 2x)/(1 + 2\gamma x)$. Note that (8.49) is an exact, general solution for the reflectance. It makes two assumptions about the medium: that the components scatter light isotropically and independently of each other, and that the optical thickness of the added layer can be made so small that its square and higher powers can be ignored.

8.7.3.3 Properties, analytic approximations, and moments of the H functions

Equation (8.49) is the exact solution for the bidirectional reflectance of a semi-infinite medium of isotropic scatterers. However, the solution is in terms of the nonlinear integral equation (8.48). Values of the H functions for isotropic scatterers have been tabulated in several places (see Chandrasekhar, 1960). They are plotted for several values of w in Figure 8.7 along with the equivalent two-stream approximation.

In this section, some of the properties of the H functions will be described and some useful analytic approximations derived. Note that $H(x) \approx 1$ for all values of x when $w << 1$. In that case, (8.49) reduces to the Lommel–Seeliger law, equations (8.14) and (8.35a) with $f(g) = p(g) = 1$ and $K_{LS} = Jw/4\pi$. This describes the reflectance of a medium in which only single isotropic scattering is important. When $x \to 0$, $H(x) \to 1$, showing that at glancing angles of incidence and emergence only single scattering is important in the reflectance, no matter what the value of w. For $w > 0$, the H functions have a logarithmically infinite slope at $x = 0$, but as x increases, the curve rapidly flattens and becomes almost a straight line whose slope increases monotonically with w.

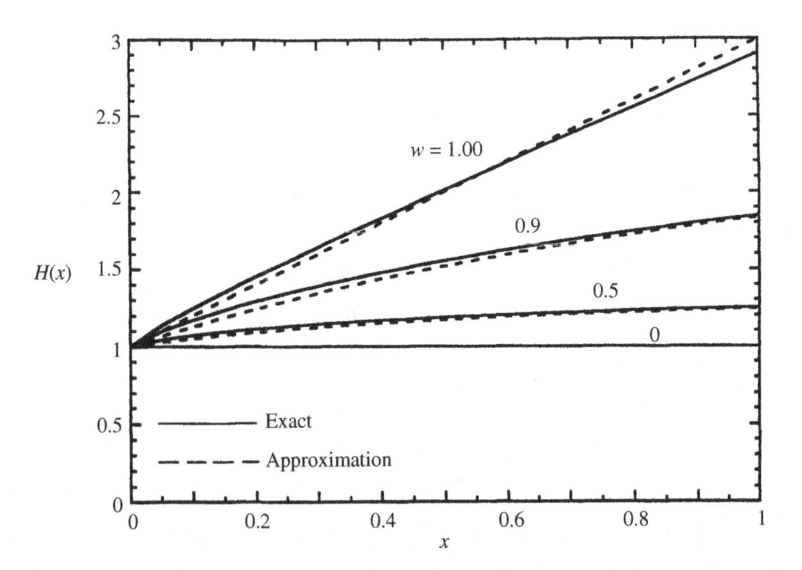

Figure 8.7 The function $H(x)$ vs. x for several values of w. Solid lines, exact solution; dashed lines, approximation (equation [8.53]). The more accurate approximation, equation (8.56), is indistinguishable from the exact solution at the resolution of this figure.

The jth moment of the H function is defined by

$$H_j = \int_0^1 H(x)x^j dx. \tag{8.50}$$

The 0th moment, which is simply the integral of the H function, and also its average value over the interval between 0 and 1, may be found from equation (8.48) as follows:

$$H_0 = \int_0^1 H(x)dx = \int_{x=0}^1 \left[1 + \frac{w}{2}\int_{x'=0}^1 H(x)H(x')\frac{x}{x+x'}dx'\right]dx$$

$$= 1 + \frac{w}{2}\int_{x=0}^1 \int_{x'=0}^1 H(x)H(x')\left(1 - \frac{x'}{x+x'}\right)dxdx'$$

$$= 1 + \frac{w}{2}\left[\int_{x=0}^1 H(x)dx\right].\left[\int_{x'=0}^1 H(x')dx'\right]$$

$$- \int_{x'=0}^1 \left[\frac{w}{2}x'H(x')\int_{x=0}^1 \frac{H(x)}{x'+x}dx\right]dx'$$

$$= 1 + \frac{w}{2}[H_0].[H_0] - [H_0 - 1].$$

Rearranging gives

$$\frac{w}{2}H_0^2 - 2H_0 + 2 = 0.$$

This quadratic equation has the roots $H_0 = 2\left(1 \pm \sqrt{1-w}\right)/w$. The minus sign must be chosen because H_0 is finite as $w \to 0$. Thus, putting $\gamma = \sqrt{1-w}$ gives

$$H_0 = \frac{2}{1+\gamma}. \tag{8.51}$$

Expressing H_0 in terms of the diffusive reflectance $r_0 = (1-\gamma)/(1+\gamma)$ gives

$$H_0 = 1 + r_0. \tag{8.52}$$

The first moment H_1 has been calculated by numerical integration by Chamberlain and Smith (1970).

We will now describe a few useful analytic approximations to the H functions. The first is the two-stream approximation from equation (8.47),

$$H(x) \simeq \frac{1+2x}{1+2\gamma x}. \tag{8.53}$$

This approximation is plotted as the dashed line in Figure 8.7. It differs by less than 4% from the exact values everywhere, and it is better than that in most places.

It can be seen from Figure 8.7 that $H(x)$ is almost linear over most of its range. This suggests that for certain purposes it may be approximated by a linear function of the form $H(x) \simeq A + Bx$, in which the integral over x is required to equal H_0 exactly. This gives the condition $A + B/2 = H_0$. Because $H(0) = 1$, a possible choice might be to take $A = 1$, giving

$$H(x) \simeq 1 + 2r_0 x. \tag{8.54}$$

Because the slope of $H(x)$ is infinite at $x = 0$, another choice is to set the slope dH/dx at $x = 0.5$ to be equal to the slope of the approximation (8.53) at the same x. This gives

$$H(x) \simeq H_0\left[1 + r_0\left(x - \tfrac{1}{2}\right)\right]. \tag{8.55}$$

By themselves, neither (8.54) nor (8.55) are very useful, because (8.53) is nearly as simple and has smaller errors. Its strength lies in the evaluation of certain integrals that involve $H(x)$. For example, an excellent approximation to $H(x)$ can be obtained by writing (8.48) in the form, $H(x) = \{1 - \frac{w}{2}x\int_0^1 \frac{H(x')}{x+x'}dx'\}^{-1}$, and substituting (8.54) for $H(x')$ in the integral. This gives

$$\begin{aligned} H(x) &\simeq \left\{1 - \frac{w}{2}x\int_0^1 \frac{(1+2r_0x')}{x+x'}dx'\right\}^{-1} \\ &= \left\{1 - wx\left[r_0 + \frac{1-2r_0x}{2}\ln\left(\frac{1+x}{x}\right)\right]\right\}^{-1}. \end{aligned} \tag{8.56}$$

This approximation has relative errors smaller than 1% everywhere, which is adequate for most applications.

Equation (8.54) or (8.55) also gives useful approximations for the moments of the H functions. Inserting (8.55) into (8.50) gives

$$H_j \approx \frac{1}{j+1}\frac{2}{1+\gamma}\left[1 + \frac{j}{2(2+j)}r_0\right]. \tag{8.57}$$

In particular,

$$H_1 \approx \frac{1}{1+\gamma}\left[1 + \frac{1}{6}r_0\right]. \tag{8.58}$$

8.7.4 Anisotropic scatterers

8.7.4.1 Exact solutions

Chandrasekhar (1960) has detailed the procedure for finding the exact solution for the bidirectional reflectance of a semi-infinite medium of nonisotropic scatterers and has carried it out for the cases of Rayleigh $[p(g) \propto 1 + \cos^2 g]$ and first-order Legendre polynomial $[p(g) \propto 1 + b_1 \cos g]$ scattering functions. The solutions are expressed in terms of several functions that satisfy nonlinear integral equations analogous to (8.50). These functions have been tabulated for certain values of b_1 and w by Chandrasekhar (1960) and Harris (1957). Other methods are discussed in many references, e.g., Sobolev (1975), Van de Hulst (1980), and Lenoble (1985). Unfortunately these solutions are complicated and inconvenient and must ultimately be evaluated numerically by computer. Thus we seek other methods that are more convenient, but sufficiently accurate for many purposes.

8.7.4.2 Similarity relations

In the two-stream solution for the diffusive reflectance it was possible to reduce the problem for nonisotropic scatterers to an equivalent problem of isotropic scatterers using the similarity relations. Unfortunately, this is not possible in the solution for the bidirectional reflectance, as was emphasized by Sobolev (1975). Although equation (8.37a) with a nonisotropic scattering function can be reduced to an equivalent isotropic form, it is found that an additional term is added to (8.37b) that cannot be removed except by making the asymmetry factor $\beta = 0$.

A common approximation in diffusion theory is to use solutions of the radiative-transfer equation for isotropic scatterers, but replace the scattering coefficient $S = N\sigma Q_S$ by the transport coefficient $S_T = S(1-\xi)$. This is equivalent to treating those photons scattered by an average particle of the medium at scattering angles $\theta <$ invcos ξ as unscattered. Then in the solutions for the reflectance the single-scattering albedo $w = S/(S+A)$ is replaced by

$$w^* = \frac{S(1-\xi)}{S(1-\xi)+A} = \frac{\frac{S}{S+A}(1-\xi)}{\frac{S+A}{S+A} - \frac{S}{S+A}\xi} = \frac{1-\xi}{1-\xi w}w, \tag{8.59a}$$

and the optical depth by

$$d\tau* = -[A+S(1-\xi)]d\tau = -[A+S]\left[1 - \frac{S}{A+S}\xi\right]d\tau = -E(1-\xi w)d\tau. \tag{8.59b}$$

These expressions have the same form as the similarity relations (8.30) derived for the diffusive reflectance, except that β is replaced by ξ.

Van de Hulst (1974) has shown that these similarity relations give excellent results when used in the expression for the bihemispherical or spherical reflectance (Chapter 11). Unfortunately, although the similarity relations are highly satisfactory for calculating integrated reflectances, they are less so when it is necessary to work with angle-resolved reflectances. The reason for this is that departures from isotropic scattering effectively transfer the scattered radiance from one direction into another. These differences are averaged out in the integrated quantities, but have a much larger effect on the angular distribution of reflectance.

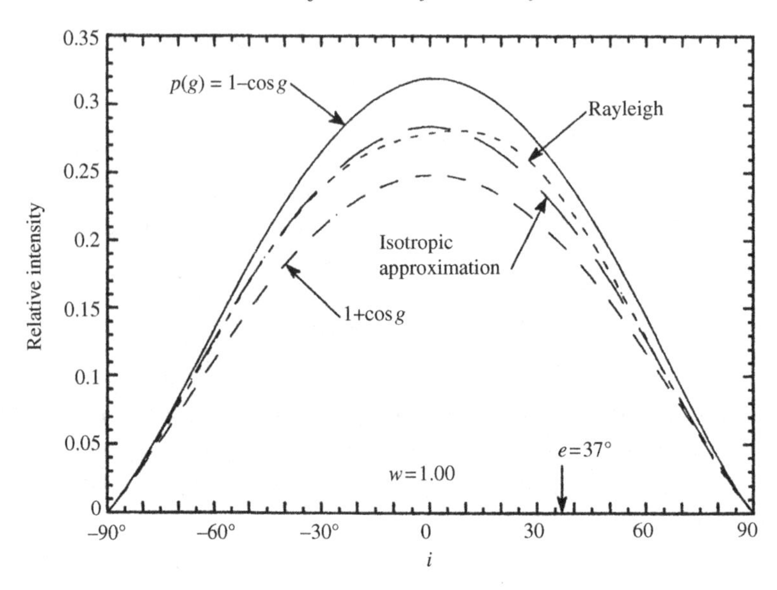

Figure 8.8 Multiply scattered component of the radiance scattered into the principal plane from media with $w = 1.00$ and single-particle scattering functions as shown; e is held constant at 37°, while i varies.

8.7.4.3 The isotropic multiple-scattering approximation (IMSA)

An approximation that is useful if the particle scattering function is not too anisotropic can be obtained by noting that most of the effects of anisotropy are carried by the single-scattering term. As emphasized by Chandrasekhar (1960) and Hansen and Travis (1974), the brighter the surface, the more times the average photon is scattered before emerging from the surface. This tends to randomize the directions of the scattered photons and average out directional effects in the multiply scattered intensity distribution, causing it to be not too different from the distribution produced by isotropically scattering particles.

The dependence of the multiply scattered component of the scattered radiance on $p(g)$ is illustrated in Figure 8.8 for the cases where $w = 1$ and $p(g) = 1$, $1 \pm \cos g$, and $\frac{3}{4}(1 + \cos^2 g)$.

Although the curve for $p(g) = 1$ is somewhat too low when $p(g) = 1 - \cos g$, and high when $p(g) = 1 + \cos g$, all three curves have similar shapes. When $p(g)$ is symmetric, even though it is not isotropic, the multiply scattered component is quite close to that for isotropic scatterers, as illustrated by the curve for Rayleigh scatterers.

The relative insensitivity of the multiply scattered term to $p(g)$ suggests that the solution for isotropic scatterers be used to approximate the multiply scattered contribution to the bidirectional reflectance of a medium of nonisotropic scatterers, while retaining the exact expression for the singly scattered contribution. The exact single-scattering contribution to the radiance at the detector for an arbitrary particle phase function is given by equation (8.35a),

$$I_{SS} = J\frac{w}{4\pi}\frac{\mu_0}{\mu_0+\mu}p(g),$$

and the contribution of multiple scattering from isotropic scatterers is the difference between (8.51) and (8.35a) with $p(g) = 1$,

$$I_{MS} = J\frac{w}{4\pi}\frac{\mu_0}{\mu_0+\mu}[H(\mu_0)H(\mu) - 1].$$

Hence, the bidirectional reflectance may be approximated by

$$r(i, e, g) = \frac{w}{4\pi}\frac{\mu_0}{\mu_0+\mu}[p(g) + H(\mu_0)H(\mu) - 1], \tag{8.60}$$

where $H(x)$ is given by (8.53) or (8.56), depending on the degree of precision required. Equation (8.60) is the isotropic multiple-scattering approximation, or IMSA model. It is widely used in planetary work to analyze the light reflected from surfaces of solar system bodies.

The adequacy of this approximation clearly depends on the single-scattering albedo and the degree of nonisotropy of the scatterers. The approximation would be poor for a medium consisting of large, weakly absorbing, widely separated particles, which have strong refractive and diffractive forward scattering. However, in planetary regoliths and laboratory powders the diffractive term is absent, some absorption is invariably present, and the irregular shapes and presence of internal scatterers cause the particle phase functions to be fairly isotropic. For these materials this approximation should be reasonably accurate. When the medium consists of large, well-separated particles, as in a cloud, the diffraction term in $p(g)$ cannot be ignored, and (8.60) will be seriously in error. In this case, Joseph *et al.* (1976) suggest treating diffraction as a delta function in the radiative-transfer equation.

Lumme and Bowell (1981a) have suggested using a polynomial fit to the exact isotropic solution, equation (8.48), for the multiple-scattering term, except that they replace w by w^*, where w^* is given by the similarity relations (8.59). However, as Sobolev (1975) has emphasized, the similarity relations are reasonably accurate only for hemispherically averaged fluxes. In particular, when $w = 1$, $w^* = w$ independently of ξ; hence, in this case the desired correction to the multiple-scattering term does not happen.

8.7.4.4 The modified IMSA model (MIMSA)

The principal difficulty with the IMSA model is that, although the general shape of a predicted curve of reflectance with angle is correct, the amplitude is slightly high compared with the exact solution if the particle phase function is forward-scattering and low if it is backscattering. The IMSA model may be modified to obtain a more accurate, but still analytic, approximate expression for the bidirectional reflectance of anisotropic scatterers. The modification uses the general invariance equation (7.57) and makes two approximations in order to evaluate the integrals: the particle angular scattering functions are replaced by their averages over the range of integration and taken out from under the integral; and the H functions for isotropic scatterers, equation (8.48), replace the L functions in the integrals. With these approximations the integrals are independent of azimuth and the invariance equation becomes

$$L(\mu_0,\mu) \approx p(\Omega_0,\Omega) + L_1(\mu_0)\frac{w}{2}\mu H(\mu)\int_0^1 \frac{H(\mu_0')}{\mu_0'+\mu}d\mu_0'$$
$$+L_1(\mu)\frac{w}{2}\mu_0 H(\mu_0)\int_0^1 \frac{H(\mu')}{\mu'+\mu_0}d\mu'$$
$$+L_2\left[\frac{w}{2}\mu_0 H(\mu_0)\int_0^1 \frac{H(\mu')}{\mu'+\mu_0}\right]\left[\frac{w}{2}\mu H(\mu)\int_0^1 \frac{H(\mu_0')}{\mu_0'+\mu}\right],$$

Where $L_1(\mu_0)$, $L_1(\mu)$, and L_2 are defined as follows.

The function $L_1(\mu_0)$ is the directionally averaged radiance scattered into the entire lower hemisphere by a particle illuminated from a single direction making an angle i with the vertical,

$$L_1(\mu_0) = \frac{1}{2\pi}\int_{e'=\pi/2}^{\pi}\int_{\psi'=0}^{2\pi} p(g')\sin e' de' d\psi'. \tag{8.61}$$

$L_1(\mu)$ is the radiance scattered by a particle into a single direction in the upper hemisphere that makes an angle e with the vertical when uniformly illuminated from the entire lower hemisphere. Since photons can travel in either direction along a ray path, $L_1(\mu)$ has the same functional dependence on μ as $L_1(\mu_0)$ does on μ_0. L_2 is the average intensity scattered back into the entire lower hemisphere by a particle uniformly illuminated from the entire lower hemisphere,

$$L_2 = \frac{1}{(2\pi)^2}\int_{i'=0}^{\pi/2}\int_{\psi_i'=0}^{2\pi}\int_{e'=0}^{\pi/2}\int_{\psi_e=0}^{2\pi} p(g')\sin e' de' d\psi_e' \sin i' di' d\psi_i'. \tag{8.62}$$

With these approximations, and using equation (8.48), the bidirectional reflectance becomes

$$r(i,e,g) = \frac{w}{4\pi}\frac{\mu_0}{\mu_0+\mu}\{p(g) + L_1(\mu_0)[H(\mu)-1] + L_1(\mu)[H(\mu_0)-1] \tag{8.63}$$
$$+L_2[H(\mu)-1][H(\mu_0)-1]\}.$$

Equation (8.63) is the modified IMSA (MIMSA) model.

In general, the quantities L_1 and L_2 must be evaluated numerically using equations (8.61) and (8.62). In the case where $p(g)$ can be represented as a sum of Legendre polynomials

$$p(g) = 1 + \sum_{n=1}^{\infty} b_n P_n(\cos g), \tag{8.64}$$

they can be found analytically by using a property of Legendre polynomials known as the addition theorem. This theorem states that

$$P_n(\cos g) = P_n(\cos i) P_n(\cos e) + 2 \sum_{m=1}^{\infty} \frac{(n-m)!}{(n+m)!} P_{nm}(\cos i) P_{nm}(\cos e) \cos m\psi, \tag{8.65}$$

where the P_{nm} are the associated Legendre polynomials (see Appendix C). Inserting (8.64) and (8.65) into (8.61), the integral over the azimuth vanishes, giving

$$L_1(\mu_0) = \frac{1}{2\pi} \int_{e'=/2}^{\pi} \int_{\psi'=0}^{2\pi} \left[1 + \sum_{n=1}^{\infty} b_n P_n(\cos i) P_n(\cos e') \right] \sin e' de' d\psi'$$

$$= 1 + \sum_{n=1}^{\infty} b_n P_n(\mu_0) \int_{\mu'=-1}^{0} P_n(\mu') d\mu'.$$

The integral may be evaluated using the recurrence relation

$$P_n(\mu') = \frac{1}{2\pi} \left[\frac{dP_{n+1}(\mu')}{d\mu'} - \frac{dP_{n-1}(\mu') d\mu'}{d\mu'} \right],$$

which gives

$$L_1(\mu_0) = 1 + \sum_{n=1}^{\infty} \frac{b_n P_n(\mu_0)}{2n+1} \{[P_{n+1}(0) - P_{n-1}(0)] - [P_{n+1}(-1) - P_{n-1}(-1)]\}.$$

Now, $P_n(-1) = (-1)^n P_n(+1)$ and $P_n(+1) = 1$; also $P_n(0) = 0$ if n is odd, and

$$P_n(0) = \frac{(-1)^{n/2}}{n+1} \frac{1 \cdot 3 \cdot 5 - - - (n+1)}{2 \cdot 4 \cdot 6 - - - n}$$

if n is even. Using these values $L_1(\mu_0)$ can be found; L_2 can be evaluated in a similar manner. After a little algebra, the final results are

$$L_1(\mu_0) = 1 + \sum_{n=1}^{\infty} A_n b_n P_n(\mu_0), \tag{8.66a}$$

$$L_1(\mu) = 1 + \sum_{n=1}^{\infty} A_n b_n P_n(\mu), \tag{8.66b}$$

$$L_2 = 1 + \sum_{n=1}^{\infty} A_n^2 b_n, \tag{8.66c}$$

Table 8.2. *Legendre scattering function coefficients*

n	A_n	A_n^2
1	−0.5000	0.2500
3	0.1250	0.0156
5	−0.0625	0.00391
7	0.0391	0.00153
9	−0.0273	0.000748
11	0.0205	0.000421
13	−0.0161	0.000259
15	0.0131	0.000172

Note: For $n =$ even, $A_n = A_n^2 = 0$.

where $A_n = 0$ if $n =$ even, and

$$A_n = \frac{(-1)^{(n+1)/2}}{n} \frac{1 \cdot 3 \cdot 5 - - - n}{2 \cdot 4 \cdot 6 - - - (n+1)} \tag{8.66d}$$

if $n =$ odd. The coefficients A_n and A_n^2 are listed to 15th order in Table 8.2. (Note: there was a typographical error in equation [8.66c] published in the original paper [Hapke, 2001].)

The procedure for calculating the bidirectional reflectance using the MIMSA is as follows. The coefficients b_n must be known. If they are not known they can be calculated from $p(g)$ using the procedure outlined in Appendix C. The b_ns are then inserted into (8.66) to find both L_1s and L_2. These are then inserted into (8.63) to give $r(i, e, g)$.

8.8 Comparison of the IMSA model with measurements

Because the isotropic multiple-scattering approximation is analytic and relatively mathematically simple it has been widely used in remote-sensing work to analyze and characterize photometric and spectroscopic observations. Hence, its ability to correctly predict and/or describe measured data has been tested extensively. Such tests are of two types: forward and reverse modeling. In forward modeling the reflectance of a material of known properties is calculated from the model and compared with the measured reflectance. It is a test of the physical correctness and absolute accuracy of the model. In reverse modeling the reflectance calculated by the model is fitted to measured reflectance by adjusting the parameters of the model until the differences between the reflectances are minimized. If the parameters are known independently comparison of the retrieved and known values tests the absolute validity of the model. If a model has been sufficiently tested parameters retrieved in this way allow values of microphysical properties of a surface to be inferred and thus, the model is useful in remote sensing. However, a word of caution concerning the accuracy of such inferences is in order. The reflectance equation contains several parameters that may have opposite effects on it, so that often unique values of all parameters are difficult to obtain. Parameter retrieval is discussed in more detail in Chapter 14.

The most stringent forward-modeling test was by Hapke *et al.* (2009) who compared the predicted and measured angular variation of reflectance of a medium of anisotropically scattering particles. The material was a white powder consisting of spheres of soda–lime ($SiO_2 + Na_2O + CaO$) glass. The spheres ranged in size from 1 to 56 μm with a distribution $N(r) \propto r^{-3}$, where r is the radius, and mean equivalent particle diameter $D = 5.1\,\mu m$. The filling factor was $\Phi = 43\%$, and the refractive index was $n = 1.51 + i3.5 \times 10^{-5}$. The bidirectional reflectance was measured in the principal plane at $i = 0$ with e varying between $0 < e = g < 80°$, and $i = 60°$ with e varying between $0 < e < 80°$ on both sides of the normal, so that $-20° < g < 140°$.

The volume-averaged extinction, scattering and absorption efficiencies, single-scattering albeo, and particle phase function were calculated as if the spheres of the medium were widely separated using Mie theory. The diffraction peak was then removed from $p(g)$, a diffraction efficiency $Q_D = 1$ subtracted from Q_S, and w recalculated. The results are shown in Figure 8.9.

These quantities, with and without diffraction, were inserted into the IMSA equation (8.60) and the bidirectional reflectance calculated. The results are shown in Figure 8.10.

The agreement between prediction and measurement is quite satisfactory when Fraunhofer diffraction is removed (solid lines), but is rather poor when the diffraction peak is retained (dashed lines). This reinforces the arguments in Chapter 7 that Fraunhofer diffraction does not exist in a closely packed medium. It must be emphasized that no parameters were adjusted in the model calculations. The largest discrepancy between theory and measurement occurs in the $i = 0$ curve at large values of g. However, the agreement there would be improved if $p(g)$ is larger at mid-phase angles and smaller at large and small phase angles from that shown by the solid line in Figure 8.9, that is, if the particles are more isotropic than predicted by Mie theory, even after diffraction is removed. We have seen in Chapter 6 that near-field and coherent interactions between adjacent particles could cause such changes in the effective particle scattering functions.

Note that spherical particles are among the most highly anisotropic scatterers likely to be found in nature. Yet the IMSA model does a credible job of matching the measured reflectance. Most particles to be found in laboratory powders and planetary regoliths are probably irregular and filled with internal scatterers, so that

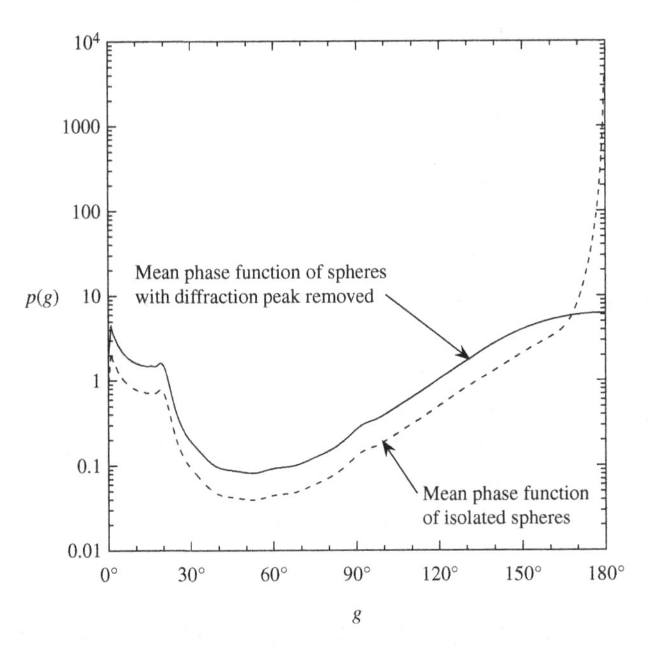

Figure 8.9 Mean single-scattering phase function vs. phase angle for a distribtion of spherical particles with the Fraunhofer diffraction peak removed (solid line) and retained (dashed line). See text for details of the properties of the spheres.

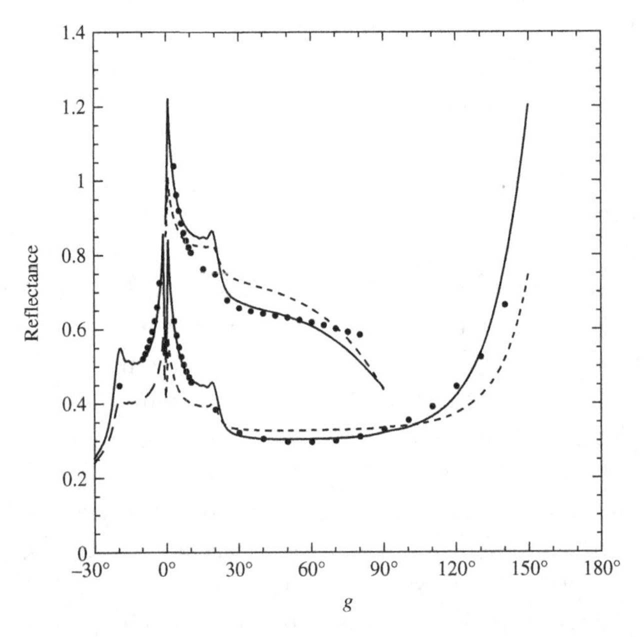

Figure 8.10 Comparison of the measured reflectance (dots) of a powder of spheres with the particle scattering function shown in Figure 8.9 with the reflectance calculated using the IMSA model with the diffraction peak removed (solid line) and retained (dashed line).

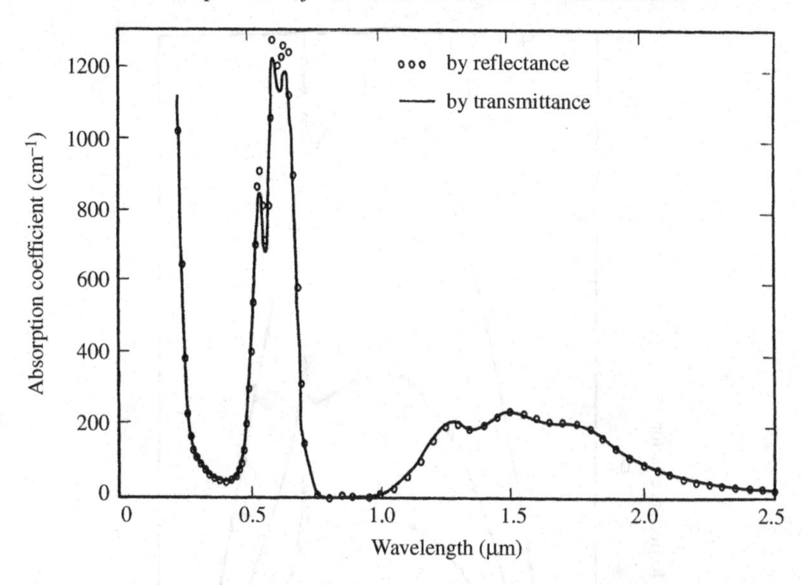

Figure 8.11 Spectral absorption coefficient of cobalt-doped silicate glass. The line shows the spectrum measured by transmission of thin sections; the circles show the spectrum calculated from the bidirectional reflectance using the espat function. (Reproduced from Hapke *et al.* [1981], copyright 1981 by the American Gephysical Union.)

their phase functions are much more isotropic than the spheres of this experiment. Hence, the IMSA model should be well able to describe such materials.

An important test of the reverse modeling type used the synthetic silicate glass containing Co^{2+} described in Chapter 6. This material was chosen because its absorption coefficient varied over a wide range of values in the visible portion of the spectrum. Part of the glass was sliced and polished into a thin section. Its spectral absorption coefficient $\alpha(\lambda)$ was measured by transmission over the wavelength range from 0.25 to 2.5 μm. The spectrum is shown as the solid line in Figure 8.11.

The remainder of the glass was ground into a powder and its bidirectional reflectance measured in the principal plane at $i = 30°$, $e = 30°$, $g = 60°$ over the same wavelengths. The particle phase function was assumed to be isotropic, $p(g) = 1$. In that case the only adjustable parameter in equation (8.47) is the single-scattering albedo, so that the measured valued of the reflectance could be solved for $w(\lambda)$ at each wavelength. Solving approximate equation (6.40) for the absorption coefficient gives $\alpha(\lambda) = [1 - w(\lambda)]/w(\lambda)D_e$, where D_e is the effective particle size. An empirical value for the effective particle size D_e was found from the value of a measured by transmission in blue wavelengths, and the absorption coefficient

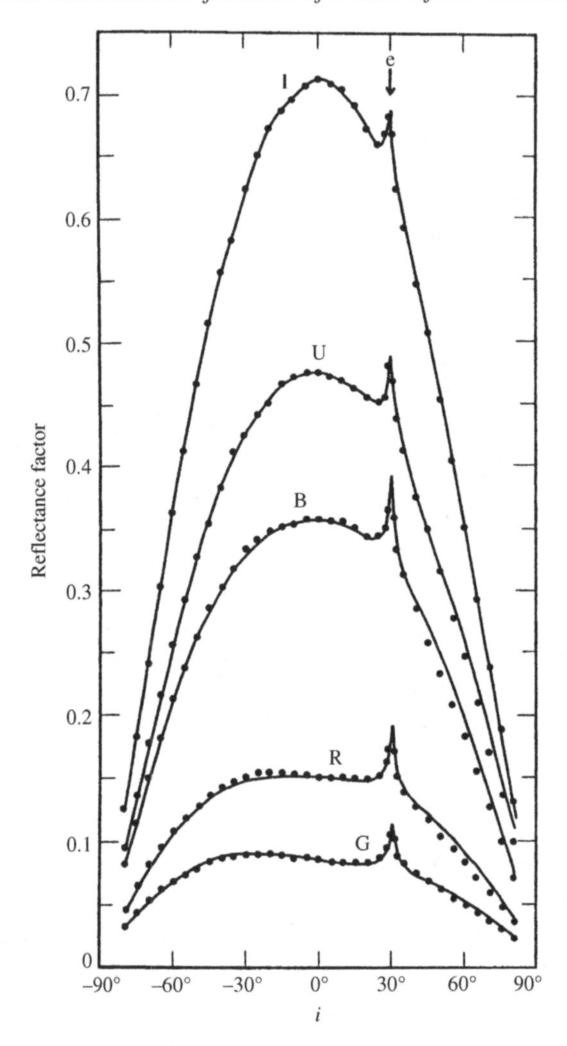

Figure 8.12 Bidirectional reflectance vs. i for a powder of size $<37\ \mu m$ made from the cobalt glass whose absorption spectrum is shown in Figure 8.11. The detector views the powder at $e = 30°$, while i is varied in the principal plane. The dots show the reflectance measured at five wavelengths in the ultraviolet (U), blue (B), green (G), red (R), and infrared (I). The lines show the IMSA model, equation (8.47), fitted to the data. The peaks at $i = 40°$ are the opposition effect, which is treated in Chapter 9. (Reproduced from Hapke and Wells [1981], copyright 1981 by the American Geophysical Union.)

was then calculated from $w(\lambda)$ for the other wavelengths. The values of $a(\lambda)$ found in this way are shown as the circles in Figure 8.11. The agreement between the two spectra is excellent, except at the highest values, where the approximate equation (6.40) for w becomes inaccurate.

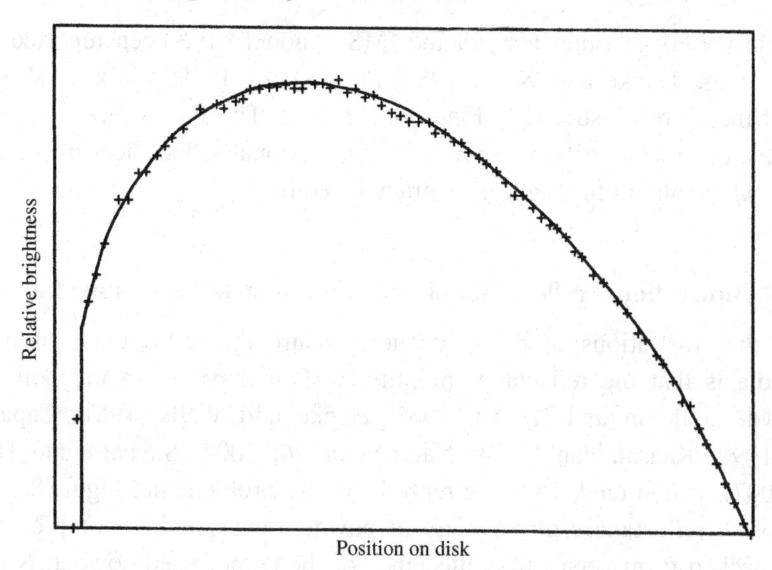

Figure 8.13 Relative brightness profile along the equator of Venus as a function of longitude. The crosses show data measured by the *Mariner 10* spacecraft. The line shows equation (8.47). (Reproduced from Hapke and Wells [1981], copyright 1981 by the American Geophysical Union.)

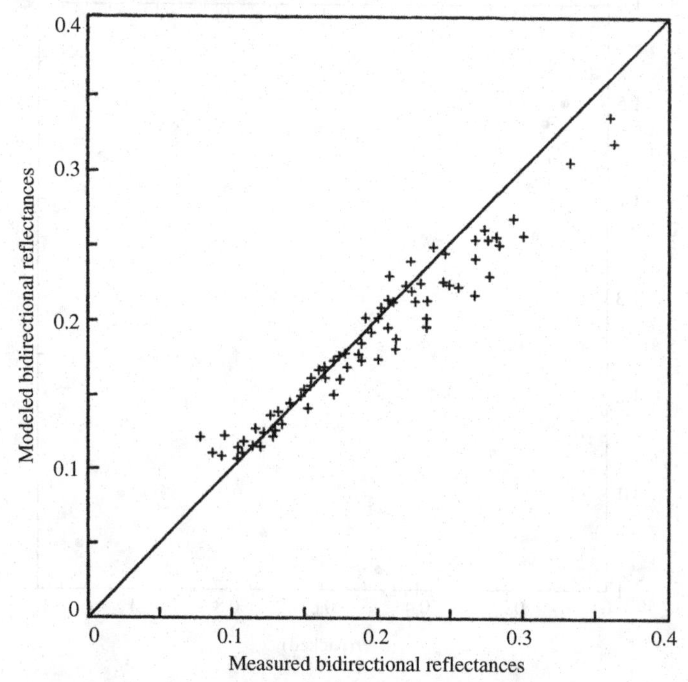

Figure 8.14 Comparison (crosses) between the reflectances in visible light of the bare soil in a plowed field predicted by the theory and measured in various geometries. If the predicted and measured values agreed exactly, the crosses would fall on the straight line. (Reproduced from Pinty *et al.* [1989], copyright 1989 with permission of Elsevier.)

A wide variety of other tests of the IMSA model have been reported in the literature (e.g., Hapke and Wells, 1981; Pinty *et al.*, 1989; Clark *et al.*, 1993). Some of these are illustrated in Figures 8.12–8.14. The IMSA model appears to be capable of satisfactorily describing the bidirectional reflectances of particulate material where all but the highest accuracy is required.

8.9 Bidirectional reflectance of a medium of arbitrary filling factor

One of the frustrations of persons who measure the reflectances of particulate media is that the reflectance amplitude often depends on the porosity of the powder (Blevin and Brown, 1967; Hapke and Wells, 1981; Capaccioni *et al.*, 1990; Kaasalainan, 2003; Naranen *et al.*, 2004; Shepard and Helfenstein, 2007), which tends to make reproducibility problematic. Figure 8.15 plots the ratios of reflectances of a variety of powders measured in compressed and loosely packed form versus the reflectance of the compressed powder. Note that compression increases the reflectance, but that the amount of increase decreases as

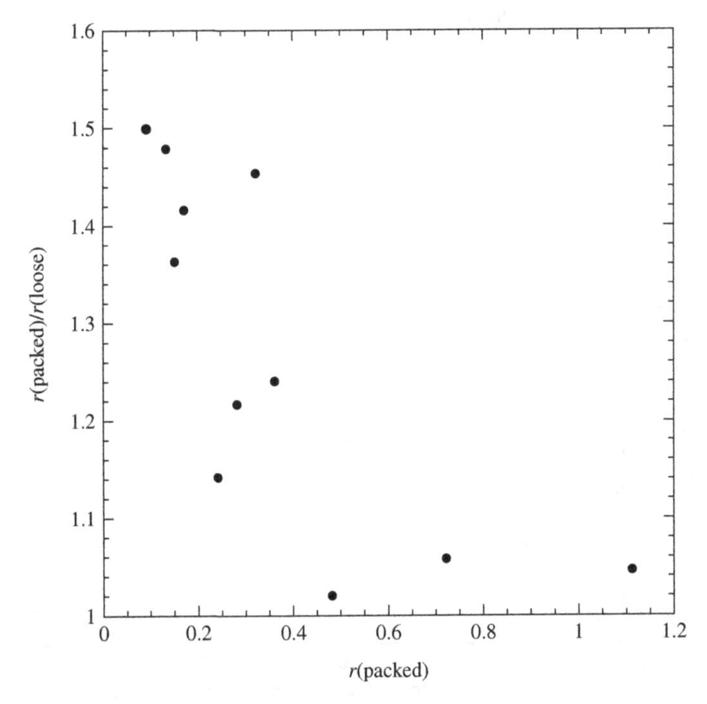

Figure 8.15 The ratios of the reflectances of a variety of powders in packed form to those of the same powders in loose packing plotted against the reflectances of the packed powders. (Reproduced from Hapke [2008], copyright 2008 with permission of Elsevier.)

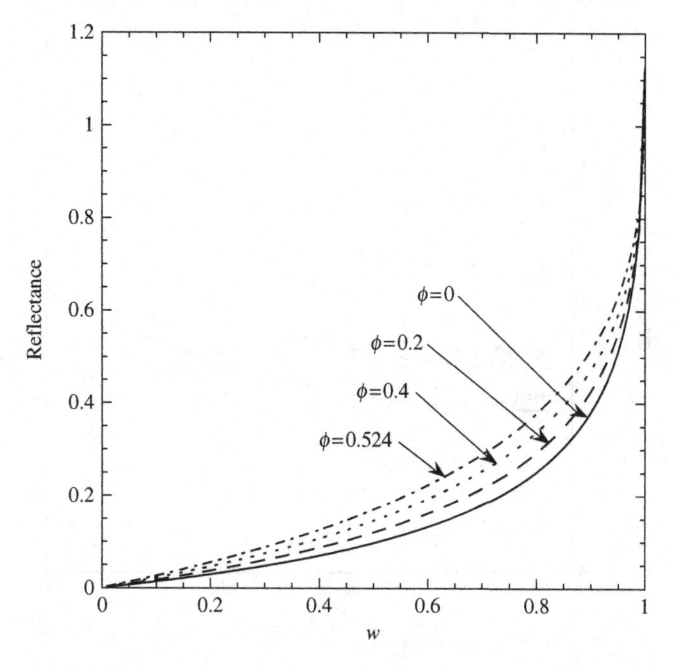

Figure 8.16 Reflectance (relative to a perfectly diffuse surface) of a medium of isotropic scatterers illuminated and viewed normally vs. single-scattering albedo w for several different values of the filling factor ϕ.(Reproduced from Hapke [2008], copyright 2008 with permission of Elsevier.)

the reflectance increases. However, models based on the equation of radiative transfer in its usual sparse packing form, equation (7.38), are independent of porosity or filling factor.

The porosity dependence arises because, as discussed in Section 7.4.3, the simple transmissivity function for well-separated particles, equation (7.25) is incorrect and equation (7.46) must be used instead in any part of the derivation involving the transmissivity. This means that in the derivation of the reflectance from the radiative transfer equation $\exp(-\tau/\mu)$ and $\exp(-\tau/\mu_0)$ must be replaced everywhere by $K\exp(-K\tau/\mu)$ and $K\exp(-K\tau/\mu_0)$, respectively. The procedure for finding the reflectance is exactly the same as in Section 8.7.3.1. The result is

$$r(i,e,g) = K\frac{w}{4\pi}\frac{\mu_0}{\mu_0+\mu}[p(g)+H(\mu_0/K)H(\mu/K)-1], \tag{8.70a}$$

where

$$H(x/K) = \frac{1+2x/K}{1+2\gamma x/K}, \tag{8.70b}$$

and $K = -\ln(1-1.209\phi^{2/3})/1.209\phi^{2/3}$ for media of equant particles.

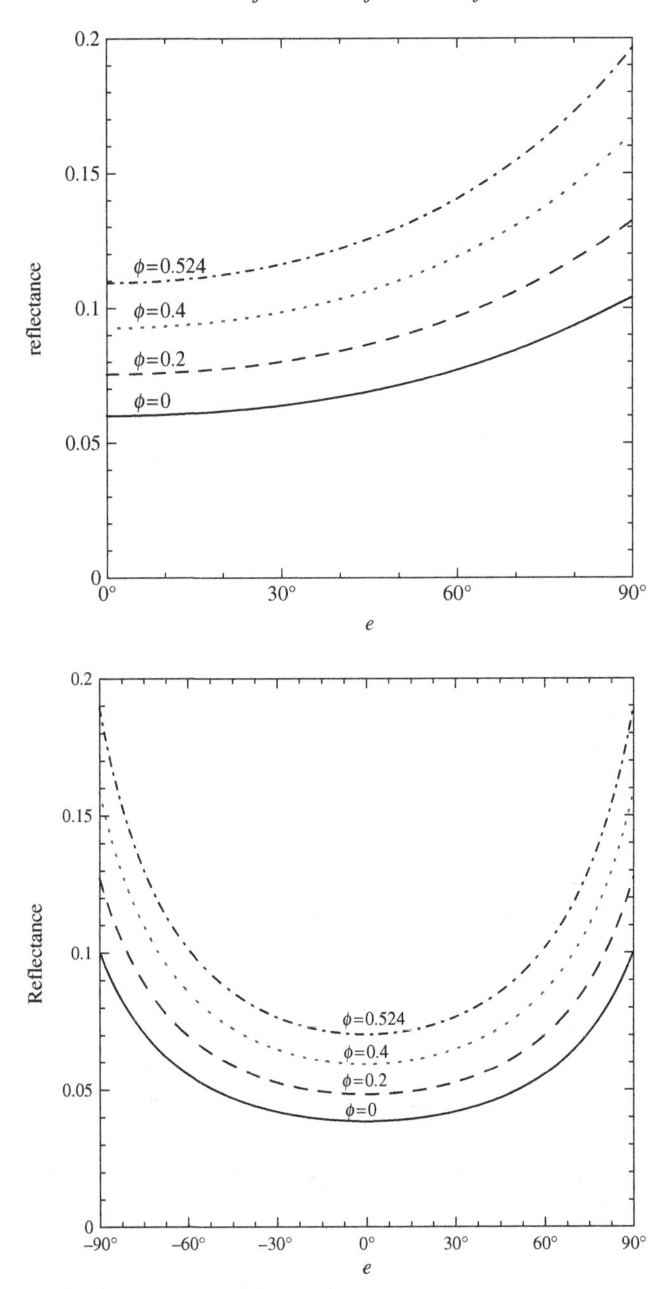

Figure 8.17 Reflectance (relative to a perfectly diffuse surface) of a medium of isotropic scatterers and $w = 0.36$ vs. viewing angle e for several different values of ϕ; (a) $i = 0$; (b) $i = 60°$. (Reproduced from Hapke [2008], copyright 2008 with permission of Elsevier.)

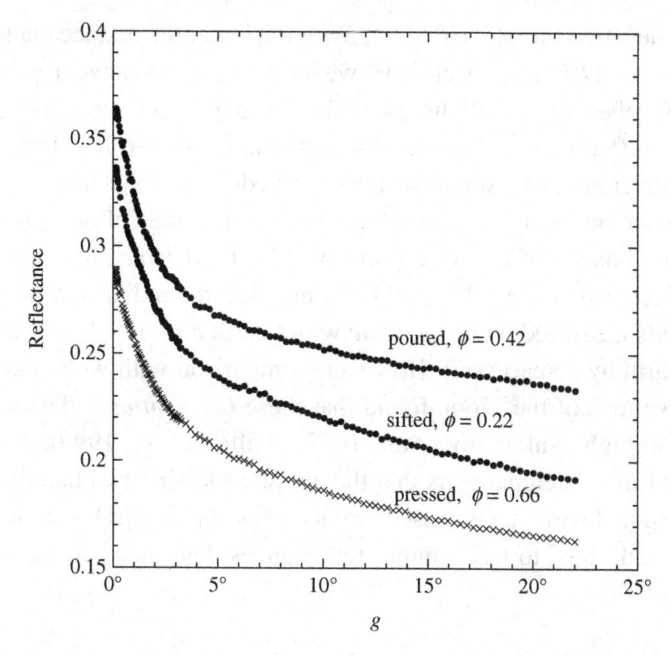

Figure 8.18 Measured reflectance of SiC powder of particles approximately 17 μm in size illuminated at $i = 0$ vs. phase angle g. All reflectances are relative to a perfectly diffuse surface.

With increasing filling factor K increases but the H functions decrease. The net result is that at low albedos r increases as ϕ increases, but the amount of increase becomes smaller as the single-scattering albedo increases, consistent with Figure 8.15, until at the highest values of w, r decreases at certain angles. Equation (8.70) of r vs. w with $p(g) = 1$ is plotted in Figure 8.16 for several values of the filling factor. The figure shows that, depending on w, compression can increase the reflectance by as much as a factor of 2 over the sparse packing values.

Figures 8.17 plots the reflectance as a function of emission angle. Note that the shapes of the curves are virtually independent of ϕ. Thus a difference in porosity can easily be misinterpreted as a difference in single-scattering albedo, as emphasized by Shepard and Helfenstein (2007). This makes it difficult to retrieve a unique value of w by reverse modeling of an observational data set.

As ϕ increases, equation (8.70) is valid only up to the critical point $\phi < 52\%$ where coherent effects become important. This avoids the difficulty that this equation predicts that the reflectance would become very large when $\phi \to 75\%$. However, this means that equation (8.70) is not applicable to solid rocks or extremely compressed powders. Figure 8.18 shows the reflectances of the same sample of SiC powder with grain sizes about 17 μm measured in three conditions of packing. The sifted sample with $\phi = 22\%$ has a lower reflectance than the poured powder with $\phi = 42\%$, as expected. However, the pressed powder with $\phi = 66\%$ has the lowest reflectance of all three samples. Evidently at large densities coherent and collective effects cause the individual grains to behave like larger particles, resulting in lower effective single-scattering albedos and reflectances.

A possible example of porosity effects occurred in the calibration of absolute reflectances of images of the Moon observed by the *Clementine* spacecraft. Initially the reflectances were calibrated by comparing the radiance of an area in the lunar highlands measured by *Clementine* with that of a sample from the same area brought to Earth by a spacecraft. However, comparison with well-calibrated telescopic observations of the Moon found that these *Clementine* reflectances were a factor of 2 too high (Shkuratov *et al.*, 1999a; Hillier *et al.*, 1999). The probable explanation of this discrepancy is that the act of collecting and handing the lunar samples, combined with the increased gravity, caused the samples in the laboratory to be denser and, thus, to have higher reflectances than their natural state on the lunar surface.

9
The opposition effect

9.1 Introduction

The *opposition effect* is a sharp surge observed in the reflected brightness of a particulate medium around zero phase angle. Its name derives from the fact that the phase angle is zero for solar-system objects at astronomical opposition when the Sun, the Earth, and the object are aligned. Depending on the material the angular width of the peak can range from about 1° to more than 20°. It has many names including the *heiligenschein* (literally *"holy glow"*), *hot spot*, *bright shadow* and *backscatter peak*. We have already encountered the opposition effect peak in Figures 8.12 and 8.18. and it is further illustrated in Figures 9.1–9.3. It should not be confused with the glory in the phase function of a sphere (Chapter 5), which is also often called the heiligenschein.

The opposition effect is a nearly ubiquitous property of particulate media, including vegetation (Hapke *et al.*, 1996), laboratory powders (Hapke and Van Horn, 1963; Oetking, 1966; Egan and Hilgeman, 1976; Montgomery and Kohl, 1980; Nelson *et al.*, 1998, 2000), and regoliths of the Moon (Gehrels *et al.*, 1964; Whitaker, 1969; Wildey, 1978; Buratti *et al.*, 1996), Mars (Thorpe, 1978), asteroids (Gehrels *et al.*, 1964; Bowell and Lumme, 1979; Belskaya and Shevchenko, 2000), satellites of the outer planets (Brown and Cruikshank, 1983; Domingue *et al.*, 1991), and the rings of Saturn (French *et al.*, 2007). On a clear day you can see it as a glow around the shadow of your head when your shadow falls on grass or soil. It is readily observed from an airplane as a bright glow around the shadow of the plane when the shadow falls on vegetation or soil. The halo disappears on pavement. It is particularly pronounced in powders with grains a few micrometers in size.

The first historical record of the opposition effect is in the autobiography of the sixteenth-century Florentine artist, sculptor, and rogue, Benvenuto Cellini. Cellini noted that he often saw a glow around the shadow of his head on grass. Since he did not see it around the shadow of anyone else, he took this as evidence that he was

Figure 9.1 The opposition effect on the surface of the Moon can be seen as the glow around the shadow of the head of the astronaut. (Courtesy of the National Aeronautics and Space Administration.)

Figure 9.2 The opposition effect on the surface of Mars is seen around the shadow of the camera on the Mars rover *Spirit*. (Courtesy of the National Aeronautics and Space Administration.)

Figure 9.3 The opposition effect in the rings of Saturn as imaged by the visual and infrared imaging spectrometer on the *Cassini* spacecraft. (Courtesy of the National Aeronautics and Space Administration.)

especially favored by God. He noted that it was more pronounced when the grass was covered with morning dew, so obviously the glory partly contributed to Cellini's observation. However, the glow persisted even after the dew had evaporated, so the opposition effect also contributed. Cellini would doubtless have been delighted to know that 500 years later the phenomenon is still sometimes referred to as "Cellini's halo."

The first modern discovery of the opposition effect was by Seeliger (1887, 1895) in the light scattered by Saturn's rings. It was later independently discovered by Gehrels on asteroids and the Moon (Gehrels *et al.*, 1964).

Several mechanisms have been suggested to explain the opposition effect observed on solar system bodies, including shadow-hiding, coherent backscatter, glories from glass beads, and crystalline corner reflectors. The last two hypotheses can be readily eliminated. The ubiquitous nature of the phenomenon in the solar system, and the fact that glass beads are not required to cause a backscatter peak in laboratory powders, makes the third explanation superfluous. In principle, the faces of crystals with a cubic structure could act as corner reflectors, which consist of three planes set at right angles to one another and have the property that a ray incident on one of the planes is redirected by multiple specular reflections back toward the source (Trowbridge, 1978; Muinonen *et al.*, 1989). However, the surfaces of all of the airless bodies on which the effect is observed are continuously abraded by micrometeorites, which would destroy the smooth surfaces necessary for retroreflection. Furthermore, the effect is observed on laboratory powders made of irregular particles. Thus, while corner reflectors may contribute to the opposition effect in certain cases, they are not the major cause. Hence we are left with shadow-hiding and coherent backscatter, and there are considerable experimental and observational indications that both processes are operating. Both will be discussed in detail in this chapter. We will use the convenient acronyms SHOE for the shadow-hiding opposition effect and CBOE for the coherent backscatter opposition effect.

9.2 The shadow-hiding opposition effect (SHOE)

9.2.1 Physical principles

The shadow-hiding backscatter surge occurs in any particulate medium in which the grains are larger than the wavelength so that they have shadows. Particles near the surface cast shadows on the deeper grains. These shadows are visible at large phase angles, but close to zero phase they are hidden by the objects that cast them.

Another way of understanding the phenomenon is to think of the interstices between the particles that make up a powder as resembling tunnels through which light penetrates. At large phase angles the interiors of these tunnels visible to the detector are in shadow, the light being blocked by the particles that make up the walls of the tunnels. However, when the phase angle is small, the sides and bottoms of the tunnels are illuminated, resulting in enhanced brightness.

The opposition effect is particularly pronounced in fine powders with a mean grain size less than about 20 μm (Figure 9.4). The shadow-hiding hypothesis accounts for this by noting that for particles as fine as this the intermolecular adhesive (electrostatic and van der Waals) forces that act between the contacting surfaces of two adjacent grains exceed the gravitational forces attracting them to the planet. Consequently, only one point of contact is necessary to stably support the weight of a small particle, instead of the minimum of three for a large particle. Because of this, the microstructures formed by fine cohesive powders can be very open, porous, and intricate, consisting of lacy towers and bridges that Hapke and Van Horn (1963) dubbed "fairy castle structures." Because the Moon has a strong opposition effect these authors argued, well before the *Surveyor* and *Apollo* missions, that the upper layers of the lunar regolith are fine-grained and have a high porosity.

The equation of radiative transfer does not account for the SHOE, so that it must be added ad hoc. This was first done by Seeliger (1887), and it has been treated theoretically by several persons in addition to Seeliger, including Bobrov (1962),

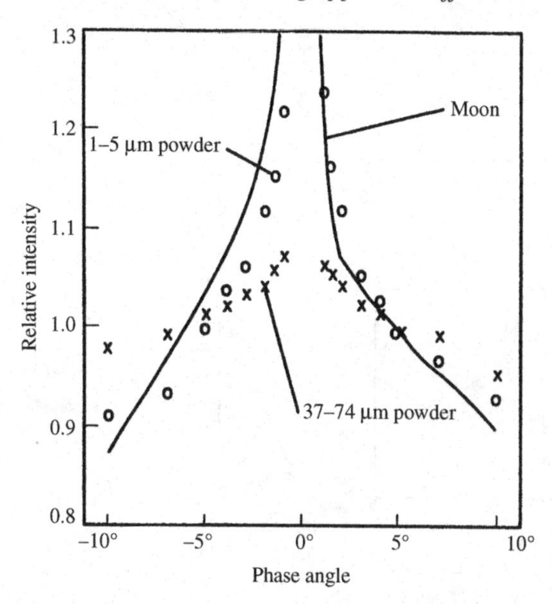

Figure 9.4 Relative brightnesses of an area on the lunar surface (the crater Copernicus, from data of Van Diggelen [1965]) and pulverized basalt of two different sizes, illustrating the opposition effect. (Reproduced from Hapke [1968], copyright 1968 with permission of Pergamon Press, Ltd.)

Hapke (1963,1986), Irvine (1966), Lumme and Bowell (1981a), and Shkuratov *et al.* (1999a). Some of the models require a computer to obtain a numerical answer; others give approximate analytic equations that are valid only when the porosity is large. In this section the general treatment of Hapke (1986) will be followed, because this gives a convenient, approximate analytic expression whose parameters can be given a straightforward interpretation in terms of physical properties of the medium. The result will be seen to be a modification to the singly scattered, Lommel–Seeliger part of the bidirectional reflectance.

An order-of-magnitude estimate of the expected half-width of the shadow-hiding peak can be made as follows. The average distance a ray travels in a particulate medium before encountering a particle and being either absorbed or scattered is the extinction length, Λ_E, so that this will also be the mean length of a shadow cast by a particle in the medium. Thus, the system may be crudely idealized as a spherical particle of radius a casting its shadow on a screen a distance Λ_E away (Figure 9.5). When the phase angle between the source and detector is zero the particle hides its own shadow. As the phase angle increases the detector sees some of the shadow. Half of the shadow will be visible when the ray from the center of the shadow to the detector just grazes the edge of the particle, so that the half-width of the peak is HWHM $\sim a/\Lambda_E$.

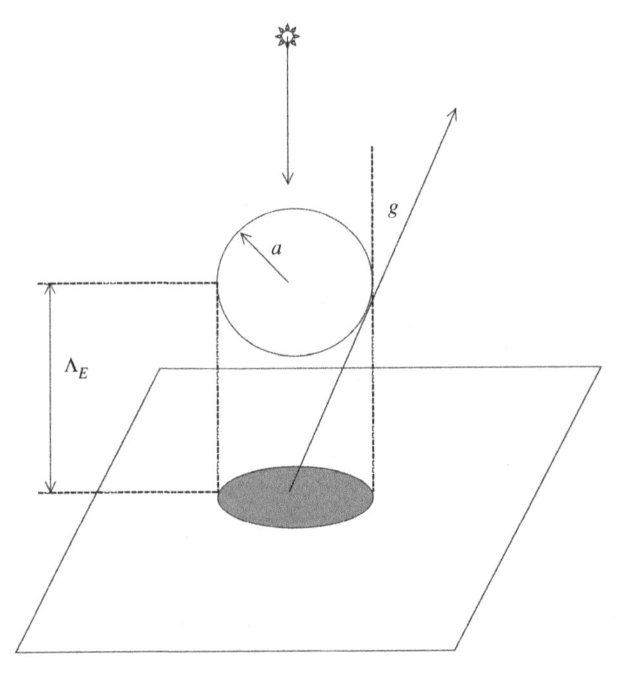

Figure 9.5 Simple model of the shadow-hiding backscatter effect.

9.2.2 Derivation

To make a more rigorous model, we begin by considering light that has been scattered only once. In Section 8.7 the following expression for the singly scattered radiance was derived:

$$I_{SS}(i, e, g) = \frac{J}{4\pi}\frac{1}{\mu}\int_0^\infty w(\tau)p(\tau, g)T_i(\tau, \mu_0)T_e(\tau, \mu)d\tau, \tag{9.1}$$

where

$$T_i(\tau, \mu_0) = Ke^{-K\tau/\mu_0} \tag{9.2a}$$

is the incident transmissivity, the probability that the incident irradiance J will penetrate to a level $\tau = \int_z^\infty E(z')dz'$ in the medium. The transmissivity has been generalized by the addition of the porosity factor K, in accordance with the discussion in Sections 7.4 and 8.9. Similarly, in Section 8.7.2 it was assumed that

$$T_e(\tau, \mu) = Ke^{-K\tau/\mu} \tag{9.2b}$$

is the exit transmissivity, the probability that the light scattered in the direction of the detector by a particle at this level will escape the medium. In this case (9.1) can be easily integrated to give the generalized Lommel–Seeliger law,

$$I_{SS}(i, e, g) = JK\frac{w}{4\pi}\frac{\mu_0}{\mu_0 + \mu}p(g).$$

Note that at zero phase,

$$I_{SS}(i, e=i, 0) = JK\frac{w}{8\pi}p(0). \tag{9.3}$$

Let

$$\sigma_E = <\sigma Q_E> = E(z)/N(z) \tag{9.4}$$

be the volume-average extinction cross section, and

$$a_E(z) = \sqrt{\sigma_E(z)/\pi} \tag{9.5}$$

be the mean extinction radius; that is, a_E is the radius of an equivalent sphere having the cross-sectional area σ_E. Note that τ is equal to the total average number of particles having their centers in a vertical cylinder of radius a_E lying above altitude z. It will be assumed that σ_E, w, and $p(g)$ are independent of τ.

Expressions (9.2) assume that the incident probability T_i and exit probability T_e are independent of each other. However, if the phase angle is small, the probabilities are not independent, and (9.3) underestimates T_e. To understand this, suppose a ray of light from the source is scattered through phase angle g at some point **P** located at altitude z in the medium. Now, T_i is the probability that no particle has its center in a cylinder of radius a_E whose axis is the ray connecting the source and **P**. If T_e were independent of T_i, then, similarly, T_e would be the probability that no particle has its center in the cylinder with radius a_E coaxial with the ray connecting **P** with the detector. However, portions of the two probability cylinders overlap, as is illustrated schematically in Figure 9.6, and the probability of extinction by any particles that are in the common volume has been counted twice. Let V_c be the common volume. In Figure 9.6 the area **APBC** is the cross section of V_c in the scattering plane containing **P**. Correcting for this effect, the escape probability can be written

$$T_i T_e = K^2 \exp[-K(\tau/\mu_0 + \tau/\mu - \tau_c)], \tag{9.6a}$$

where

$$\tau_c = \int_{V_c} n_e(z) dV, \tag{9.6b}$$

and V denotes volume.

The overlap can be calculated exactly at zero phase. Then the direction to the detector coincides with the direction to the source, and the two cylinders overlap perfectly, so that $\tau_c = \tau/\mu_0$, and

$$I_{SS}(i, e=i, 0) = -J\frac{w}{4\pi}p(0)\frac{1}{\mu}\int_0^\infty K^2 e^{-K\tau/\mu} d\tau = JK\frac{w}{4\pi}p(0), \tag{9.7}$$

which is exactly twice (9.3). In effect, the incident ray has preselected a preferential escape path for rays leaving the medium at small phase angles. That is, if a ray is

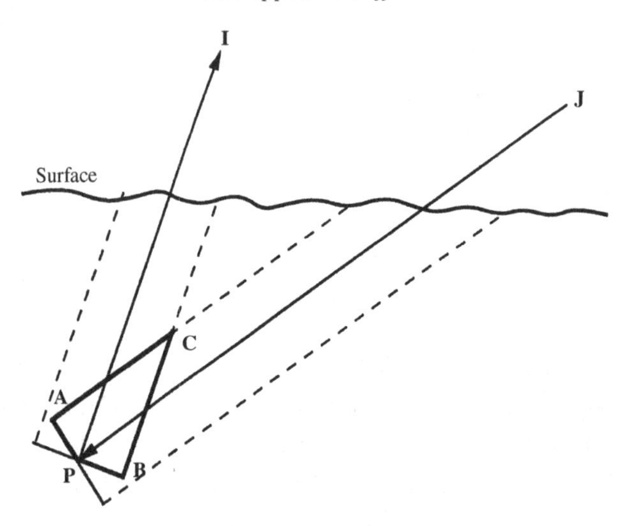

Figure 9.6 Cross section through the scattering plane containing point **P**. The polygon **APBC** is the cross section in the scattering plane of the common volume V_c.

able to penetrate to a given level in the medium and illuminate a point there without being extinguished, then rays scattered from that same point exactly back toward the source are able to escape without being blocked by any particle.

It is sometimes erroneously stated that the opposition effect requires opaque particles. However, the blocking is by extinction, not absorption, and occurs whether the particles are transparent or opaque. It is important to note the clear difference between the scattering properties of a continuous medium and a particulate medium. In a continuous medium $a_E = 0$ so that exponential attenuation occurs along the entire incident and exit paths of the rays, and a shadow-hiding opposition effect cannot arise. The common volume is the intersection of two circular cylinders, of which **APBC** is the cross section in the scattering plane containing **P**, and its calculation is mathematically cumbersome (e.g., Irvine, 1966; Goguen, 1981). An approximate value for V_c may be derived as follows.

Let $\sigma_E = \sigma Q_E$, the mean particle extinction cross section. Now, the area **APBC** consists of two right triangles with sides a_E and $a_E \cot(g/2)$, common hypotenuse $\zeta = \mathbf{PC} = a_E \csc(g/2)$ and area $a_E{}^2 \cot(g/2)$. Let z_1 be the projection of ζ onto the vertical axis. Then z_1 is given by the following system of simultaneous equations

$$q^2 = \frac{z_1^2}{\mu^2} + a_E{}^2 \csc^2\frac{g}{2} - 2\frac{z_1 \cos\frac{g}{2}}{\mu} a_E \csc\frac{g}{2} = \left(\frac{z_1}{\mu} - a_E \cot\frac{g}{2}\right)^2 + a_E{}^2,$$

$$q_0^2 = \frac{z_1^2}{\mu_0^2} + a_E{}^2 \csc^2\frac{g}{2} - 2\frac{z_1 \cos\frac{g}{2}}{\mu_0} a_E \csc\frac{g}{2} = \left(\frac{z_1}{\mu_0} - a_E \cot\frac{g}{2}\right)^2 + a_E{}^2,$$

$$(q+q_0)^2 = z_1^2(\tan^2 i + \tan^2 e - 2\tan i \tan e \cos\psi)\cos g,$$
$$\cos g = \cos i \cos e + \sin i \sin e \cos\psi$$

where q is the distance from **C** to the intersection of the exit ray with the horizontal plane containing **C**, q_0 is the corresponding distance for the incident ray, and ψ is the azimuth angle between the projections of the incident and exit rays on the horizontal plane.

For small phase angles, $\cot(g/2) \gg 1$, so that

$$q \simeq \pm[z_1/\mu - a_E \cot(g/2)],$$
$$q_0 \simeq \mp[z_1/\mu_0 - a_E \cot(g/2)],$$

where the positive sign is to be used for q and the negative sign for q_0 if $e > i$, and oppositely if $i > e$. When g is small, each term on the right-hand side of the last two equations is large, but both q and q_0 are small. Hence, the difference between q and q_0 will be small also, and to a sufficient approximation,

$$|q - q_0| \simeq z_1/\mu + z_1/\mu_0 - 2a_E \cot(g/2) \simeq 0,$$

so that

$$z_1 \simeq \langle\mu\rangle a_E \cot(g/2), \tag{9.8}$$

where

$$\frac{1}{\langle\mu\rangle} = \frac{1}{2}\left(\frac{1}{\mu_0} + \frac{1}{\mu}\right) \text{ or } \langle\mu\rangle = \frac{2\mu_0\mu}{\mu_0 + \mu} \tag{9.9}$$

is the average secant of the incident and exit ray paths. Expression (9.8) for z_1 is a good approximation when g is small. It is poor when g is large and the incident and exit rays are on opposite sides of the normal. However, when g is large, the opposition effect makes only a small contribution to the brightness. Hence, using (9.8) will cause a negilgible error in the total radiance for any value of g.

At any altitude z' between z and $z + z_1$ a horizontal cut through V_c consists of the common area between two overlapping ellipses. The thickness of this area in the direction perpendicular to the scattering plane is much less sensitive to z' than is the width in the parallel direction. Hence, only a small error will result if V_c is approximated by a volume of constant thickness u whose cross section in any plane parallel to the scattering plane is a triangle. This approximation essentially involves replacing the ellipses by rectangles. We require the area of the triangles to have the same area $a_E{}^2\cot(g/2)$ and projected altitude z_1 as **APBC**, and its base is required to lie in the horizontal plane containing **P**. Then

$$V_c \simeq u a_E{}^2 \cot(g/2). \tag{9.10}$$

The portion of V_c lying above any plane at z' between z and $z + z_1$ is $V_{ca}(z') = V_c[1 - (z' - z)/z_1]^2$.

Differentiating,

$$dV = (dV_{ca}/dz')dz' = -2V_c[1-(z'-z)/z_1]dz'/z_1$$
$$= -(2ua_E/\mu)[1-(z'-z)/z_1]dz'.$$

Thus, the value of τ_c in (9.6) is

$$\tau_c = \int_{V_c} N(z')dV \simeq -\int_z^{z+z_1} N(z')\frac{2ua_E}{\langle\mu\rangle}\left(1-\frac{z'-z}{z_1}\right)dz'. \tag{9.11}$$

The thickness u of the approximation to common volume is determined by requiring that $\tau_c = \tau/\mu$, at $g=0$, when $\langle\mu\rangle = \mu = \mu_0$ and $z_1 \to \infty$; or

$$\int_z^\infty N(z')\frac{2ua_E}{\langle\mu\rangle}dz' = \frac{1}{\mu}\int_z^\infty E(z')dz' = \frac{1}{\mu}\int_z^\infty N(z')\sigma_E dz'.$$

Thus,

$$u = \sigma_E/2a_E = \pi a_E/2, \tag{9.12}$$

so that

$$\tau_c = -\frac{1}{\langle\mu\rangle}\int_z^{z+z_1} N_E(z)\sigma_E\left(1-\frac{z'-z}{z_1}\right)dz' = -\frac{1}{\langle\mu\rangle}\int_{\tau(z)}^{\tau(z+z_1)}\left(1-\frac{z'-z}{z_1}\right)d\tau(z') \tag{9.13}$$

Integrating by parts gives

$$\tau_c = \tau/\langle\mu\rangle - \tau'/\langle\mu\rangle, \tag{9.14}$$

where

$$\tau' = \frac{1}{z_1}\int_z^{z+z_1}\tau(z')dz'. \tag{9.15}$$

Inserting this result into (9.6) gives

$$\tau/\mu_0 + \tau/\mu - \tau_c = (\tau+\tau')/\langle\mu\rangle. \tag{9.16}$$

Thus, the singly scattered radiance can be written

$$I_{SS}(i,e,g) = J\frac{w}{4\pi}p(g)\int_0^\infty K^2 e^{-K(\tau+\tau')/\langle\mu\rangle}\frac{d\tau}{\mu}. \tag{9.17}$$

Equations (9.15) and (9.17) can be readily evaluated for a slab particle density distribution. Let

$$N(z) = \begin{cases} 0 \text{ for } z < -D \text{ and } z > 0, \\ N = \text{constant} > 0 \text{ for } -D < z < 0, \end{cases}$$

Where D is the thickness of the slab,

$$\tau_1 = E z_1, \tag{9.18a}$$

and z_1 is defined by (9.8), and

$$\tau_D = ED. \tag{9.18b}$$

Then

$$\tau' = \tau^2/2\tau_1 \text{ for } -D < z < 0, \quad \text{if } z_1 > D,$$

$$\tau' = \tau^2/2\tau_1 \text{ for } -D < z < 0, \quad \text{if } z_1 < D \text{ and } -z_1 < z < 0,$$

$$\tau' = \tau - \tau_1/2 \text{ for } -D, z < 0, \quad \text{if } z_1 < D \text{ and } -D < z < -z_1,$$

$$\tau' = 0 \text{ for } z < -D \quad \text{or } z > 0.$$

Equation (9.17) can be readily evaluated to give

$$I_{SS}(i,e,g) = JK\frac{w}{4\pi}p(g)\frac{\mu_0}{\mu_0+\mu}B(g), \tag{9.19}$$

where

$$U(g) = \sqrt{\frac{4\pi}{y}}\exp\left(\frac{1}{y}\right)\left[erf\left(\frac{K\tau_D\sqrt{y}}{2<\mu>}+\frac{1}{\sqrt{y}}\right) - erf\left(\frac{1}{\sqrt{y}}\right)\right],$$

$$\text{if } y \le \frac{2<\mu>}{K\tau_D}, \tag{9.20a}$$

$$U(g) = \sqrt{\frac{4\pi}{y}}\exp\left(\frac{1}{y}\right)\left[erf\left(\sqrt{\frac{4}{y}}\right) - erf\left(\sqrt{\frac{1}{y}}\right)\right]$$

$$+\exp(-3/y) - \exp\left(-\frac{2K\tau_D}{<\mu>}+\frac{1}{y}\right), \quad \text{if } y \ge \frac{2<\mu>}{K\tau_D}, \tag{9.20b}$$

$$y = 2\frac{\langle\mu\rangle}{K\tau_1} = 2\frac{\tan(g/2)}{KN\sigma_E a_E}, \tag{9.20c}$$

and $erf(x) = (2/\sqrt{\pi})\int_0^x \exp(-t^2)dt$ is the error function of argument x.

(The error function is tabulated in many places. For those persons whose computers may not have a program that evaluates the error function, a useful approximation is

$$erf(\pm x) \approx \pm\left\{1 - \frac{\exp(-x^2)}{2|x|/\sqrt{\pi} + \left[(\sqrt{\pi} - 2/\sqrt{\pi})^2 x^2 + 1\right]^{1/2}}\right\}.$$

This expression is accurate to better than 1% everywhere.)

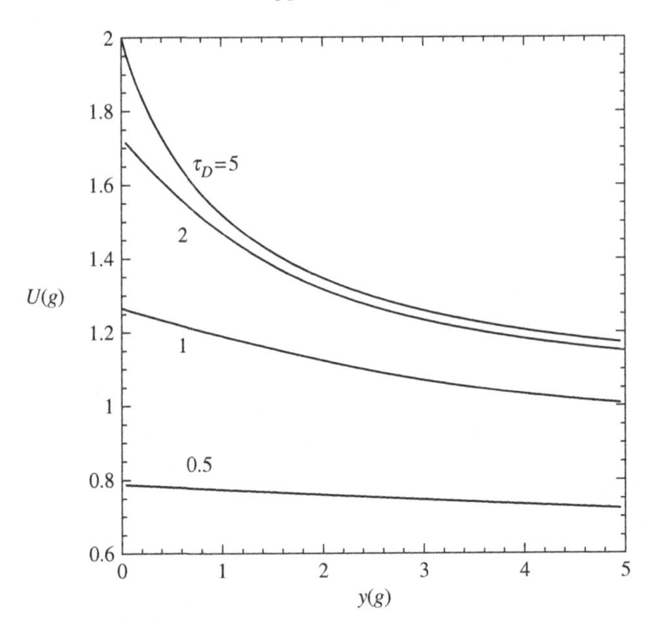

Figure 9.7 The opposition function $U(g)$ (equation 9.20) plotted against the parameter $y(g)$ for several values of the optical thickness of the scattering layer.

Equations (9.20) are plotted versus y in Figure 9.7 for several values of the optical thickness τ_D of the slab. When the optical thickness is small, $U(g)$ is low because of the fewer particles scattering incident light into the detector. There is only a slight increase in brightness as g decreases toward zero phase. As the optical thickness increases the general reflectance level increases and the zero phase peak becomes narrower and higher. When the medium completely fills the lower half-space, $\tau_D \to \infty$ and the singly scattered reflectance is equal to the Lommel–Seeliger law, except for a sharp spike near $g = 0$. Thus, for the important case of an optically thick medium, $D \to \infty$, we write

$$U(g) = 1 + B_S(g), \tag{9.21a}$$

where $B_S(g)$ is the shadow-hiding opposition function of a semi-infinite medium

$$B_S(g) = \sqrt{\frac{4\pi}{y}} \exp\left(\frac{1}{y}\right)\left[erf\left(\sqrt{\frac{4}{y}}\right) - erf\left(\sqrt{\frac{1}{y}}\right)\right] + \exp(-3/y) - 1. \tag{9.21b}$$

Equation (9.21b) is plotted as the solid line in Figure 9.8.

Although equation (9.21) is analytic, it is cumbersome and inconvenient. To good accuracy it may be approximated by the function

$$B_S(g) \simeq (1+y)^{-1} = \left(1 + \frac{1}{h_S}\tan\frac{g}{2}\right)^{-1}, \tag{9.22}$$

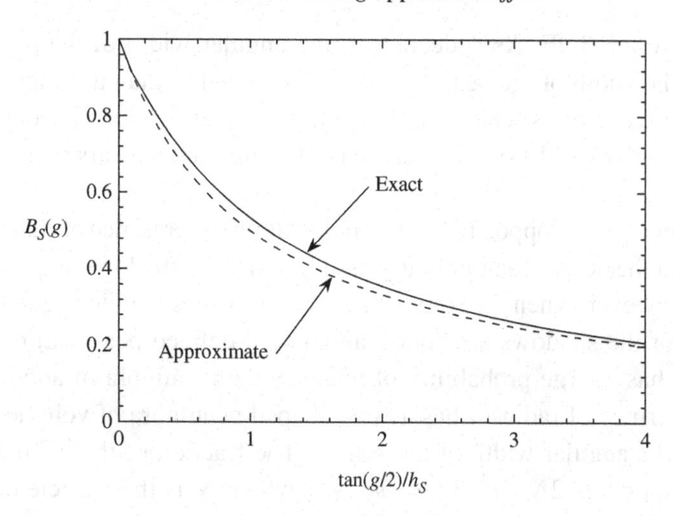

Figure 9.8 The shadow-hiding opposition effect function vs. $\tan(g/2)/h_S$ for a semi-infinte medium. The solid line is the exact equation (9.21). The dashed line is the approximate equation (9.22).

where

$$h_S = \frac{1}{2} KN\sigma_E a_E = \frac{KEa_E}{2}. \tag{9.23}$$

Expression (9.22) is shown as the dashed line in Figure 9.8. It is much more convenient than (9.21b), and the difference between the two curves probably cannot be distinguished observationally. It will be used to describe the SHOE in the rest of this book.

9.2.3 The angular width of the SHOE

The angular half-width at half-maximum of the opposition-effect peak is

$$\text{HWHM} \simeq 2h_S = KEa_E = a_E/\Lambda_E. \tag{9.24}$$

Thus, the rough estimate made in Section 9.2.1 turns out to be surprisingly accurate. In Hapke (1986) the equations of the previous section were also evaluated for a density distribution that was exponential with altitude and one that had a hyperbolic-tangent altitude distribution. In these cases it was found that the behavior of $B_S(g)$ was very similar to that predicted by (9.22), except that the angular-width parameter h_S refers to the density $N(z)$ at the altitude where the slant-path optical depth $\tau(z)/\langle\mu\rangle \simeq 1$.

Equation (9.24) shows that h_S increases as ϕ increases and becomes infinite as $\phi \to 0.752$, so that the peak is infinitely wide. This states that a surface in which the particles are so closely packed that light cannot penetrate between the particles does not have a SHOE. As ϕ decreases, the angular width of the peak narrows. Eventually the width of the peak becomes much smaller than the angular width of the source or detector as seen from the surface. In that case it is averaged over the combined angular widths of the source and detector and appears as a lower, wider peak.

The existence of the opposition effect depends on the presence of sharp extinction shadows, and the derivation implicitly assumes that the shadows are infinitely long cylinders. However, when the source of illumination has a finite angular width, the penumbras of the shadows are cones, and a well-defined peak will occur only if one particle has a large probability of being in the penumbra of another particle. An equant particle of radius a has a cone-shaped penumbra of volume $2\pi a^3/3\theta_s$, where θ_s is the angular width of the source. The fraction of the volume occupied by the penumbras is $2N\pi a^3/3\theta_s = \phi/2\theta_s$, where N is the particle density. The opposition effect will be negligible if this fraction is small, say ≤ 0.1. Thus, the condition for the opposition effect to occur is

$$\phi \geq \theta_s/5. \tag{9.25}$$

At 1 astronomical unit (AU) from the Sun, $\theta_s \sim 0.01$ so that the critical filling factor is $\phi \geq 0.002$; that is, the particles must be separated by less than about seven times their diameter. Thus, most clouds would not be expected to cause an opposition effect, but planetary regoliths would.

If the particles are larger than the wavelength and are equant, and the particle size distribution is narrow, then $Q_E \simeq 1, a_E \simeq a, \sigma_\mathrm{E} \simeq \pi a^2$, and $\phi \simeq N\frac{4}{3}\pi a^3$, so that equation (9.23) is

$$h_S = 3K\phi/8. \tag{9.26}$$

For a powder with $\phi = 0.4$, $K = 1$, and a narrow size distribution, $h_S = 0.24$, and the predicated half-width is about 28°. However, if the material has a wide range of sizes, then h_S can be much smaller. In Table 9.1, h_S is evaluated for a variety of particle size distributions, including a power law of the form $N(a) \propto \alpha^{-\nu}$. Figure 9.9 illustrates the behavior of h_S when the size distribution is a power law and the ratio of the largest to smallest particle sizes is $a_l/a_s = 1000$. The case when $v = 4$ is of special interest because this distribution characterizes products of a comminution process. Lunar regolith approximates this size distribution (McKay *et al.*, 1974) because it is the product of comminution or grinding by meteorite impacts. For $v = 4$,

$$h_S = \frac{3\sqrt{3}}{8}\frac{K\phi}{\ln(a_l/a_s)}, \tag{9.27}$$

Table 9.1. *Shadow-hiding opposition effect width factors for various particle size distributions*

$N(a)$	$8hs/3K\phi$
$\delta\,(a-\bar{a})$	1
$ae^{-a/\bar{a}}$	$\sqrt{3/8}$
a^{-v}:	
$v=0$	$4/3\sqrt{3}$
$v=1$	$3/\sqrt{8\ln(a_l/a_s)}$
$v=2$	$2\sqrt{a_s/a_l}$
$v=3$	$\sqrt{2}[\ln(a_l/a_s)]^{3/2}(a_s/a_l)$
$v=4$	$\sqrt{3}/\ln(a_l/a_s)$
$v=5$	$1/\sqrt{2}$

Note: a_l and a_s are the radii of the largest and smallest particles in the distribution, respectively; $\bar{a}$ is the average particle radius.

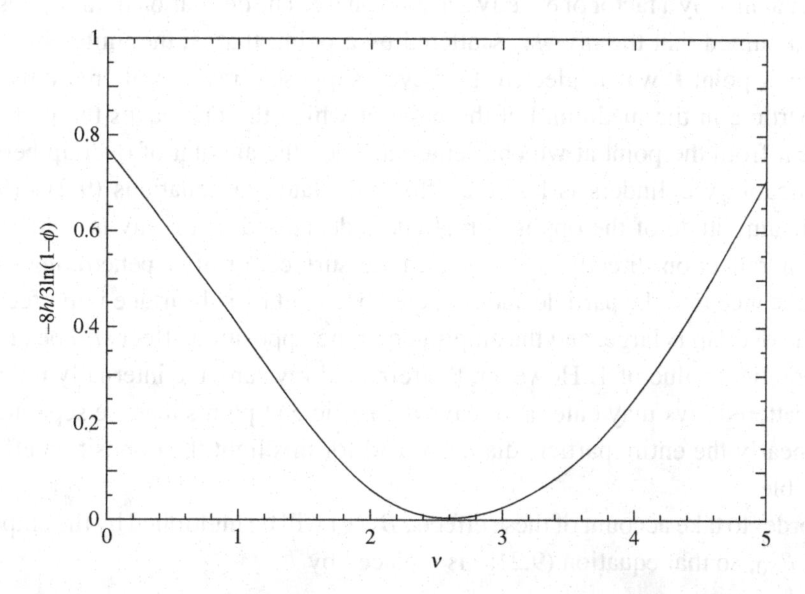

Figure 9.9 Opposition effect angular width parameter for a power-law particle size distribution of the form $n(a)\propto a^{-}\nu$ vs. ν. The ratio of the largest to the smallest particle is 1000.

where a_l and a_s are the largest and smallest particle sizes, respectively, in the distribution. If $a_l/a_s \simeq 1000$ and $\phi = 0.4$, then $h_S = 0.061$, corresponding to a half-width of about 7°.

9.2.4 The amplitude of the SHOE

9.2.4.1 The contribution of multiply scattered light

It is easy to show that the shadow-hiding effect is negligible for the multiply scattered component. The enhanced brightness is caused by the overlap of the incident and emergent probability cylinders associated with a ray scattered from a single point within the medium. If the ray is scattered between two or more particles before emerging, the probability that the incident and emergent cylinders will overlap is very small. Esposito (1979) carried out a numerical calculation of the opposition effect including doubly scattered light and found a negligible addition to the reflectance. The main effect of multiple scattering is to reduce the height of the peak relative to the total continuum reflectance, which includes both singly and multiply scattered light.

9.2.4.2 Effects of finite particle size

Theoretically, the opposition effect should increase the singly scattered component of the radiance by a factor of exactly 2 at zero phase. The derivation of the opposition effect assumed that the ray was scattered by a point; that is, the finite size of the particle at point **P** was neglected. However, suppose a ray is scattered only once by a particle in the medium, but the point at which the ray leaves the particle is different from the point at which it entered. Then the amount of overlap between the probability cylinders is less than that calculated in equations (9.21)–(9.23), and the amplitude of the opposition effect is decreased. If the ray is scattered by specular reflection directly from the particle surface, or after penetrating only a short distance into the particle and scattered back out by subsurface imperfections, then the overlap is large, and the amplitude of the opposition effect will be close to its theoretical value of 1. However, the refracted rays and the internally reflected, backscattered rays may enter and leave the particle at points that are separated by up to nearly the entire particle diameter, and for this light the opposition effect is negligible.

In order to take account of these effects, $B_S(g)$ will be multiplied by the empirical factor B_{S0}, so that equation (9.21a) is replaced by

$$B(g) = 1 + B_{S0} B_S(g).$$

For a SHOE, $0 \leq B_{S0} \leq 1$, and B_{S0} is the ratio of the light scattered from near the illuminated portion of the surface of the particle to the total amount of light scattered at zero phase:

$$B_{S0} \simeq S(0)/wp(0), \tag{9.28}$$

where $S(0)$ is the fraction of incident light scattered at or close to the illuminated part of the surface of the particle facing the source, and $wp(0)$ is the total amount of light scattered by the particle at zero phase. A lower limit to $S(0)$ occurs when the particle is opaque. Then $S(0)$ is the specular component of the particle scattering function,

$$S(0) = R(0) = \frac{(n_r - 1)^2 + n_i^2}{(n_r + 1)^2 + n_i^2}, \tag{9.29}$$

the Fresnel reflection coefficient at normal incidence. However, subsurface scattering (Chapter 6) will increase $S(0)$ above this value for most particles except true opaques.

If a particle is sufficiently large and complex the sub-particle structures can also cast shadows on one another, so that the individual particle scattering functions can have their own SHOE. This would properly be considered to be part of $p(g)$, but could have the effect of making B_{S0} appear to be > 1. Because particles with size parameter $X \lesssim 1$ do not have well-defined shadows, it might be expected that a SHOE does not occur in the reflectance of a medium made up of small particles. However, small particles tend to cohere to one another and form clumps. Large clumps of small particles can cast shadows, allowing a SHOE in such media.

9.3 The coherent backscatter opposition effect (CBOE)

9.3.1 Physical principles

An entirely different phenomenon can cause a surge in the brightness of a disordered medium at small phase angles whether the particles are larger or smaller than the wavelength. The phenomenon is known variously as *coherent backscatter, time-reversal symmetry,* and *weak photon localization.* It was first discussed by Watson (1969) in connection with radar observations of plasmas. The first laboratory identification was in the reflectance of a particulate medium in visible light by Kuga and Ishimaru (1984). At optical wavelengths it has been investigated both experimentally and theoretically by several other researchers, including Wolf and Maret (1985), Akkermans *et al.* (1986), MacKintosh and John (1988), Hapke (1990), Mishchenko (1995), Nelson *et al.* (1998, 2000), Muinonen (2004), and many others. It was independently suggested as a cause of the opposition effect observed on solar system bodies by Shkuratov (1988), Muinonen (1990), and Hapke (1990).

The term *weak localization* comes from an analogy with the transport of electrons through conducting and semiconducting media. There a similar phenomenon occurs in the transition region between the conditions where the electrons may be described as waves propagating through the medium and where the electron wave functions become localized on individual atoms. Although a photon never becomes permanently confined to an atom it can temporarily follow looping, nearly closed, multiply scattered paths and, hence, be "weakly" localized.

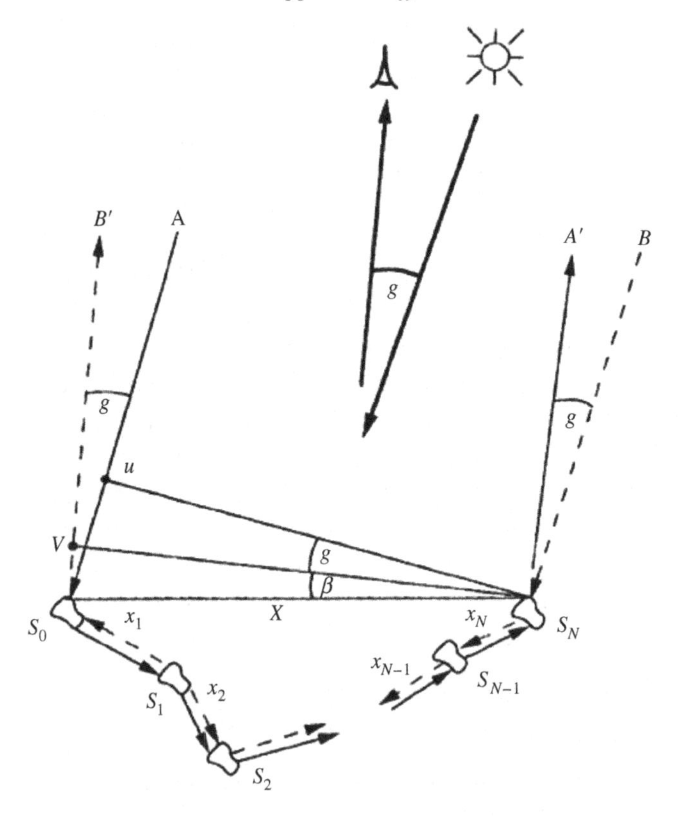

Figure 9.10 Schematic diagram of the coherent backscatter opposition effect.

The principle of coherent backscatter is illustrated in Figure 9.10. Part of a wave front incident on a particulate medium encounters a scattering center and is scattered two or more times before exiting the medium at a small phase angle. For every such path another portion of the same wave front will traverse the same path within the medium but in the opposite direction. At large phase angles there is no coherence between the two wavelets and the total intensity is twice the individual intensities. At zero phase angle the two wavelets will be in phase upon emerging from the medium and will combine coherently so as to interfere with each other positively (Chapter 2), and the total intensity is quadruple the individual intensities.

In Figure 9.10 S_0 and S_N are the initial and final scatterers, respectively, encountered by part of the wave and are separated by distance X. There can be any number of additional scatterers between S_0 and S_N. Denote the incident wave front by $S_N - U$ and the emergent wave front by $S_N - V$. Consider two wave paths in the principal plane $A - S_0 - - - - -S_N - A'$ (shown as the solid line in the figure) and its reverse $B - S_N - - - S_0 - B'$ (dashed line). Let $L = \sum_1^N X_N$ be the total distance traveled by the two partial waves between S_0 and S_N. Then just after emerging from the medium wavelet $A - A'$ has traveled a distance $X\sin(\beta + g) + L$ and wavelet $B - B'$ has traveled a distance $X\sin\beta + L$, so that the path difference is $\Delta X = X\sin(\beta + g) - X\sin\beta$. The difference in phase between the two waves is $\Delta\phi = 2\pi\,\Delta X/\lambda = (2\pi/X\beta\lambda)[\sin(\beta + g) - \sin a] = (2\pi X/\lambda)(\sin\beta\cos g + \cos\beta\sin g - \sin\alpha) \approx 2\pi Xg\cos\beta/\lambda$ to first order in g.

If the amplitude of each of the wavelets is E_0 and if the relative phases of the waves are random, then the combined intensity is proportional to $|E_0|^2 + |E_0|^2 = 2|E_0|^2$. However, in the exact backscatter direction, $\Delta\phi = 0$, and the amplitudes add coherently, so that the combined intensity is proportional to $|E_0 + E_0|^2 = 4|E_0|^2$. Thus, the intensity of the multiply scattered contribution to the reflectance is doubled.

9.3.2 Derivation

It must be emphasized that no rigorous theory of the CBOE in a closely packed, semi-infinite medium of complex particles large compared to the wavelength currently exists. An approximate model of the CBOE for isotropic scatterers has been derived by Akkermans *et al.* (1986) and an empirical model by Shkuratov and Ovcharenko (1998). The exact solution for Rayleigh scatterers that takes account of the vector nature of light has been published by Ozrin (1992) and a numerical simulation by Mishchenko, (2000). We will follow the general scheme of Akkermans *et al.* (1986) because it is relatively mathematically simple and analytic, yet gives surprisingly good agreement with a number of observations.

The general procedure will be as follows. The medium is assumed to consist of point particles that scatter light isotropically. Only the case of vertically incident light will be treated. Consider two portions of an incident wave that enter the medium through two infinitesimal surface regions of area ΔA separated by a horizontal distance r. We will use the two-stream approximation to the radiative transfer equation to find the directionally averaged radiance emerging from each area. The radiance emerging in a given direction from the two areas will be combined coherently and then integrated over the entire surface to give the reflectance as a function of phase angle.

To find the intensity in a semi-infinite medium we start by solving the two-stream approximation to the radiative-transfer equation in an infinite medium from a single point source located at the origin. We will then add a sink of photons in such a manner that the boundary conditions appropriate to a semi-infinite medium are approximately satisfied. Although this solution has no physical meaning for the part of the medium above the surface, this does not concern us because it does satisfy both the radiative-transfer equation and the boundary conditions for the portion of the medium at and below the surface.

The expression governing the distribution of the directionally averaged radiance φ is given by equation (8.38),

$$\frac{1}{4}\frac{d^2\varphi}{d\tau^2} = \frac{1}{4E^2}\frac{d^2\varphi}{dz^2} = -\gamma^2\varphi + source,$$

where z is the vertical direction. Now, the operator d^2/dz^2 is actually the *del-squared* operator ∇^2 (see Appendix C) in rectangular coordinates for a horizontally stratified medium. Our initial goal is to find the intensity from a point source located at the origin in an infinite uniform medium, which is a problem with spherical symmetry. In spherical coordinates the appropriate equation is

$$\frac{1}{4E^2}\nabla^2\varphi = \frac{1}{4E^2R^2}\frac{d}{d}\left(R^2\frac{d\varphi}{dR}\right) = -\gamma^2\varphi + S\delta(R), \tag{9.30}$$

where R is the distance to the source, S is the radiant power emitted per unit solid angle by the point source, and $\delta(R)$ is the Dirac delta function (Chapter 5). The solution is (hint: substitute $\varphi = F/R$)

$$\varphi = C\exp(-2\gamma ER)/R.$$

The constant C is found by requiring that the total power $4\pi S$ emitted by the source is equal to the power absorbed per unit volume, $4\pi\varphi A$ integrated over all space, where A is the volume-averaged absorption coefficient, $A = (E - S) = (1 - S/E)E = (1 - w)E = \gamma^2 E$. This gives

$$4\pi S = \int_0^\infty 4\pi\varphi A 4\pi R^2 dR = 4\pi^2 C/E,$$

Hence, $C = ES/\pi$ and

$$\varphi = \frac{ES}{\pi}\frac{e^{-2\gamma ER}}{R}. \tag{9.31}$$

To find the solution for a semi-infinite medium we change to cylindrical coordinates (r, z, φ), where r is the distance from the z-axis, φ is the azimuth angle, and z is the vertical distance. Let the surface of the medium be the $z = 0$ plane. The source is placed at a point on the vertical axis, $r = 0$, a distance z_1 below the surface. The distance from any point to the source is $R = [r^2 + (z_1 + z)^2]^{1/2}$.

The solution must satisfy the boundary condition (equation 8.44),

$$\varphi = (1/2)(d\varphi/d\tau) = -(1/2E)(d\varphi/dz)$$

at the surface $z = 0$ of the medium. We can try to accomplish this by mathematically adding a virtual sink of photons (that is, a negative source) on the z-axis at some distance above the surface. Unfortunately, it turns out to be mathematically difficult to satisfy the boundary condition exactly for all values of r on the surface $z = 0$. However, we can satisfy it approximately as follows.

It is desired to force the radiance to have a slope at the surface $d\varphi(0)/dz = -2E\varphi(0)$. Suppose this slope is linearly extrapolated from the surface to the distance z_E above the surface where the extrapolated radiance becomes zero. Then $d\varphi(0)/dz = -\varphi(0)/z_E = -2E\varphi(0)$, so that the extrapolation length is given by $z_E = 1/2E$. Hence, we will replace equation (8.40) by the approximate boundary condition $\varphi(z_E) = 0$. This condition can be satisfied by placing a sink of strength $-S$ at the mirror point at $r = 0$, $z = z_1 + 2z_E$. Then the combined directionally averaged intensity from the source and sink is

$$\varphi(r,z,\psi) = \frac{ES}{\pi}\left[\frac{\exp\left(-2\gamma E\sqrt{r^2+(z_1+z)^2}\right)}{\sqrt{r^2+(z_1+z)^2}} - \frac{\exp\left(-2\gamma E\sqrt{r^2+(z_1+2z_E-z)^2}\right)}{\sqrt{r^2+(z_1+2z_E-z)^2}}\right], \tag{9.32}$$

where $z < 0$.

For the CBOE calculation we have an area ΔA on the surface of a semi-infinite medium illuminated vertically by irradiance J, which penetrates to a depth z_1. Therefore, we let the source be the light scattered per unit solid angle by an incremental volume of area ΔA and thickness dz_1 located a distance z_1 below the surface:

$$S = JwET(z_1)\Delta A dz_1/4\pi,$$

where $T(z_1) = K\exp(-K\tau_1/\mu_0) = K\exp(KEz_1)$, since $\mu_0 = 1$. This scattered light diffuses through the medium and illuminates another infinitesimal volume $\Delta A dz_2$ at the point (r_2, z_2, ψ_2) with total radiant power $4\pi\varphi(r_2, z_2, \psi_2)$. A fraction $wE/4\pi$ per unit solid angle of this is directed toward the surface and is attenuated by an amount $T(z_2) = K\exp(-K\tau_2/\mu) = K\exp(KEz_2/\mu)$ before exiting in a direction making a small phase angle with the vertical. Since the problem involves only small phase angles we may take $\mu \approx 1$ to a sufficient approximation. Then the increment of radiant power leaving the surface above point 2 due to a source at point 1 is

$$dI_{1-2} = \frac{JwEK\exp(KEz_1)}{4\pi}\Delta A dz_1 \frac{E}{\pi}\varphi(r_2, z_2, \psi_2)\frac{wEK\exp(KEz_2)}{4\pi}\Delta A dz_2. \tag{9.33}$$

It is trivial to show that the incremental radiant power dI_{2-1} leaving the surface above point 1 from a source located at point 2 is the same: $dI_{2-1} = dI_{1-2}$. These will interfere coherently, depending on the phase difference between them. The geometry is shown schematically in Figure 9.11. From the law of cosines, $\cos\beta = \cos(\pi/2 - g)\cos\psi_2 + \sin(\pi/2 - g)\sin\psi_2\cos(\pi/2) = \sin g\cos\psi_2$, $\approx g\cos\psi_2$ for small g, and from the figure, $\Delta L = r_2\cos\beta = r_2 g\cos\psi_2$. Let the field of the wave emerging at the point $(r_2, 0, \psi_2)$ on the surface be E_0 and from the origin be $E_0\exp(\Delta\varphi)$, where $\Delta\varphi = 2\pi\Delta L/\lambda$ is the phase difference. Then the combined

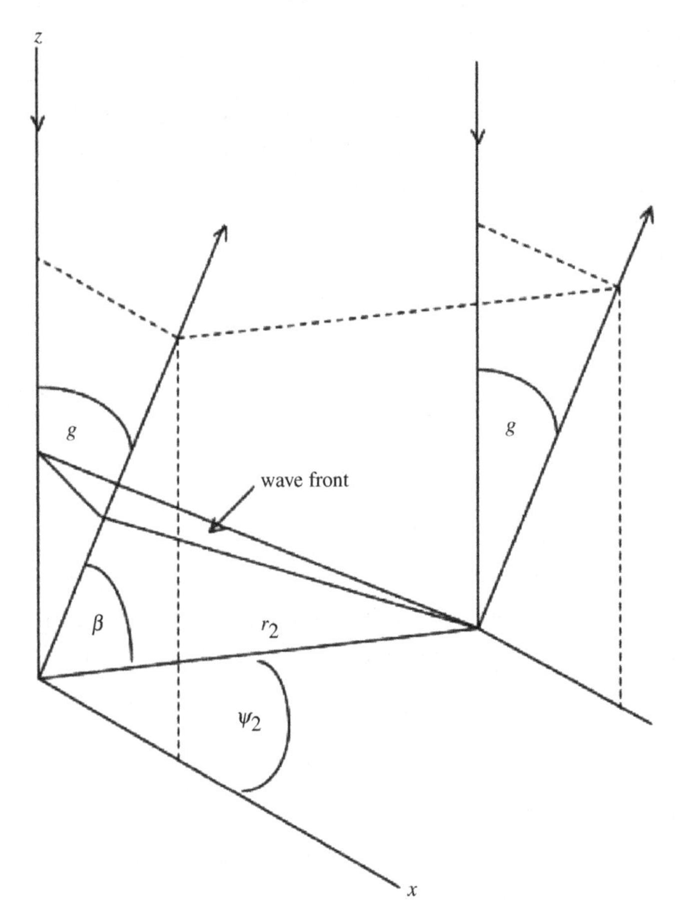

Figure 9.11 Schematic diagram of the wave front leaving the surface above source points 1 and 2.

radiant power from the two reverse-path pairs is

$$dI \propto |E_0 + E_0 e^{i\Delta\phi}|^2 = (E_0 + E_0 e^{i\Delta\phi})(E_0 + E_e^{-i\Delta\phi})$$
$$= E_0^2\left(1 + 2\frac{e^{i\Delta\phi} + e^{-i\Delta\phi}}{2} + 1\right) = 2E_0^2(1+\cos\Delta\phi) \propto dI_{1-2}(1+\cos\Delta\phi). \tag{9.34}$$

The integral of dI over the entire medium is the power per unit solid angle emerging from the area ΔA on the axis at $r = 0$ combined with the power emerging from all its reverse-pair areas around it,

$$I = \int_{z_2=-\infty}^{0}\int_{z_1=-\infty}^{0}\int_{r_2=0}^{\infty}\int_{\psi_2=0}^{2\pi} 2(1+\cos\Delta\phi)dI_{1-2}d\psi_2 r_2 dr_2 dz_2 dz_1. \tag{9.35}$$

The integration over azimuth can be carried out using the relation (Jahnke and Emde, 1945)

$$J_{2j}(v) = (-1)^j \frac{2}{\pi} \int_0^{\pi/2} \cos(v \cos\psi) \cos(2j\psi) d\psi,$$

where $J_{2j}(v)$ is the Bessel function of the first kind of argument v and order $2j$. Putting $j = 0$ and $v = 2\pi r_2 g/\lambda$ gives

$$\int_0^{2\pi} [1 + \cos\Delta\phi] d\psi_2 = 4 \int_0^{\pi/2} \left[1 + \cos\left(\frac{2\pi g r_2}{\lambda} \cos\psi_2\right)\right] d\psi_2$$
$$= 2\pi \left[1 + J_0\left(\frac{2\pi g r_2}{\lambda}\right)\right]. \tag{9.36}$$

The emerging radiance is radiant power per unit area $I/\Delta A$, and the reflectance is the radiance divided by the incident power $J\Delta A$. Thus the reflectance is

$$r_M = \frac{w^2 K^2 E^3}{8\pi^2} \int_{z_2} \int_{z_1} \int_{r_2} \left[1 + J_0\left(\frac{2\pi g r_2}{\lambda}\right)\right]$$
$$\left[\frac{\exp\left(-2\gamma E\sqrt{r_2^2 + (z_1 + z_2)^2}\right)}{\sqrt{r_2^2 + (z_1 + z_2)^2}} - \frac{\exp\left(-2\gamma E\sqrt{r_2^2 + (z_1 + 2z_E - z_2)^2}\right)}{\sqrt{r_2^2 + (z_1 + 2z_E - z_2)^2}}\right]$$
$$\exp(KEz_2)\exp(KEz_1) r_2 dr_2 dz_1 dz_2, \tag{9.37}$$

where the subscript M is to indicate that this is the multiply scattered part of the reflectance r, since the singly scattered light does not contribute to the CBOE.

Non-absorbing particles If the particles of the medium do not absorb light and are perfect scatterers, $w = 1$ and $\gamma = 0$. In this case the integration can be performed analytically. The integrals over r_2 can be evaluated using the relation (Bracewell, 2000)

$$\int_0^{\infty} J_0(fr) \frac{1}{\sqrt{r^2 + b^2}} r dr = \frac{\exp(-fb)}{f},$$

and putting $f = 2\pi g/\lambda$, $r = r_2$, and $b = z_1 + z_2$ or $b = z_1 + z_E - z_2$. The rest of the integrations are straightforward, but tedious. The result is

$$r_M = \frac{w^2}{K\pi} \left\{1 + 2z_E KE + \frac{1}{(1 + 2\pi g/KE\lambda)^2} \left[1 + \frac{1 - \exp(4\pi g z_E/\lambda)}{2\pi g/KE\lambda}\right]\right\}. \tag{9.38}$$

This is of the same form as that obtained by Akkermans *et al.* (1986), except that they considered only well-separated particles, so $K = 1$ in their expression. Since the particles are assumed to be nonabsorbing, $w = 1$ and $E = S = 1/\Lambda_S$, so these authors made the approximation, commonly used in diffusion theory to correct for anisotropic particle phase functions, of replacing Λ_S by the transport mean free path Λ_T. They also used a more accurate value for the extrapolation length z_E from diffusion theory of $z_E = 0.71\Lambda_T$.

With these substitutions, but retaining K, r_M can be written

$$r_M = r_{M0}[1 + B_c(g)] \tag{9.39}$$

where $r_{M0} = (1 + 1.42K)/K\pi$ is the multiply scattered background intensity and

$$B_C(g) = \frac{1}{(1+1.42K)(1+2\pi\Lambda_T g/\lambda)^2}\left[1 + \frac{1-\exp(1.42K2\pi\Lambda_T g/\lambda)}{2\pi\Lambda_T g/\lambda}\right] \tag{9.40}$$

is the coherent backscatter angular function for a nonabsorbing scattering medium. Recall that in deriving (9.40) sin g was replaced by g. When using a computer to calculate the intensity over a wide range of phase angles, if sin g is used $B_S(g)$ will decrease and then increase again when $g > 90°$. In order to prevent this from happening it is often convenient to replace sin g or g by $2\tan(g/2)$. Then (9.40) can be written

$$B_C(g) = \frac{1}{1+1.42K}\frac{1}{\left(1+\frac{1}{h_C}\tan\frac{g}{2}\right)^2}\left[1 + \frac{1-\exp(-1.42K\frac{1}{h_C}\tan\frac{g}{2})}{\frac{1}{h_C}\tan\frac{g}{2}}\right], \tag{9.41a}$$

where

$$h_C = \frac{\lambda}{4\pi\Lambda_T}. \tag{9.41b}$$

Equation (9.41a) for $B_C(g)$ is plotted as the solid lines in Figure 9.12 as a function of $\tan(g/2)/h_C$ for $K = 1$ and 1.9. Like the SHOE it is sharply peaked peak at $g = 0$ and falls rapidly as g increases. However, it should be noted that it does not fall off as rapidly as an exponential function. The angular half-width at half-maximum occurs at

$$\mathrm{HWHM} = 0.33 \cdot 2h_c = 0.33\lambda/2\pi\Lambda_T. \tag{9.42}$$

An empirical approximation to $B_C(g)$ that is much more convenient than (9.41a) is

$$B_C(g) \approx \left\{1 + [1.3 + K]\left[\left(\frac{1}{h_C}\tan\frac{g}{2}\right) + \left(\frac{1}{h_C}\tan\frac{g}{2}\right)^2\right]\right\}^{-1}. \tag{9.43}$$

This is plotted as the dashed lines in Figure 9.12.

Absorbing media The integration over the radial distance cannot be carried out analytically when $\gamma > 0$ in (9.35), corresponding to particles with $w < 1$ in a quasi-continuous approximation to a particulate medium. It is clear from the discussion

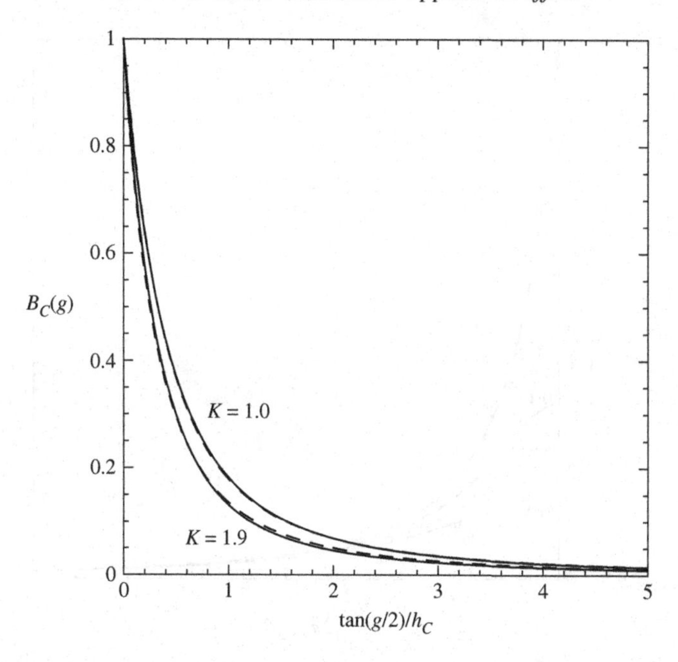

Figure 9.12 The coherent backscatter opposition effect function vs. $\tan(g/2)/h_C$ for two values of the parameter K. The solid lines are the exact equation (9.41) and the dashed lines are the approximate equation (9.43).

in Section 9.3.1 that photons that travel only a short distance before emerging will combine coherently over a wide range of phase angles and, thus, contribute to the wide base of the CBOE peak. Photons that travel a long distance before emerging are much fewer in number and combine coherently only over a short range of phase angles, thus forming the narrow top of the peak. Although the contribution from each radial distance is a rounded hump, the cumulative effect of all paths is a sharp peak when $w = 1$. However, if the photons are absorbed as they propagate between the scatterers, few will be able to travel a long distance, so that the peak would be expected to be lower and rounded off.

Akkermans *et al.* (1988) studied the interaction of a short pulse of light as it propagated and spread out through a medium, so that they were able to ascertain the effect of eliminating the long paths. They found that the coherent backscatter peak of an absorbing medium is an equation of the same form as (9.38) for $B_C(g)$, except that $2\pi \Lambda_T g/\lambda$ is replaced by

$$\frac{2\pi g \Lambda_T}{\lambda} \to \sqrt{\left(\frac{2\pi g \Lambda_T}{\lambda}\right)^2 + \frac{3\Lambda_T}{\Lambda_A}}. \tag{9.44}$$

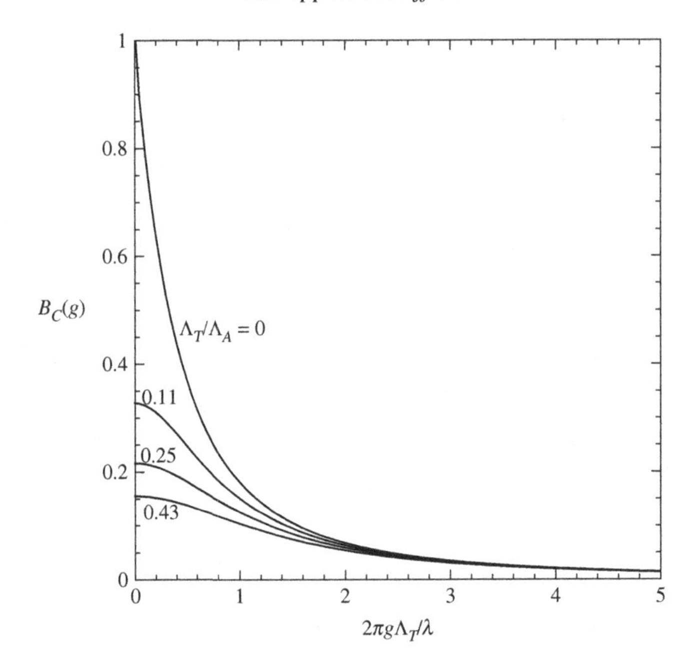

Figure 9.13 CBOE peak shape predicted for scatterers embedded in an absorbing medium for several values of Λ_T/Λ_A.

Where Λ_A is the absorption mean free path. This expression for $B_C(g)$ is plotted in Figure 9.13 for several values of Λ_T/Λ_A. When this parameter is not zero the backscatter peak is rounded instead of sharply peaked. As the value of the parameter increases the peak height is predicted to decrease and the angular width to increase. When $\Lambda_T/\Lambda_A = 0.70$ the HWHM is $1.45\lambda/2\pi\Lambda_T$, over four times as wide as when $\Lambda/\Lambda_A = 0$.

9.3.3 Comparison of CBOE models with experiment

9.3.3.1 Media of particles smaller than the wavelength

The backscatter peak has been studied in a large variety of colloidal suspensions of polystyrene spheres and TiO_2 particles smaller than the wavelength (Van Albada *et al.*, 1988, 1990; Wolf *et al.*, 1988). For the suspensions of small particles, the relatively simple model of Akkermans *et al.* (1986) seems to work remarkably well. Figure 9.14 from Van der Mark *et al.* (1988) shows a plot of the angular widths of the peaks in colloidal suspensions of a variety of nonabsorbing particles versus the ratio of the transport mean free path to the wavelength. The measured data agree quite well with predictions of theory over a wide range.

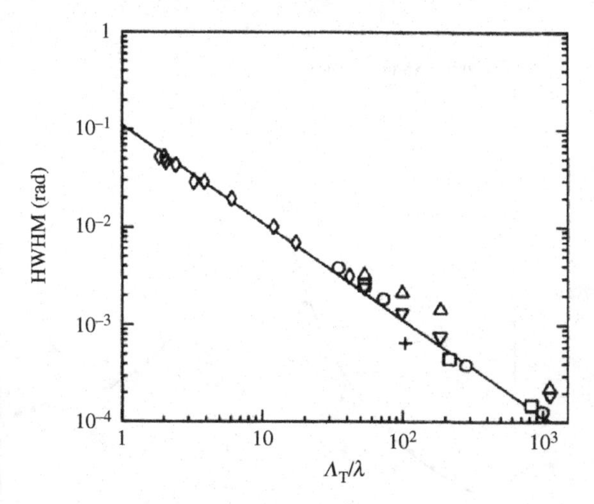

Figure 9.14 Measured full widths of the coherent backscatter peaks of optically thick colloidal suspensions versus the ratio of the transport mean free path to the wavelength. Diamonds: 220 nm TiO_2 in 2-methylpentane-2,4-diol. Circles: 1091-nm polystyrene spheres in water. Squares: 482-nm polystyrene spheres in water. Triangles (point up): 214-nm polystyrene spheres in water, polarized light, polarizer parallel to plane of emergence. Triangles (point down): 214-nm polystyrene spheres in water, polarized light, polarizer perpendicular to plane of emergence. Plus signs: 2020-nm polyvinyltoluene spheres in water. Line: theoretical prediction. (Reproduced from Van der Mark *et al.* [1988], copyright 1988 by the American Physical Society.)

The effect of eliminating the long paths that make the CBOE peak high and sharp were studied experimentally in two ways: by adding absorbing dyes to the suspending media, and using optically thin layers so that photons that were scattered more than a few times escaped in a downward direction. The peaks in these media were found to become low and rounded, in agreement with theory.

9.3.3.2 The shape of the CBOE peak in powders of particles larger than the wavelength

When the model developed in the preceding section is extrapolated to media of large particles in contact it is found that the line shape for a nonabsorbing medium, equation (9.41), can be fitted empirically to the data with good agreement. However, the predicted shape and dependence on wavelength, transport mean free path, and absorption does not agree with experimental results.

The two-stream diffusion model of the CBOE derived in Section 9.3.2 predicts that the peak should be sharp only when the particles are pure scatterers, because absorption cuts off the long diffusion paths that control the top of the peak. However,

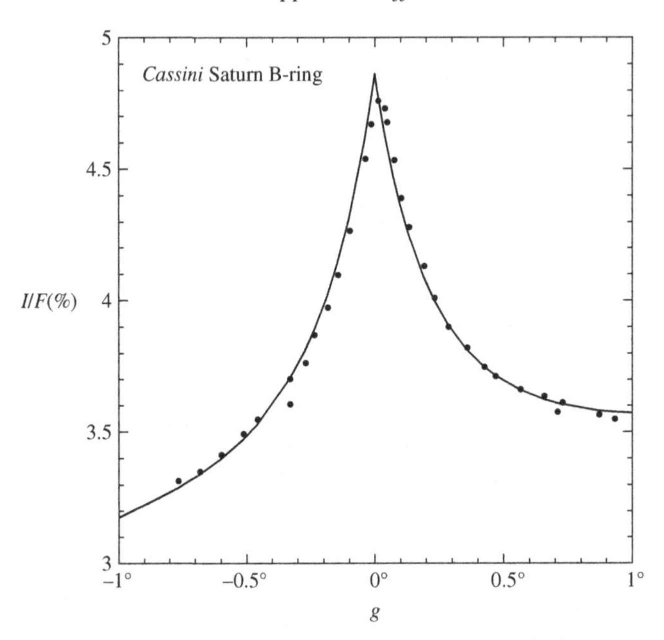

Figure 9.15 The opposition effect shown in Figure 9.3 in a ringlet of the B-ring of Saturn measured along the ringlet a constant distance from the planet. Dots: measured data from *Cassini* spacecraft. Line: theoretical prediction (equation [9.41] with the addition of a linear slope to account for detector nonuniformity).

measurements on particulate media of a wide variety of particle sizes and compositions (Hapke *et al.*, 1993; Shkuratov *et al.*, 1997; Nelson *et al.*, 1998, 2000), find that the peaks are invariably sharp and are well described by equation (9.41) This is illustrated by Figure 8.18 for commercial SiC abrasive particles approximately 17 μm in size measured using a He–Ne laser at $\lambda = 633$ nm. The samples have reflectances at 10° ranging from only 18% to 26%, yet there is no sign of a rounding off. The same is true of the opposition effect in the rings of Saturn shown in Figure 9.3. The relative radiance along the ring through the zero phase point is plotted in Figure 9.15, where the slight rounding can be accounted for entirely by the finite angular size of the Sun at Saturn (about 0.05°).

In the experiments where a dye was continuously distributed between the scatterers the long scattering paths were absorbed. However, in powders of absorbing particles in clear media the absorber is confined entirely to the scatterers. The sharp peak implies that long nonabsorbing paths between scattering locations exist, even in low-albedo powders. This would not be the case for photons that are forward-scattered multiple times deeply into the medium, but it would be true the if the photons that contribute significantly to the CBOE peak are those that, upon the first encounter with the medium, are scattered sideways, more or less parallel to the irregular surface, so that they travel several particle diameters before scattering again. This would explain why equation (9.41) for nonabsorbing particles also correctly describes the shapes of the coherent backscattering peaks of powders of absorbing particles embedded in a nonabsorbing medium. The sharp peak observed in low-albedo powders is another case where attempts to model a system of discrete particles by a quasi-continuous medium breaks down. This problem has also been discussed by Weaver (1993).

9.3.3.3 The height of the CBOE peak in powders of particles large compared to the wavelength

The fact that coherent backscatter operates only on the multiply scattered portion of the reflectance of a particulate medium might lead the reader to predict that low-albedo media should not possess a CBOE, and that the amplitude of the peak should increase monotonically as the albedo increases. However, samples of lunar soil with albedos of the order of 10% possess well-developed opposition effects which have been shown to be CBOEs, Hapke *et al.* (1993). Shkuratov and Ovcharenko (1998) discovered that mixtures of MgO and amorphous C powder have strong CBOEs, whereas the opposition effects of the pure end members are weak. There are two reasons for this behavior.

To understand the first reason we must carefully be aware of what is meant by "scattering" in the context of the CBOE. In Chapters 5–8 the term referred to the interaction of a light wave with an entire particle. However, in the CBOE the term refers to any event that changes the direction of a wave. This can occur by refraction, reflection, or internal scattering within a single particle, as well as from a particle as a whole. If a particle is complex, as most large particles are, multiple scattering can occur between parts of the surface or between internal scatterers. Thus, depending on the nature of the particle, the CBOE can amplify much of the singly scattered portion of the reflectance in addition to all of the multiply scattered terms, so that even low-albedo media can exhibit a CBOE. All of the light in the SHOE that is multiply scattered within or on the surface of the particle is also amplified by coherent backscatter. These arguments are supported by the paper by Zubko *et al.* (2008) who studied computer models of scattering by complex single particles and found that they exhibited coherent backscatter.

The second reason arises because of the vector nature of light. Each scattering event tends to rotate the plane of polarization by an amount that depends on the nature of the event. The initial direction of polarization becomes randomized after a few scatterings and the reverse pairs can no longer combine coherently (Van Albada *et al.*, 1988). Thus, the longer multiple scattering paths that occur in high-albedo media only contribute to the background reflectance, but not to the coherent backscatter peak. The CBOE angular function equation (9.41) was derived using

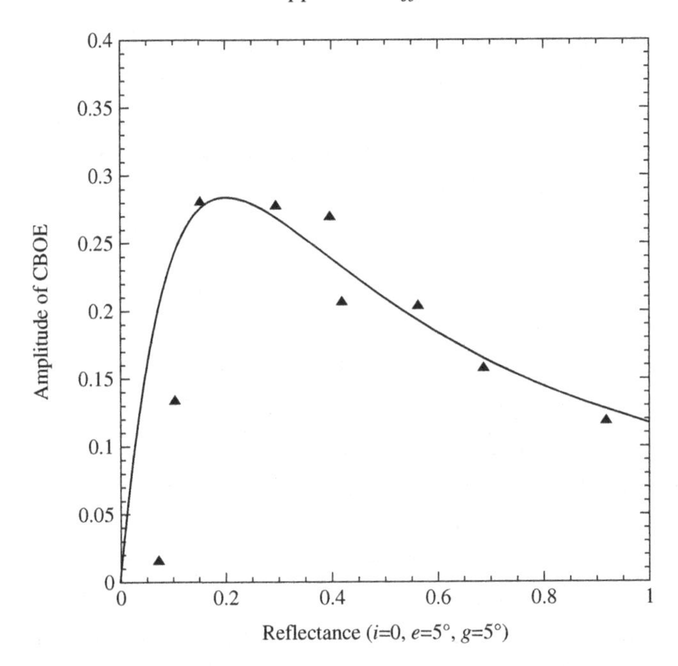

Figure 9.16 Height of the CBOE peak relative to the continuum reflectance for mixtures of Al_2O_3 and B_4C abrasive powders. Triangles: data. Line: theoretical prediction.

a scalar treatment that did not take polarization randomization into account and so overestimated the height of the peak.

To study these effects Nelson *et al.* (2004) measured the height of the backscatter peak, relative to the background reflectance, of mixtures of 22-μm abrasive powders of Al_2O_3, which has a high albedo, and B_4C, which has a low albedo. The results are plotted versus reflectance as the points in Figure 9.16. When the reflectances are low the peak amplitude first increases as the reflectance increases, but goes through a maximum at a reflectance around 20% and then decreases.

To analyze their data Nelson *et al.* assumed that the background could be modeled to sufficient accuracy by equation (8.47) for the bidirectional reflectance of a medium of isotropic scatterers. This equation was expanded in powers of w up to w^4, which gave the contribution of each order of scattering to the reflectance. These contributions, relative to the entire reflectance, are plotted in Figure 9.17. Each of these curves was multiplied by an adjustable weighting factor between 0 and 1 and then added together, and the sum compared with the data. The best fit, shown as the solid line in Figure 9.16, was obtained if singly scattered light and light scattered more than four times made negligible contributions to the coherent backscatter peak, while the peak consisted of 75% of the light scattered

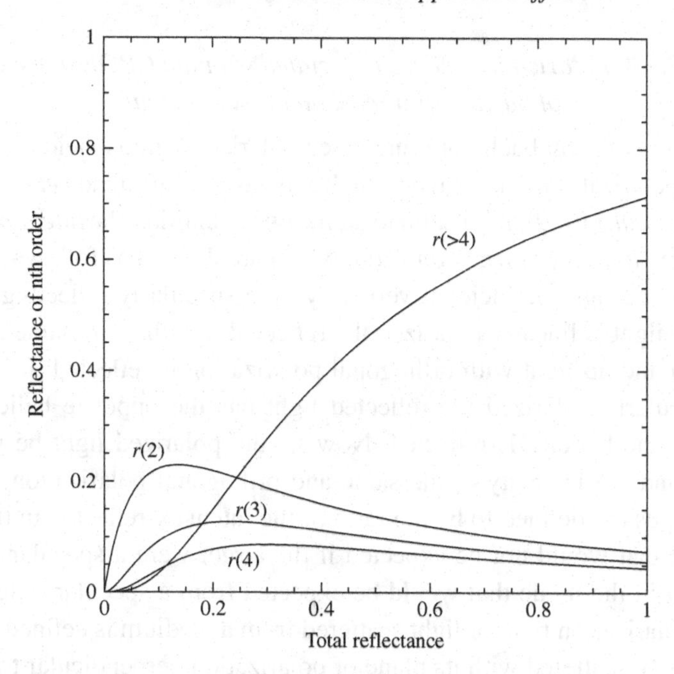

Figure 9.17 Ratio of each order of scattering to the total reflectance of a medium of isotropic scatterers plotted against the total reflectance.

twice, 65% of the light scattered three times, and 50% of the light scattered four times.

The main discrepancy occurs at small values of the reflectance, where the theoretical curve is too high. The reason is that this crude model averages the albedos of all the particles, so that the weighting factors are assumed to be independent of the mixing ratio. However, in the low-reflectance mixtures most of the second-order scattering occurs between the abundant low-albedo B_4C particles, so that a photon has to be scattered more than twice before encountering a high-albedo Al_2O_3 particle. In real media of low reflectance this effect decreases the relative importance of second-order scattering below the predicted curve.

These effects can be taken into account by multiplying $B_C(g)$ by an amplitude factor B_{C0}, so that the reflectance is amplified by the coherent backscattering factor

$$1 + B_{C0} B_C(g), \tag{9.45}$$

where B_{C0} is an empirical quantity that cannot be predicted exactly at the present time, but is related to the albedo.

9.3.3.4 Polarization effects associated with the CBOE in media of particles larger than the wavelength

The effects of coherent backscattering when polarized light is incident depends on the type of polarization. They have been discussed by Van Albada *et al.* (1990) and MacKintosh *et al.* (1989). It is useful to define two quantities, the *linear polarization ratio* and the *circular polarization ratio*, which are defined as follows.

Let polarized light be incident vertically on a specularly reflecting surface. If the incident light is linearly polarized the reflected light has the same direction of polarization and no light with orthogonal polarization is reflected. If the incident light is circularly polarized the reflected light has the opposite helicity and no light with same helicity is reflected. Now let the polarized light be incident on a medium and the intensity in the same and orthogonal polarizations measured. Then the ratios are defined to be the ratio of the intensity reflected in the mode of polarization that would not be expected if the target were a specular reflector to the intensity in the mode that would be expected from a specular reflector. Thus, the linear polarization ratio of light scattered from a medium is defined as the ratio of the intensity scattered with its plane of polarization perpendicular to that of the incident light to the intensity scattered with its plane of polarization parallel to that of the incident light. Similarly, the circular polarization ratio of light scattered from a medium is defined as the ratio of the intensity scattered with the same direction of helicity as that of the incident light to the intensity scattered with the opposite direction of helicity as that of the incident light.

If the polarization is completely randomized the polarization ratios are equal to 1. The polarization ratios are a measure of the degree to which the incident light is depolarized by multiple scattering within the medium. For that reason they are sometimes referred to as *depolarization ratios.*

Because multiple scattering tends to randomize the polarization, the lowest orders of scattering contribute the most toward the polarization effects associated with coherent backscatter. Large particles scatter light primarily by two processes: specular reflection from the surface and transmission. If the particle is opaque all of the light is scattered by surface reflection. If the particle is not opaque, the transmitted light is forward-scattered deeper within the medium, where on average it has to undergo several scatterings to escape in the backward direction, and so contributes relatively little to the lower orders. Thus, light that is scattered only a few times by specular reflection with intermediate paths in directions roughly parallel to the surface makes the largest contribution to the polarization effects.

When circularly polarized light is incident the light reflected by single scattering directly back toward the source has helicity opposite to that of the incident light. Light that is doubly scattered has its helicity reversed twice, so that light of the same helicity as incident is backscattered. The doubly reflected light is slightly elliptically polarized because the Fresnel reflection coefficients at oblique incidence for the times when the electric vectors are instantaneously perpendicular or parallel to the local scattering plane are not equal. Since elliptically polarized light can be decomposed into two circularly polarized components of opposite helicity (Chapter 4), the light has been partially depolarized. However, the amount of depolarization is fairly minor.

To understand the circular polarization effects it will be sufficient to consider only singly and doubly scattered light. Let the intensity of the singly scattered light be $2I_1$, let the unenhanced intensity of the doubly scattered light with the same helicity as incident be I_{2S} and of the doubly scattered light with the opposite heliciy be I_{2O}. Then outside of the coherent peak the circular polarization ratio is $I_{2S}/(I_1 + I_{2O})$. Suppose that inside the coherent peak each component of the doubly scattered light is enhanced by a factor of E. Then inside the peak the circular polarization ratio is $EI_{2S}/(I_1 + EI_{2O}) = I_{2S}/(I_1/E + I_{2S})$, which is greater than the ratio outside the peak. Thus, the circular polarization ratio is *increased* inside the coherent peak.

MacKintosh *et al.* (1989) have pointed out another mechanism that preserves helicity in a medium in which large particles are sufficiently separated that diffraction is important. In this case a significant amount of light can be forward-scattered by diffraction at each scattering, which does not change the helicity. Eventually by repeated scattering through several small angles the light can be scattered back out of the medium with its helicity intact.

Another way of thinking about helicity preservation is that multiple scatterings depolarize by rotating the electric vectors. However, in circularly polarized light the electric vector is already rotating. Hence, the original direction of helicity tends to be preserved through many scatterings to a much greater extent than either the direction of polarization of linearly polarized light or the direction of propagation.

Van Albada *et al.* (1990) have discussed the case when linearly polarized light is incident on a medium of particles smaller than the wavelength. They showed that the linear polarization ratio decreases in the coherent backscatter peak. Here we show how their analysis can be extended to the case of large particles. When linearly polarized light is incident the singly backscattered light has its electric vector parallel to that of the incident radiation. However, it is not coherently amplified. The doubly scattered light has its electric vector rotated by an amount that depends on the angle between the direction of the line connecting the two scatterers and the incident electric vector and is strongly randomized. Coherent backscattering enhances both the polarized and depolarized components of the doubly scattered light.

However, let us consider triple scattering in detail, as illustrated schematically in Figure 9.18. We are looking down vertically, parallel to the direction of incidence,

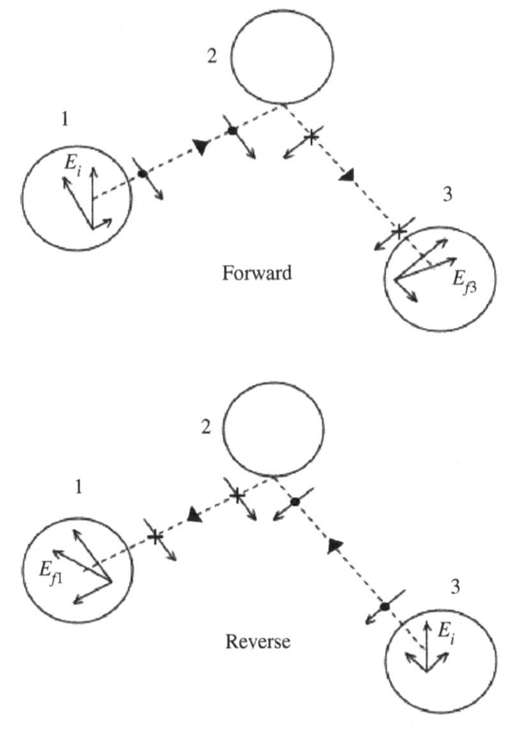

Figure 9.18 Schematic diagram of the reflections and relative phases of triply scattered light.

at three particles near the surface of the medium. For purposes of illustration, the figure shows the situation where the scattering facets on the surfaces of the particles are separated by integral multiples of the wavelength, but the phase relationships between the final forward and reverse states are the same for arbitrary separations. The upper figure shows the case where part of the incident wave is scattered from particle 1 to 2 to 3 and then to the detector. The electric vector of the incident light is labeled E_i. Part of the light incident on particle 1 is specularly reflected from a surface facet of particle 1 towards particle 2. The wave reflected from particle 1 has two components parallel and perpendicular to the scattering plane formed by the direction of incidence and the direction from particle 1 to 2. These components are shown either as arrows, or as dots if the electric vector points up, or as plus signs if the field points down. As discussed in Chapter 4, the directions of both vectors are reversed at each reflection. The light is reflected from a facet on particle 2 towards particle 3 and finally reflected from a facet on particle 3 towards the observer with electric vector E_{f3}.

Now consider the light that travels the reverse path, shown schematically in the lower figure. Light with electric vector E_i incident on particle 3 is reflected to 2 and 1 and then reflected upward towards the observer as E_{f1}. Note that E_{f3} points toward the top of the page and to the right, while E_{f1} points to the top and to the left. That is, the components of the triply scattered vectors parallel to the incident radiation point in the same direction, but the components perpendicular to E_i point oppositely to one another. Outside of the coherent backscatter peak the scatterings are uncorrelated. However, inside the peak the components of the triply backscattered light parallel to E_1 tend to add coherently, but the components perpendicular to E_i tend to cancel each other. Thus coherence tends to increase the polarized component of the scattered light, but reduce the depolarized component, so that the linear polarization ratio *decreases* in the peak.

By contrast, the SHOE is dominated by single reflections from the surface of a particle. This preserves the direction of linearly polarized light, but reverses the helicity of circularly polarized light. Thus, both the linear and the circular polarization ratios *decrease* in the SHOE peak.

These effects are illustrated in Figure 9.19, which shows the polarization ratios of the light scattered by 17-μm SiC abrasive powder illuminated by linearly and circularly polarized light. Note the dip in the linear polarization ratio and the peak in the circular polarization ratio of about the same angular width as the intensity surge. Based on these polarization relationships, there is little doubt that the narrow peak in the reflectance of the SiC powder is caused by coherent backscatter, rather than shadow hiding.

Since this book is concerned mainly with particles larger than the wavelength we will not consider small particles in detail. It turns out that the linear polarization ratio decreases in the coherent backscatter peak in media of small particles, just as for large particles. However, the increase in the circular polarization ratio in small particles is much smaller than is the case for large particles, and it may even decrease slightly in certain cases (MacKintosh *et al.*, 1989; Van Albada *et al.*, 1990).

9.3.3.5 *The angular width of the CBOE peak in close-packed powders as a function of particle size and wavelength*

The model for the shape of the peak derived in Section 9.2 predicts that the angular width of the peak should be proportional to λ/Λ_T. This is also predicted by every theory of the CBOE published to date, and implies that the width should depend strongly on particle size, filling factor and wavelength. For equant particles of diameter D the reciprocal of the transport mean free path is given by

$$\Lambda_T = [KN\sigma Q_S(1-\xi)]^{-1}$$

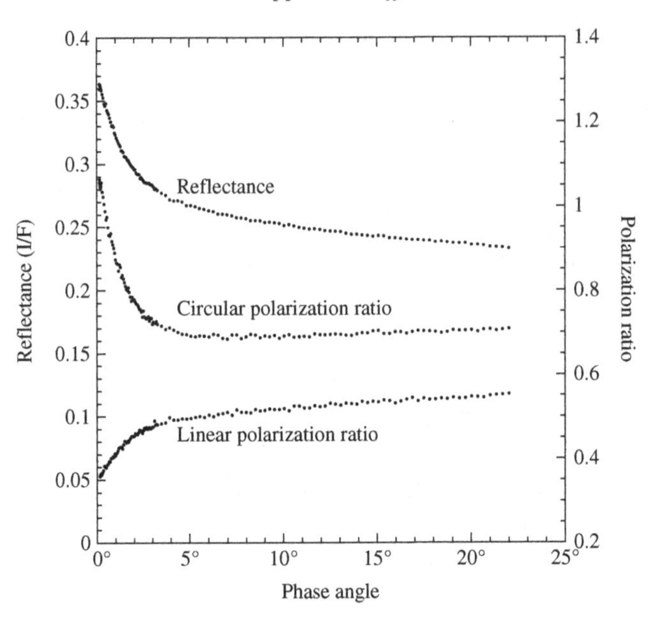

Figure 9.19 Measured reflectance and polarization ratios of a powder of 17-μm SiC abrasive particles.

$$= [K\frac{\phi}{v}\sigma Q_S(1-\xi)]^{-1} \approx \left[\frac{K\phi}{\pi D^3/6}(\pi D^2/4)Q_S(1-\xi)\right]^{-1}$$

$$= \frac{2}{3}\frac{D}{K\phi Q_s(1-\xi)}. \tag{9.46}$$

If the size parameter $X < 1$, then $Q_S \propto X^4 \propto D^4$, and $\xi \approx 0$, so the peak width is predicted to be proportional to D^3 for constant filling factor. On the other hand, if $X > 1$, then $Q_S = wQ_E$, where w decreases as D increases for constant absorption coefficient. Now, $Q_E \approx 1$ for close-packed powders, and ξ will not change very much with size. Thus, for constant φ the peak width decreases proportional to or faster than $1/D$ as D increases. Starting with particles smaller than the wavelength, as the size increases, the peak width is predicted to first increase and then decrease, with the maximum occuring around the wavelength, as illustrated by the line in Figure 9.20 for a constant w. If the particle size is much larger or smaller than λ the peak width should be so small as to be unobservable.

If Λ_T is held constant, the peak width is predicted to be directly proportional to the wavelength.

However, none of these predictions are borne out by measurements on close-packed particles. The observed widths of the opposition effect peaks are too wide by orders of magnitude and are insensitive to particle size, filling factor, or wavelength.

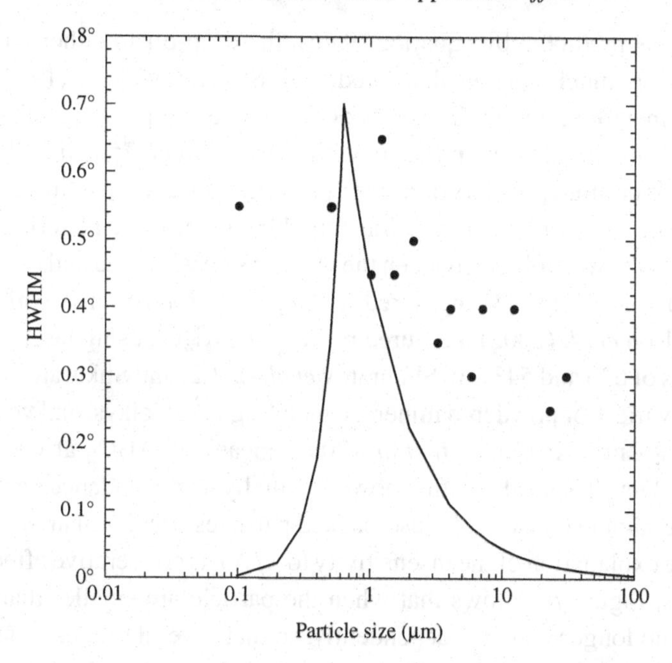

Figure 9.20 Width of the CBOE peak of a particulate medium plotted against particle size. Dots: measured widths for Al_2O_3 abrasive powders. Line: predicted widths for particles of constant single-scattering albedos.

For example, the narrow peak observed for the SiC powder of Figure 9.19 has a HWHM of 1.4°. This figure shows that the circular polarization ratio also has a peak of about the same width as the intensity peak, while the linear polarization ratio simultaneously decreases. No explanation other than coherent backscatter has been advanced for this behavior of the polarization ratios, so it is highly likely that the peak is indeed a CBOE. The transport mean free path can be calculated from equation (9.46) using the known size, single scattering albedo, and filling factor to be $\Lambda_T \sim 175\,\mu\text{m}$. The peak width predicted from equation (9.42) is only 0.01°.

Figure 8.18 shows the curves of reflectance versus phase angle for the 17-μm SiC powder samples measured at different filling factors. The widths of the opposition peaks are virtually indistinguishable, even though the porosity changes by a factor of 3.

Nelson *et al.* (1998, 2000) and Piatek *et al.* (2004) have measured the reflectances of powders of varying composition over a wide range of sizes in both linearly and circularly polarized light. The light source was a He–Ne laser emitting a wavelength of 633 nm. Representative results are shown as the dots in Figure 9.20. All the peaks had increased circular polarization ratios associated with them and are undoubtedly caused by coherent backscatter. Although the peak width is a maximum around the wavelength, as predicted by equation (9.46), the drop-off to either large or small particle sizes is much smaller than predicted. Shkuratov *et al.* (1997) measured powders of metallic Fe with diameters of 5, 40, and 100 μm. Kaasalainan (2003) measured basalt and alumina powders with sizes between 75 and 500 μm. Psarev *et al.* (2007) measured powders of olivine particles 3, 5, and 8 μm in size. No strong dependence of peak widths on size was found in any of these materials.

Similar results were obtained when the particle size was fixed and the wavelength varied. Hapke *et al.* (1993) measured a sample of *Apollo* lunar soil at 633 and 442 nm. Nelson *et al.* (2002) measured powders of Al_2O_3 using laser sources with wavelengths of 633 and 543 nm. Shkuratov *et al.* (2002) and Shkuratov *et al.* (2004) measured a variety of powdered minerals, metals, glasses, clays, and volcanic ashes at 633 and 442 nm. Kaasalainen *et al.* (2005) measured Al_2O_3 at wavelengths of 633 and 1024 nm. The peak widths showed virtually no dependence on wavelength.

When the sample consists of close-packed particles smaller than the wavelength the probable explanation of the insensitivity to λ/Λ_T is cooperative effects between the particles. Figure 7.5 shows that when the particle are smaller than the wavelength they no longer scatter independently, so that several of them tend to behave like a single particle. In such media amplitude and wavelength effects are probably determined more by the statistics of the clumps within the medium than the properties of the individual particles. However, this needs to be investigated.

No adequate explanation has been advanced for the behavior of large particles. If the medium consists of complex particles the CBOE could be dominated by multiple scattering between internal and surface irregularities of complex particles rather than between particles. While this could account for the insensitivity to particle size in low-albedo media it does not explain the lack of wavelength dependence. Neither does it explain the wide, readily observable CBOE peak in high-albedo Al_2O_3 powders (Nelson *et al.*, 1998), nor in media of smooth uniform spherical particles (Hapke *et al.*, 2007).

Interestingly, when Van der Mark *et al.* (1988) measured the transport mean free path of suspensions of TiO_2 as the filling factor was varied they noted that Λ_T appeared to saturate when $\phi > \sim 15\%$. They did not offer any explanation for this behavior. Hence, equation (7.24b) for Λ_T may not be valid for media of large particles in contact.

The observed lack of dependence on λ/Λ_T could be understood if, instead of Λ_T, a parameter of the order of λ is the critical distance in close-packed media of large particles. In that case the ratio would be independent of particle size or wavelengh separately. For example, it is possible that the CBOE in close-packed media is dominated by evanescent waves, which can transfer energy between particles when their surfaces are within about a wavelength of each other (Petrova, 2007), so that the separation distance is the critical length that determines the peak width.

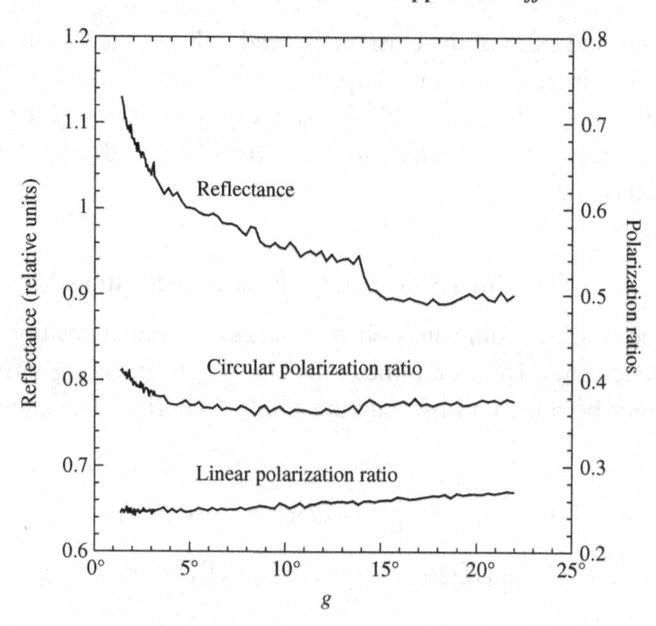

Figure 9.21 Reflectance and polarization ratios of the CBOE peak of the fractured surface of a solid basaltic rock.

Further complicating this discussion is the surprising discovery by Shepard and Arvidson (1999) that the fractured surface of a basaltic rock had a weak opposition effect. Subsequent polarized reflectance measurements showed that the peak was a CBOE (Figure 9.21). It is probably caused by scattering between imperfections inside the rock.

Of course, it is possible that the opposition effect in powders of large close-packed particles is dominated by shadow-hiding rather than coherent backscatter. This would explain the insensitivity to wavelength and particle size. However, it would not account for the lack of dependence on porosity or the strong peaks observed in high-albedo materials. Moreover, it would not be consistent with the increased circular polarization ratio in the peaks, for which no explanation other than coherent backscatter has ever been advanced.

At the present time we do not understand the coherent backscatter opposition effect in media of touching particles larger than the wavelength. In principle, techniques like the T-matrix method and the discrete dipole approximation could be used to study the problem, but the CPU and memory requirements to model realistic semi-infinite media are well beyond present technology. Present capabilities allow the calculation of scattering by clusters of only a few particles, but these are poor simulations of semi-infinite media because major amounts of light leak out of the fronts and sides of the clusters, instead of being backscattered. As computer capabilities increase we may expect to see improvements in such models that may resolve the problem. It is likely that the correct explanation will again be found to be the result of a breakdown of approximating a discrete medium by a quasi-continuous one.

9.4 Combined SHOE, CBOE, and IMSA models

Combining the shadow-hiding and coherent backscattering opposition effects with the IMSA model, the SHOE amplifies only the single scattering term, while the CBOE amplifies both the multiple and the single scattering. Thus, the reflectance becomes

$$r(i,e,g) = K\frac{w}{4\pi}\frac{\mu_0}{\mu_0+\mu}\{p(g)[1+B_{S0}B_S(g)] \\ +[H(\mu_0/K)H(\mu/K)-1]\}[1+B_{C0}B_S(g)] \qquad (9.47)$$

where $B_S(g)$, h_S, $B_C(g)$, and h_C are given respectively by (9.22), (9.23), (9.43), and (9.41b), K by (7.44) and (7.45), and $H(x)$ by (8.70b).

The shapes of $B_C(g)$ and $B_S(g)$ are compared in Figure 9.22. In these curves the angular width parameters have been adjusted so that the curves are equal at the half-maximum points. Also shown for comparison is the exponential function, which is often used empirically to describe the opposition effect. The curves are indistinguishable between $g = 0$ and the HWHM. However, the SHOE and CBOE functions fall off more slowly than the exponential function at larger phase angles.

A question of interest in remote sensing is how to tell what type of opposition effect is being observed in the reflectance of a medium. Shortly after the CBOE was discovered, it was thought that the two types could be distinguished by their wavelength dependence, but subsequent laboratory investigations have shown that this test is unreliable. The most definitive test seems to be the behavior of the polarization ratios. As the phase angle decreases the circular polarization ratio increases and the linear ratio decreases if the opposition peak is a CBOE. If the peak is a SHOE both ratios decrease. These ratios can be measured for samples in the laboratory, but usually not in remote-sensing situations, where the illumination is natural sunlight.

Another test, but one that is less stringent, is the angular width. A CBOE peak tends to be one or two degrees wide, while a SHOE peak tends to be several degrees wide if the medium has a wide particle size distribution, and 10°–20° wide, or more, if the size distribution is narrow. Thus the CBOE peak can easily be distinguished from the rest of the phase curve of a medium by its narrow width. However, when the SHOE peak is narrow it is difficult to separate it from the CBOE because of the

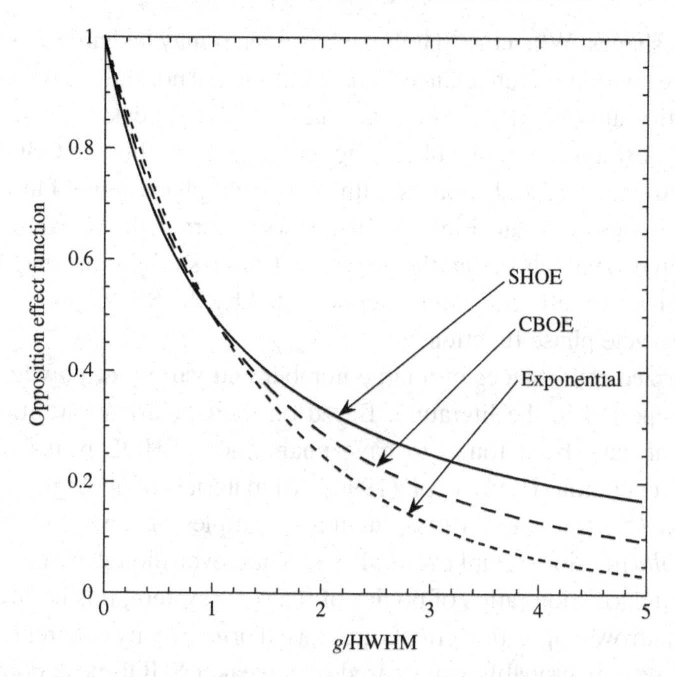

Figure 9.22 The CBOE and SHOE functions. For comparison, the exponential function is also shown. The angular width parameters have been adjusted so that the curves are equal at the half-power point.

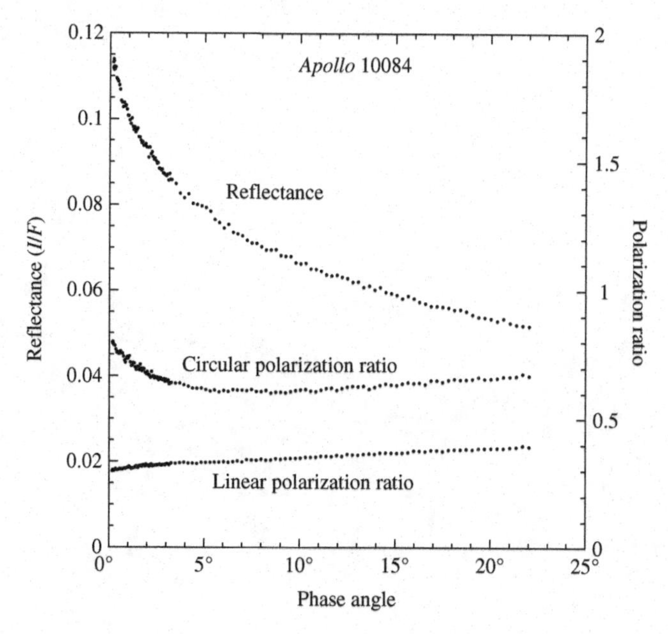

Figure 9.23 Measured reflectance and polarization ratios of a sample of lunar soil (*Apollo* 10084).

similarity in shapes. When the SHOE peak is wide it may be hard to distinguish its base from the continuum reflectance. When illuminated normally the phase function of most particulate materials has a high, narrow CBOE peak, outside of which it decreases almost linearly as the phase angle increases. Within the CBOE peaks the circular polarization ratio decreases with increasing phase angle, but in the linear region it increases or is flat (Figure 9.19). At least part of the linear portion of the reflectance may be a SHOE, but the departure from a straight line may be so slight that it is difficult to tell how much is contributed by the SHOE and how much by the single-particle phase function.

The polarized reflectances of a large number and variety of powdered samples have been reported in the literature. Based on their polarization ratios, the only materials that have been found to have unamiguous SHOE peaks were certain varieties of vegetation (Hapke *et al.* (1996). All materials of geological interest had CBOE peaks (Nelson *et al.*, 1998), including samples of lunar soil (Figure 9.23) from the *Apollo* missions (Hapke *et al.*, 1993). Thus, even though we cannot measure the circular polarization ratios of bodies of the solar system, it is highly likely that their sharp, narrow, opposition effects are caused primarily by coherent backscatter. This CBOE peak is probably superposed on a weaker SHOE peak combined with the wide backscattered lobe of $p(g)$.

10

A miscellany of bidirectional reflectances and related quantities

10.1 Introduction

Expressions for the bidirectional reflectance of a particulate medium were derived in Chapter 8. In this chapter several quantities that are frequently encountered in applications of bidirectional reflectance are discussed, including some variants in other geometries. Equations are also derived for the calculation of the reflectances of media consisting of two layers with different properties and mixtures of different kinds of particles.

10.2 Some commonly encountered bidirectional reflectance quantities

The *bidirectional reflectance r(i,e,g)* is the ratio of the radiance scattered from the surface of a medium into a given direction to the collimated power incident per unit area perpendicular to the direction of incidence. The following equations omit the opposition effect. It may be added using the formalism of equation (9.47).

The *bidirectional-reflectance distribution function* (BRDF) is defined as the ratio of the radiance scattered from the surface of a medium into a given direction to the collimated power incident on a unit area of the surface. The incident radiant power per unit area of surface is $J\mu_0$, and the scattered radiance is $Jr(i, e, g)$. Thus,

$$\mathrm{BRDF}(i, e, g) = r(i, e, g)/\mu_0. \tag{10.1}$$

If r is described by (8.70),

$$\mathrm{BRDF}(i, e, g) = K\frac{w}{4\pi}\frac{1}{\mu_0 + \mu}[p(g) + H(\mu_0/K)H(\mu/K) - 1] \tag{10.2}$$

The *reflectance factor*, denoted by REFF, of a surface is defined as the ratio of the reflectance of the surface to that of a perfectly diffuse surface under the same conditions of illumination and measurement. The reflectance factor is sometimes also called the *reflectance coefficient*. The bidirectional reflectance of a perfect Lambert surface is $r_L = \mu_0/\pi$, so the reflectance factor of a surface with bidirectional reflectance $r(i, e, g)$ is

$$\mathrm{REFF}(i, e, g) = \pi r(i, e, g)/\mu_0. \tag{10.3}$$

If r is given by (8.70a), then

$$\mathrm{REFF}(i, e, g) = K\frac{w}{4\pi}\frac{1}{\mu_0 + \mu}[p(g) + H(\mu_0/K)H(\mu/K) - 1]. \tag{10.4}$$

The *radiance factor*, denoted by RADF, is similar to the reflectance factor, except that it is defined as the ratio of the bidirectional reflectance of a surface to that of a perfectly diffuse surface illuminated at $i = 0$, $r_L(0) = 1/\pi$, rather than at the same angle of illumination as the sample. In his classic treatise on radiative transfer Chandrasekhar (1960) introduced the notation $F = J/\pi$. Hence, a commonly used symbol for the radiance factor is I/F. The radiance factor is given by

$$\mathrm{RADF}(i, e, g) = \frac{I}{F}(i, e, g) = \pi r(i, e, g),$$

$$= K\frac{w}{4}\frac{\mu_0}{\mu_0 + \mu}[p(g) + H(\mu_0/k)H(\mu/k) - 1]. \tag{10.5}$$

The *relative reflectance*, denoted by Γ, of a particulate sample is defined as the reflectance relative to that of standard surface consisting of an infinitely thick particulate medium of nonabsorbing, isotropic scatterers, with negligible opposition effect, and illuminated and viewed at the same geometry as the sample.

If the bidirectional reflectances of the sample and the standard are given by (8.70), the relative bidirectional reflectance is

$$\Gamma(i, e, g) = \frac{K_{Sa}}{K_{St}}w\frac{p(g) + H(\mu_0/K_{Sa}, w)H(\mu/K_{Sa}, w) - 1}{H(\mu_0/K_{St}, w = 1)H(\mu/K_{St}, w = 1)}. \tag{10.6}$$

where K_{Sa} and K_{St} are the porosity coefficients for the sample and standard, respectively. If $K_{Sa} = K_{St} \approx 1$, the relative bidirectional reflectance of a sample of isotropic scatterers measured outside of the opposition peak is given by

$$\Gamma(i, e, g \gg h) = \frac{w}{(1 + 2\gamma\mu_0)(1 + 2\gamma\mu)} \tag{10.7}$$

where approximation (8.53) has been used for the H functions.

The reduced reflectance, denoted by $r_r(i, e, g)$, is defined as the radiance factor divided by the Lomme I–Seeliger factor: $r_R(i, e, g) = (I/F)/[\mu_0(\mu_0 + \mu)]$.

10.3 Reciprocity

A powerful and useful theorem in reflectance work is the *principle of reciprocity*, which was first formulated by Helmholtz (Minnaert, 1941). The principle may be stated as follows: suppose a uniform surface is illuminated by light from a collimated source making an angle i with the surface normal, and observed by a detector that measures the light emerging from a small part of the illuminated surface at an angle e from the normal. Let the bidirectional reflectance of the surface under these conditions be $r(i,\ e,\ g)$. Next, interchange the positions of the source and detector, and denote the new reflectance by $r(e,\ i,\ g)$. The reciprocity principle states that the reflectances must satisfy the following relation:

$$r(i,e,g)/\cos i = r(e,i,g)/\cos e, \tag{10.8}$$

The proof is intuitive. Note that $Jr(i,e,g)\cos e$ is the radiance scattered by unit area of the surface when the source is at angle i and the detector at angle e with the normal. Similarly, $Jr(e,i,g)\cos i$ is the radiance scattered by a unit area in the reciprocal geometry. Imagine that the source and detector are connected by ray bundles of light, which emerge from the source, pass through this unit area of surface, are scattered within the medium, and pass into the detector. Although the rays follow a multitude of complex paths within the medium, these paths remain the same when the positions of the source and detector are interchanged. The property of reciprocity follows from the fact that photons can travel in either direction along the ray paths.

Equivalent statements of the principle of reciprocity are

$$\mathrm{BRDF}(i,e,g) = \mathrm{BRDF}(e,i,g), \tag{10.9a}$$

or

$$\mathrm{REFF}(i,e,g) = \mathrm{REFF}(e,i,g). \tag{10.9b}$$

The principle of reciprocity is useful for testing proposed scattering laws. If they do not obey reciprocity, they are incorrect. Note that (8.70) is reciprocal, as are the Lambert and Minnaert laws. Equation (10.8) may also be used to calculate the brightness of a surface at a given set of angles from measurements made at the reciprocal angles.

One caveat must be emphasized: an important assumption underlying the principle is that the surface is laterally uniform. If the measured brightness of a surface is found to actually violate reciprocity, the likely cause is a lateral inhomogeneity, which causes the detector to view two different kinds of surfaces when source and detector are interchanged.

10.4 Diffuse reflectance from a medium with a specularly reflecting surface

The upper surfaces of many particulate materials may be sufficiently smooth on a scale comparable to the wavelength that light is scattered both specularly from the surface and diffusely from below the surface. The specular component is known as *regular reflection.* Because a surface effectively becomes more optically smooth at large angles of incidence (Chapter 6), the regular component may become especially important at large phase angles.

The most familiar example of the combination of diffuse reflection and regular reflection is water containing suspended solids, as in rivers, lakes, and oceans. If a body of water is examined in the geometry for specular reflection of sunlight from the surface, a bright glare is seen, which is the reflected image of the Sun. However, if the same body is examined in an off-specular configuration, it looks dark and may be colored blue, brown, or green, depending on the nature of the suspended solids. Combinations of specular and diffuse reflectances are encountered in many other materials, including glossy paints, glazed ceramic tiles, glazed paper, polished wood, and leaves with waxy surfaces. In some commercial reflectance spectrometers the sample is held vertically or upside down, necessitating the use of a specularly reflecting cover glass if the material is a powder.

A first-order expression for the specular-diffuse, bidirectional reflectance $r_{sd}(i, e, \psi)$ of such types of surfaces may be obtained by simply combining the specular- and bidirectional-reflectance laws, after correcting for transmission through the specular surface:

$$r_{sd}(i, e, \psi) = R(i)\delta(e - i)\delta(\psi - \pi) + [1 - R(i)][1 - R(e')]r(i', e', \psi). \tag{10.10}$$

Here $R(x)$ is the Fresnel reflection coefficient averaged over the two directions of polarization; $\delta(x)$ is the delta function, $r(i', e', \psi)$ is the bidirectional reflectance of the scattering medium under the smooth surface, and i' and e' are given by the Snell relations, $\sin i = n \sin i'$ and $\sin e = n \sin e'$, where n is the index of refraction.

Equation (10.10) is an oversimplification in several respects. First, it assumes that the regular component remains collimated after reflection, whereas most natural surface have some roughness structure, in which to first order the surface may be treated as made of up smooth facets tilted in a variety of directions. The roughness spreads the specularly reflected light out into an elliptical cone whose angular half-width in the principal plane is approximately equal to twice the mean surface tilt.

Second, at small phase angles the radiance reflected from a rough surface is maximum at the specular angle, but at large angles of incidence and reflection the maximum intensity is skewed toward phase angles larger than specular. The reason for this is that the Fresnel reflection coefficients are rising rapidly and nonlinearly at large angles, so that light reflected from surface facets tilted away from the source makes a much larger contribution to the radiance than that from the facets tilted toward the source.

Third, equation (10.10) ignores the diffuse light that is multiply internally scattered between the surface and the scatterers. However, the lowest-order term of this component is of order $[1 - R]^2 Rr^2$ and, because R is usually small, can be ignored in most applications.

A particularly interesting application of combined specular and diffuse reflectances is the scattering of radar waves from a planetary surface. If an inner planet, such as the Moon, is illuminated by radar, most of the power is scattered back to the antenna by specular reflection from the surface, which is relatively smooth on the scale of radar wavelengths (Evans, 1962; Evans and Hagfors, 1971). However, the return also displays a diffuse background (Figure 10.1) whose size, relative to the specular component, depends on the planet and the wavelength. The nature of this diffuse component is uncertain. It is often attributed to specular reflection from surface facets tilted towards the antenna on exceptionally rough portions of the planet. However, if the regolith is sufficiently inhomogeneous, then some or all of the diffuse component must come from radiation that has been transmitted through the surface and diffusely scattered by the inhomogeneities (Thompson *et al.*, 1970; Pollack and Whitehill, 1972). Certain satellites of the outer planets

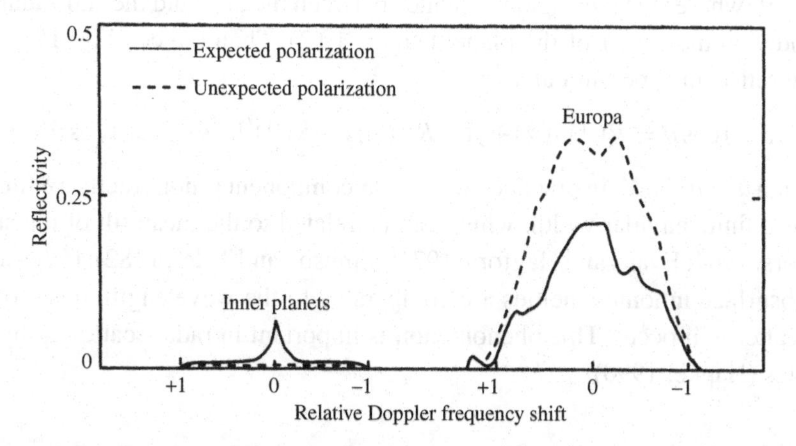

Figure 10.1 Typical radar signals received from solar system objects in circularly polarized light. The inner bodies, such as the Moon, have a strong specular component and weak diffuse component, whereas the situation is reversed for outer satellites, such as Europa. The terms "expected" and "unexpected" refer to the circular polarization ratios defined in Chapter 9. (Reproduced from Ostro [1982], copyright 1982 with permission of the University of Arizona Press.)

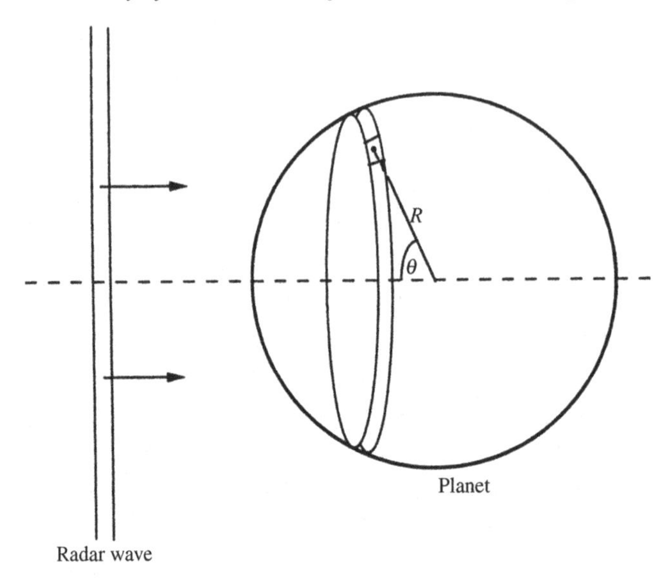

Figure 10.2 Geometry of radar scattering from a planet.

have strong diffuse components (Ostro, 1982) and it has been suggested that subsurface scattering is the dominant process in the reflectance of radar waves from these bodies (Hapke, 1990; Ostro and Shoemaker, 1990).

If the diffuse radar component is due to subsurface scattering then equation (10.10) should apply. Monostatic radar observes an area on a planet at $g = 0$ and $i = e = \vartheta$, where ϑ is the angular distance between the area and the sub-radar point subtended at the center of the planet (Figure 10.2). Then the combined specular–diffuse reflectance per unit area is

$$r_{sd}(\vartheta, \vartheta, 0)\cos\vartheta = R(\vartheta)\delta(\vartheta) + [1 - R(\vartheta)][1 - R(\vartheta)']r(\vartheta', \vartheta', 0)\cos\vartheta', \quad (10.11)$$

where $\sin\vartheta = n\sin\vartheta'$. In practice, the regular component is not exactly a δ function, but has a finite angular width, which can be related to the mean tilt of the surface (Hagfors, 1964; Evans and Hagfors, 1971; Simpson and Tyler, 1982). If the sizes of the subsurface inhomogeneities are comparable to the wavelength, then coherent backscatter will occur. This phenomenon is important in radar scattering by outer satellites (Hapke, 1990).

10.5 Oriented scatterers: applications to vegetation canopies

Some media consist of asymmetric particles with orientations in a preferred direction. An important example is vegetation, in which the "particles" are leaves, which are flattened structures whose faces are sometimes oriented with respect to the direction of sunlight or prevailing wind. A second example is a sediment in which the deposition process has preferentially oriented the particles. A third example occurs when a powder is subjected to such a high pressure that the grains are reoriented or even deformed so that they tend to lie with flat sides parallel to the surface of the medium. Such a medium will have a strong specular component. Kortum (1969) gives several examples of pressure-induced specular reflection.

If the particles are not randomly oriented, all of the scattering parameters defined in Chapter 7 will depend on the mean orientations of the particles with respect to the directions of incidence and emergence. In general, preferentially orienting the particles will completely change the angular pattern of the singly scattered component of the reflectance. Instead of the bidirectional reflectance being described by (8.47) or (8.70), the single-scattering term preserves the angular-scattering properties of the individual scatterers. For example, a semi-infinite medium consisting of parallel mirrors would have a specular, rather than diffuse, single-scattering law. It is only the multiply scattered component that tends to diffuse and randomize the light scattered by the individual particles.

Scattering by vegetation canopies is a major topic in terrestrial remote sensing. It has been discussed by many authors using a variety of approaches, including Suits (1972), Kimes and Kerchner (1982), Otterman (1983), Ross and Marshak (1984), Gerstl and Zardecki (1985b), Camillo (1987), Dickinson *et al.* (1990), and Pinty *et al.* (1989, 1990). See also Smith (1983).

Woessner and Hapke (1987) found that the IMSA model, which was derived for soils, gave a reasonably good fit to observational data on clover (Figure 10.3). Verstraete *et al.* (1990) and Pinty *et al.* (1990) have extended the IMSA model to the case of preferentially oriented scatterers. Although their model was derived specifically for applications to vegetation, a similar analysis applies to any medium in which the scatterers are not randomly oriented.

Suppose the particles are not equant, but are relatively flattened disks with mean cross-sectional area σ and with a preferred orientation. For instance, the disk could represent leaves or portions of leaves. Then, in the terms involving the incident light in the radiative-transfer equation, σ must be replaced by σ_i, where σ_i is the average of the projected cross-sectional areas of the particles onto a plane perpendicular to the direction of incidence. Let

$$\mu_i = \sigma_i/\sigma. \tag{10.12}$$

Similarly, in the terms involving the emerging rays, σ must be replaced by σ_e, the mean projected area of the scatterers onto a plane perpendicular to the direction to

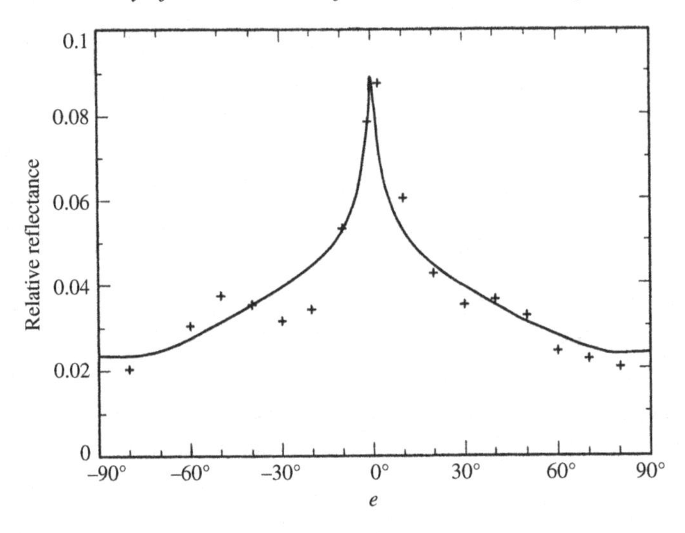

Figure 10.3 Scattering by clover. The crosses are values measured in the principal plane as a function of e with $i = 0$; the line is the theoretical IMSA bidirectional-reflectance function, equation (8.47). (Reproduced from Woessner and Hapke [1987], copyright 1987 with permission of Elsevier.)

the detector. Let

$$\mu_e = \sigma_e/\sigma. \tag{10.13}$$

In vegetation, μ_i and μ_e involve the *leaf droop*, which is a measure of the tilt of the average leaf surface from horizontal.

The multiple-scattering contribution to the scattered radiance inside the medium is much less sensitive to both the particle angular-scattering function and orientation. Hence, the same simplifying approximation may be made as in the derivation of (8.47), namely, that the radiance multiply scattered within the medium is similar to that which would occur if the particles were randomly oriented, isotropic scatterers. Then it is straightforward to show that these modifications result in an equation almost identical with (8.47), except that μ_0 and μ are replaced by μ_0/μ_i and μ/μ_e, respectively.

As with other particulate media, a vegetation canopy exhibits a strong opposition effect, which is known as the *hot spot*. Hapke *et al.* (1996) measured the circularly polarized reflectances of a number of types of vegetation and found that in most cases the hot spot is dominated by shadow hiding. Hence, the hot spot may be described by an equation of the form of (9.22), except that the interpretation of some of the parameters is changed.

With these modifications, the generalization of the bidirectional-reflectance model to the case of oriented scatterers, including vegetation canopies, is

$$r(i,e,g) = \frac{w}{4^1} \frac{\mu_0/\mu_i}{\mu_0/\mu_i + \mu/\mu_e} \times \{[1 + B_{SO}B_S(g)]p(i,e,g) + H(\mu_0/\mu_i)H(\mu/\mu_e) - 1\}. \quad (10.14)$$

Leaves have complex structures that act as internal scatterers. Hence, much of the light backscattered at small phase angles comes from the parts of the leaf surfaces close to the points at which the rays entered, so that the opposition-effect amplitude parameter B_{SO} should not be very different from 1. In the angular-width parameter $h_s = KN\sigma_E a_E/2$ the leaf filling factor is sufficiently low that $K = 1$. Because the leaves are large and have a fairly narrow size distribution, $a = (\sigma/\pi)^{1/2}$ and $\sigma a = \sigma^{3/2}/\pi^{1/2}$. However, σ must be corrected for leaf orientation. Within the opposition peak, $\mu_i \simeq \mu_e$; hence, to a sufficient approximation, σ may be replaced by $\sigma\overline{\mu}$, where $\overline{\mu} = (\mu_i + \mu_e)/2$. Thus,

$$h_s \simeq \frac{(N\sigma\bar{\mu})^{3/2}}{2^{1/2}}. \quad (10.15)$$

Note that the angular width of the hot spot decreases as the leaves get farther apart and that h_s also depends on the leaf droop.

The *leaf-area index* LAI(z) at an altitude z is defined as the total fraction of the area above that altitude occupied by leaves:

$$\mathrm{LAI}(z) = \int_z^\infty N\sigma\mu_h dz', \quad (10.16)$$

where μ_h is the cosine of the angle between the perpendicular to σ and the vertical. Hence, for horizontally oriented leaves the leaf-area index equals the optical depth: $\mathrm{LAI}(z) = \tau(z)$. If the integrand is approximately constant, $\mathrm{LAI}(z) = N\sigma\mu_h z$, where z is the depth below the top of the vegetation canopy. Thus, the hot-spot angular-width parameter is proportional to the leaf-area index:

$$h_s = \mathrm{LAI}\Delta z, \quad (10.17a)$$

where

$$\Delta z = \frac{\sigma^{1/2}\bar{\mu}^{3/2}}{2^{1^{1/2}}\mu_h}. \quad (10.17b)$$

For example, if the leaves all lie horizontally and the canopy is illuminated vertically, h is equal to the leaf-area index of a layer whose thickness Δz is approximately one-quarter of the mean leaf diameter.

Verstraete *et al.* (1990) have suggested an alternative expression to describe the hot spot. However, it is not clear that their expression is more accurate than (10.17).

Before leaving this topic, it must be emphasized that these equations are only a beginning. The general problem of the scattering of radiation by vegetation is formidable indeed, and it requires treating at least four layers of materials: the atmosphere, the vegetation canopy, the floor litter, and the soil. Additionally, the plants may not be randomly positioned, but aligned in rows. The treatment of these problems is beyond the scope of this book.

10.6 Reflectance of a layered medium

10.6.1 The diffusive reflectance and transmittance of layered systems

10.6.1.1 The diffusive radiance in a layer of finite thickness

In this section we will use the diffusive reflectance formalism of Section 8.6 to find approximate solutions for the reflectance and transmittance of layered media. We begin by considering a layer consisting of isotropically scattering particles with density N, and volume-average cross-sectional area σ, extinction efficiency Q_E, single-scattering albedo w and albedo factor $\gamma = \sqrt{1-w}$. As is usual in the diffusive-reflectance approximation, any light crossing the top or bottom surface of the layer is assumed to be immediately converted into an isotropic radiance. Let the total radiant power incident per unit area on the top surface be πI_0 so that the radiance (power per unit area per unit solid angle) entering the top surface and illuminating the medium is I_0.

Let the optical thickness of the layer be $\tau_0 = \int_{-\infty}^{\infty} E dz$, where $E = N\sigma Q_E$ is the extinction coefficient, and z is the altitude. Obviously, in order that τ_0 be finite, N must approach zero above and below certain altitudes that may be described as the top and bottom of the layer, respectively.

In the diffusive two-stream solution to the equation of radiative transfer the radiance was shown to have the general solution of the form given by equations (8.23),

$$I_1(\tau) = A(1-\gamma)e^{-2\gamma} + B(1+\gamma)e^{2\gamma\tau}, \tag{10.18a}$$

$$I_2(\tau) = A(1+\gamma)e^{-2\gamma} + B(1-\gamma)e^{2\gamma\tau}, \tag{10.18b}$$

where I_1 and I_2 are, respectively, the upward- and downward-going radiances in the layer, τ is the optical depth, and A and B are constants to be determined by appropriate boundary conditions. Approximate solutions for nonisotropic scatterers may be found by replacing w and τ by w^* and τ^*, respectively, using equations (8.59).

We will find the solutions to these equations for several layered problems of interest in the sections below.

10.6.1.2 Diffusive reflectance of an optically thick layer

Equations (10.18) were solved in Chapter 8 for the case of an infinitely thick layer. The reflectance was shown to be the diffusive expression (8.25), which is repeated here for completeness,

$$r_0 = \frac{1-\gamma}{1+\gamma}. \tag{10.19}$$

10.6.1.3 Diffusive reflectances of two-layer media

Consider the case where one layer with albedo factor γ lies on top of a second material that is infinitely thick and has diffusive reflectance $r_L = (1-\gamma_L)/(1+\gamma_L)$ in the absence of the upper layer. This might describe, for instance, a cloud over the ground or a snow pack over soil. If the upper layer is infinitely thick it has a reflectance $r_U = (1-\gamma)/(1+\gamma)$. The boundary conditions are that the radiance must be continuous across the top and bottom surfaces of the upper layer. At the top of this layer, this requires that

$$I_2(0) = A(1+\gamma) + B(1-\gamma) = I_0. \tag{10.20}$$

At the interface between the two layers, $\tau = \tau_0$, the upwelling radiance must be equal to the fraction of the downwelling radiance reflected back up from the lower layer:

$$\begin{aligned} I_1(\tau_0) &= A(1-\gamma)e^{-2\gamma\tau_0} + B(1+\gamma)e^{2\gamma\tau_0}. \\ &= r_L I_2(\tau_0) = r_L[A(1+\gamma)e^{-2\gamma\tau_0} + B(1-\gamma)e^{2\gamma\tau_0}] \end{aligned} \tag{10.21}$$

Solving equations (10.20) and (10.22) for A and B gives

$$A = \frac{I_0}{1+\gamma}\frac{1-r_L r_U}{(1-r_L r_U)+r_U(r_L-r_U)e^{-4\gamma\tau_0}}, \tag{10.22a}$$

$$B = \frac{I_0}{1-\gamma}\frac{r_U(r_L-r_U)^{-4\gamma\tau_0}}{(1-r_L r_U)+r_U(r_L-r_U)e^{-4\gamma\tau_0}} \tag{10.22b}$$

The diffusive reflectance r_0 of the system is the fraction of the incident power emerging from the upper layer,

$$r_0 = \frac{\pi I_1(0)}{\pi I_0} = \frac{A(1-\gamma)+B(1+\gamma)}{I_0},$$

which can be put into the form

$$r_0 = r_U\left[\left(1+\frac{1}{r_U}\frac{r_L-r_U}{1-r_L r_U}e^{-4\gamma\tau_0}\right)\Big/\left(1+r_U\frac{r_L-r_U}{1-r_L r_U}e^{-4\gamma\tau_0}\right)\right] \tag{10.23}$$

Let us explore some of the properties of this solution for r_0. The radiance in the upper layer and the degree to which the reflectance of the lower layer influences the total reflectance are governed by the quantity $4\gamma\tau_0$. As $\tau_0 \to \infty$, $r_0 \to r_U$, the diffusive reflectance of the upper layer, but the layer may be considered to be optically thick when $\gamma\tau_0 \gtrsim r_1$. When $\tau_0 \to 0$, $r_0 \to r_L$, the diffusive reflectance of the lower layer.

The diffusive transmittance t_0 of the upper layer in the system is the fraction of the incident power incident on the bottom surface of the upper layer,

$$t_0 = \frac{\pi I_2(\tau_0)}{\pi I_0} = \frac{A(1+\gamma)e^{-2\gamma\tau_0} + B(1-\gamma)e^{-2\gamma\tau_0}}{I_0},$$

which gives

$$t_0 = \left[\left(1 + r_U \frac{r_L - r_U}{1 - r_L r_U}\right) e^{2\gamma\tau_0}\right] \Big/ \left[1 + r_U \frac{r_L - r_U}{1 - r_L r_U} e^{-4\gamma\tau_0}\right]. \tag{10.24}$$

Values of the transmittance range from $t_0 = 1$ at $\tau_0 = 0$ to $t_0 = 0$ as $\tau_0 \to \infty$. Equation (10.24) is useful for estimating the radiance at the interface between two media, such as at the base of an atmosphere.

An interesting situation occurs when there is negligible absorption in the upper layer. This might approximate clouds or snow covering the surface of a planet. When $w \simeq 1$, $\gamma = 1$. Assume that γ is so small that $e^{-4\gamma\tau_0} \approx 1 - 4\gamma\tau_0$. Then expression (10.23) becomes

$$r_0 = \frac{r_L(1 - r_U^2) - 4\gamma\tau_0(r_L - r_U)}{(1 - r_U^2) - 4\gamma\tau_0 r_U(r_L - r_U)}.$$

To first order in γ, $r_U \simeq 1 - 2\gamma$ and $1 - r_U^2 = 4\gamma/(1+\gamma)^2 \simeq 4\gamma$, so that

$$r_0 \simeq \frac{r_L + (1 - r_L)\tau_0}{1 + (1 - r_L)\tau_0}. \tag{10.25}$$

Similarly,

$$t_0 \simeq \frac{1}{1 + (1 - r_L)\tau_0}. \tag{10.26}$$

When $\tau_0 = 0$, $r_0 = r_L$, and $\tau_0 = 1$; when $\tau_0 \gg 1$, $r_0 \simeq 1$; and $t_0 \simeq 0$; when $t_0 \simeq 0$; and $r_L = 0$, $r_0 = \tau_0/(1+\tau_0)$ and $t_0 = 1/(1+\tau_0)$.

Although the diffusive-reflectance model is not exact, nevertheless it is capable of giving answers that are sufficiently accurate for many applications. This is illustrated by Figure 10.4, which shows the two-layer diffusive-reflectance model, equation (10.23), fitted to a series of measurements by Wells *et al.* (1984). Thin layers of fine volcanic dust 1–5 μm in diameter were deposited on top of surfaces covered with white paint and black paint. The bidirectional reflectance $r(6°, 0°, 6°)$

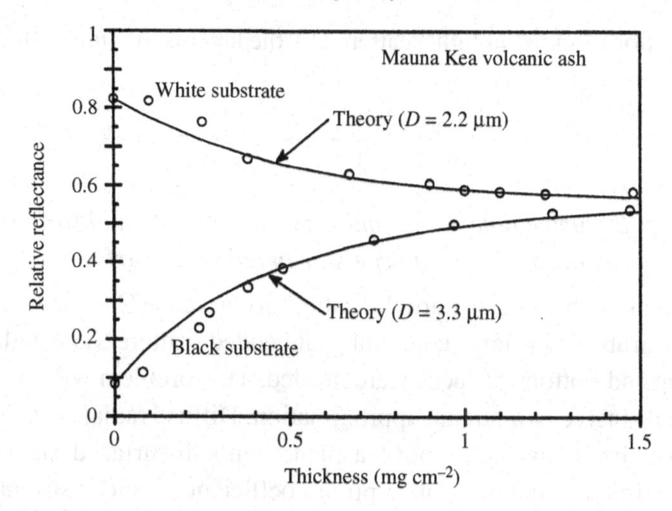

Figure 10.4 Reflectances of thin uniform layers of fine volcanic ash on bright and dark substrates plotted against the thicknesses of the layers. The points are values measured by Wells *et al.* (1984); the lines are equation (10.23) with the particle sizes that give the best fits.

was measured as a function of dust thickness in units of mass per unit area, M in mg/cm^2.

Now, for a uniform upper layer of thickness L, $\tau_0 = N\sigma Q_E L$. For equant particles of diameter $D \gg \lambda$ and density ρ, $Q_E = 1$ $M = N(\pi/6)D^3\rho L$, and $\sigma = (\pi/4)D^2$. Hence, $\tau_0 = \frac{3}{2}(M/\rho D)$. The only adjustable parameter in these fits is the mean dust-particle diameter. The sizes that gave the best fits were 2.0–3.5 μm, which are comfortably within the measured size range.

The effects of layering on spectra are discussed in Chapter 14.

10.6.1.4 *Diffusive reflectance and transmittance of an isolated layer*

The reflectance and transmittance of a layer suspended in space or lying on a material of very low albedo may be found from (10.23) and (10.24) by setting $r_L = 0$. This gives

$$r_0 = r_U \frac{1 - e^{-4\gamma\tau_0}}{1 - r_U^2 e^{-4\gamma\tau_0}}, \tag{10.27a}$$

$$t_0 = \frac{(1 - r_U^2)e^{-2\gamma\tau_0}}{1 - r_U^2 e^{-4\gamma\tau_0}}. \tag{10.27b}$$

The total fraction of incident light scattered by the layer is the sum of the reflectance and transmittance,

$$r_0 + t_0 = \frac{r_U + e^{-2\gamma\tau_0}}{1 + r_U e^{-2\gamma\tau_0}}. \tag{10.28}$$

10.6.1.5 Diffusive reflectance and transmittance of an absorbing and scattering slab with specularly reflecting surfaces

In the equivalent-slab model of particle scattering derived in Chapter 6 the scattering properties of an absorbing layer containing embedded scatterers and with specularly reflecting top and bottom surfaces were needed. This problem will now be solved by using the diffusive two-stream approximation. Diffuse radiance I_0 is uniformly incident from the hemisphere above a plane, smooth-surfaced slab of material having real refractive index n, absorption coefficient α, and distributed internal scattering coefficient s. The thickness of the slab is L. The reflection coefficients of the surface of the slab for diffuse light externally incident is S_e, and for diffuse light internally incident is S_i, where S_e and S_i are functions of m and were defined in Chapter 6. The light is assumed to be scattered isotropically by the distributed internal scatterers and by the surfaces of the slab.

Inside the slab, $A = \alpha$, $S = s$ and $E = \alpha + s$; hence, $w = S/E = s/(\alpha + s)$ and $\gamma = \sqrt{1-w} = \sqrt{\alpha/(\alpha+s)}$. The general solution for the radiation field inside the slab is given by equations (8.23), where $\tau = -Ez = -(\alpha+s)z$. Let the upper surface of the slab be at $z = 0$, and the lower surface at $z = -L$. The optical thickness of the slab is $\tau_0 = (\alpha+s)L$, and $2\gamma\tau_0 = 2\sqrt{\alpha(\alpha+s)}L$. The boundary conditions are as follows: at the lower surface, $\tau = \tau_0$, of the slab a fraction S_i of the downgoing radiance is reflected by the interface back into the upward direction. That is,

$$I_1(\tau_0) = S_i I_2(\tau_0),$$

or

$$A(1-\gamma)e^{-2\sqrt{\alpha(\alpha+s)}L} + B(1-\gamma)e^{2\sqrt{\alpha(\alpha+s)}L}$$
$$= S_i[A(1+\gamma)e^{-2\sqrt{\alpha(\alpha+s)}L} + B(1-\gamma)e^{2\sqrt{\alpha(\alpha+s)}L}].$$

At the upper surface, a fraction $(1 - S_e)$ of the incident radiance I_0 is transmitted through the surface into the slab, and a fraction S_i of the upgoing radiance is reflected by the surface back into the downward direction inside the slab. Thus,

$$I_2(0) = (1 - S_e)I_0 + S_i I_1(0)$$

or

$$A(1+\gamma) + B(1-\gamma) = (1 - S_e)I_0 + S_i[A(1-\gamma) + B(1+\gamma)].$$

Solving these equations for A and B gives

$$A = \frac{1-S_e}{1+\gamma}\frac{1-S_i r_i}{(1-S_i r_i)^2 - (S_i - r_i)^2 \exp(-4\sqrt{\alpha(\alpha+s)}L)} I_0, \tag{10.29a}$$

$$B = \frac{1-S_e}{1+\gamma}\frac{(S_i - r_i)\exp(-4\sqrt{\alpha(\alpha+s)}L)}{(1-S_i r_i)^2 - (S_i - r_i)^2 \exp(-4\sqrt{\alpha(\alpha+s)}L)} I_0, \tag{10.29b}$$

where

$$r_i = \frac{1-\gamma}{1+\gamma} = \frac{1-\sqrt{\alpha/(\alpha+S)}}{1+\sqrt{\alpha/(\alpha+S)}}$$

is the equivalent diffusive reflectance of a medium with $A=\alpha$, $S=s$, and $E=\alpha+s$. The external radiant power incident per unit area on the upper surface of the slab is πI_0. The power per unit area emerging from the upper surface is $^1[S_e I_0 + (1-S_i)I_1(0)]$, and that from the lower surface is $\pi(1-S_i)I_2(\tau_0)$. Hence, the reflectance of the slab is

$$\begin{aligned} r_0 &= S_e + (1-S_i)I_1(0)/I_0 = S_e + (1-S_i)[A(1-\gamma)+B(1+\gamma)]/I_0 \\ &= S_e + (1-S_e)(1-S_i)\frac{r_i(1-S_i r_i) + (S_i - r_i)\exp(-4\sqrt{\alpha(\alpha+S)}L)}{(1-S_i r_i) + (S_i - r_i)^2 \exp(-4\sqrt{\alpha(\alpha+S)}L)}. \end{aligned} \tag{10.30}$$

and the transmittance is

$$\begin{aligned} t_0 &= (1-S_i)I_2(\tau_0)/I_0 \\ &= (1-S_i)\left[A(1+\gamma)e^{-2\sqrt{\alpha(\alpha+S)}L} + B(1-\gamma)e^{2\sqrt{\alpha(\alpha+S)}L}\right]/I_0 \\ &= (1-S_i)\left[A(1+\gamma)e^{-2\sqrt{\alpha(\alpha+S)}L} + B(1-\gamma)e^{2\sqrt{\alpha(\alpha+S)}L}\right]/I_0 \\ &= (1-S_e)(1-S_i) \\ &\quad \times \frac{(1-S_i r_i)\exp(-2\sqrt{\alpha(\alpha+S)}L) + r_i(S_i - r_i)\exp(-2\sqrt{\alpha(\alpha+S)}L)}{(1-S_i r_i)^2 - (S_i - r_i)^2 \exp(-4\sqrt{\alpha(\alpha+S)}L)}. \end{aligned} \tag{10.31}$$

The total fraction of incident light scattered by the slab is

$$r_0 + t_0 = S_e + (1-S_e)(1-S_i)\frac{r_i + \exp(-2\sqrt{\alpha(\alpha+s)}L)}{(1-S_i r_i) + (r_i - S_i)\exp(-2\sqrt{\alpha(\alpha+S)}L)}.$$

If these expressions are taken to represent scattering by a particle, then $r_0 + t_0 = Q_s$, the nondiffractive part of the scattering efficiency. After a little algebra, the last equation may be written in the form

$$Q_s = S_e + (1-S_e)\frac{(1-S_i)\Theta}{1-S_i\Theta}, \tag{10.32a}$$

where

$$\Theta = \frac{r_i + \exp(-2\sqrt{\alpha(\alpha+s)}L)}{1 + r_i \exp(-2\sqrt{\alpha(\alpha+s)}L)}. \tag{10.32b}$$

Define the scattering difference ΔQ_S as the difference between the reflectance and transmittance,. Then $\Delta Q_s = r_0 - t_0$ can be put into the form

$$\Delta Q_S = S_e + (1 - S_e)(1 - S_i)\frac{\Psi}{1 - S_i \Psi}, \tag{10.33a}$$

where

$$\Psi = \frac{r_i - \exp(-2\sqrt{\alpha(\alpha+s)}L)}{1 - r_i \exp(-2\sqrt{\alpha(\alpha+s)}L)} \tag{10.33b}$$

Then

$$r_0 = (Q_S + \Delta Q_s)/2,$$

and

$$t_0 = (Q_S - \Delta Q_s)/2.$$

Equations (10.32) and (10.33) were the expressions used in the equivalent-slab model of a particle in Chapter 6.

If the slab is optically thick, $\sqrt{2\alpha(\alpha+s)}L \gg 1$, and

$$r_0 = S_e + (1 - S_e)(1 - S_i)\frac{r_i}{1 - S_i r_i}. \tag{10.34}$$

If we allow $\tau_0 \to \infty$, equation (10.34) is the diffusive reflectance of an optically thick layer of powder underneath a cover glass.

If the internal scatterers are not isotropic, but can be described by the asymmetry factor ξ, the reflectance and transmittance of the slab can be found to a rough approximation from the similarity equations (8.30) by substituting $w^* = (1-\xi)w/(1-\xi w)$ and $L^* = (1 - \xi w)L$ for w and L, respectively.

10.6.2 The bidirectional reflectance of a two-layer medium

Although the diffusive reflectance is convenient for estimating the radiance in a double-layer geometry, a more accurate model for the bidirectional reflectance is desirable. In this section we will find the two-stream solution for the combined bidirectional reflectance of an optically thin layer overlying an optically thick medium. For simplicity it will be assumed that $K = 1$ and the opposition effect will be ignored.

Consider an optically thin layer that contains particles of single-scattering albedo w_U and angular phase function $p_U(g)$, and with extinction coefficient E_U, lying on top of a second material of very large optical thickness, and containing particles of single-scattering albedo w_L and phase function $p_L(g)$, and with extinction

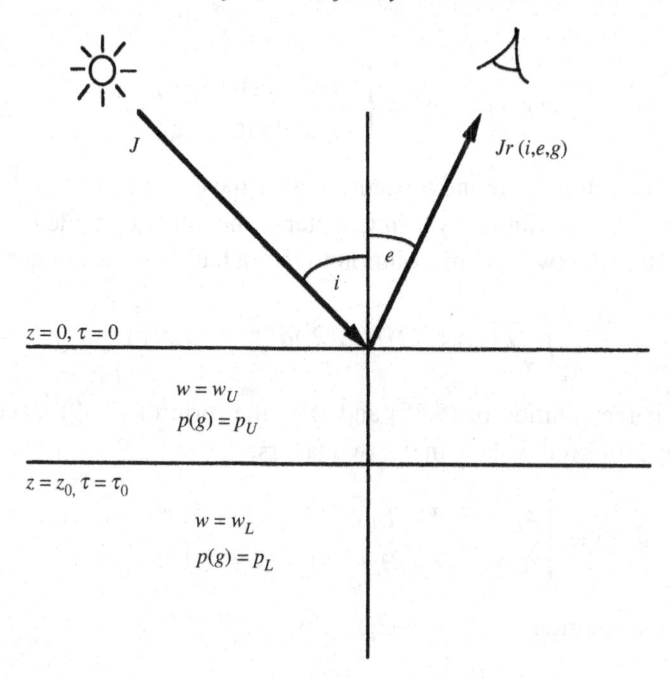

Figure 10.5 Schematic diagram of the two-layer bidirectional-reflectance problem.

coefficient E_L (see Figure 10.5). The optical thickness of the top layer is τ_0, and z_0 is the altitude of the interface. The medium is illuminated from above by a distant collimated source of irradiance J incident from a direction making an angle i with the vertical, and viewed by a distant detector from a direction making an angle e with the vertical.

Using the same procedure as in Chapter 8, the exact expression will be found for the singly scattered radiance for particles of arbitrary phase function. An approximate expression for the multiply scattered radiance will be found using the two-stream approximation with collimated source, and assuming that the particles scatter isotropically. From equation (8.34), the radiance at the detector can be written in the form

$$I_D(\Omega) = -\int_{\infty}^{0} \frac{1}{\mu}\left[F(\tau,\Omega) + \frac{w(\tau)}{4\pi}\int_{4\pi} p(\tau,\Omega',\Omega)I(\tau,\Omega')d\Omega'\right]e^{-\tau/\mu}d\tau. \tag{10.35}$$

For this problem,

$$F(\tau,\Omega) = J\frac{w(\tau)}{4\pi}p(\tau,g)e^{-\tau/\mu_0}, \tag{10.36a}$$

where

$$w(\tau), p(\tau, g) = \begin{cases} w_U, P_U(g)(\tau < \tau_0) \\ w_L, P_L(g)(\tau > \tau_0). \end{cases} \tag{10.36b}$$

Using the same assumptions and procedures as in the derivation of the bidirectional reflectance of a semi-infinite layer in Chapter 8, the integral inside the brackets is approximated by the two-stream solution to the radiative-transfer equation

$$\int_{4\pi} p(\tau, \Omega', \Omega) I(\tau, \Omega') d\Omega' \simeq 4\pi\varphi(\tau), \tag{10.37}$$

where $\varphi(\tau)$ is the solution of (8.38) and is of the form of (8.39), except that the constants have different values in the two layers:

$$\varphi(\tau) = \begin{cases} A_U e^{-2\gamma_U \tau} + B_U e^{2\gamma_U \tau} + C_U e^{-\tau/\mu_0}(\tau > \tau_0) \\ A_L e^{-2\gamma_U \tau} + B_L e^{2\gamma_U \tau} + C_L e^{-\tau/\mu_0}(\tau > \tau_0). \end{cases} \tag{10.38}$$

In addition, the solution requires the quantity

$$\Delta\varphi(\tau) = \frac{1}{2}\frac{d\varphi(\tau)}{d\tau}.$$

The constants C_U and C_L can be found by inserting (10.38) into the differential equation for $\varphi(\tau)$ and equating coefficients of $e^{-\tau/\mu_0}$. The remaining four constants are found from the boundary conditions as follows: (1) $\varphi(\tau)$ must remain finite as $\tau \to \infty$. This requires that $B_L = 0$. (2) The upward- and downward-going radiances must be continuous across the interface at τ_0. This is equivalent to requiring that φ and $\Delta\varphi$ be continuous at $\tau = \tau_0$. (3) At the top surface, from equation (8.44), $\varphi(0) = \frac{1}{2} d\varphi(0)/d\tau$. Conditions (2) and (3) give three simultaneous equations that can be solved for A_U, B_U, and C_L. The algebra is straightforward, but tedious.

The final result for the bidirectional reflectance is

$$r(i, e, g) = I_D/J = r_s(i, e, g) + r_m(i, e, g), \tag{10.39}$$

where r_s and r_m, are, respectively, the singly and multiply scattered components of the reflectance and are given by the following expressions:

$$r_s(i, e, g) = \frac{\mu_0}{\mu_0 + \mu} \times \left\{ \frac{W_U}{4\pi} p_U(g)[1 - e^{-1(1/\mu_0 + 1/\mu)\tau_0}] + \frac{w_L}{4\pi} P_L(g) e^{-(1/\mu_0 + 1/\mu)\tau_0} \right\}, \tag{10.40a}$$

$$r_m(i,e,g) = \frac{w_U}{4\pi}\left\{\frac{A_U}{1+2\gamma_U\mu}[1-e^{-(1/\mu+2\gamma_U)\tau_0}]\right.$$

$$\left.+\frac{B_U}{1-2\gamma_U\mu}[1-e^{-(1/\mu+2\gamma_U)\tau_0}]+\frac{C_U}{1+\mu/\mu_0}[1-e^{-(1/\mu_0+1/\mu)\tau_0}]\right\}$$

$$+\frac{w_L}{4^1}\left\{\frac{A_L}{1+2\gamma_L\mu}e^{-(1/\mu+2\gamma_L)\tau_0}+\frac{C_L}{1+\mu/\mu_0}e^{-(1/\mu_0+1/\mu)\tau_0}\right\},$$

(10.40b)

$$C_U = \frac{1-\gamma_U^2}{1/4\mu_0^2-\gamma_U^2}, \tag{10.41a}$$

$$C_L = -\frac{1-\gamma_U^2}{1/4\mu_0^2-\gamma_U^2}, \tag{10.41b}$$

$$B_U = \left\{\left[\frac{C_U}{1+\gamma_U}\right]\left(1+\frac{1}{2_{\mu_0}}\right)(\gamma_L-\gamma_U)e^{-2\gamma_\mu I_0}\right.$$

$$\left.+(C_L-C_U)\left(\gamma_L-\frac{1}{2\mu_0}\right)e^{-2\tau_0/\mu_0}\right\}$$

$$X\left\{(\gamma_L+\gamma_U)e^{2\gamma_U\tau_0}-[(1-\gamma_U)/(1+\gamma_U)](\gamma_L+\gamma_U)e^{2\gamma_U\tau_0}\right\}^{-1} \tag{10.41c}$$

$$A_U = -[(1-\gamma_U)/(1-\gamma_U)]B_U-[(1+1/2\mu_0)/(1+\gamma_U)]C_U, \tag{10.41d}$$

$$A_L = \frac{2\gamma_U e^{2\gamma_U}\tau_0 B_U-(C_L-C_U)(\gamma_U-1/2\mu_0)e^{-\tau_0/\mu_0}}{\gamma_L-\gamma_U}e^{2\gamma_L\tau 0}. \tag{10.41e}$$

Note that equations (10.40) and (10.41) for $r(i,e,g)$ include the specific bidirectional scattering properties of both the lower and upper layers, all orders of multiple scattering within both the lower and upper layers, and all orders of multiple scattering between the two layers.

Johnson *et al.* (2004) used the equations developed in this section to model dust-covered rocks on Mars and compared predictions of the model with laboratory measurements, with generally good agreement. These expressions should also be useful for calculating the combined bidirectional reflectance of such systems as a hazy atmosphere over a planetary regolith and the reflectance of a vegetation canopy over soil. For the latter application, the fact that the scatterers may not be randomly oriented may have to be taken into account, as discussed in Section 9.3.

10.7 Mixing formulas

10.7.1 Area mixtures

Few materials encountered in nature consist of only one type of particle. Hence, it is necessary to be able to calculate the reflectances of mixture of different particle types. In remote-sensing applications two kinds of mixtures are of interest: areal and intimate. *Areal mixtures* are also known as *inhomogeneous*, *linear*, *macroscopic*, and *checkerboard mixtures*. In an areal mixture, the surface area viewed by the detector consists of several unresolved, smaller patches, each of which consists of a pure material. In this case the total reflectance is simply the linear sum of each reflectance weighted by area. That is,

$$r = \Sigma_j F_j r_j \tag{10.42}$$

Here r may represent any type of reflectance, as appropriate, r_j is the same type of reflectance for the jth area, and F_j is the fraction of the area viewed by the detector that is occupied by the jth area.

10.7.2 Intimate mixtures

Intimate mixtures are also called *homogeneous*, *microscopic,* and *nonlinear mixtures.* In an intimate mixture the medium consists of different types of particles mixed homogeneously together in close proximity. In this case the averaging process is on the level of the individual particle, and the parameters that appear in the equation of radiative transfer are the volume averages of the properties of the various types of particles in the mixture weighted by cross-sectional area. Because the parameters appear in the reflectance equation nonlinearly, the reflectance of an intimate mixture is a nonlinear function of the reflectances of the pure end members, as has been noted empirically by several authors (e.g., Nash and Conel, 1974).

The formulas for intimate mixtures have another important use: calculating the effects on the reflectance of small asperities and subsurface fractures at the surfaces of grains. It was shown in Chapter 6 that such structures may be treated as small particles mixed with the large ones. If the surface structures are smaller than the wavelength, they can act as Rayleigh scatterers and absorbers. If the imaginary part of the refractive index is small, they act primarily as scatterers and increase the reflectance. However, if k is large, as for a metal, the asperities are efficient Rayleigh absorbers and decrease the reflectance.

The mixing formula for each parameter follows from the definition of that parameter, as given in Chapter 7. Because these formulae follow from the definitions in the radiative-transfer equation, they are independent of the method or type of approximation used to solve that equation.

Mixtures are often specified by the fractional mass of each component, rather than by the number of particles. Assume that the particles are equant, and let the subscript j refer to any property of the particles, such as size, shape, or composition. Then the cross-sectional area of the jth type of particle is

$$\sigma_j = \pi a_j^2 \tag{10.43}$$

where a_j is the radius of the jth type of particle. The bulk density of the jth component is

$$M_j = N_j \frac{4}{3} \pi a_j^3 \rho_j \tag{10.44}$$

where N_j is the number of particles of type j per unit volume, and ρ_j is their solid density. Then

$$N_j \sigma_j = \frac{3}{4} \frac{M_j}{\rho_j a_j} = \frac{3}{2} \frac{M_j}{\rho_j D_j} \tag{10.45}$$

where $D_j = 2a_j$ is the average size of the jth type of particle.

From the definitions is Section 7.3 the volume average single scattering albedo is

$$w = \frac{S}{E} = \frac{\sum_j N_j \sigma_j Q_{Sj}}{\sum_j N_j \sigma_j Q_{Ej}} = \left(\sum_j \frac{M_j Q_{Sj}}{\rho_j D_j} \right) \Big/ \left(\sum_j \frac{M_j Q_{Ej}}{\rho_j D_j} \right), \tag{10.46}$$

and the volume-average particle scattering function is

$$p(g) = \frac{G(g)}{S} = \frac{\sum_j N_j Q_{sj} p(g)}{\sum_j N_j \sigma_j Q_{sj}}$$

$$= \left(\sum_j \frac{M_j Q_{sj}}{\rho_j D_j} p_j(g) \right) \Big/ \left(\sum_j \frac{M_j Q_{sj}}{\rho_j D_j} \right). \tag{10.47}$$

Because, $w_j = Q_{Sj}/Q_{Ej}$ the last two expressions may be written in the alternative forms

$$w = \frac{\sum_j N_j \sigma_j Q_{Ej} w_j}{\sum_j N_j \sigma_j Q_{Ej}} = \left(\sum_i \frac{M_j Q_{Ej}}{\rho_j D_j} w_j \right) \Big/ \left(\sum_i \frac{M_j Q_{Ej}}{p \rho_j D_j} \right). \tag{10.48}$$

$$p(g) = \frac{\sum_j N_j \sigma_j Q_{Ej} w_j p_j(g)}{\sum_j N_j \sigma_j Q_{Ej} w_j} = \left(\sum_i \frac{M_j Q_{Ej}}{\rho_j D_j} w_j p_j(g) \right) \Big/ \left(\sum_i \frac{M_j Q_{Ej}}{\rho_j D_j} w_j \right). \tag{10.49}$$

If there is a continuous distribution of properties, the summations in these definitions are replaced by integrations.

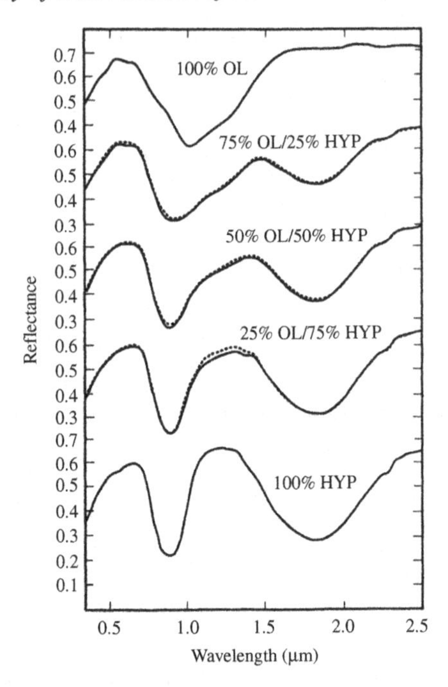

Figure 10.6 Comparison of the measured (solid lines) and calculated (dotted lines) spectra of 0.25/0.75, 0.5/0.5, and 0.75/0.25 intimate mixtures of olivine (OL) and hypersthene (HYP). The spectra are vertically offset to avoid confusion. (Reproduced from Johnson *et al.* [1983], copyright 1983 by the American Geophysical Union.)

If (10.49) is multiplied by $(1/4\pi)\cos g$ and integrated over solid angle, the following equation for the volume-average cosine asymmetry factor is obtained:

$$\xi = \frac{\sum_j N_j \sigma_j Q_{Ej} w_j \xi_j}{\sum_j N_j \sigma_j Q_{Ej} w_j} = \left(\sum_j \frac{M_j Q_{Ej}}{\rho_j D_j} w_j \xi_j \right) \Big/ \left(\sum_j \frac{M_j Q_{Ej}}{\rho_j D_j} w_j \right) \tag{10.50}$$

where ξ_j is the cosine asymmetry factor of the *j*th type of particle. The SHOE and CBOE parameters are

$$B_{so} = \frac{S(0)}{wp(0)} = \frac{\sum_j N_j \sigma_j Q_{Ej} S_j(0)}{\sum_j N_j \sigma_j Q_{Ej} w_j P_j(0)}$$

$$= \left(\sum_j \frac{M_j Q_{Ej}}{\rho_j D_j} S_j(0) \right) \Big/ \left(\sum_j \frac{M_j Q_{Ej}}{\rho_j D_j} w_j P_j(0) \right), \tag{10.51}$$

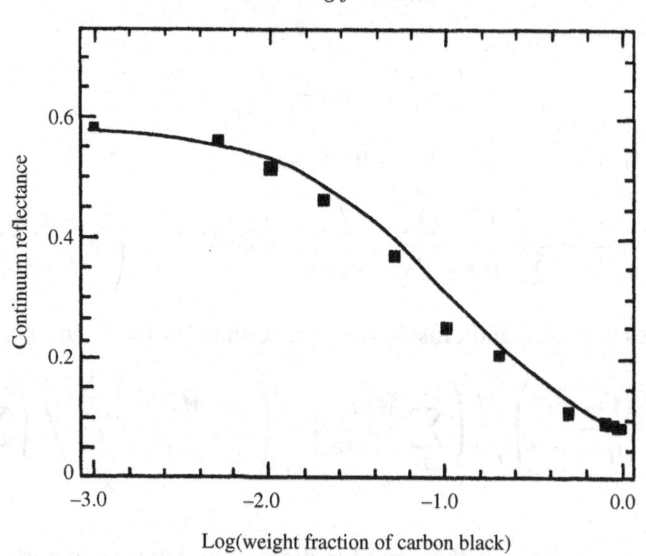

Figure 10.7 Reflectances of intimate mixtures of montmorillonite and charcoal. The points are the measured values. The line shows the values calculated from mixing theory. (Reproduced from Clark [1983], copyright 1983 by the American Geophysical Union.)

where $S_j(0)$ is the amount of light scattered into zero phase from the back surface of the *j*th particle, and

$$h_s = \frac{1}{2}K(\phi)Ea_E = \frac{3}{8}K(\phi)\left(\sum_j \frac{M_j Q_{Ej}}{\rho_j D_j}\right)^{3/2} \Bigg/ \left(\sum_j \frac{M_j}{\rho_j D_j^3}\right)^{1/2}, \tag{10.52a}$$

where the filling factor is

$$\phi = \sum_j N_j \frac{4}{3}\pi a_j^3 = \sum_j \frac{M_j}{\rho_j}, \tag{10.52b}$$

$$h_C = \frac{\lambda}{4\pi \Lambda_T} = \frac{\lambda}{4\pi}S(1-\xi). \tag{10.53}$$

If the mixture is a binary one, a useful formula due to Helfenstein and Veverka (private communication, 1990) may be derived as follows. For a mixture of two types of particles, equations (10.50) and (10.51) may be written

$$w = \frac{A_1 w_1 + A_2 w_2}{A_1 + A_2} \quad \text{and} \quad \xi = \frac{A_1 w_1 \xi_1 + A_2 w_2 \xi_2}{A_1 w_1 + A_2 w_2},$$

where $A_1 = N_1\sigma_1 Q_{E1}$ and $A_2 = N_2\sigma_2 Q_{E2}$. These equations can be solved for A_1 and A_2, and the resulting expressions substituted back into the expression for ξ, to give

$$\xi = \frac{(w - w_2)w_1\xi_1 - (w - w_1)w_1\xi_2}{(w_1 - w_2)w} \tag{10.54}$$

Finally, the espat function of a mixture is

$$W = \frac{1-w}{w} = \frac{k}{E} = \frac{\sum_j N_j\sigma_j Q_{Aj}}{\sum_j N_j\sigma_j Q_{Sj}} = \left(\sum_j \frac{M_j Q_{AJ}}{\rho_j D_j}\right) \Big/ \left(\sum_j \frac{M_j Q_{Sj}}{\rho_j D_j}\right) \tag{10.55}$$

If the particles are large and closely packed, so that $Q_{Ej} = 1$, and $Q_{sj} = w_j$

$$w = \left(\sum_j \frac{M_j(1-w_j)}{\rho_j D_j}\right) \Big/ \left(\sum_j \frac{M_j w_j}{\rho_j D_j}\right) = \left(\sum_j \frac{M_j w_j}{\rho_j D_j} W_j\right) \Big/ \left(\sum_j \frac{M_j w_j}{\rho_j D_j}\right), \tag{10.56}$$

where $W_j = (1 - w_j)/w_j$ is the espat function of the jth type of particle. Note that the weighting parameter of W_j in W is $M_j w_j/\rho_j D_j$, not $M_j/\rho_j D_j$.

The mixing equations have been verified experimentally by several workers. Clark (1983) and Johnson *et al.* (1983) measured the reflectances of mixtures and compared them with the values predicted from the end members. Comparisons between their predicted and measured reflectances are given in Figures 10.6 and 10.7.

11

Integrated reflectances and planetary photometry

11.1 Introduction

The basic expression for the bidirectional reflectance of a semi-infinite medium of isotropically scattering particles was derived in Chapter 8 and is given by equation (8.48). This expression was further refined to include an approximate correction for anisotropic scatterers (8.60) and considerations of the effects of porosity (8.70) and the opposition effect in Chapter 9 (9.47). Variants in different geometries were discussed in Chapter 10. In this chapter we will derive expressions that involve integration of the bidirectional reflectance over one or both hemispheres. The remote sensing of bodies of the solar system has its own nomenclature because of the special problems of astronomical observation. Expressions for quantities commonly encountered in planetary spectrophotometry will be derived. Only media for which the porosity factor $K = 1$ will be treated. Approximate expressions when $K > 1$ may be readily obtained by substituting one of the linear approximations (8.54 or 8.55) for the H function in the integrand.

11.2 Integrated reflectances

11.2.1 Biconical reflectances

If the source and detector do not occupy negligibly small solid angles as seen from the surface, appropriate expressions for the reflectances may be found by numerically integrating one of the above equations over the angular distribution of the radiance from the source and the angular distribution of the response of the detector. In general, such reflectances will be biconical. However, because they would be specific to each particular system, it would not be particularly useful to discuss biconical reflectances in detail.

11.2.2 The hemispherical reflectance (directional–hemispherical reflectance)

11.2.2.1 Introduction

The directional–hemispherical reflectance, or, more simply, the hemispherical reflectance, is denoted by r_h. It is defined as the ratio of the power scattered into the entire upper hemisphere by a unit area of the surface of the medium to the collimated power incident on the unit surface area. The hemispherical reflectance is also called the hemispherical albedo or the plane albedo in planetary photometric work, where it is denoted by A_h (see Section 11.3.4).

The hemispherical reflectance is important for two reasons. First, it is the quantity that is measured by many commercial reflectance spectrometers. Second, it is one of the properties of a material that determines the radiative equilibrium temperature.

The power incident on a unit area of the surface of a medium is $J\mu_0$. The power emitted into unit solid angle per unit area of the surface is $Y = Jr(i,e,g)\mu$. The total power emitted per unit area of the surface into the entire hemisphere above the surface is obtained by integrating the emitted power per unit area over the entire hemisphere into which the radiance is emitted. Hence, the general expression for the hemispherical reflectance is

$$r_h(i) = \frac{1}{J\mu_0}\int_{2\pi} Y(i,e,g)d\Omega_e = \frac{1}{\pi_0}\int_{e=0}^{\pi/2}\int_{\Psi=0}^{2\pi} r(i,e,g)\mu d\omega_e, \tag{11.1}$$

where $d\omega_e = \sin e\, de\, d\Psi = -d\,\mu d\Psi$.

The general bidirectional reflectance includes the opposition effect. However, note that the opposition effect typically has an angular half-width of about 0.1 radian or less. Hence, its relative contribution to the integral over a hemisphere is $\leq \pi(0.1)^2/2\pi \ll 1$, so that it makes a negligible contribution to the integrated reflectances. Thus, to a sufficient approximation, the opposition effect may be ignored in the integrand.

11.2.2.2 The hemispherical reflectance of a medium of isotropic scatterers

Well-separated scatterers The hemispherical reflectance r_h is obtained by inserting equation (8.49) for $r(i,e,g)$ when the particles are well separated into (11.1), which gives

$$r_h = \frac{w}{4\pi}\int_{\psi=0}^{2\pi}\int_{e=0}^{\pi/2}\frac{\mu}{\mu_0+\mu}H(\mu_0)H(\mu)\sin e\, de\, d\psi.$$

$$= \frac{w}{2}\int_{\mu=0}^{1}\frac{\mu}{\mu_0+\mu}H(\mu_0)H(\mu)d\mu,$$

which can be readily evaluated by writing $\mu/(\mu_0+\mu) = 1-\mu_0/(\mu_0+\mu)$,

$$\begin{aligned} r_{hi} &= \frac{w}{2}H(\mu_0)\int_0^1 H(\mu)d\mu - \frac{w}{2}\mu_0 H(\mu_0)\int_0^1 \frac{H(\mu)}{\mu_0+\mu}d\mu \\ &= \frac{1-\gamma^2}{2}H(\mu_0)\frac{2}{1+\gamma} - [H(\mu_0)+1] \\ &= 1-\gamma H(\mu_0). \end{aligned} \tag{11.2}$$

Using the two-stream approximation for $H(\mu_0)$ gives

$$r_h(i) = \frac{1-\gamma}{1+2\gamma\mu_0}. \tag{11.3}$$

An alternate derivation of (11.3) for isotropic scatterers can be found from equations (8.43)–(8.45). In the two-stream approximation, the upward-going flux at the surface is

$$I_1(0) = A(1-\gamma) + C\frac{2\mu_0 - 1}{2\mu_0} = \frac{J}{2\pi}2\mu_0\frac{1-\gamma}{1+2\gamma\mu_0}.$$

Hence, the total power per unit area leaving the surface in the upward direction is

$$\int_0^{\pi/2} I_1(0)\cos e 2\pi \sin e de = J\mu_0\frac{1-\gamma}{1+2\gamma\mu_0},$$

from which (11.3) follows. This expression was first obtained by Reichman (1973).

Arbitrary separations between scatterers The hemispherical reflectance for isotropically scattering particles with arbitrary separations can be similarly obtained from the two-stream approximation. The result is

$$r_h = \frac{1-\gamma}{1+2\gamma\mu_0/K}. \tag{11.4}$$

Equation (11.3) for $r_h(i)$ is plotted versus w in Figure 11.1 for several values of i for the case of isotropic scatterers.

11.2.2.3 The IMSA model for the hemispherical reflectance

Putting equation (8.60) for r in (11.1) gives

$$r_h(i) = \frac{w}{4\pi}\int_{\Psi=0}^{2\pi}\int_{\pi/2}^{e=0}\frac{\mu}{\mu_0+\mu}[p(g)+H(\mu_0)H(\mu)-1]\sin e de d\Psi \tag{11.5}$$

Separating the integral into two parts,

$$r_h = r_{hi} + r_{ha},$$

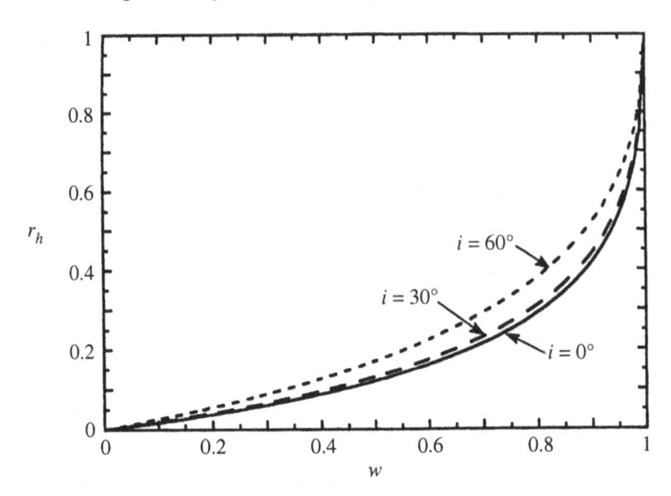

Figure 11.1 Hemispherical reflectance r_h (also called directional–hemispherical reflectance, hemispherical albedo, and plane albedo) for isotropic scatterers plotted against the single-scattering albedo for several values of the angle of incidence. The curve for $i = 60°$ is identical with r_0. The curves of the hemispherical–directional reflectance are identical if i is replaced by e.

where r_{hi} is the contribution to r_h by an equivalent medium of isotropic scatterers and has been evaluated in equation (11.3), and

$$r_{ha} = \frac{w}{4\pi} \int_{\Psi=0}^{2\pi} \int_{\mu=0}^{1} \frac{\mu}{\mu_0 + \mu} [p(g) - 1] \, d\mu d\Psi. \tag{11.6}$$

Equation (11.6) can be evaluated analytically if $p(g)$ can be expressed as a Legendre polynomial series,

$$\begin{aligned} p(g) &= 1 + \sum_{n-1}^{\infty} b_n p_n(\cos g) \\ &= 1 + \sum_{n-1}^{\infty} b \left[p_n(\mu_0) p_n(\mu) + \sum_{m-1}^{n} \frac{(n-m)!}{(n+m)!} p_{nm}(\mu_0) p_{nm}(\mu) \cos \Psi \right], \end{aligned} \tag{11.7}$$

where we have used the addition theorem for Legendre polynomials (Appendix C), in which $P_{nm}.(x)$ is an associated Legendre polynomial. Inserting (11.7) into (11.6), the integral over cos ψ vanishes and (11.6) becomes

$$r_{ha} = \frac{w}{2} \sum_{n=1}^{\infty} b_n p_n(\mu_0) \int_0^1 \frac{\mu}{\mu_0 + \mu} p_n(\mu) d\mu.$$

Since a Legendre polynomial can be expressed as a sum of cosines, the integral in the last equation consists of terms of the form $\int_0^1 \mu^j(\mu_0+\mu)^{-1}d\mu$. These can be evaluated using

$$\int \frac{x^j}{a+bx} = \frac{1}{b^{j+1}}\left[(-a)^j \ln(a+bx) + \sum_{k=0}^{j-1} \frac{j!(-a)^k(a+bx)^{j-k}}{(j-k)!(j-k)k!}\right]. \tag{11.8}$$

We will find r_h for the case where $p(g)$ consists of a first-order Legendre series, $p(g) = 1 + b_1 \cos g$. Then the integral over μ is

$$\int_0^1 \mu^2(\mu_0+\mu)^{-1}d\mu = 1/2 - \mu_0 + \mu_0^2 \ln(1/\mu_0 + 1).$$

Combining r_{hi} and r_{ha} gives

$$r_h(i) = 1 - \gamma H(\mu_0) + \frac{w}{2} b_1 \mu_0 \left[\frac{1}{2} - \mu_0 + \mu_0^2 \ln \frac{\mu_0+1}{\mu_0}\right]. \tag{11.9}$$

If we substitute the two-stream approximation (8.53) for $H(\mu_0)$ and use the approximation $\mu_0 \ln[(\mu_0+1)/\mu_0] \approx 2\mu_0/(1+2\mu_0)$, which has about the same accuracy as (8.53), equation (11.9) becomes

$$r_h(i) = \frac{1-\gamma}{1+2\gamma\mu_0} + b_1 \frac{w}{4} \frac{\mu_0}{1+2\mu_0}. \tag{11.10}$$

11.2.2.4 The modified IMSA model for the hemispherical reflectance

The directional–hemispherical reflectance of a medium of nonabsorbing particles must equal unity. However, equation (11.10) predicts that when $w = 1$, $r_h(i) = 1 + b_1\mu_0/4(1+2\mu_0)$. For example, when $i = 0$, $r_h = 1 + b_1/12$, so that if b_1 is as large as 1, the discrepancy is 8%. This is an indication of the magnitude of the general errors inherent in the approximation that places all the effects of particle anisotropy in the single-scattering term of the bidirectional reflectance. The errors can be reduced by using the MIMSA model.

Inserting equation (8.63), in which $p(g)$ is expressed in Legendre polynomials, into equation (11.1) gives

$$r_h(i) = \frac{w}{4\pi} \int_{\psi=0}^{2\pi} \int_{e=0}^{\pi/2} \frac{\mu}{\mu_0+\mu} \{p(g) + L_1(\mu_0)[H(\mu)-1] + L_1(\mu)[H(\mu_0)-1] + L_2[H(\mu)-1][H(\mu_0)-1]\} d\mu, \tag{11.11}$$

where $L_1(\mu_0)$, $L_1(\mu)$, and L_2 have been defined in Section 8.7.4.4. Inserting the expressions for the Ls, with the Legendre polynomial addition theorem for $p(g)$, and carrying out the integration over azimuth, after some algebra (11.11) can be put into the form

$$r_h(i) = \frac{w}{2}\int_0^1 \frac{\mu}{\mu_0+\mu} H(\mu_0)H(\mu)d\mu + \sum_{n=1}^{\infty} b_n\{P_n(\mu_0)+A_n[H(\mu_0)-1]\}$$
$$\times\left\{\frac{w}{2}\int_0^1 \frac{\mu}{\mu_0+\mu}P_n(\mu)d\mu\right.$$
$$\left.+A_n\frac{w}{2}\int_0^1\frac{\mu}{\mu_0+\mu}H(\mu)d\mu - A_n\frac{w}{2}\int_0^1\frac{\mu}{\mu_0+\mu}d\mu\right\}, \quad (11.12)$$

where A_n has been defined in (8.66d). The first term on the right of (11.12) is the hemispherical reflectance for isotropic scatterers, equation (11.2). The first and third integrals inside the curly brackets are all of the form $\int_0^1 \mu^j(\mu_0+\mu)^{-1}d\mu$ and, as discussed in the preceding section, can be evaluated using (11.8). The second integral inside the curly brackets is

$$A_n\frac{w}{2}\int_0^1\frac{\mu}{\mu_0+\mu}H(\mu)d\mu = A_n\frac{w}{2}\int_0^1\left(1-\frac{\mu_0}{\mu_0+\mu}\right)H(\mu)d\mu$$
$$= A_n\left[(1-\gamma)-\frac{H(\mu_0)-1}{H(\mu_0)}\right] = A_n\left[\frac{1}{H(\mu_0)}-\gamma\right].$$

For the case where $p(g)$ is a first-order Legendre polynomial,

$$r_h(i) = 1-\gamma H(\mu_0)+b_1\left\{\mu_0-\frac{1}{2}[H(\mu_0)-1]\right\}$$
$$\times\left\{\frac{w}{2}\left[\frac{1}{2}-\mu_0+\mu_0^2\ln\frac{1+\mu_0}{\mu_0}\right]\right.$$
$$\left.+\frac{1}{2}\left[\frac{1}{H(\mu_0)}-\gamma\right]-\frac{w}{4}\left[1-\mu_0\ln\frac{1+\mu_0}{\mu_0}\right]\right\}. \quad (11.13)$$

The improvement of the MIMSA over the IMSA model may be judged by the fact that when $w=1$, conservation of energy requires that $r_h(i)=0$. When $i=0$ equation (11.13) predicts that $r_h(0)=1+0.0088b_1$. Since $|b_j|\le 1$, the error is $< 1\%$.

11.2.3 The hemispherical–directional reflectance

The hemispherical–directional reflectance is defined as the ratio of (1) the radiance scattered into a particular direction from the surface of a medium that is being uniformly illuminated from all directions in the hemisphere above the surface to (2) the incident radiance. To calculate the hemispherical–directional reflectance, the incident irradiance J must be replaced by $I_0 d\Omega_i$, where I_0 is the incident radiance, assumed independent of direction, and $d\Omega_i = \sin i\, di\, d\psi$. The radiance scattered from the surface into a given direction is the integral of $I_0 r(i, e, g) d\Omega_i$ over the hemisphere from which the incident radiance comes. Hence, the hemispherical–directional reflectance is

$$r_{hd}(e) = \frac{1}{I_0} \int_{2\pi} I_0 r(i, e, g) d\Omega_i = \int_{2\pi} r(i, e, g) d\Omega_i. \tag{11.14}$$

By the same arguments as used in deriving the hemispherical reflectance, the contribution of the opposition effect to the total scattered radiance is negligible. Because the bidirectional reflectance must obey reciprocity (Section 10.3),

$$r_{hd} = \int_{2\pi} r(i, e, g) d\Omega_e = \int_{2\pi} r(e, i, g) \frac{\mu}{\mu_0} d\Omega_e = r_h(e). \tag{11.15}$$

Hence, the hemispherical–directional reflectance has the same functional dependence on e that the directional–hemispherical reflectance has on i. For isotropic scatterers we can write, by inspection,

$$r_{hd}(e) = 1 - \gamma H(\mu) \tag{11.16a}$$

Using the two-stream approximation for $H(\mu)$,

$$r_{hd}(e) \frac{1-\gamma}{1+2\gamma\mu} \tag{11.16b}$$

If i is replaced by e in Figure 11.1, the curve shows r_{hd} as a function of w for several values of e.

Expressions for the IMSA and MIMSA models can be found by replacing i with e in the appropriate equations in Section 11.2.2.

11.2.4 The spherical reflectance (bihemispherical reflectance)

The bihemispherical reflectance, or, more simply, the spherical reflectance, is denoted by r_s. It is the ratio of (1) the total power scattered into the upward hemisphere from unit area of a surface that is being uniformly illuminated by diffuse radiance from the entire upper hemisphere to (2) the total power incident on unit area of the surface. In Section 11.3.7 this quantity will be shown to be equivalent to the Bond or spherical albedo A_s of a spherical planet.

The total power incident on unit area of the surface is $\int_{2\pi} I_0 \mu_0 2\pi \sin i \, di = I_0 \pi$, where I_0 is the incident radiance. The total power scattered into the upper hemisphere is the integral of $I_0 d\Omega_i r \mu d\Omega_e$ over both the direction from which the radiance comes and the direction into which it is scattered. Hence, the spherical reflectance is

$$r_s = \frac{1}{\pi} \int_{2\pi} \int_{2\pi} r(i, e, g) \mu d\Omega_e d\Omega_i. \tag{11.17}$$

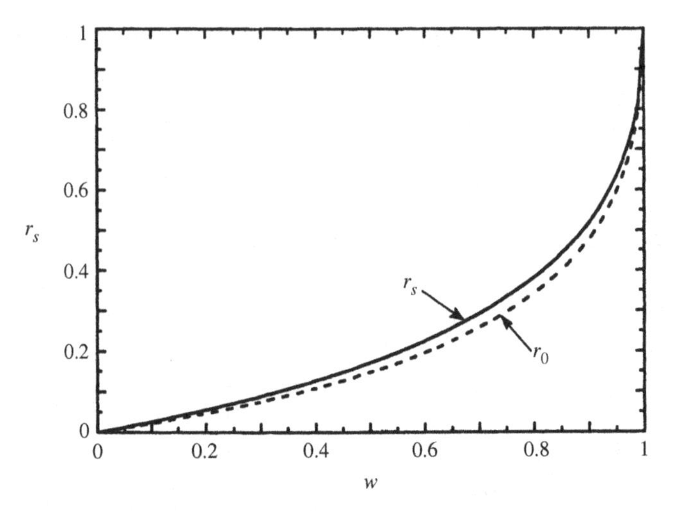

Figure 11.2 Spherical reflectance r_s (also called bihemispherical reflectance and Bond albedo) vs. w for isotropic scatterers. The exact and approximate expressions are indistinguishable on this scale. Also shown is the diffusive reflectance r_0.

Equation (11.17) will be evaluated first for the case of isotropic scatterers. The opposition effect can be ignored. Then r is independent of azimuth and we may put $d\Omega_i = 2\pi d\mu_0$ and $d\Omega_e = 2\pi d\mu$. Inserting (8.49) for r, and using the result from (11.2),

$$r_s = \frac{1}{\pi}\int_0^1 \int_0^1 \frac{w}{4\pi}\frac{\mu_0\mu}{\mu_0+\mu} H(\mu_0)H(\mu)2\pi d\mu_0 = 2\int_0^1 [1-\gamma H(\mu_0)]\,\mu_0 d\mu_0.$$

Hence,

$$r_s = 1 - 2\gamma H_1, \tag{11.18}$$

where H_1 is the first moment of the H function (equation [8.50]). The quantities H_1 and r_s have been calculated by numerical integration by Chamberlain and Smith (1970). The spherical reflectance is plotted versus w in Figure 11.2.

An approximate expression for r_s may be obtained by using (8.58) for H_1 in (11.18) to obtain

$$r_s \simeq r_0\left(1 - \frac{1}{3}\frac{\gamma}{1+\gamma}\right). \tag{11.19}$$

This approximation has relative errors of less than 2.0%.

Note that r_s and the diffusive reflectance r_0 are both types of bihemispherical reflectances and have very similar functional dependences on w. However, they are not equal. The reason has to do with the way the incident radiance is assumed to interact with the medium in the derivations of the two expressions. In the diffusive reflectance r_0 the incidence radiance is assumed to become uniformly distributed with angle as soon as it crosses the mathematical upper surface. However, in the more physically realistic solution for r_s, the interaction takes place via the source function within a layer a few times $1/E$ thick below the upper surface. This alters the distribution with depth of the radiance within the medium, and hence alters the reflectance.

It was emphasized in Chapter 8 that the diffusive reflectance is a useful, mathematically simple quantity that gives surprisingly accurate first-order estimates of reflectance. As a demonstration of the power of the diffusive reflectance, it is left as an exercise for the reader to show that the following identities hold for media of isotropic scatterers:

$$\text{(1)} \qquad \Gamma(60^\circ, 60^\circ, g \gg h) = r_0, \tag{11.20a}$$

$$\text{(2)} \qquad r_h(60^\circ) = r_0, \tag{11.20b}$$

$$\text{(3)} \qquad r_{dh}(60^\circ) = r_0. \tag{11.20c}$$

Because of the similar behaviors of the diffusive and spherical reflectances, it is reasonable to ask if similarity relations can be found that will convert the expression for r_s for isotropic scatterers to one applicable to nonisotropic scatterers. Although a general answer cannot be given, this question has been investigated numerically by Van de Hulst (1974). He calculated the spherical reflectance of a medium of scatterers having an angular function that can be described by the Henyey–Greenstein function,

$$p(\theta) = \frac{1-\xi^2}{(1-2\xi\cos\theta+\xi^2)^{3/2}},$$

where ξ is the cosine asymmetry factor. Van de Hulst finds that replacing w and γ by w^* and γ^*, respectively, in (11.18), where these quantities are given by the similarity relations

$$w^* = \frac{1-\xi}{1-\xi w} w, \tag{11.21a}$$

$$\gamma^* = \left[\frac{1-w}{1-\xi w}\right]^{1/2}, \tag{11.21b}$$

gives an approximation for r_s that is accurate to 0.002 for all values of w and ξ.

11.3 Planetary photometry

11.3.1 Introduction

Planetary scientists usually describe the photometric properties of solar-system bodies by several different kinds of reflectances known as albedos and photometric functions. The word *albedo* comes from the Latin word for whiteness. Just as there are many reflectances, so there are several different kinds of albedos, depending on the geometry. Historically, most of these quantities arose in attempts to characterize the scattering properties of the surface of the Moon as observed from the Earth, and the definitions were then extended to other bodies of the solar system.

In addition to defining the various quantities for a planet whose surface has arbitrary photometric properties, the following sections will derive approximate analytic expressions for these quantities on a spherical body covered with an optically thick, uniform, particulate regolith whose bidirectional reflectance can be described by the IMSA model with opposition effect, equation (9.47) with $K = 1$. As usual, $r(i, e, g)$ will denote the bidirectional reflectance of a surface, and J the incident collimated irradiance. The radiance $I(i, e, g)$ interacting with the eye produces the sensation of brightness, and the terms *radiance* and *brightness* are often used interchangeably in planetary work.

11.3.2 The normal albedo

The *normal albedo* A_n is the ratio of the brightness of a surface observed at zero phase angle from an arbitrary direction to the brightness of a perfectly diffuse surface located at the same position, but illuminated and observed perpendicularly. That is, the normal albedo is the radiance factor of the surface at zero phase.

Thus,

$$A_n = [Jr(e, e, 0)]/[J/\pi] = \pi r(e, e, 0). \tag{11.22}$$

Using (9.47) with $K = 1$,

$$A_n = \frac{w}{8}\{p(0)(1 + B_{S0}) + [H(\mu)^2 - 1][1 + B_{C0}]\}. \tag{11.23}$$

In general, the normal albedo is a function of e. However, for dark bodies of the solar system, including the Moon, Mercury, and many asteroids, $[H(\mu)^2 - 1] \ll 1$, so that to a first approximation for these bodies

$$A_n \approx \frac{w}{8} p(0)(1 + B_{S0})(1 + B_{C0}) \tag{11.24}$$

which is constant, independent of e. Thus, the normal albedos of different areas on the surfaces of these bodies may be intercompared even though they are observed at different angles. Hence, the normal albedo is a useful parameter to characterize their surfaces.

11.3.3 The photometric function

The ratio of the brightness of a surface viewed at a fixed e, but varying i and g, to its value at $g = 0$ is called the *photometric function* of the surface:

$$f(i,e,g) = r(i,e,g)/r(e,e,0). \tag{11.25}$$

Then the radiance scattered by the surface is given by

$$I(i,e,g) = Jr(i,e,g) = Jr(e,e,0)f(i,e,g) = \frac{J}{\pi}A_n f(i,e,g). \tag{11.26}$$

For a Lambert surface, $A_n = A_L$, and $f(i,e,g) = \mu_0$. For a particulate medium that obeys (9.47),

$$f(i,e,g) = 2\frac{\mu_0}{\mu_0+\mu}\frac{\{p(g)[1+B_{S0}B_S(g)]+[H(\mu_0)H(\mu)-1]\}[1+B_{C0}B_C(g)]}{\{p(0)(1+B_{S0})+[H(\mu)^2-1]\}[1+B_{C0}]}. \tag{11.27}$$

The Moon keeps the same face toward the Earth; hence, a given area will always be viewed from the Earth at nearly the same e. Because the lunar surface has a low albedo, $H(\mu_0)H(\mu)-1 \ll 1$, and $H(\mu)^2-1 \ll 1$. Hence, to a first appoximation, for areas on the Moon observed from Earth,

$$f(i,e,g) \simeq 2\frac{\mu_0}{\mu_0+\mu}\frac{p(g)}{p(0)}\frac{[1+B_{S0}B_S(g)][1+B_{C0}B_C(g)]}{(1+B_{S0})(1+B_{C0})}. \tag{11.28}$$

This expression has the same functional dependence on i, e, and g as does the generalized Lommel–Seeliger law, equation (8.35a). KenKnight, Rosenberg, and Wehner (1967) have shown that the relative brightness of different areas on the Moon are described to a good approximation by the Lommel–Seeliger law.

11.3.4 The hemispherical albedo (plane albedo)

The *hemispherical albedo* A_h is the total fraction of collimated irradiance incident on unit area of surface that is scattered into the upward hemisphere. It is also known as the plane albedo. The hemispherical albedo is the same as the directional–hemispherical or hemispherical reflectance; that is,

$$A_h(i) = r_h(i). \tag{11.29}$$

Explicit expressions for this quantity have been defined, and analytic approximations derived in Section 11.2.2. It is plotted versus w and i in Figure 11.1.

11.3.5 The physical albedo (geometric albedo)

The *physical albedo* is also known as the *geometric albedo*. It is defined as the ratio of the integral brightness of a body at $g = 0$ to the brightness of a perfect Lambert disk of the same radius and at the same distance as the body, but illuminated and observed perpendicularly. The physical albedo is the weighted average of the normal albedo over the illuminated area of the body.

For a spherical body of radius R the physical albedo is given by

$$A_p = \left[\int_{A(i)} Jr(e,e,0)\mu dA\right] / \left[(J/\pi)\pi R^2\right] = R^{-2}\int_{A(i)} r(e,e,0)\mu\, dA, \quad (11.30)$$

where $dA = 2\pi R^2 \sin e\, de = -2\pi R^2 d\mu$ is the increment of area on the surface of the body, and $A(i)$ is the area of the illuminated hemisphere. Using (9.47),

$$\begin{aligned} A_p &= \frac{w}{4}\int_0^1 \left\{p(0)(1+B_{S0}) + [H^2(\mu) - 1]\right\}[1 + B_{C0}]\mu d\mu \\ &= \frac{w}{8}[p(0)(1+B_{S0}) - 1](1+B_{C0}) + \frac{w}{4}(1+B_{C0})\int_0^1 H^2(\mu)\mu d\mu. \end{aligned} \quad (11.31)$$

If the opposition effect is ignored, the last term on the right-hand side of (11.31) is the physical albedo of a planet covered with isotropically scattering particles. Denote this quantity by A_{pi}. It is plotted in Figure 11.3. An approximate expression for A_{pi} may be found be using the linear approximation for the H function, equation (8.55). If terms of order $r_0^3/24$ are neglected, the expression

$$A_{pi} = \frac{w}{4}\int_0^1 H^2(\mu)\mu d\mu \approx \frac{1}{2}r_0 + \frac{1}{6}r_0^2 \quad (11.32)$$

is obtained. This has relative errors of less than 3% everywhere.

Alternatively, empirically fitting a second-order polynomial to A_{pi} gives an expression for A_{pi} that is accurate to better than 0.4% everywhere:

$$A_{pi} \simeq 0.49r_0 + 0.196r_0^2. \quad (11.33)$$

Then (11.31) becomes

$$A_p; \left\{\frac{w}{8}[p(0)(1+B_{s0}) - 1] + (0.49r_0 + 0.19r_0^2)\right\}(1+B_{s0}). \quad (11.34)$$

Note that A_p may exceed unity if the opposition effect is large or if the particles of the medium are strongly backscattering. This simply means that the bidirectional reflectance of the medium is more backscattering than that of a Lambert surface and does not violate any restrictions imposed by conservation of energy.

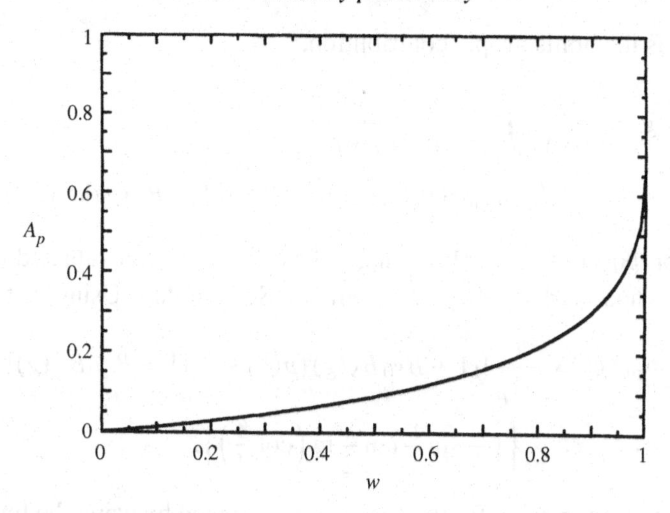

Figure 11.3 Physical albedo A_p (also called geometric albedo) of a planet covered with a medium of isotropic scatterers vs. w. The opposition effect is ignored. The exact and approximate expressions are indistinguishable on this scale.

11.3.6 The integral phase function

The *integral phase function* of a body is the relative brightness of the entire planet seen at a particular phase angle, normalized to its brightness at zero phase angle. Thus,

$$\Phi_p(g) = \left[\int_{A(i,v)} Jr(i,e,g)\mu dA\right] \Big/ \left[\int_{A(i)} Jr(e,e,0)\mu\, dA\right], \tag{11.35}$$

where $A(i, v)$ is the area of the planet that is both illuminated and visible at phase angle g, and $A(i)$ is the area of the illuminated part. If the body is a sphere of radius R, then, from (11.30),

$$\Phi_p(g) = \frac{1}{R^2 A_p}\int_{A(i,v)} r(i,e,g)\mu dA. \tag{11.36}$$

We will derive an approximate expression for a spherical planet covered by a soil whose bidirectional reflectance can be described by (9.47). The phase function can be written in the form

$$\Phi_p(g) = \Phi_{pi}(g) + \Phi_{pn}(g) \tag{11.37}$$

where $\Phi_{pi}(g)$ is the isotropic portion of the phase function,

$$\Phi_{pi}(g) = \frac{1}{R^2 A_p}\int_{A(i,v)} \frac{w}{4\pi}\frac{\mu_0\mu}{\mu_0+\mu} H(\mu_0)H(\mu)[1+B_{S0}B_S(g)][1+B_{C0}B_C(g)]dA, \tag{11.38}$$

and $\Phi_{pn}(g)$ is the nonisotropic contribution,

$$\Phi_{pn}(g) = \frac{1}{R^2 A_p} \int_{A(i,v)} \frac{w}{4\pi} \frac{\mu_0 \mu}{\mu_0 + \mu} \times \{[1 + B_{S0} B_S(g)] p(g) - 1\}[1 + B_{C0} B_C(g)] dA. \tag{11.39}$$

A quantity proportional to the last integral has already been evaluated in Chapter 6 for a sphere whose surface obeys the Lommel–Seeliger law. Using that result gives

$$\Phi_{pn}(g) = \frac{w}{2A_p} \{[1 + B_{S0} B_S(g)] p(g) - 1\}[1 + B_{C0} B_C(g)] \times \left\{1 - \sin\frac{g}{2} \tan\frac{g}{2} \ln\left(\cot\frac{g}{4}\right)\right\}. \tag{11.40}$$

An approximate expression for $\Phi_{pi}(g)$ may be obtained by using the linear approximation for the H functions, equation (8.55), and ignoring terms of order $r_0^3/24$, as was done in the derivation of the approximation for A_p. The integration is straightforward and gives

$$\Phi_p(g) = \frac{1}{2A_p} r_0 \left\{[1 - r_0]\left[1 - \sin\frac{g}{2} \tan\frac{g}{2} \ln\left(\cot\frac{g}{4}\right)\right] + \frac{4}{3} r_0 \frac{\sin g + (\pi - g)\cos g}{\pi}\right\} \times \{1 + B_{S0} B_S(g)\}\{1 + B_{C0} B_C(g)\}. \tag{11.41}$$

Combining equations (11.37)–(11.41) gives

$$\Phi(g) = \frac{r_0}{2A_p} \left\{\left[\frac{(1+\gamma)^2}{4} \{[1 + B_{S0} B_S(g)] p(g) - 1\} + [1 - r_0]\right] \times \left[1 - \sin\frac{g}{2} \tan\frac{g}{2} \ln\left(\cot\frac{g}{4}\right)\right] + \frac{4}{3} r_0 \frac{\sin g + (\pi - g)\cos g}{\pi}\right\} [1 + B_{c0} B_c(g)], \tag{11.42}$$

where A_p is given by (11.34). The first term of (11.42) describes a sphere covered by a Lommel–Seeliger scattering surface, modified by the particle phase function and the opposition effect. This term will be dominant for low-albedo bodies, such as the Moon. The second term describes a sphere covered by a Lambert scattering surface. This term will dominate for high-albedo bodies, such as icy satellites.

Equations based on (11.42) have been used to describe the integral phase functions of a wide variety of bodies in the solar system (Hapke, 1984; Buratti, 1985; Helfenstein and Veverka, 1987; Herbst *et al.*, 1987; Simonelli and Veverka, 1987; Veverka *et al.*, 1988; Bowell *et al.*, 1989; Domingue *et al.*, 1991). In order to model light scattering from planetary bodies more realistically, equation (11.42) must be modified to include the shadowing and other phenomena caused by large-scale roughness. These modifications will be described in Chapter 12.

11.3.7 The spherical albedo (Bond albedo)

The *spherical albedo*, also called the *Bond albedo* and *global albedo*, is the total fraction of incident irradiance scattered by a body into all directions. It bears the same relation to a planet as the single-scattering albedo w does to a particle. The spherical albedo may be found by integrating the fraction of power incident on unit area of a planet emitted into all directions $P_{dh}(i)$ over the illuminated area $A(i)$ of the body. For a sphere,

$$A_s = \left[\int_{A(i)} P_{dh}(i)dA\right] / \left[J\pi R^2\right]$$
$$= \left[\int_{A(i)} \int_{\Omega_e=2\pi} Jr(i,e,g)\mu d\Omega_e dA\right] / \left[J\pi R^2\right]. \quad (11.42)$$

Because the increment of the illuminated half of a sphere is

$$dA = R^2 \sin i \, di \, d\Psi = R^2 d\Omega_i,$$
$$A_s = \frac{1}{\pi}\int_{2\pi}\int_{2\pi} r(i,e,g)\mu d\Omega_i d\Omega_e. \quad (11.43)$$

This is identical with equation (11.17) for the bihemispherical or spherical reflectance r_s. Thus,

$$A_s = r_s, \quad (11.44)$$

as stated without proof in Section 11.2.4. In particular, the spherical albedo of a medium of isotropic scatterers is

$$A_s = 1 - 2\gamma H_1 \quad (11.45a)$$

and may be approximated by

$$A_s \simeq \frac{1-\gamma}{1+\gamma}\left(1 - \frac{1}{3}\frac{\gamma}{1+\gamma}\right) = r_0\left(1 - \frac{1-r_0}{6}\right). \quad (11.45b)$$

In Figure 11.2 A_s is plotted as a function of w. For nonisotropic scatterers the similarity relations (11.21) may be used. Conservation of energy requires that the spherical albedo cannot exceed unity.

11.3.8 The bolometric albedo

The bolometric albedo A_b is also called the *radiometric albedo*. It is defined as the average of the spectral spherical albedo $A_s(\lambda)$ weighted by the spectral irradiance of the Sun $J_s(\lambda)$:

$$A_b = \left[\int_0^\infty A_s(\lambda) J_s(\lambda) d\lambda\right] / \left[\int_0^\infty J_s(\lambda) d\lambda\right], \tag{11.46}$$

where $J_s(\lambda)$ is approximately the Planck function of a black body at a temperature of 5770 K. The bolometric albedo is the primary quantity that determines the mean temperature of a planet, along with distance from the sum.

11.3.9 The phase integral

The *phase integral* q of a body is defined as

$$q = \frac{1}{\pi} \int_{4\pi} \Phi_p(g) d\Omega_g = 2 \int_0^\pi \Phi_p(g) \sin g \, dg, \tag{11.47}$$

where $d\Omega_g = 2\pi \sin g dg$ is the increment of solid angle associated with the phase angle.

An important relation involving q, A_s, and A_p may be derived as follows. From (11.36),

$$q = \left[\int_{4\pi} \int_{A(i,v)} Jr(i, e, g) \mu dA d\Omega_g\right] / \left[J\pi R^2 A_p\right].$$

Now, the integral in this last expression for q is the total amount of light scattered into all directions by the entire surface of the body. But, by the definition of the spherical albedo, this is just $J\pi R^2 A_s$. Hence,

$$q = A_s / A_p. \tag{11.48}$$

Russell (1916) has noted the following empirical relation, which is known as *Russell's rule:*

$$q \approx 2.17 \Phi_p(54°).$$

12

Photometric effects of large-scale roughness

12.1 Introduction

The expressions for reflectance developed in previous chapters of this book implicitly assume that the apparent surface of the particulate medium is smooth on scales large compared with the particle size. Although that assumption may be valid for surfaces in the laboratory, it is certainly not the case for planetary regoliths. In this chapter the expressions that were derived in Chapters 8–10 to describe the light scattered from a planet with a smooth surface will be modified so as to be applicable to surfaces with large-scale roughness. By "large-scale roughness" is meant that areas of the surface larger than the particle size but smaller than the detector footprint are tilted with an irregular distribution of slopes. Persons uninterested in the details of the derivation may wish to jump directly to the Summary Section 12.D, after reading this introductory section.

In calculations of this type we are immediately faced with the problem of choosing an appropriate geometric model to describe roughness. Some authors have chosen specific shapes, such as hemispherical cups (Van Diggelen, 1959; Hameen-Anttila, 1967), that approximate impact craters on the surface of a planet. However, such models may not be applicable to other geometries such as hills or dunes. To make the expressions to be derived as general as possible, it will be assumed that the surfaces are randomly rough. There is a large body of literature that treats shadowing on such surfaces – see, for example, Muhleman (1964), Saunders (1967), Wagner (1967), Hagfors (1968), Lumme and Bowell (1981a,b), Simpson and Tyler (1982), Van Ginneken *et al.* (1988), Shepard and Campbell (1998), as well as the references cited in those papers. Many of these papers deal only with specular reflection, such as is involved in analyses of sea glitter or backscattered lunar radar signals. In order to treat diffuse bidirectional-reflectance functions, the approach of Hapke (1984) will be followed. As in the other parts of this book, the emphasis will be on the development of useful approximate analytic expressions, rather than perfect mathematical rigor.

The derivation is based on the following assumptions:

(1) Geometric optics is valid. If the medium is composed of particles smaller than the wavelength, the objects that control the scattering are large clumps rather than individual particles.

(2) The macroscopically rough surface is considered to be made up of small, locally smooth facets that are large compared with the mean particle size and are tilted at a variety of angles. The normals to the facets are described by a two-dimensional slope distribution function $a(\vartheta, \zeta)d\vartheta d\zeta$, where ϑ is the zenith angle between a facet normal and the vertical, and ζ is the azimuth angle of the facet normal.

(3) The distribution function of the facet orientations is independent of azimuth angle, so that $a(\vartheta, \zeta)$ can be written simply as $a(\vartheta)$. The assumption that the slope distribution function is independent of azimuth will certainly be true on the average for surfaces made up of craters and hills, and it also appears to be reasonable for the surface of the ocean (Cox and Munk, 1954). Its validity may be questioned for morphologies with preferred orientations, like folded mountain ranges and fields of parallel sand dunes. However, on the small scales that dominate the distribution function, the slopes are likely to be caused by such erosive agents as microscopic impacts, eolian gusting, and fluvial action, which are roughly isotropic in azimuth. Hence, the assumption appears to be reasonable.

If the slope distribution function is independent of azimuth then, in general (Hagfors, 1968), if $a'(\vartheta)$ is the one-dimensional function that describes the distribution of slopes on any vertical cut through the surface made at an arbitrary azimuth angle, the corresponding two-dimensional, azimuth-independent distribution function can be written in the form

$$a(\vartheta)d\vartheta d\zeta = a'(\vartheta)\sin\vartheta d\vartheta d\zeta. \tag{12.1}$$

(4) It will be assumed that $a'(\vartheta)$ can be described by a Gaussian distribution of the form

$$a'(\vartheta)d\vartheta = A\exp[-B\tan^2\vartheta]d(\tan\vartheta), \tag{12.2}$$

where A and B are constants to be determined. Then,

$$a(\vartheta) = A\exp[-B\tan^2\vartheta]\sec^2\vartheta\sin\vartheta. \tag{12.3}$$

The slope distribution function is normalized such that

$$\int_0^{\pi/2} a(\vartheta)d\vartheta = 1 \tag{12.4}$$

and is characterized by a mean slope angle $\overline{\theta}$ defined by

$$\tan\overline{\theta} = \frac{2}{\pi}\int_0^{\pi/2} a(\vartheta)\tan\vartheta d\vartheta. \tag{12.5}$$

Inserting (12.3) into (12.4) and (12.5) gives

$$A = 2/\pi \tan^2\overline{\theta}, \tag{12.6a}$$

and

$$B = 1/\pi \tan^2\overline{\theta}. \tag{12.6b}$$

(5) The mean slope angle $\overline{\theta}$ is assumed to be small. Vertical scarps and overhangs are assumed to constitute a negligible part of the surface. Although the general equations for the roughness effects will be derived for an arbitrary $\overline{\theta}$, their analytic solutions are greatly simplified if terms of order $\overline{\theta}^3$ and higher can be ignored.

(6) Light multiply scattered from one surface facet to another is neglected. However, radiance that is multiply scattered from one particle to another within each surface facet is included in the derivation. This is the most restrictive assumption of the model. The limitation that this assumption imposes can be estimated by the following calculation: consider a depression in the shape of a sector of a sphere of radius x and with maximum slope ϑ_M. The inside is covered with a Lambert surface with albedo A_L. The depression is illuminated vertically by irradiance J (see Figure 12.1). Then the radiance scattered once from a small area ΔA at the bottom of the cup is

$$I_1 = \frac{J}{\pi}A_L\Delta A.$$

What is the radiance I_2 due to light scattered onto ΔA from the rest of the inside of the cup?

Consider an increment of area $dA = x^2\sin\vartheta d\vartheta d\psi$ located at zenith angle ϑ and azimuth ψ on the inside of the cup. Then the doubly scattered radiance of light scattered from dA to ΔA and then to the detetector is

$$dI_2 = \frac{J}{\pi}dA\cos\vartheta A_L\cos\vartheta'\frac{\Delta A\cos\vartheta'}{y^2}\frac{1}{\pi}A_L,$$

where ϑ' and y are defined in Figure 12.1. Now, $\vartheta' = (\pi - \vartheta)/2$, and $y = 2x\sin(\vartheta/2)$. Hence, the radiance from ΔA due to light scattered from the entire inside of the cup is

$$I_2 = \int_{\vartheta=0}^{\vartheta M}\int_{\zeta=0}^{2\pi}\frac{J\Delta A}{4\pi^2}A_L^2\sin\vartheta\cos\vartheta d\vartheta d\psi = \frac{J}{4}\Delta A A_L^2\sin^2\vartheta_M,$$

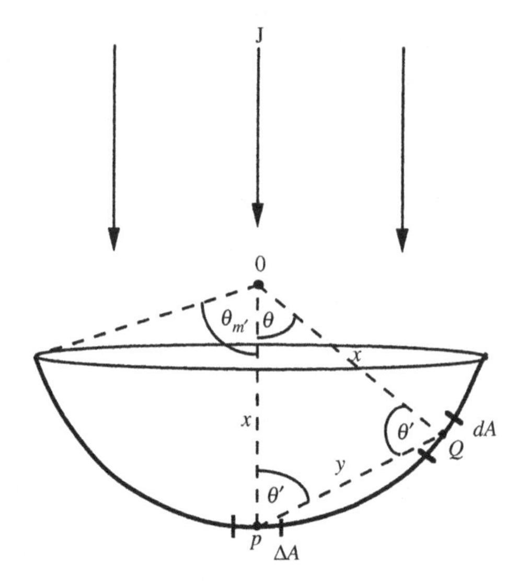

Figure 12.1

and the ratio of doubly to singly scattered radiance is

$$I_2/I_1 = \frac{1}{4} A_L \sin^2 \vartheta_M .$$

For example, if $A_L = 0.5$ and $\vartheta_M = 45°$, $I_2/I_1 = 6\%$. The assumption that interfacet scattering can be neglected is seen to be consistent with assumption (3). Light multiply scattered from one facet to another usually can be ignored if either the albedo or mean slope is small. The major exception to this statement occurs for high-albedo surfaces at large phase angles. Then most of the visible facets may be in the shadow of the direct irradiance from the source, but will still be illuminated by light scattered from adjacent surfaces not in shadow.

The major effect of multiple scattering is to fill in the shadows. The shadowed surfaces tend to be located at the bottoms of hills or depressions, where the slopes tend to be small. Illuminating these areas tends to increase the contribution of the surfaces with smaller slopes. Hence, neglect of facet-to-facet multiple scattering means that a value of $\overline{\theta}$ obtained by fitting the model to observations underestimates the actual mean slope. so that the photometric $\overline{\theta}$ is a lower limit. An empirical correction for this effect will be added later in Section 12.3.3.

Buratti and Veverka (1985) investigated the effects of neglecting multiple scattering between facets experimentally. They measured the reflectances of plaster-of-Paris models with normal albedos of 2.1% and 95% with and without craters of hemi-elliptical cross section with a depth-to-diameter ratio of 0.4. The fractional differences in reflectance at $e = 0$ as i is varied with and without the craters are shown

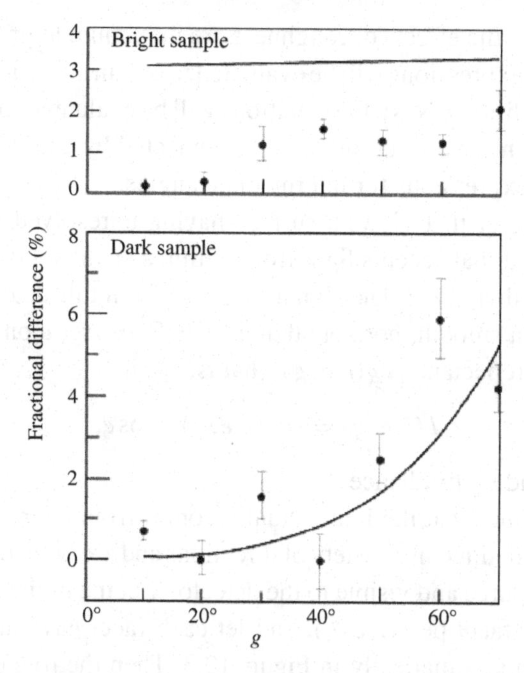

Figure 12.2 Comparison between predicted and measured changes caused in the reflectance by roughness for a bright ($r = 0.95$) and a dark ($r = 0.02$) surface. The dots are the measured fractional changes in percentage, and the lines are the predicted changes. (Reproduced from Buratti and Veverka [1985], copyright 1985 with permission of Elsevier.)

as the dots in Figure 12.2. Buratti and Veverka also calculated the reflectances of the models assuming that the surface facets scattered light according to the Lommel–Seeliger law for the dark surface, and Lambert's law for the bright surface, but neglecting interfacet scattering. The theoretical changes are shown as the solid lines in Figure 12.2. Theoretical and measured changes were similar for the dark surface, but differed by about 2% for the bright surface, which can be attributed to the effects of shadow-filling. Errors of 2% are within the accuracy with which absolute reflectances can be measured in most situations.

12.2 Derivation

12.2.1 Derivation of the general equations

The general scheme of the derivation will be as follows. We will seek a formalism by which the bidirectional reflectance of a medium having a smooth surface can be corrected to one describing the same medium, but with a surface roughness characterized by a mean slope angle $\bar{\theta}$. General equations that are mathematically rigorous will be derived first, and the parameters necessary for their evaluation will be defined. Because the effects of roughness are maximum at grazing illumination and viewing, these expressions will be evaluated to obtain analytic functions that are exact for these conditions. Next, the equations will be evaluated for vertical viewing and illumination. The two solutions will be connected by analytic interpolation to give approximate expressions for intermediate angles.

Consider a detector that views a surface having unresolved roughness from a large distance R and that accepts light from within a small solid angle $\Delta\omega$ about a direction with zenith angle e. The signal $I(i, e, g)$ from this detector is interpreted as if it came from a smooth, horizontal area $A = R^2 \Delta\omega \sec e$ on the mean surface with bidirectional reflectance $r_R(i, e, g)$; that is,

$$I(i, e, g) = J r_R(i, e, g) A \cos e, \tag{12.7}$$

where J is the incident irradiance.

The model assumes that the light actually comes from a large number of unresolved facets that are tilted in a variety of directions and are both directly illuminated by light from the source and visible to the detector. Let the bidirectional reflectance of each individual facet be $r(i, e, g)$, and let each facet have area $A_f \ll A$. The geometry is shown schematically in Figure 12.3. Then the true expression for the light reaching the detector is

$$I(i, e, g) = J \int_{A(i,v)} r(i_t, e_t, g) \cos e_t dA_t, \tag{12.8}$$

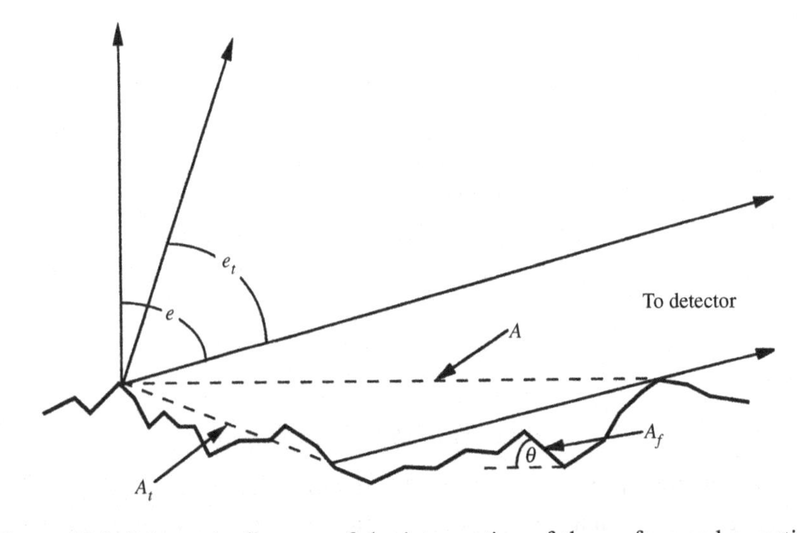

Figure 12.3 Schematic diagram of the intersection of the surface and a vertical plane containing the detector. Shown are the actual surface, consisting of a multitude of unresolved facets A_f, the nominal surface A, and the effective tilted surface A_t. A cut by a vertical plane containing the source would be similar.

where the subscript t (standing for "tilted") on i, e, and A denotes values appropriate to an incremental surface of area

$$dA_t = A_f a(\vartheta) d\vartheta d\zeta, \tag{12.9a}$$

whose normal points in direction (ϑ, ζ), and the symbol $A(i, v)$ indicates that the integration is to be taken only over those surface facets within A that are both directly illuminated by the source and visible to the detector. Using the law of cosines it may be readily shown that the angles $i_t, e_t, \vartheta, \zeta$, and ψ, where ψ is the azimuth angle between the source and detector planes, are related by

$$\cos i_t = \cos i \cos \vartheta + \sin i \sin \vartheta \cos \zeta, \tag{12.9b}$$

$$\cos e_t = \cos e \cos \vartheta + \sin e \sin \vartheta \cos(\zeta - \psi). \tag{12.9c}$$

Note that g is the same for all facets within the area viewed by the detector. The objective of this chapter is to find the relation between $r(i, e, g)$ and $r_R(i, e, g)$ as a function of $\overline{\theta}$.

There are three important effects of macroscopic roughness that will modify the reflectance. (1) Scattering of light from one facet to another will increase the reflectance and decrease the apparent mean slope. This effect will be small if either the albedo or the mean slope is small, as argued in the preceding section, and will be ignored. (2) Unresolved shadows cast on one part of the surface by another will decrease the reflectance. (3) As the surface is viewed and illuminated at increasing zenith angles, the facets that are tilted away from the observer or source will tend to be hidden or in shadow, so that the surfaces that are visible and illuminated will tend to be those that are tilted preferentially toward the detector or source.

To account for the latter two effects, we will try to write the rough-surface bidirectional reflectance $r_R(i, e, g)$ as the product of a shadowing function $S(i, e, g)$ and the bidirectional reflectance $r(i_e, e_e, g)$ of a smooth surface of effective area A_e tilted so as to have effective angle of incidence i_e and angle of emergence e_e, and with the same phase angle g. That is, we will seek expressions for $i_e(i, e, g)$, $e_e(i, e, g)$, and $S(i, e, g)$ that will make the following equation true:

$$r_R(i, e, g) = r(i_e, e_e, g) S(i, e, g) \tag{12.10}$$

From Figure 12.3, A and A_e are related by

$$A_e \cos e_e = A \cos e. \tag{12.11}$$

Let

$$\mu_e = \cos e_e, \tag{12.12a}$$

$$\mu_{0e} = \cos i_e, \tag{12.12b}$$

$$\mu_t = \cos e_t, \tag{12.12c}$$

$$\mu_{0t} = \cos i_t. \tag{12.12d}$$

Denote the reflectance per unit area of surface by Y; that is, let

$$Y_R(\mu_0, \mu, g) = r_R(i, e, g)\cos e, \tag{12.13a}$$

$$Y(\mu_{0e}, \mu_e, g) = r(i_e, e_e, g)\cos e_e, \tag{12.13b}$$

$$Y(\mu_{0t}, \mu_t, g) = r(i_t, e_t, g)\cos e_t. \tag{12.13c}$$

Combining (12.7) – (12.9),

$$\begin{aligned} I(i, e, g) &= JAY_R(i, e, g) = JA_eY(\mu_{0e}, \mu_e, g)S(\mu_0, \mu, g) \\ &= J\int_{A(i,v)} Y(\mu_{0t}, \mu_t, g)dA_t. \end{aligned} \tag{12.14}$$

Assume that Y is mathematically well behaved so that it can be expanded in a Taylor series about μ_0 and μ. Doing this on both sides of the third equal sign in (12.14), and using (12.11), gives

$$\begin{aligned} &\frac{A\mu}{\mu_e}S(\mu_0, \mu, g)\Big[Y(\mu_{0e}, \mu_e, g)|_{(\mu_0,\mu)} \\ &\qquad + \frac{\partial Y}{\partial \mu_{0e}}(\mu_{0e}, \mu_e, g)\,|_{(\mu_0,\mu)}(\mu_{0e} - \mu_0) \\ &\qquad + \frac{\partial Y}{\partial \mu_e}(\mu_{0e}, \mu_e, g)\,|_{(\mu_0,\mu)}(\mu_e - \mu) + \cdots\Big] \\ &= \int_{A(i,v)} Y(\mu_{0t}, \mu_t, g)|_{(\mu_0,\mu)}dA_t \\ &\qquad + \int_{A(i,v)} \frac{\partial Y}{\partial \mu_{0t}}(\mu_{0t}, \mu_t, g)|_{(\mu_0,\mu)}(\mu_{0t} - \mu_0)dA_t \\ &\qquad + \int_{A(i,v)} \frac{\partial Y}{\partial \mu_t}(\mu_{0t}, \mu_t, g)|_{(\mu_0,\mu)}(\mu_t - \mu)dA_t + \cdots, \end{aligned}$$

or

$$\frac{A\mu}{\mu_e}S(\mu_0,\mu,g)\left[Y(\mu_0,\mu,g)+\frac{\partial Y}{\partial\mu_0}(\mu_0,\mu,g)(\mu_{0e}-\mu_0)\right.$$
$$\left.+\frac{\partial Y}{\partial\mu}(\mu_0,\mu,g)(\mu_e-\mu)+\cdots\right]$$
$$=Y(\mu_0,\mu,g)\int_{A(i,v)}dA_t+\frac{\partial Y}{\partial\mu_0}\int_{A(i,v)}(\mu_{0t}-\mu_0)dA_t$$
$$+\frac{\partial Y}{\partial\mu}(\mu_0,\mu,g)\int_{A(i,v)}(\mu_t-\mu)dA_t+\cdots. \tag{12.15}$$

Because μ_0 and μ are independent variables and Y can be an arbitrary function of these variables, equation (12.15) will be satisfied if the coefficients of Y and its partial derivatives are separately equal on both sides of the equality. This gives

$$S(i,e,\psi)=\frac{\mu_e}{A\mu}\int_{A(i,v)}dA_t, \tag{12.16}$$

$$\mu_{0e}(i,e,\psi)=\frac{\int_{A(i,v)}\mu_{0t}dA_t}{\int_{A(i,v)}dA_t}, \tag{12.17}$$

$$\mu_e(i,e,\psi)=\frac{\int_{A(i,v)}\mu_t dA_t}{\int_{A(i,v)}dA_t}. \tag{12.18}$$

In general, there will be two types of shadows. Some of the facets will not contribute to the scattered radiance because their normals will be tilted by more than 90° to the direction from the source or detector; such facets will be said to be in a *tilt shadow*. Some facets will not contribute because other parts of the surface will obstruct either the view of the detector or the light from the source; such facets will be said to be in a *projected shadow*. We shall follow Saunders (1967) and assume that any facet that is not in a tilt shadow has a statistical probability of being in a projected shadow that is independent of the slope or azimuth angle of its tilt.

Let P be the probability that a facet is not in a projected shadow. Then in equation (12.15), $Y(\mu_{0t},\mu_t,g)$ can be multiplied by P, if at the same time the boundaries of the integration are replaced by the tilt-shadow boundaries. This has the effect of multiplying both the numerators and denominators in (12.17) and (12.18) by P, which thus cancels out in these equations. Therefore, inserting (12.9) into (12.17) and (12.18) and writing the latter out explicitly, these equations become

$$\mu_{0e}(i,e,\psi) = \frac{\cos i \int_{(A(\text{tilt})} \cos\vartheta a(\vartheta) d\vartheta d\zeta + \sin i \int_{A(\text{tilt})} \sin\vartheta \cos\zeta a(\vartheta) d\vartheta d\zeta}{\int_{A(\text{tilt})} a(\vartheta) d\vartheta d\zeta}, \tag{12.19}$$

$$\mu_e(i,e,\psi) = \frac{\cos e \int_{A(\text{tilt})} \cos\vartheta a(\vartheta) d\vartheta d\zeta + \sin e \int_{A(\text{tilt})} \sin\vartheta \cos(\zeta - \psi) a(\vartheta) d\vartheta d\zeta}{\int_{A(\text{tilt})} a(\vartheta) d\vartheta d\zeta}, \tag{12.20}$$

where A(tilt) denotes the boundaries of the tilt shadows in A. These boundaries will depend on i, e, and ψ.

Let A_T be the total area of all the facets within the nominal area A, whether visible and illuminated or not:

$$A_T = \int_{\zeta=0}^{2\pi} \int_{\vartheta=0}^{\pi/2} dA_t = 2\pi \int_0^{\pi/2} A_f a(\vartheta) d\vartheta = 2\pi A_f,$$

because of azimuthal symmetry and the normalization condition on $a(\vartheta)$. Now, A is just the projection of all the facets onto the horizontal plane:

$$A = \int_{\zeta=0}^{2\pi} \int_{\vartheta=0}^{\pi/2} \cos\vartheta dA_t = 2\pi A_f \int_{\vartheta 0}^{\pi/2} a(\vartheta) \cos\vartheta d\vartheta.$$

Hence,

$$A/A_T = \langle \cos\vartheta \rangle,$$

where

$$\langle \cos\vartheta \rangle = \int_0^{\pi/2} a(\vartheta) \cos\vartheta d\vartheta. \tag{12.21}$$

Thus, expression (12.16) for S can be written

$$S(\mu_0,\mu,\psi) = \frac{\mu_e}{\mu} \frac{A_T}{A} \frac{1}{A_T} \int_{A(i,v)} dA_t = \frac{\mu_e}{\mu \langle \cos\vartheta \rangle} F_{(i,v)}(\mu_0,\mu,\psi), \tag{12.22}$$

where

$$F_{(i,v)}(\mu_0,\mu,\psi) = \frac{1}{A_T} \int_{A(i,v)} dA_t \tag{12.23}$$

is the probability that a facet is both illuminated and visible.

Let $F_i(i)$ and $F_e(e)$ be the fraction of the facets that are illuminated and the fraction that are visible, respectively. Because of azimuthal symmetry both $F_i(i)$ and $F_e(e)$ are independent of ψ, and furthermore, $F_i(i)$ has the same functional dependence on i as $F_e(e)$ has on e.

The solutions for **S**, μ_{0e}, and μ_e have different forms depending on whether i is larger or smaller than e. Suppose $i \leq e$. Then an illumination shadow cast by a given object is always smaller than its visibility shadow, and the illumination shadow may be regarded as partially hidden in the visibility shadow. When $\psi = 0$ the illumination shadow is completely hidden, so that $F_{(i,v)}(\mu_0, \mu, 0) = F_e(e)$. But when $\psi = 0$ and $i \leq e$, no shadows are visible: the detector's field of view is completely filled by surfaces that are both visible and totally illuminated. Thus,

$$S(\mu_0, \mu, 0) = \frac{\eta_e(e)}{\mu\langle\cos\vartheta\rangle} F_e(e) = 1,$$

where

$$\eta_e(e) = \mu_e(i, e, \psi = 0), i \leq e. \tag{12.24}$$

Hence,

$$F_e(e) = \frac{\mu\langle\cos\vartheta\rangle}{\eta_e(e)}. \tag{12.25}$$

Now suppose that $e \leq i$. Then by exactly the same arguments,

$$F_i(i) = \frac{\mu_0\langle\cos\vartheta\rangle}{\eta_{0e}(i)}. \tag{12.26}$$

where

$$\eta_{0e}(i) = \mu_{0e}(i, e, \psi = 0), e \leq i. \tag{12.27}$$

12.2.2 The case when $i \leq e$

We will derive the equations for the case when $i \leq e$ first. Then when $\psi \neq 0$ the visibility shadow partially hides the illumination shadow. Let $f(\psi)$ be the fraction of the illumination shadow that is hidden in the visibility shadow. Let

$$A_e/A_T = 1 - F_e(e)$$

be the fraction of the facets in the visibility shadows, and let

$$A_i/A_T = 1 - F_i(i)$$

be the fraction of the facets in the illumination shadows. These include both the tilt and projected shadows. As we have seen, when $\psi = 0$, all of the illumination shadows are hidden in the visibility shadows. The visibility and illumination shadows are perfectly correlated, so that $f(0) = 1$ and

$$F_{(i,v)} = F_e = 1 - A_e/A_T.$$

As ψ increases, a fraction $1 - f(\psi)$ of the illumination shadows will be exposed. When $\psi = \pi$, $f(\pi) = 0$, and the two types of shadows are completely uncorrelated, so that

$$\begin{aligned} F_{(i,v)} &= 1 - A_e/A_T - A_i/A_T + (A_e/A_T)(A_i/A_T) \\ &= (1 - A_e/A_T)(1 - A_i/A_T) = F_e F_i, \end{aligned}$$

where the term $A_e A_i / A_T^2$ corrects for the amount of random overlap. This last expression states that when the two types of shadows are completely uncorrelated, the probability that a facet will be both illuminated and visible is the product of the separate probabilities.

When $0 < \psi < \pi$, A_i in the last expression for $F_{(i,v)}$ must be replaced by $(1 - f)A_i/(A_T - fA_i)$. This accounts for the fact that only a portion $1 - f$ of the illumination shadow is randomly exposed, and only an area $A_T - fA_i$ is available to be occupied by the uncorrelated part of the illumination shadow. Thus, the general expression for the probability that a facet will be both illuminated and visible is

$$\begin{aligned} F_{(i,v)} &= [1 - A_e/A_T][1 - (1 - f)A_i/(A_T - fA_i)] \\ &= [1 - A_e/A_T][1 - A_i/A_T]/[1 - fA_i/A_T] \\ &= F_e F_i/(1 - f + fF_i). \end{aligned}$$

Combining this result with (12.22), (12.25), and (12.26) gives

$$S(i, e, \psi) = \frac{\mu_e}{\mu_e(0)} \frac{\mu_0}{\mu_{0e}(0)} \frac{\langle \cos\vartheta \rangle}{1 - f(\psi) + f(\psi)[\mu_0/\mu_{0e}(0)]\langle \cos\vartheta \rangle}. \tag{12.28}$$

Thus far, the derivation has been rigorous. In the remainder of this section we will derive approximate analytic expressions for μ_{0e}, μ_e, and S that are suitable for practical calculations. First, an expression for $f(\psi)$, the fraction of the illumination shadow hidden in the visibility shadow, will be found. It will be assumed that $f(\psi)$ is a function of ψ only and is independent of i and e, which is reasonable if i and e are near 90°.

Recall that the shadows have two components, tilt and projected shadows. When i and e are near 90°, the contributions of the two components are roughly equal. As ψ increases from zero, the fraction of the tilt component of the illumination shadows that are exposed increases approximately linearly with ψ, and the exposure is complete when $\psi = \pi$. Hence, the contribution of the tilt component to f will be $\sim \frac{1}{2}(\psi/\pi)$. Now, the surface can be considered as consisting of depressions and protuberances of mean width Δd and mean height $(\Delta d/2)\tan\overline{\theta}$. When i and e are near 90°, the projected shadows are cast by objects of width $\sim\Delta d$ onto surfaces a distance $\sim\Delta d$ away, so that this component of the illumination shadows is nearly completely exposed when $\psi \gtrsim 1$ radian. At $\psi \approx 1$, the fraction of each type of shadow exposed is approximately $\frac{1}{2} \times (1/\pi) + \frac{1}{2} \times 1 \simeq \frac{2}{3}$. Hence, $f(\psi)$ may be described by a function that decreases linearly from a value of $f(0) = 1$ to $f(1) \simeq 1 - \frac{2}{3} = \frac{1}{3}$, and then decreases to zero as $\psi \to \pi$. A simple function with the required properties is

$$f(\psi) = \exp\left(-2\tan\frac{\psi}{2}\right), \tag{12.29}$$

and this will be adopted for $f(\psi)$.

Returning to μ_{0e} and μ_e, only the tilt shadows affect these quantities. The boundary of the tilt illumination shadow can be found by putting $i_t = \pi/2$ in (12.9a). This gives

$$\cos\zeta = -\cot\vartheta \cot i. \tag{12.30}$$

This equation has no solution when $0 \le \vartheta \le \pi/2 - i$, but $\pi/2 \le \zeta \le 3\pi/2$ when $\pi/2 - i \le \vartheta \le \pi/2$. Similarly, the boundary of the tilt visibility shadow is given by putting $e_t = \pi/2$ in (12.9b),

$$\cos(\zeta - \psi) = -\cot\vartheta \cot e, \tag{12.31}$$

which has no solution when $0 \le \vartheta \le \pi/2 - e$, but $\pi/2 + \psi \le \zeta \le 3\pi/2 + \psi$ when $\pi/2 - e \le \vartheta \le \pi/2$.

The integrals in equations (12.19) and (12.20) are readily evaluated when $i = e = 0$ and $i = e = \pi/2$. For vertical illumination and viewing, when there are no shadows, the integrals over $\cos\zeta$ vanish because of azimuthal symmetry, so that

$$i = e = 0 \sim \mu_{0e} = \mu_e = \langle\cos\vartheta\rangle. \tag{12.32}$$

If the tilt shadows are represented in a polar diagram with ϑ as the radial variable and ζ as the angular variable, then at grazing incidence and viewing the tilt-shadow boundaries are the straight radial lines $\zeta = \psi - \pi/2$ and $\zeta = \pi/2$, and the limits on ϑ are 0 to $\pi/2$. Hence

$$i = e = \pi/2 : \mu_{0e} = \mu_e = (1 + \cos\psi)\frac{\langle\sin\vartheta\rangle}{\pi - \psi}, \tag{12.33}$$

where

$$\langle\sin\vartheta\rangle = \int_0^{\pi/2} \sin\vartheta a(\vartheta) d\vartheta.$$

Equation (12.33) shows that when $\psi = 0$ the effective surface is tilted toward the source and detector by an angle $\cos^{-1}[(2/\pi)\langle\sin\vartheta\rangle]$.

For intermediate values of i and e these integrals are much more difficult to evaluate, because the shadow boundaries (12.30) and (12.31) are overlapping curves in the (ϑ, ζ) polar diagram. Approximate expressions may be obtained for these integrals as follows. First the separate effects of the illumination and viewing shadows will be found. Next, the effect of overlapping the two shadows will be estimated by replacing the curved boundary lines by straight lines of constant ζ and circles of constant ϑ in the (ϑ, ζ) plane. This approximation will then be improved by substituting the results from the solutions for the separate shadows. Finally, the integrals will be evaluated.

The effect of the illumination shadow alone on μ_{0e} may be seen by setting $e = 0$ and using (12.30) as the A(tilt) boundary in (12.19). Then the integration over ζ in the first integral in the numerator of (12.19) may be carried out exactly to give

$$\int_{A(\text{tilt})} \cos\vartheta a(\vartheta) d\vartheta d\zeta$$
$$= \int_0^{\pi/2-i} 2\pi \cos\vartheta a(\vartheta) d\vartheta$$
$$+ \int_{\pi/2-i}^{\pi/2} 2\sin^{-1}(1 - \cot^2\vartheta \cot^2 i)^{1/2} \cos\vartheta a(\vartheta) d\vartheta, \qquad (12.34a)$$

where the value of the $\sin^{-1}$ lies between $\pi/2$ and π. Now, the factor $(1 - \cot^2\vartheta \cot^2 i)^{1/2}$ has the following properties: it rises with infinite slope from 0 at $\vartheta = \pi/2 - i$, then levels off to 1 at $\vartheta = \pi/2$, where it has slope 0. Thus, as a first approximation, this factor may be replaced by a unit step function at $\vartheta = \pi/2 - i$. Then equation (12.34a) becomes

$$\int_{A(\text{tilt})} \cos\vartheta a(\vartheta) d\vartheta d\zeta$$
$$\simeq \int_0^{\pi/2-i} 2\pi \cos\vartheta a(\vartheta) d\vartheta + \int_{\pi/2-i}^{\pi/2} \pi \cos\vartheta a(\vartheta) d\vartheta. \qquad (12.34b)$$

Similarly, the effect of the visibility shadow alone may be ascertained by setting $f = 0$ and using (12.31) in the integrals in (12.19).

Then the first integral in the numerator of (12.19) is

$$\int_{A(\text{tilt})} \cos\vartheta a(\vartheta) d\vartheta d\zeta$$
$$= \int_0^{\pi/2-e} 2\pi \cos\vartheta a(\vartheta) d\vartheta$$
$$+ \int_{\pi/2-e}^{\pi/2} 2\sin^{-1}(1 - \cot^2\vartheta \cot^2 e)^{1/2} \cos\vartheta a(\vartheta) d\vartheta, \qquad (12.34c)$$

which, upon approximating $(1 - \cot^2\vartheta \cot^2 e)^{1/2}$ by a unit step function at $\vartheta = \pi/2 - e$, becomes

$$\int_{A(\text{tilt})} \cos\vartheta a(\vartheta) d\vartheta d\zeta$$
$$\simeq \int_0^{\pi/2-e} 2\pi \cos\vartheta a(\vartheta) d\vartheta + \int_{\pi/2-e}^{\pi/2} \pi \cos\vartheta a(\vartheta) d\vartheta. \qquad (12.34d)$$

As shown in Figure 12.4, these approximations are equivalent to replacing the curved illumination shadow boundary in the (ϑ, ζ) diagram by a square-cornered

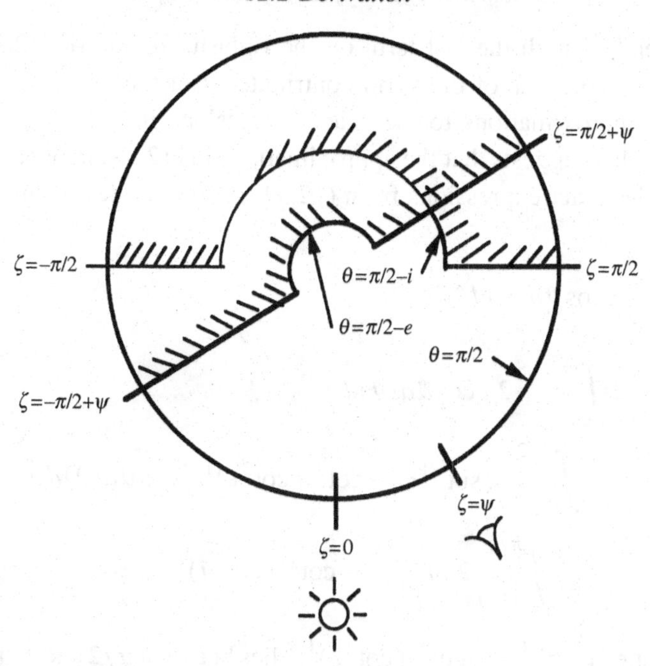

Figure 12.4 Schematic diagram showing the square-cornered approximation to the tilt shadows in (ϑ, ζ) space for the case when $i \leq e$. The projected shadows are assumed to be randomly distributed over the part of (ϑ, ζ) space not in tilt shadows.

shadow bounded by the radial lines $\zeta = \pi/2$ and $3\pi/2$ and the circles $\vartheta = \pi/2 - i$ and $\pi/2$, and replacing the visibility shadow by a square-cornered shadow bounded by the radial lines $\zeta = \pi/2 + \psi$ and $3\pi/2 + \psi$ and the circles $\vartheta = \pi/2 - e$ and $\pi/2$. At grazing illumination and viewing these approximations become exact. With the simplified boundaries of the squared-shadow approximation, the first integral in (12.19) may be evaluated for the case in which neither i nor e nor ψ is zero to give

$$
\begin{aligned}
f_{A(\text{tilt})} \cos\vartheta\, a(\vartheta) d\vartheta\, d\zeta \simeq & \int_0^{\pi/2-e} 2\pi \cos\vartheta\, a(\vartheta) d\vartheta \\
& + \int_{\pi/2-e}^{\pi/2} \pi \cos\vartheta\, a(\vartheta) d\vartheta \\
& - \frac{\psi}{\pi} \int_{\pi/2-i}^{\pi/2} \pi \cos\vartheta\, a(\vartheta) d\vartheta. \qquad (12.34e)
\end{aligned}
$$

Now, the first two terms on the right-hand side of (12.34e) are the same as those of (12.34d). The third term on the right-hand side of (12.34e) corrects for the fraction of the illumination shadow sticking out from behind the visibility shadow and is almost identical with the last term on the right-hand side of (12.34b), except that only a fraction ψ/π of this term contributes. But the terms in (12.34b) and (12.34d) are approximations to the exact expressions in (12.34a) and (12.34c), respectively. This suggests that the approximations in (12.38) may be improved by substituting the exact expressions from (12.34a) and (12.34c). If this is done, we obtain

$$\begin{aligned}\int_{A(\text{tilt})} & \cos\vartheta\, a(\vartheta) d\vartheta d\zeta \\ &\simeq \int_0^{\pi/2-e} 2\pi \cos\vartheta\, a(\vartheta) d\vartheta \\ &\quad + \int_{\pi/2-e}^{\pi/2} 2\sin^{-1}(1-\cot^2\vartheta\cot^2 e)^{1/2} \cos\vartheta\, a(\vartheta) d\vartheta \\ &\quad - \frac{\psi}{\pi}\int_{\pi/2-i}^{\pi/2} 2\sin^{-1}(1-\cot^2\vartheta\cot^2 i)^{1/2} \cos\vartheta\, a(\vartheta) d\vartheta, \end{aligned} \qquad (12.35)$$

where the value of $\sin^{-1}(1-\cot^2\vartheta\cot^2 e)^{1/2}$ lies between $\pi/2$ and π, but the value of $\sin^{-1}(1-\cot^2\vartheta\cot^2 i)^{1/2}$ lies between 0 and $\pi/2$.

By an identical argument the following expression for the integral in the denominator of (12.19) is obtained:

$$\begin{aligned}\int_{A(\text{tilt})} a(\vartheta) d\vartheta d\zeta &\simeq \int_0^{\pi/2-e} 2\pi\, a(\vartheta) d\vartheta \\ &\quad + \int_{\pi/2-e}^{\pi/2} 2\sin^{-1}(1-\cot^2\vartheta\cot^2 e)^{1/2} a(\vartheta) d\vartheta \\ &\quad - \frac{\psi}{\pi}\int_{\pi/2-i}^{\pi/2} 2\sin^{-1}(1-\cot^2\vartheta\cot^2 i)^{1/2} a(\vartheta) d\vartheta. \end{aligned} \qquad (12.36)$$

An approximate expression for the second integral in the numerator of (12.19) may be found using similar arguments. When $e=0$,

$$\begin{aligned}\int_{A(\text{tilt})} & \sin\vartheta\cos\zeta\, a(\vartheta) d\vartheta d\zeta \\ &= \int_{\pi/2-i}^{\pi/2} 2(1-\cot^2\vartheta\cot^2 i)^{1/2} \sin\vartheta\, a(\vartheta) d\vartheta. \end{aligned} \qquad (12.37a)$$

Approximating $(1-\cot^2\vartheta\cot^2 i)^{1/2}$ by a step function, this becomes

$$\int_{A(\text{tilt})} \sin\vartheta\cos\zeta\, a(\vartheta) d\vartheta d\zeta \simeq \int_{\pi/2-i}^{\pi/2} 2\sin\vartheta\, a(\vartheta) d\vartheta. \qquad (12.37b)$$

Setting $i = 0$, the integral is

$$\int_{A(\text{tilt})} \sin\vartheta \cos\zeta\, a(\vartheta) d\vartheta d\zeta$$

$$= \int_{\pi/2-e}^{\pi/2} 2\cos\psi (1 - \cot^2\vartheta \cot^2 e)^{1/2} \sin\vartheta\, a(\vartheta) d\vartheta, \tag{12.37c}$$

which is approximately

$$\int_{A(\text{tilt})} \sin\vartheta \cos\zeta\, a(\vartheta) d\vartheta d\zeta \simeq \int_{\pi/2-e}^{\pi/2} 2\cos\psi \sin\vartheta\, a(\vartheta) d\vartheta. \tag{12.37d}$$

Combining both shadows and using the square-boundary approximation gives

$$\int_{A(\text{tilt})} \sin\vartheta \cos\zeta\, a(\vartheta) d\vartheta d\zeta$$

$$\simeq \int_{\pi/2-e}^{\pi/2} 2\cos\psi \sin\vartheta\, a(\vartheta) d\vartheta + \int_{\pi/2-i}^{\pi/2} (1 - \cos\psi) \sin\vartheta\, a(\vartheta) d\vartheta. \tag{12.37e}$$

The first term on the right-hand side of (12.37e) is the same as the right-hand side of (12.37d), which is an approximation to the right-hand side of (12.37c). Similarly, the second term on the right-hand side of (12.37e) is the same as the right-hand side of (12.37b), except for the coefficient that accounts for the fact that only part of the illumination shadow contributes to the integral, and may be replaced by the right-hand side of (12.37a). Thus, we obtain

$$\int_{A(\text{tilt})} \sin\vartheta \cos\zeta\, a(\vartheta) d\vartheta d\zeta$$

$$\simeq \int_{\pi/2-e}^{\pi/2} 2\cos\psi (1 - \cot^2\vartheta \cot^2 e)^{1/2} \sin\vartheta\, a(\vartheta) d\vartheta$$

$$+ \int_{\pi/2-i}^{\pi/2} (1 - \cos\psi)(1 - \cot^2\vartheta \cot^2 i)^{1/2} \sin\vartheta\, a(\vartheta) d\vartheta, \tag{12.38}$$

where the value of $\sin^{-1}(1 - \cot^2\vartheta \cot^2 e)^{1/2}$ lies between $\pi/2$ and π, and that of $\sin^{-1}(1 - \cot^2\vartheta \cot^2 i)^{1/2}$ lies between 0 and $\pi/2$.

The next step is to carry out the integration over ϑ in equations (12.35), (12.36), and (12.38). Let the average value of any function $G(\vartheta)$ be defined as

$$\langle G(\vartheta) \rangle = \int_0^{\pi/2} G(\vartheta) a(\vartheta) d\vartheta. \tag{12.39}$$

We will obtain approximate analytic expressions that are valid for i and e near 90° by expanding the integrals in Taylor series in $\cot e$ and $\cot i$. This gives, for the integral in (12.35),

$$\int_{A(\text{tilt})} \cos\vartheta\, a(\vartheta) d\vartheta d\zeta$$
$$\simeq \pi \langle\cos\vartheta\rangle + 2\langle\cot\vartheta\cos\vartheta\rangle \cot e$$
$$-\frac{\psi}{\pi}(\pi\langle\cos\vartheta\rangle - 2\langle\cot\vartheta\cos\vartheta\rangle\cot i)$$
$$= \pi\langle\cos\vartheta\rangle\left[2 - \left(1 - \frac{2}{\pi}\frac{\langle\cot\vartheta\cos\vartheta\rangle}{\langle\cos\vartheta\rangle}\cot e\right)\right.$$
$$\left. - \frac{\psi}{\pi}\left(1 - \frac{2}{\pi}\frac{\langle\cot\vartheta\cos\vartheta\rangle}{\langle\cos\vartheta\rangle}\cot i\right)\right]$$
$$\simeq \pi\langle\cos\vartheta\rangle\left[2 - \exp\left(-\frac{2}{\pi}\frac{\langle\cot\vartheta\cos\vartheta\rangle}{\langle\cos\vartheta\rangle}\cot e\right)\right.$$
$$\left. - \frac{\psi}{\pi}\exp\left(-\frac{2}{\pi}\frac{\langle\cot\vartheta\cos\vartheta\rangle}{\langle\cos\vartheta\rangle}\cot i\right)\right]. \quad (12.40)$$

The derivation up to this point has not depended on $\overline{\theta}$ being small. This assumption will now be used for the first time. Using the slope distribution function (12.3) and applying (12.4) and (12.5), it is found that, to second order in $\overline{\theta}$,

$$\langle\cos\vartheta\rangle = 1/(1+\pi\tan^2\overline{\theta})^{1/2}, \quad (12.41a)$$

$$\langle\cot\vartheta\rangle = \cot\overline{\theta}, \quad (12.41b)$$

$$\langle\sin\vartheta\rangle = \frac{\pi}{2}\langle\cos\vartheta\rangle\tan\overline{\theta}, \quad (12.41c)$$

$$\langle\cot\vartheta\cos\vartheta\rangle = \cot\overline{\theta}\langle\cos\vartheta\rangle = \frac{2}{\pi}\cot^2\overline{\theta}\langle\sin\vartheta\rangle. \quad (12.41d)$$

Then (12.40) becomes

$$\int_{A(\text{tilt})} \cos\vartheta\, a(\vartheta) d\vartheta d\zeta$$
$$\simeq \left[\pi/(1+\pi\tan^2\overline{\theta})^{1/2}\right]$$
$$\times\left[2 - \exp\left(-\frac{2}{\pi}\cot\overline{\theta}\cot e\right) - \frac{\psi}{\pi}\exp\left(-\frac{2}{\pi}\cot\overline{\theta}\cot i\right)\right]. \quad (12.42)$$

Expressions (12.36) and (12.38) may be evaluated in a similar way to give

$$\int_{A(\text{tilt})} a(\vartheta) d\vartheta d\zeta$$
$$\simeq \pi\left[2 - \exp\left(-\frac{2}{\pi}\cot\overline{\theta}\cot e\right) - \frac{\psi}{\pi}\exp\left(-\frac{2}{\pi}\cot\overline{\theta}\cot i\right)\right]. \quad (12.43)$$

and

$$\int_{A(\text{tilt})} \sin\vartheta \cos\zeta a(\vartheta) d\vartheta d\zeta = \frac{\pi \tan\overline{\theta}}{(1+\pi\tan^2\overline{\theta})^{1/2}}$$
$$\times\left[\cos\psi \exp\left(-\frac{1}{\pi}\cot^2\overline{\theta}\cot^2 e\right) + \sin^2\frac{\psi}{2}\exp\left(-\frac{1}{\pi}\cot^2\overline{\theta}\cot^2 i\right)\right]. \quad (12.44)$$

Let

$$\chi(\overline{\theta}) = \langle\cos\vartheta\rangle = 1/(1+\pi\tan^2\overline{\theta})^{1/2}, \quad (12.45a)$$

$$E_1(x) = \exp\left(-\frac{2}{\pi}\cot\overline{\theta}\cot x\right), \quad (12.45b)$$

and

$$E_2(x) = \exp\left(-\frac{1}{\pi}\cot^2\overline{\theta}\cot^2 x\right). \quad (12.45c)$$

Then the approximate analytic expression for (12.19) is

$$\mu_{0e}(i,e,\psi) = \chi(\overline{\theta})\left[\cos i + \sin i \tan\overline{\theta}\frac{\cos\psi E_2(e) + \sin^2(\psi/2)E_2(i)}{2 - E_1(e) - (\psi/\pi)E_1(i)}\right]. \quad (12.46)$$

Using the identical procedure to evaluate (12.20) gives

$$\mu_e(i,e,\psi) = \chi(\overline{\theta})\left[\cos e + \sin e \tan\overline{\theta}\frac{E_2(e) - \sin^2(\psi/2)E_2(i)}{2 - E_1(e) - (\psi/\pi)E_1(i)}\right]. \quad (12.47)$$

If desired, expressions correct to higher order in $\overline{\theta}$ may be found, but such a complication probably would not be justified because of the assumption that interfacet scattering can be ignored.

Equations (12.46) and (12.47) describe the effective tilt of the surface when $i \le e$. Although they are somewhat mathematically complicated, their behavior is relatively simple. When i and e are smaller than about $\pi/2 - \overline{\theta}$, $\mu_{0e} \simeq \mu_0\chi(\overline{\theta})$ and $\mu_e \simeq \mu\chi(\overline{\theta})$. However, when either i or e exceeds about $\pi/2 - \overline{\theta}$, the effective surface tilts toward the source or detector by about $\overline{\theta}$, except that if both i and e are large and ψ is close to π the effective tilt angle goes to zero.

Finally, from (12.46) the effective cosines at $\psi = 0$, which appear in the shadow factor $\mathcal{S}$, can be evaluated:

$$\eta_e(e) \simeq \chi(\overline{\theta})\left[\cos e + \sin e \tan\overline{\theta}\frac{E_2(e)}{2 - E_1(e)}\right], \quad (12.48)$$

and, by symmetry,

$$\eta_{0e}(i) \simeq \chi(\overline{\theta})\left[\cos i + \sin i \tan\overline{\theta}\frac{E_2(i)}{2 - E_1(i)}\right]. \quad (12.49)$$

(This can also be derived by letting $\psi = 0$ in [12.52] for μ_{0e} when $e \le i$ below.) Hence, from (12.28),

$$S(i,e,\psi); \frac{\mu_e}{\eta_e(e)} \frac{\mu_0}{\eta_{0e}(i)} \frac{\chi(\overline{\theta})}{1 - f(\psi) + f(\psi)\chi(\overline{\theta})[\mu_0/\eta_{0e}(i)]}, \tag{12.50}$$

where

$$f(\psi) = \exp\left(-2\tan\frac{\psi}{2}\right). \tag{12.51}$$

12.2.3 The case when $e \le i$

Similar reasoning when $e < i$ leads to the following expressions:

$$\mu_{0e}(i,e,\psi) \simeq \chi(\overline{\theta})\left[\cos i + \sin i \tan\overline{\theta}\,\frac{E_2(i) - \sin^2(\psi/2)E_2(e)}{2 - E_1(i) - (\psi/\pi)E_1(e)}\right], \tag{12.52}$$

$$\mu_e(i,e,\psi) \simeq \chi(\overline{\theta})\left[\cos e + \sin e \tan\overline{\theta}\,\frac{\cos\psi\, E_2(i) + \sin^2(\psi/2)E_2(e)}{2 - E_1(i) - (\psi/\pi)E_1(e)}\right], \tag{12.53}$$

$$\mathbf{S}(i,e,\psi) \simeq \frac{\mu_e}{\eta_e(e)} \frac{\mu_0}{\eta_{0e}(i)} \frac{\chi(\overline{\theta})}{1 - f(\psi) + f(\psi)\chi(\overline{\theta})[\mu/\eta_e(e)]}, \tag{12.54}$$

where $\eta_{0e}(i)$, $\eta_e(e)$, and $f(\psi)$ are the same as for the $i \le e$ case and are given by (12.48), (12.49), and (12.51), respectively, except that $f(\psi)$ is now to be interpreted as the fraction of the visibility shadow hidden in the illumination shadow.

12.2.4 The physical meaning of $\bar{\theta}$

The approximate formalism developed in this chapter has been compared with detailed calculations of the brightness of an artificial rough surface generated by a computer model (Helfenstein, 1988), and its predictions have been found to be generally consistent with the model. However, an important question is the physical meaning of $\overline{\theta}$. This parameter is the mean slope angle averaged according to equation (12.5) over all distances on the surface between upper and lower limits that are determined by the angular resolution of the detector and the physics of the radiative-transfer equation. The upper limit is the footprint of the detector on the surface of the planet, which in planetary remote sensing is typically meters to kilometers.

It might be thought that the lower limit is given by the sizes of the particles making up the surface. However, it must be remembered that the radiative-transfer equation for a particulate medium, on which the solutions for the reflectance are based, implicitly averages the radiance over a distance that not only is larger than the distances between the particles, but also contains a representative distribution of the various types of particles. Thus the lower limit is several times the mean particle separation, and is typically of the order of 100–1000 μm.

Moreover, the maximum slopes that can occur on natural surfaces are determined by the strengths of materials and the cohesiveness of the soil. The effects of these properties are strongly size-dependent, such that the small-scale slopes tend to be the highest. Thus, the slope distribution functions tend to be dominated by millimeter-scale roughness.

The physical meaning of $\overline{\theta}$ has been investigated by several workers. Shepard and Campbell (1998) found that the shadowing behavior of this model was similar to a fractal surface they generated by computer modeling. By varying the parameters of their synthetic surface they concluded that the scale that dominates the photometric roughness, and thus determines $\overline{\theta}$, is the smallest scale at which well-defined shadows exist. Helfenstein and Shepard (1999) argued that for lunar regolith this scale is ~100 μm.

The main deficiency of the roughness model presented in this chapter is its neglect of interfacet multiple scattering, which by filling in shadows has the effect of making a surface appear to be photometrically smoother. This means that the photometric roughness will appear to decrease as the albedo increases. An empirical correction for this is given in Section 12.3.3. Since the actual roughness angle is a physical property of the surface it should, in principle, be independent of the wavelength at which the reflectance is measured. However, Cord *et al.* (2003) and others have pointed out that if the reflectance changes with wavelength the photometrically measured $\overline{\theta}$ will also depend on wavelength.

12.3 Applications to planetary photometry

12.3.1 Disk-resolved photometry

The equations derived in Section 12.2 can be used to calculate the effects of macroscopic roughness on light scattered by a surface having an arbitrary diffuse-reflectance function. These results will now be applied to the approximate analytic IMSA bidirectional-reflectance equation (9.47). For a surface characterized by a mean roughness slope angle $\overline{\theta}$, this equation becomes

$$r_R(i,e,g) = K\frac{w}{4\pi}\frac{\mu_{0e}}{\mu_{0e}+\mu_e}\{[p(g)[1+B_{S0}B_S(g)] \\ +[H(\mu_{0e}/K)H(\mu_e/K)-1]\}[1+B_{C0}B_C(g)]S(i,e,g). \quad (12.55)$$

Without loss of generality, it may be assumed that $g \geq 0$. When $i \leq e$, or luminance longitude $\Lambda \leq -g/2$, μ_{0e}, μ_e, and $\mathbf{S}(i, e, g)$ are given by equations (12.45) – (12.51); when $i \geq e$, or $\Lambda \geq -g/2$, μ_{0e}, μ_e, and $\mathbf{S}(i, e, g)$ are given by equations

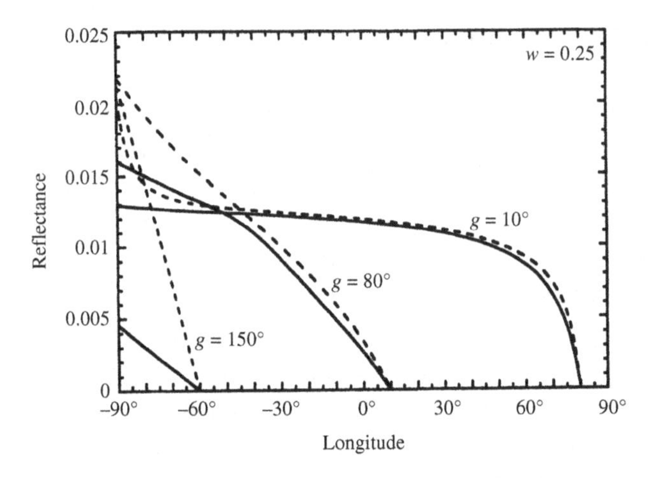

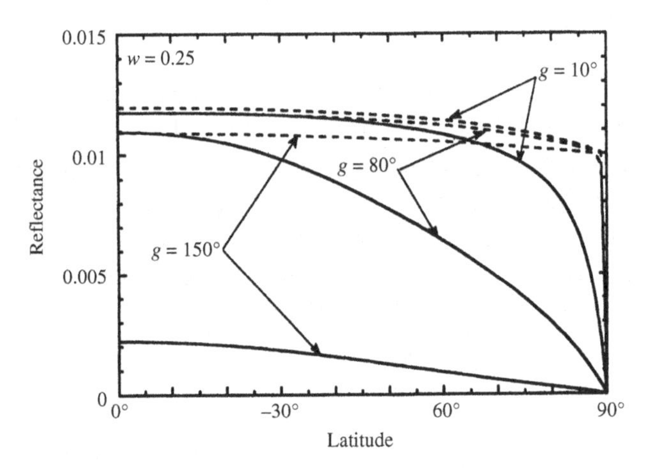

Figure 12.5 Effect of macroscopic roughness on the brightness profiles of a low-albedo ($w = 0.25$) planet at three phase angles g. For simplicity, the surface is assumed to be covered with isotropically scattering particles, and the opposition effect is neglected. Solid lines, radiance factor for $\overline{\theta} = 25°$; dashed lines, radiance factor for $\overline{\theta} = 0$. (Top) Profiles along the luminance equator. (Bottom) Profiles along the central meridian of the illuminated part of the disk. (Reproduced from Hapke [1984], copyright 1984 with permission of Elsevier.)

(12.45) and (12.48) – (12.54); the mean slope angle $\overline{\theta}$ is defined in (12.5), and the other quantities are defined in Chapters 8 and 9.

Equation (12.55) satisfies the reciprocity requirement (Section 10.3), as may be verified by writing down the detailed expressions for the various terms in the reflectance. If this is done, it must be remembered that if $i < e$ for a given set of angles, then the reciprocal configuration will have $i > e$.

In order to illustrate the effects of roughness on the reflectance, let us see how the distribution of brightness across the surface of a planet is altered by increasing

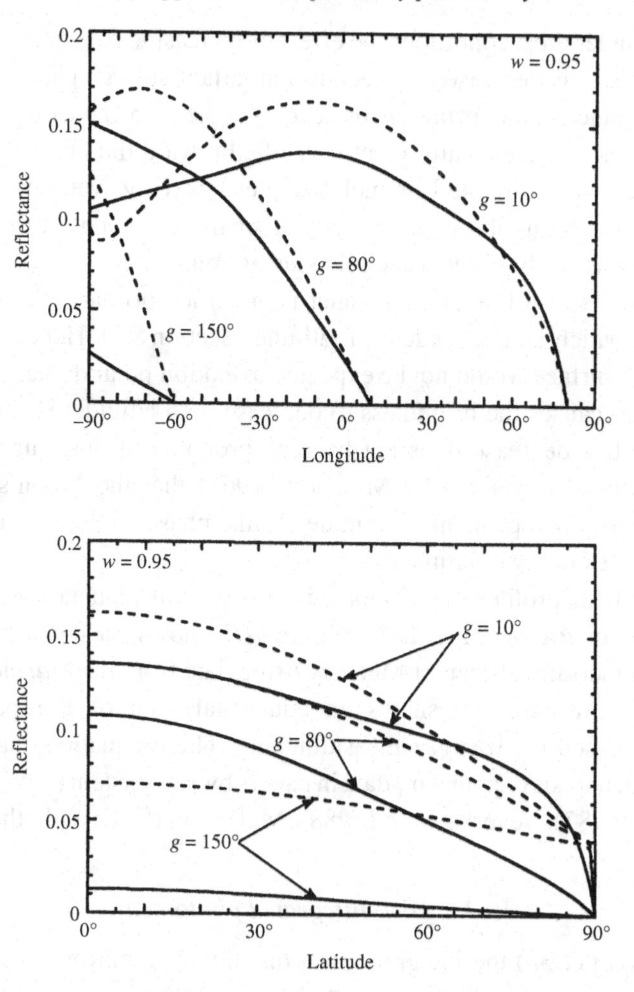

Figure 12.6 Same as Figure 12.5 for a high-albedo ($w = 0.95$) planet. (Reproduced from Hapke [1984], copyright 1984 with permission of Elsevier.)

$\overline{\theta}$. These changes are illustrated in Figures 12.5 and 12.6, which show relative reflectance profiles across the disks of hypothetical planets of varying roughnesses and albedos. Figure 12.5 (top) shows the brightness as a function of longitude along an equatorial scan on a dark, uniform planet whose regolith is composed of particles of mean single-scattering albedo $w = 0.25$. Profiles for three phase angles are given for two values of $\overline{\theta}$. Figure 12.5 (bottom) shows the brightness as a function of latitude along the central meridian of the illuminated crescent of this planet for the same phase angles and slope angles. Figure 12.6 gives similar profiles for a bright planet with $w = 0.95$. For simplicity, the profiles were calculated from equation (12.55) with $K = 1$, $p(g) = 1$ and neglecting the opposition effect.

Note that under most conditions the effect of increasing roughness is to decrease the reflectance. This decrease is especially important at large phase angles, where shadows hide much of the surface. It also occurs at the limb of the low-albedo planet, where the surface elements are selectively tilted toward the observer, eliminating the limb spike caused by the Lommel–Seeliger law. However, near the limb and terminator of the high-albedo planet seen at small phase angles, the effective tilt causes an increase in brightness with increasing roughness.

The brightness of a low-albedo planet is governed primarily by the Lommel–Seeliger law, which is independent of latitude (Section 8.7). Hence, a dark planet with a smooth surface would not be expected to exhibit polar darkening. Roughening the surface causes the brightness to decrease with latitude. However, at small phase angles this decrease does not become pronounced until high latitudes, in agreement with observations by Minnaert (1961) that the Moon's photometric function is nearly independent of latitude. As the phase angle increases, the polar darkening moves to lower latitudes.

The theoretical profiles are compared with several planetary data sets taken by spacecraft in Figures 12.7–12.9. Figure 12.7 shows an equatorial scan, and Figure 12.8 a meridional scan of Mercury using data from the *Mariner 10* mission (Hapke, 1984). Figure 12.9 shows an equatorial scan of Europa (Domingue *et al.*, 1991) based on *Voyager* measurements. These equations have also been applied to the analysis of planetary data in papers by Helfenstein (1986), Helfenstein and Veverka (1987), Veverka *et al.* (1988), McEwan (1991), and others.

12.3.2 Disk-integrated photometry

From equation (11.35) the integral phase function of a uniform, rough-surfaced, spherical planet of radius R observed at phase angle g is

$$\begin{aligned}\Phi(g,w,\overline{\theta}) &= \frac{1}{JR^2A_p}\int_{A(i,v)} Jr_R(i,e,g)\mu dA\\ &= \frac{1}{JR^2A_p}\int_{\Lambda=-\pi/2}^{\pi/2-g}\int_{L=-\pi/2}^{\pi/2} Jr_R(\Lambda,L,g)\mu R^2\cos L dL d\Lambda\\ &= \frac{4}{A_p}\int_{\Lambda=-\pi/2}^{-g/2}\int_{L=0}^{\pi/2} r_R(\Lambda,L,g)\cos\Lambda\cos^2 L dL d\Lambda, \qquad (12.56)\end{aligned}$$

where

$$\begin{aligned}A_p(w,\overline{\theta}) &= \frac{1}{JR^2}\int_{A(i)} Jr_R(i,e,0)\mu dA\\ &= 4\int_{\Lambda=-\pi/2}^{0}\int_{L=0}^{\pi/2} r_R(\Lambda,L,0)\cos\Lambda\cos^2 L dL d\Lambda \qquad (12.57)\end{aligned}$$

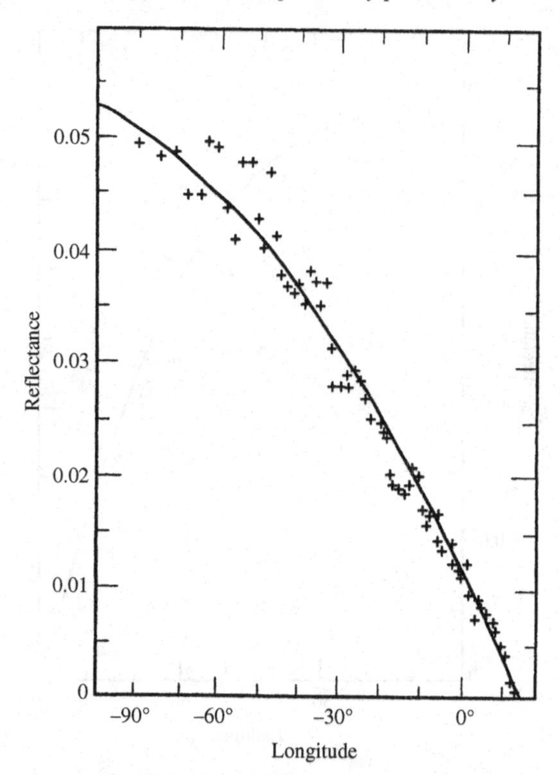

Figure 12.7 Measured and predicted brightness distributions along the equator of Mercury at $g = 77°$. The line is the theoretical brightness for $\overline{\theta} = 20°$; the crosses are data obtained by the *Mariner 10* spacecraft. (Reproduced from Hapke [1984], copyright 1984 with permission of Elsevier.)

is the physical albedo. In these equations we have used the fact that the bidirectional reflectance must be symmetric with respect to the northern and southern hemispheres and reciprocal with respect to the $\Lambda = -g/2$ meridian.

Equation (12.55) was inserted into (12.56) and (12.57) and integrated numerically for values of $\overline{\theta}$ up to 60°. It was found that to a good approximation the physical albedo can be written

$$A_p(w, \overline{\theta}) = \frac{w}{8}\left\{p(0)[1 + B_{S0}] - 1 + U(w, \overline{\theta})\frac{r_0}{2}\left(1 + \frac{r_0}{3}\right)\right\}[1 + B_{C0}], \quad (12.58)$$

and

$$\Phi(g, w, \overline{\theta}) = K(g, \overline{\theta})\Phi(g, w, 0), \quad (12.59)$$

where r_0 is the diffusive reflectance, and $\Phi(g, w, 0)$ is given by (11.42); in which $A_p(w, 0)$ is given by (11.34). The quantities $U(w, \bar{\theta})$ and $K(g, \bar{\theta})$ are, respectively,

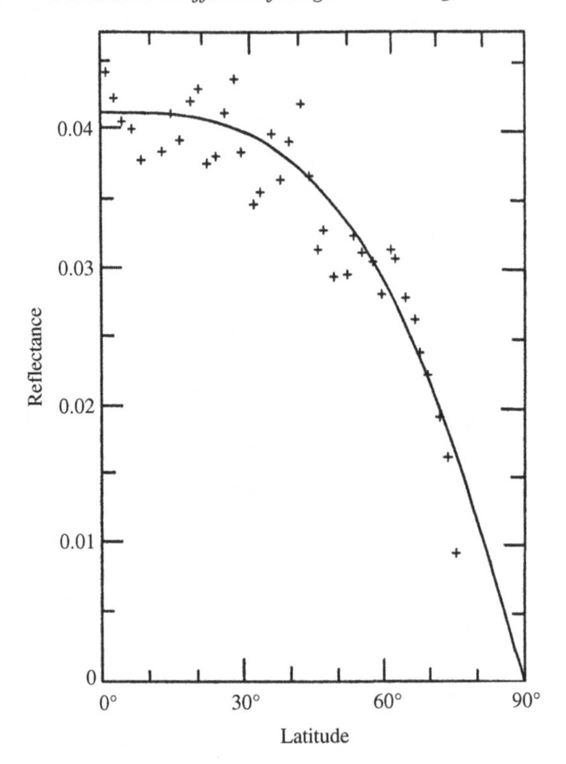

Figure 12.8 Measured and predicted brightness distributions along the $\Lambda = -50°$ meridian of luminance latitude of Mercury at $g = 77°$. The line is the theoretical brightness for $\overline{\theta} = 20°$; the crosses are data obtained by the *Mariner 10* spacecraft. (Reproduced from Hapke [1984], copyright 1984 with permission of Elsevier.)

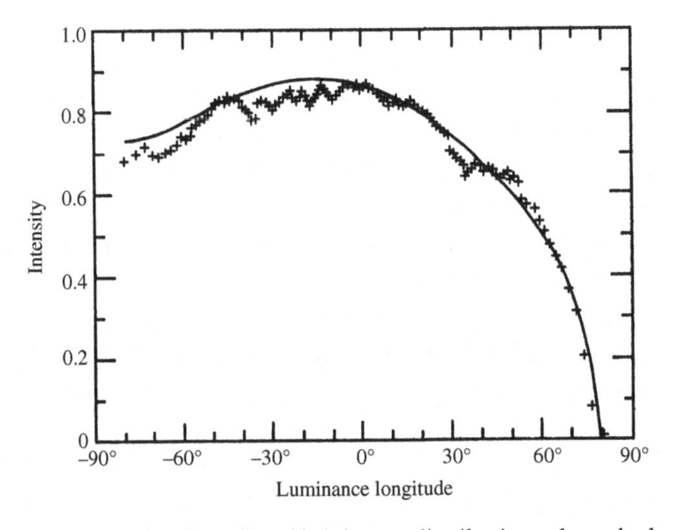

Figure 12.9 Measured and predicted brightness distributions along the luminance equator of Europa at $g = 10.5°$. The line is the theoretical brightness in relative units for $\overline{\theta} = 10°$; the crosses are data obtained by the *Voyager* spacecraft. (Reproduced from Domingue *et al.* [1991], copyright 1991 with permission of Elsevier.)

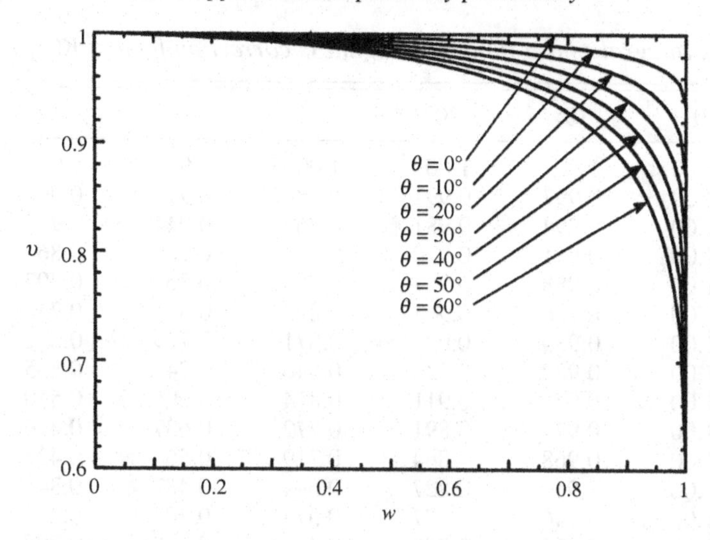

Figure 12.10 Physical-albedo correction factor $U(w, \overline{\theta})$ plotted against the single-scattering albedo for several values of $\overline{\theta}$. (Reproduced from Hapke [1984], copyright 1984 with permission of Elsevier.)

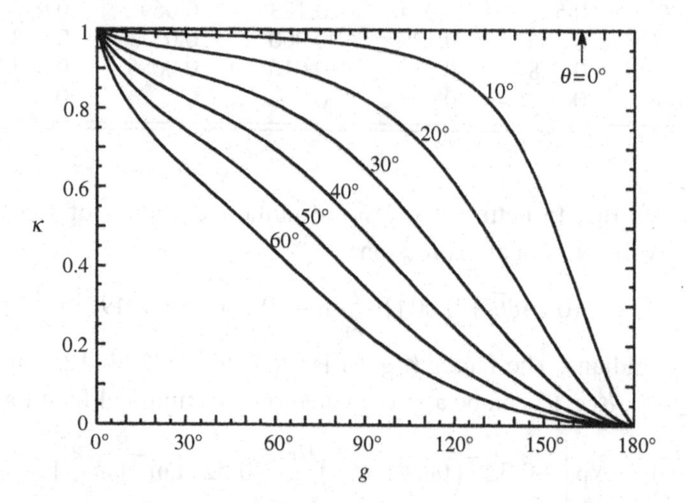

Figure 12.11 Integral-photometric-function correction factor $K(g, \overline{\theta})$ plotted against phase angle for several values of $\overline{\theta}$. This quantity is the ratio of the normalized integral brightness of a planet with a rough surface to that of one with a smooth surface. (Reproduced from Hapke [1984], copyright 1984 with permission of Elsevier.)

the factors by which the physical albedo and integral phase function of a planet with a smooth surface may be corrected for effects of macroscopic roughness. These correction factors are plotted in Figures 12.10 and 12.11.

Table 12.1. *Integral-phase-function roughness correction factors* $K(g, \bar{\theta})$

g	0°	10°	20°	30°	40°	50°	60°
0	1.00	1.00	1.00	1.00	1.00	1.00	1.00
2	1.00	0.997	0.991	0.984	0.974	0.961	0.943
5	1.00	0.994	0.981	0.965	0.944	0.918	0.881
10	1.00	0.991	0.970	0.943	0.909	0.866	0.809
20	1.00	0.988	0.957	0.914	0.861	0.797	0.715
30	1.00	0.986	0.947	0.892	0.825	0.744	0.644
40	1.00	0.984	0.938	0.871	0.789	0.692	0.577
50	1.00	0.982	0.926	0.846	0.748	0.635	0.509
60	1.00	0.979	0.911	0.814	0.698	0.570	0.438
70	1.00	0.974	0.891	0.772	0.637	0.499	0.366
80	1.00	0.968	0.864	0.719	0.566	0.423	0.296
90	1.00	0.959	0.827	0.654	0.487	0.346	0.231
100	1.00	0.946	0.777	0.575	0.403	0.273	0.175
110	1.00	0.926	0.708	0.484	0.320	0.208	0.130
120	1.00	0.894	0.617	0.386	0.243	0.153	0.094
130	1.00	0.840	0.503	0.290	0.175	0.107	0.064
140	1.00	0.747	0.374	0.201	0.117	0.070	0.041
150	1.00	0.590	0.244	0.123	0.069	0.040	0.023
160	1.00	0.366	0.127	0.060	0.032	0.018	0.010
170	1.00	0.128	0.037	0.016	0.0085	0.0047	0.0026
180	1.00	0	0	0	0	0	0

It was found that to better than 1% the numerical values of $U(w, \bar{\theta})$ can be represented by the empirical expression

$$U(w, \bar{\theta}) = 1 - (0.048\bar{\theta} + 0.0041\bar{\theta}^2)r_0 - (0.33\bar{\theta} - 0.0049\bar{\theta}^2)r_0^2, \tag{12.60}$$

where $\bar{\theta}$ is in radians. The factor $K(g, \bar{\theta})$ is tabulated in Table 12.1. It was found that for $g \leq 60°$, $K(g, \bar{\theta})$ can be approximated by the empirical function

$$K(g, \bar{\theta}) = \exp\left[-0.32\bar{\theta}\left(\tan\bar{\theta}\tan\frac{g}{2}\right)^{1/2} - 0.52\bar{\theta}\tan\bar{\theta}\tan\frac{g}{2}\right]. \tag{12.61}$$

If $g > 60°$ this equation overestimates K, and Table 12.1 should be used.

Note that macroscopic roughness causes a broad opposition effect in the integral phase function. This opposition effect is caused by macroscopic shadow hiding between the surface facets. It is superposed onto the microscopic shadow hiding and coherent backscatter peaks. This effect is also responsible for a surge in the thermally emitted radiance of many solar system bodies near zero phase in the infrared, where it is known as "thermal beaming."

At small phase angles the regions that are strongly shadowed tend to be close to the limb, where their effects are minimized by the factor μ in the integrand for

$\Phi(g, w, \overline{\theta})$. Over most of the disk, decreases in brightness of those facets that are oriented away from the Sun are largely compensated by increases in the brightness of other facets that are oriented more directly toward the Sun. The major effects of roughness are seen when $g > 90°$, when the heavily shadowed areas are near the center of the disk. At very large phase angles the brightness of a rough planet is only a few percent of that of a corresponding smooth planet, thus accounting for the observation, known since antiquity, that the Moon is invisible when less than about 1 day from new.

The theoretical integral phase function is compared with observations of Mercury, a low-albedo planet, in Figure 12.12 (Hapke, 1984) and Europa, a high-albedo body, in Figure 12.13 (Domingue *et al.*, 1991). Equations (12.56) – (12.61) have been widely used in a number of analyses to describe the integral scattering properties of objects in the solar system, including studies by Hapke (1984), Buratti (1985), Helfenstein and Veverka (1987), Simonelli and Veverka (1987), Herbst *et al.*, (1987), Veverka *et al.* (1988), Bowell *et al.* (1989), and Domingue *et al.* (1991).

12.3.3 An empirical correction for multiple interfacet scattering

Mutiple scattering between the facets that are assumed to make up the surface should be similar to interparticle multiple scattering in the sense that it is small when the albedos of the facets are small, and increases nonlinearly as the albedo becomes close to 1. This suggests that, as an approximate correction for multiple scattering, $\overline{\theta}$ be replaced in all equations by

$$\overline{\theta}_p = (1 - r_0)\overline{\theta}. \tag{12.62}$$

where r_0 is the diffusive reflectance and $\overline{\theta}_p$ is the effective value of the roughness that is measured photometrically. Thus, $\overline{\theta} = \overline{\theta}_p / (1 - r_0)$ should approximate the actual roughness. The correction has only a minor effect when the albedo is small, but reduces the photometric roughness $\overline{\theta}_p$ to nearly zero when the albedo is high. This also means that the calculated spherical reflectance of a medium of non-absorbing particles will be close to 1, as conservation of energy requires.

12.4 Summary of the roughness correction model

Suppose a semi-infinite, plane-parallel, particulate medium has a surface that is smooth on a scale larger than the particles, and has a bidirectional reflectance $r(i, e, g, w)$. Suppose also that a spherical planet covered with an optically thick layer of this medium has a physical albedo $A_p(w)$ and integral phase function $\Phi(g, w)$. Suppose now that the surface of this medium is warped into a series of elevations and/or depressions whose geometry is unspecified, except that it is independent of

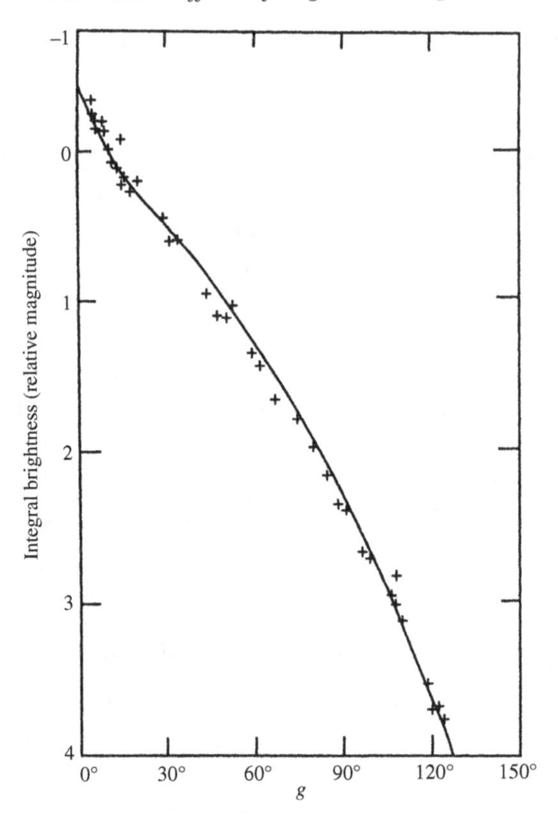

Figure 12.12 Measured and predicted integral phase functions of Mercury. The line is the theoretical phase function for $\overline{\theta} = 20°$; the crosses are data of Danjon (1949) in magnitudes. (Reproduced from Hapke [1984], copyright 1984 with permission of Elsevier.)

the azimuth angle ψ, and is characterized by a distribution of slope angles relative to the horizontal by the Gaussian function

$$a(\vartheta) = \frac{2}{\pi \tan^2 \overline{\theta}} \exp\left(-\frac{\tan^2 \vartheta}{\pi \tan^2 \overline{\theta}}\right) \sec^2 \vartheta \sin \vartheta,$$

where

$$\tan \overline{\theta} = [(< \cos \vartheta >^{-2} - 1)/\pi]^{1/2}.$$

and $< \cos \vartheta >$ is the mean cosine of the slope angle.

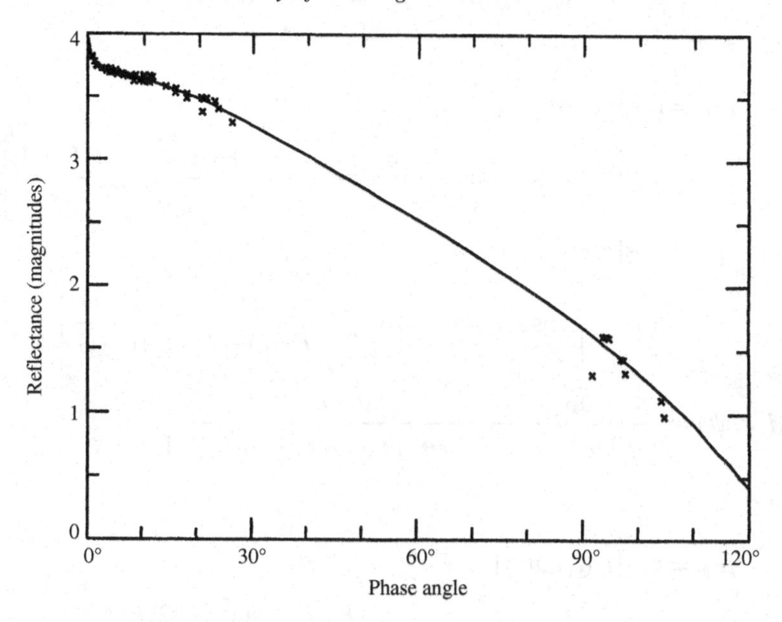

Figure 12.13 Measured and predicted integral phase functions of Europa. The line is the theoretical phase function for $\overline{\theta} = 10°$; the crosses are combined telescopic and *Voyager* data. (Reproduced from Domingue *et al.* [1991], copyright 1991 with permission of Elsevier.)

Let

$$\overline{\theta}_p = (1 - r_0)\overline{\theta},$$

$$f(\psi) = \exp\left(-2\tan\frac{\psi}{2}\right),$$

$$\chi(\overline{\theta_p}) = 1/(1 + \pi \tan\overline{\theta_p}^2)^{1/2},$$

$$E_1(y) = \exp\left(-\frac{2}{\pi}\cot\overline{\theta}_p \cot y\right),$$

$$E_2(y) = \exp\left(-\frac{1}{\pi}\cot^2\overline{\theta_p}\cot^2 y\right),$$

$$\eta(y) = \chi(\overline{\theta_p})\left[\cos y + \sin y \tan\overline{\theta_p}\frac{E_2(y)}{2 - E_1(y)}\right],$$

where y = either i or e.

Then the bidirectional reflectance of the rough surface is

$$r_R(i, e, g, w, \overline{\theta}) = r(i_e, e_e, g, w)S(i, e, g) \tag{12.63}$$

where the effective angles of incidence i_e and emergence e_e and the shadowing function $S(i,e,g)$ are given by the following relations.

When $i \le e$,

$$\mu_{0e} = \cos[i_e(i,e,\psi)]$$
$$= \chi(\overline{\theta}_p)\left[\cos i + \sin i \tan\overline{\theta}_p \frac{\cos\psi E_2(e) + \sin^2(\psi/2)E_2(i)}{2 - E_1(e) - (\psi/\pi)E_1(i)}\right],$$

$$\mu_e = \cos[e_e(i,e,\psi)]$$
$$= \chi(\overline{\theta}_p)\left[\cos e + \sin e \tan\overline{\theta}_p \frac{E_2(e) - \sin^2(\psi/2)E_2(i)}{2 - E_1(e) - (\psi/\pi)E_1(i)}\right],$$

$$S(i,e,\psi) = \frac{\mu_e}{\eta(e)}\frac{\mu_0}{\eta(i)}\frac{\chi(\overline{\theta}_p)}{1 - f(\psi) + f(\psi)\chi(\overline{\theta}_p)[\mu_0/\eta(i)]},$$

where $e \le i$,

$$\mu_{0e} = \cos[i_e(i,e,\psi)]$$
$$= \chi(\overline{\theta}_p)\left[\cos i + \sin i \tan\overline{\theta}_p \frac{E_2(i) - \sin^2(\psi/2)E_2(e)}{2 - E_1(i) - (\psi/\pi)E_1(e)}\right],$$

$$\mu_e = \cos[e_e(i,e,\psi)]$$
$$= \chi(\overline{\theta_p})\left[\cos e + \sin e \tan\overline{\theta}_p \frac{\cos\psi E_2(i) + \sin^2(\psi/2)E_2(e)}{2 - E_1(i) - (\psi/\pi)E_1(e)}\right],$$

$$S(i,e,\psi) = \frac{\mu_e}{\eta(e)}\frac{\mu_0}{\eta(i)}\frac{\chi(\overline{\theta_p})}{1 - f(\psi) + f(\psi)\chi(\overline{\theta_p})[\mu/\eta(e)]}.$$

The planetary physical albedo is

$$A_p(w,\overline{\theta}_p) = \frac{w}{8}\left\{p(0)[1 + B_{S0}] - 1 + U(w,\overline{\theta_p})\frac{r_0}{2}\left(1 + \frac{r_0}{3}\right)\right\}[1 + B_{C0}],$$

and the integral phase function is

$$\Phi(g,w,\overline{\theta}_p) = K(g,\overline{\theta_p})\Phi(g,w,0),$$

where $\Phi(g,w,0)$ is given by (11.42), $A_p(w,0)$ by (11.34),

$$U(w,\overline{\theta}) = 1 - (0.048\overline{\theta} + 0.0041\overline{\theta}^2)r_0 - (0.33\overline{\theta} - 0.0049\overline{\theta}^2)r_0^2,$$

and $K(g,\overline{\theta_p})$ is tabulated in Table 12.1; for $g \le 60^\circ$ a good approximation for $K(g,\overline{\theta_p})$ is

$$K(g,\overline{\theta}) \approx \exp\left[-0.32\overline{\theta}\left(\tan\overline{\theta}\tan\frac{g}{2}\right)^{1/2} - 0.52\overline{\theta}\tan\overline{\theta}\tan\frac{g}{2}\right].$$

12.5 Other planetary photometric models

12.5.1 Introduction

There are a number of other models that attempt to describe the scattering of sunlight by bodies of the solar system. In this section we describe three of them in wide use, the Lumme–Bowell, Buratti–Veverka, and Shkuratov models. They will be given and discussed briefly without deriving them. For details of their derivation the oiginal references should be consulted.

12.5.2 The Lumme–Bowell model

The Lumme–Bowell model is derived and described in three papers: Lumme and Bowell (1981a, b) and Bowell *et al.* (1989). The bidirectional reflectance is given by

$$r(i,e,g) = r_1(i,e,g) + r_m(i,e),$$

where

$$r_m(i,e) = \frac{w*}{4\pi} H(\mu_0, w*) H(\mu, w*) - 1$$

is the contribution of mutiply scattered light. Thus, multiple scattering is described in a manner similar to the IMSA model, except that the reduced albedo w^* is used in the H functions.

The term $r_1(i,e,g)$ is the singly scattered contribution to the reflectance,

$$r_1(i,e,g) = \frac{w}{4\pi} \frac{\mu_0}{\mu_0 + \mu} p(g) \frac{{}_1F_1(1, 1+2x, x)}{2} \left(\frac{f}{1 + s\varsigma} + 1 - f \right).$$

where $p(g)$ is the volume-average particle phase function. The Lumme–Bowell model assumes that all particulate media have the same particle scattering function

$$p(g) = 0.95 \exp[-0.4g] + 16.11 \exp[-4.0(\pi - g)],$$

which was obtained by empirically fitting the phase functions of a number of particles measured in the laboratory.

The factor ${}_1F_1(1, 1+2x, x)/2$ describes the opposition surge, where ${}_1F_1$ is the degenerate hypergeometric function. The factor x in the arguments of ${}_1F_1$ is

$$x = \frac{\cos\Lambda + \cos(\Lambda - g)}{2.4 \sin g} \ln \frac{1}{1-\phi},$$

where Λ is the luminance longitude, and ϕ is the filling factor. The opposition effect factor amplifies only the single-scattering term and, thus, describes the SHOE. The model does not include the CBOE.

The last factor (in parentheses) in the singly scattered radiance accounts for the effects of roughness: the surface is assumed to be partially covered with holes whose sides cast shadows on their bottoms; F is the fraction of the surface occupied by the holes; s is a function of the depth to diameter ratio of the holes, and thus is a measure of the mean slope angle of the roughness; and

$$\varsigma = \frac{\left(\mu_0^2 - 2\mu_0\mu + \mu^2\right)^{1/2}}{\mu_0\mu}.$$

The roughness factor is assumed to affect only the singly scattered light. The Lumme–Bowell model justifies this by asserting that the multiple-scattering term describes light that is scattered several times between macroscopic tilted facets, as well as between individual grains. This assumption is debatable and is one of the uncertainties of the model.

The bidirectional reflectance contains four adjustable parameters: w, ϕ, f, and s.

The disk-integrated reflectance scattered by a spherical body is given by integrating the bidrectional reflectance over the surface. Lumme and Bowell fitted approximate analytic functions to the resulting expressions. They find that the scattered radiance relative to a perfect Lambert disk expressed in astronomical magnitudes m is:

$$I/F = 10^{-0.40m(g)} = a_1 F_1 + a_2 F_2 + a_3 F_1 F_3,$$

where

$$F_1 = \exp\left[-3.33\left(\tan\frac{g}{2}\right)^{0.63}\right],$$

$$F_2 = \exp\left[-1.87\left(\tan\frac{g}{2}\right)^{1.22}\right],$$

$$F_3 = \exp\left(-\frac{g}{0.333 + 2.31g}\right),$$

$$a_1 = (1-c)(1-Q)10^{-0.4m(0)},$$

$$a_2 = Q10^{-0.4m(0)},$$

$$a_3 = c(1-Q)10^{-0.4m(0)}.$$

These equations contain three adjustable parameters: $m(0)$ the magnitude at zero phase angle, Q the ratio of multiple to single scattering, and c the opposition effect parameter.

A simplified empirical version of the integral Lumme–Bowell equation has been adopted by the International Astromonmical Union to describe the phase function of asteroids:

$$H(g) = H(0) - 2.5\ln[(1-G)F_1(g) + GF_2(g)],$$

where $\mathcal{H}(g)$ is the visual magnitude at phase angle g reduced to unit heliocentric and geocentric distances. Thus the brightness in magnitdes is characterrized by two parameters: $\mathcal{H}(0)$ the absolute magnitude at zero phase, and the so-called slope parameter G.

12.5.3 The Buratti–Veverka model

The Buratti–Veverka model (Buratti and Veverka, 1985) is a combination of the generalized Lommel–Seeliger law, Lambert's law, and the Hameen-Anttila (1967) roughness model. A smooth surface is assumed to have a radiance factor given by

$$\frac{I}{F} = A\frac{\mu_0}{\mu_0 + \mu}p(g) + (1 - A)\mu_0.$$

Following Hameen-Anttila, this surface is then warped into a depression described by a paraboloid of revolution in order to account for the effects of roughness. The roughness corection is carried out by computer, so the model is not fully analytic.

The model contains three or more adjustable parameters: A, the depth-to-diameter ratio of the depressions, plus as many parameters as necessary to describe $p(g)$. The model assumes that the second term on the right describes interfacet scattering as well as interparticle scattering.

12.5.4 The Shkuratov reflectance model

The Shkuratov model (Shkuratov, 1989) is a semi-empirical description of the photometric function of a surface (Section 10.5.3). The bidirectional reflectance is

$$r(i, e, g) = \frac{A_n}{\pi} f(\Lambda, L, g) = \frac{A_n}{\pi} f_1(g) f_2(g) f_3(\Lambda, L, g)$$

where A_n is the normal albedo, $f = f_1 f_2 f_3$ is the photometric function, Λ is the luminance longitude, $\mathcal{L}$ is the luminance latitude, and g is the phase angle.

The first factor in the photometric function

$$f_1(g) = \exp(-kg)$$

describes the SHOE, where $\mathcal{K}$ is an empirical parameter that decreases with increasing albedo and depends on the geometry of the surface of the medium.

The second factor

$$f_2(g) = \frac{1}{2 + \exp(-d/\Lambda_E)}\left\{2 + \frac{\exp(-d/\Lambda_E)}{\sqrt{1 + [(4\pi\Lambda_E/\lambda)\sin(g/2)]^2}}\right\}$$

describes the CBOE, where Λ_E is the extinction mean free path, and d is a distance within the medium that is so small that coherent backscattering cannot occur within it. If the particles are larger than the wavelength, then d is of the order of the radius of a particle; if the particles are smaller than the wavelength, then d is of the order of λ.

The third factor

$$f_3(\Lambda, L, g) = \frac{\cos[\frac{\pi}{\pi-g}(\Lambda - \frac{g}{2})]}{\cos\Lambda}(\cos L)^{g/(\pi-g)}$$

describes the angular distribution of brightness at larger angles than the opposition effects. It was first suggested by L. Akimov and is sometines known as the Akimov formula. It is based on considerations of shadowing on fractal-like surfaces.

The reflectance contains four adjustable parameters: $A_n, \mathcal{K}, \Lambda_E$, and d.

The integral brightness of a body relative to a Lambert disk is

$$A_P\Phi_P = A_P f_1(g) f_2(g) f_4(g),$$

where A_P is the physical albedo, $\Phi_p = f_1 f_2 f_4$ is the integral phase function,

$$f_4(g) = \frac{2}{\sqrt{\pi}}\left(1 - \frac{g}{\pi}\right)\frac{\Gamma\left(\frac{3\pi-g}{2(\pi-g)}\right)}{\Gamma\left(\frac{4\pi-3g}{2(\pi-g)}\right)},$$

and $\Gamma(z)$ is the gamma function

$$\Gamma(z) = \int_0^\infty t^{z-1}e^{-t}dt.$$

The disk-integrated function has four adjustable parameters: $A_P, \mathcal{K}, \Lambda_E$, and d.

There are a number of difficulties with the Shkuratov model. The Shkuratov model is a useful way of describing the reflectance empirically, but the physical interpretation of the parameters is unclear. The assumption that the surface of a body covered with craters of all sizes can be described by fractals is probably reasonable down to the scale of the particles of the regolith, but the geometry of the surface changes drastically at this size, so the validity of this assumption is uncertain. It has been shown earlier in this chapter that individual particles can have a CBOE, so it is not at all clear that a finite value of d exists, particularly in view of the large number of observations of CBOEs exhibited by colloidal suspensions of particles smaller than the wavelength. Shkuratov points out that the observed independence of the opposition effect on wavelength can be explained if d and Λ_E are both proportional to λ. However, the assumption that d is of the order of λ for particles smaller than λ does not seem to be valid. Also, there is no reason why the extinction mean free path should be proportional to wavelength.

13

Polarization of light scattered by a particulate medium

13.1 Introduction

In the radiative-transport equations for the reflectance and emittance of a particulate medium developed in Chapter 7 it was assumed that polarization can be neglected. For irregular particles that are large compared with the wavelength of the observation, this assumption is justified on the grounds that the light scattered by such particles is relatively weakly polarized (e.g., Figure 6.7b). However, even though it may be small, the polarization of the light scattered by a medium does contain information about the medium and, thus, is potentially a useful tool for remote sensing. One of the advantages of using polarization is that it does not require absolute calibration of the detector, but only a measurement of the ratio of two radiances.

The discovery that sunlight scattered from a planetary regolith was polarized was made as early as 1811 by Arago, who noticed that moonlight was partially linearly polarized and that the dark lunar maria were more strongly polarized than the lighter highlands. Subsequent observations of planetary polarization were made by several persons, including Lord Rosse in Ireland. However, the quantitative measurement of polarization from bodies of the solar system was placed on a firm foundation in the 1920s by the classical studies of Lyot (1929). This work was later continued by Dollfus (1956, 1998) and his colleagues. For a more detailed historical account, the reviews by Dollfus (1961, 1962) and Gehrels (1974) should be consulted. Other important contributions have been made by Gehrels and his co-workers (Gehrels and Teska, 1963; Gehrels *et al.*,, 1964), who emphasized the importance of the variation of polarization with wavelength, Shkuratov and his colleagues (Shkuratov *et al.*, 1992a, b, 2007) and many others. Egan (1985) has discussed applications of polarization to terrestrial remote sensing.

One of the triumphs of photopolarization in planetary remote sensing occurred when Hansen and Arking (1971) were able to account for the observed variation of the polarization of Venus as a function of wavelength and phase angle by a cloud of spherical particles of refractive index $n = 1.44$ and radius 1.1 μm. This was the key observation that led to the identification of the composition of the clouds as sulfuric acid by Young (1973).

13.2 Linear polarization of particulate media

13.2.1 The Jones and Mueller matrices

We saw in Chapter 2 that electromagnetic radiation consists of a transverse wave of electric and magnetic fields vibrating perpendicularly to the direction of propagation, and that the wave can be described by Jones and Stokes vectors. The Jones vector representation is most useful when a completely polarized wave interacts with an optical device, such as a mirror, a polarizer, or a quarter wave plate. Then the interaction of the device with the electric field can be written in matrix form as $\mathbf{E} = \mathbf{J}\mathbf{E}_i$, where $\mathbf{E}_i$ and $\mathbf{E}$ are the Jones vectors that describe the electric fields of the incident and exit radiation, respectively, and $\mathbf{J}$ is a 2×2 matrix called the *Jones matrix*.

When the radiation is only partially polarized or is unpolarized the intensity can be described by the Stokes vector. In that case the general scattering process of light interacting with a medium can be written as a matrix equation

$$\mathbf{I} = \mathbf{M}\mathbf{I}_i, \tag{13.1}$$

where $\mathbf{I}_i$ and $\mathbf{I}$ are the 1×4 Stokes vectors that describe the incident and scattered radiances, respectively, and $\mathbf{M}$ is a 4×4 matrix called the *Mueller matrix* or *scattering matrix*. Thus, when the light is polarized the reflectance is actually a Mueller matrix. In general, $\mathbf{M}$ contains 16 elements. However, if the scattering medium consists of randomly oriented particles, and if each particle has a mirror particle, not all of the elements are independent and some are zero. These requirements are probably satisfied by most laboratory powders and planetary regoliths, at least in the first approximation. In that case the Mueller matrix can be written (Mishchenko *et al.*, 2000a)

$$\mathbf{M} = \begin{pmatrix} m_{11} & m_{12} & 0 & 0 \\ m_{12} & m_{22} & 0 & 0 \\ 0 & 0 & m_{33} & m_{34} \\ 0 & 0 & -m_{34} & m_{44} \end{pmatrix}, \tag{13.2}$$

which has only six non-zero independent elements.

Thus, in the case of unpolarized sunlight with Stokes vector

$$\mathbf{I}_i = \begin{pmatrix} I_i \\ 0 \\ 0 \\ 0 \end{pmatrix}$$

incident on a planetary regolith with reflectance $\mathbf{r} = \mathbf{M}$, the Stokes vector of the scattered radiance is

$$\mathbf{I} = \begin{pmatrix} m_{11} I_i \\ m_{12} I_i \\ 0 \\ 0 \end{pmatrix}.$$

13.2.2 The polarization

There are many situations in remote sensing when both the incident and scattered radiances are polarized. The light scattered from the atmosphere is partially linearly polarized (e.g., Coulson, 1971), and this may need to be taken into account in measurements of the reflectance of the surface of the Earth that are sensitive to polarization. Another important application in which the incident radiance is polarized is radar. The transmitted radio frequency pulse is usually completely polarized, either linearly or circularly, and the difference between the fractions of the power returned with the same and opposite senses of polarization as transmitted gives additional information about the medium (Evans and Hagfors, 1968). In some laboratory studies the incident irradiance is completely polarized, either linearly or circularly, and the polarized reflectances are of interest.

A general discussion of polarized reflectance and the Mueller matrix is beyond the scope of this book. Fortunately, this degree of complexity often is not needed in the interpretation of remote-sensing observations. In most field measurements the illumination is sunlight, which is essentially unpolarized and the scattered radiance is only weakly linearly polarized. Circular polarization can usually be ignored, because the polarization in the sunlight scattered from bodies of the solar system is extremely small, on the order of $10^{-4} - 10^{-5}$ (Kemp, 1974).

This chapter will concentrate on the linear polarization in the radiance at optical wavelengths scattered by a medium of irregular particles large compared with the wavelength when illuminated by an unpolarized incident irradiance. Unfortunately, the discussion will necessarily be largely qualitative, because even this apparently simple case is still poorly understood theoretically 80 years after Lyot's pioneering investigations.

It is found observationally that when the incident light is unpolarized, the radiance scattered from a particulate medium is partially linearly polarized, and the plane of the polarized component is either parallel or perpendicular to the scattering plane. This allows the polarized light, which is a vector quantity, to be represented by a scalar, the polarization. In the discussion of Stokes vectors in Chapter 2 the coordinate system was unspecified. If the direction of propagation is taken to be the positive z-axis and the scattering plane is taken to be the x–z plane then the polarization is defined as

$$P(i, e, g) = -\frac{Q}{I} = \frac{I_\perp - I_{||}}{I_\perp + I_{||}}, \tag{13.3}$$

where I and Q are the first two quantities in the Stokes vector, $I_\perp(i, e, g)$ is the component of intensity with its electric vector pointing perpendicular to the scattering plane (y direction), and $I_{||}(i, e, g)$ is the component parallel to the scattering plane (x direction).

Thus, if the radiance scattered with the electric vector perpendicular to the scattering plane is larger than the radiance parallel, the polarization is said to be *positive;* if the opposite is true, the polarization is *negative*. With this choice of sign of equation (13.3), light specularly reflected from a plane surface (Fresnel reflection), and also light scattered by a small particle (Rayleigh scattering), will always have positive polarization.

The curves of Lyot (1929) for the Moon, shown in Figure 13.1, are typical of the variation of polarization with phase angle of a particulate medium. The curves for most of the airless bodies of the solar system and for particulate media in the laboratory are similar. Although these particular curves are for the integrated light from the entire body, the curves for individual areas are virtually identical (Dollfus, 1956). At zero phase angle the polarization is zero. The polarization is negative for small phase angles and goes through a minimum of $P \sim -1\%$ at

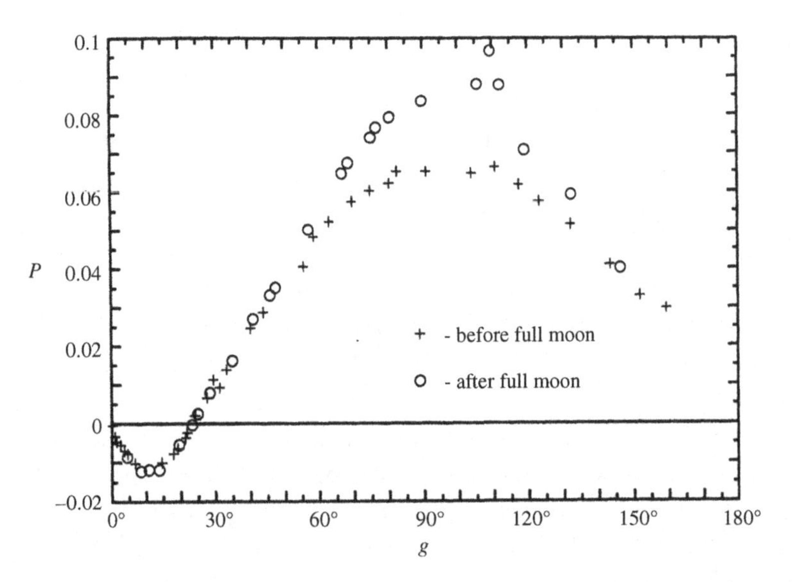

Figure 13.1 Polarization versus phase angle for the integrated light from the Moon. The crosses give the polarization before full moon, and the circles after (Lyot, 1929).

$g \sim 10°$. The negative branch of polarization may or may not be bimodal. The polarization increases to zero at phase angles between about 15° and 30°, where the plane of polarization abruptly rotates and P becomes positive. It then goes through a maximum whose amplitude depends on the material, but is typically $P \sim 10\%$ at $g \sim 100°$, after which P decreases to zero again at large phase angles.

The parameters commonly used to describe the polarization phase curve are the amount P_p and phase angle g_p at which the positive maximum occurs, the amount P_n and phase angle g_n at which the negative minimum occurs, the phase angle g_i at which the inversion occurs and the slope $h_i = dP/dg$ at g_i. If the negative branch of polarization is bimodal, P_n and g_n refer to the broad minimum at the larger phase angle. All of these parameters seem to depend on particle size, composition, albedo, and porosity in nontrivial ways.

If the polarization of light scattered from the surface of a nonopaque solid or liquid, such as a mineral or the surface of water, is examined, it is found that when the source and detector are in the vicinity of the specular configuration the polarization is large and positive, as predicted by the Fresnel reflection coefficients. However, away from the specular region the electric vector lies in the plane formed by the surface normal and the emergent ray, and the polarization is not zero at $g = 0$ (Figure 13.2). This residual negative polarization is caused by the light that has been volume-scattered within the material and is polarized by refraction as it leaves the surface, as described by the Fresnel transmission coefficients.

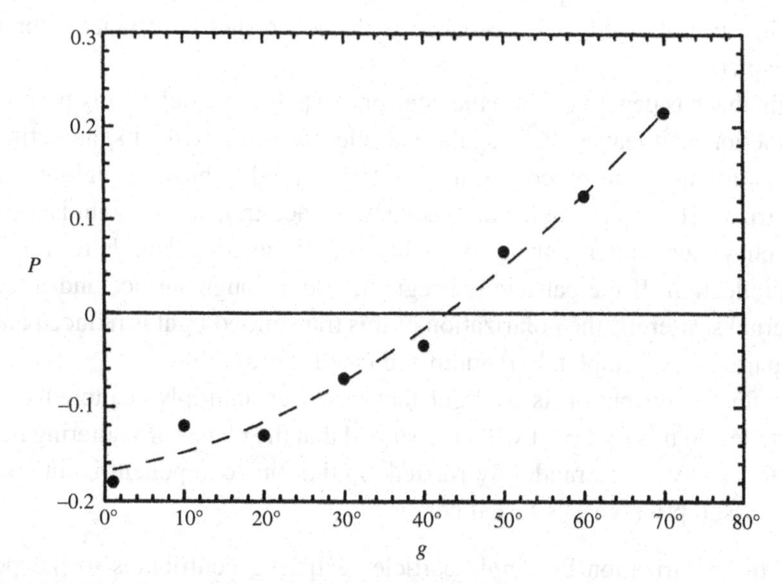

Figure 13.2 Polarization curve of light scattered from a glass plate viewed at $e = 60°$.

However, the light scattered from a particulate medium almost always has its plane of polarization parallel or perpendicular to the scattering plane, rather than with respect to the plane containing the exit ray and the surface normal. Furthermore, for low-albedo surfaces, such as lunar regolith, the polarization is a function only of phase angle and is independent of i and e. However, in materials with higher albedos the polarization phase curve is somewhat dependent on i and e also. These observations suggest that the phenomena contributing to $P(g)$ are primarily governed by single scattering by individual particles of the medium. In order to try to understand the polarization phase curve, at least qualitatively, the positive and negative branches of the curve will be discussed separately in the following sections.

13.3 The positive branch of polarization

13.3.1 Factors affecting the positive polarization

We start by hypothesizing that the positive branch of the polarization phase curve is controlled primarily by the properties of the individual particles of the medium. Recall that the light scattered from a particulate medium of particles large compared with the wavelength is the result of four phenomena:

(1) Light passing near the particle is diffracted. However, it was shown in Chapter 7 that single-particle Fraunhofer diffraction does not exist in a medium where the particles are close together.
(2) Light that has been specularly reflected from the surface of the particle is positively polarized, in accordance with the Fresnel reflectance expressions (Chapter 4).
(3) Light that has penetrated into the interior of a particle is negatively polarized by refraction as it leaves. In a regular particle, such as a perfect sphere, the plane of scattering is preserved, so that the transmitted light is strongly negatively polarized. If the particle is not opaque, this once-transmitted light is important for phase angles larger than $2\vartheta_C$, where ϑ_C is the critical angle for total internal reflection. If the particle is irregular, with a rough surface and filled with internal scatterers, the polarization of this transmitted light is reduced and may be partially or completely randomized (see Figure 6.7b).
(4) The final contribution is by light that has been multiply scattered by many particles. In this section it will be assumed that the planes of scattering between particles have been randomly rotated, so that this component is unpolarized. This assumption will be tested below.

If only the polarization by single-particle scattering contributes to the positive branch, then a theoretical expression for P may be derived as follows. The opposition effect is negligible for phase angles larger than about 20°. Therefore in the positive branch equation (12.55) becomes

$$I(i,e,g) = JK\frac{1}{4\pi}\frac{\mu_{0e}}{\mu_{0e}+\mu_e}\left\{wp(g)+w\left[H\left(\frac{\mu_{0e}}{K}\right)H\left(\frac{\mu_e}{K}\right)-1\right]\right\}S(i,e,g), \quad (13.4)$$

where the quantities in this equation are defined in Chapters 8 and 12.

In equation (13.4) the radiance that has been scattered only once by an average particle of the regolith is described by the term proportional to $Jp(g)$. Let the portion of $Jp(g)$ scattered with its electric vector perpendicular to the scattering plane of the medium be $J[p(g)]_\perp$, and let that parallel to the scattering plane be $J[p(g)]_\parallel$. Let

$$\Delta[p(g)] = [p(g)]_\perp - [p(g)]_\parallel. \quad (13.5)$$

Then if the incident radiance is unpolarized the two components of the scattered radiance are

$$I_\perp(i,e,g) = K\frac{w}{4\pi}\frac{\mu_{0e}}{\mu_{0e}+\mu_e}\left\{J[p(g)]_\perp + \frac{J}{2}[H(\mu_{0e}/K)H(\mu_e/K)-1]\right\}S(i,e,g) \quad (13.6a)$$

and

$$I_\parallel(i,e,g) = K\frac{w}{4\pi}\frac{\mu_{0e}}{\mu_{0e}+\mu_e}\left\{J[p(g)]_\parallel + \frac{J}{2}[H(\mu_{0e}/K)H(\mu_e/K)-1]\right\}S(i,e,g). \quad (13.6b)$$

If the assumption that the multiply scattered radiance has no net polarization is valid the polarization is

$$P(i,e,g) = \frac{JK\frac{w}{4\pi}\frac{\mu_{0e}}{\mu_{0e}+\mu_e}\{[p(g)]_\perp - [p(g)]_\parallel\}S(i,e,g)}{JK\frac{w}{4\pi}\frac{\mu_{0e}}{\mu_{0e}+\mu_e}\{p(g)+[H(\mu_{0e}/K)H(\mu_e/K)-1]\}S(i,e,g),}, \quad (13.7)$$

which after clearing common factors becomes

$$P(i,e,g) \simeq \frac{\Delta[p(g)]}{p(g)+[H(\mu_{0e}/K)H(\mu_e/K)-1]}. \quad (13.8)$$

Note that P is independent of the shadowing function $S(i,e,g)$. Thus, macroscopic roughness affects the polarization only through the effective tilt angles in the multiple-scattering contribution.

If w is so small that the H functions are not very different from 1 the polarization function of low-albedo objects is approximately

$$P(i,e,g) \simeq \frac{\Delta[p(g)]}{p(g)}, \quad (13.9a)$$

which is independent of i and e, as is approximately true for the Moon (Dollfus, 1961, 1962). It is also independent of surface roughness and filling factor. Note that this expression is valid for very-low-albedo materials even if the multiply scattered light is not randomly polarized. Thus, for dark surfaces the polarization of the light scattered by the medium is approximately equal to that of an average single particle of the medium, which from equation (5.16) is

$$P(i,e,g) \approx -S_{12}(g)/S_{11}(g). \tag{13.9b}$$

If the polarization curve, Figure 6.7b, of a single particle of olivine (a common mineral on the Moon) is compared with the lunar polarization curve, Figure 13.1, the two are indeed seen to be similar. However, expressions (13.9) are independent of albedo, whereas the polarization of moonlight varies inversely with the albedo; hence, the contribution of the multiply scattered light cannot be ignored.

Similarly, from equations (12.56) – (12.61), if the integral light from a body is being observed, the two components of its integral phase function outside of the opposition peak are

$$I_{\perp}(g) = \frac{J}{\pi}\pi R^2 A_p(w,\overline{\theta})\Phi_{\perp}(g,\overline{\theta}) \tag{13.10a}$$

$$= J R^2 \frac{A_p(w,\overline{\theta})}{A_p(w,0)} \left\{ \left[\frac{w}{4}([p(g)]_{\perp} - 1) + \frac{1}{2}r_0(1-r_0) \right] \right. \tag{13.10b}$$

$$\times \left[1 - \sin\frac{g}{2}\cos\frac{g}{2}\ln\left(\cot\frac{g}{4}\right) \right] \tag{13.10c}$$

$$\left. + \frac{2}{3}\frac{r_0^2 \sin g + (\pi - g)\cos g}{\pi} \right\} K(g,\overline{\theta}) \tag{13.10d}$$

and

$$I_{\parallel}(g) = \frac{J}{\pi}\pi R^2 A_p(w,\overline{\theta})\Phi_{P}(g,\overline{\theta}) \tag{13.10e}$$

$$= J R^2 \frac{A_p(w,\overline{\theta})}{A_p(w,0)} \left\{ \left[\frac{w}{4}([p(g)]_{\parallel} - 1) + \frac{1}{2}r_0(1-r_0) \right] \right. \tag{13.10f}$$

$$\times \left[1 - \sin\frac{g}{2}\cos\frac{g}{2}\ln\left(\cot\frac{g}{4}\right) \right] \tag{13.10g}$$

$$\left. + \frac{2}{3}\frac{r_0^2 \sin g + (\pi - g)\cos g}{\pi} \right\} K(g,\overline{\theta}), \tag{13.10h}$$

where R is the radius of the body. Thus,

$$P(g) = \frac{w\Delta[p(g)]}{w[p(g)-1] + 4r_0(1-r_0) + \frac{16}{3\pi}r_0^2 \frac{\sin g + (\pi - g)\cos g}{1 - \sin\frac{g}{2}\tan\frac{g}{2}\ln\left(\cot\frac{g}{4}\right)}} \tag{13.11}$$

Note that to the extent that the approximations of Chapter 12 are valid, this expression is independent of macroscopic roughness. If $w = 1$ so that terms of order r_0^2 can be ignored, $r_0 \approx w/4$, and (13.11) reduces to (13.9), the polarization of an average regolith particle.

For applications of polarization to remote sensing we wish to know what properties of the particles or the medium control P_p and g_p. Now, the scattering by an individual particle consists of two parts: Fresnel reflection from the surface, and light that has been refracted and scattered from the interior. Many theoretical models of polarization assume that the refracted light is randomly polarized and does not contribute to $\Delta[p(g)]$. Because g_p occurs in the vicinity of $2\vartheta_B$, where ϑ_B is the Brewster angle (Chapter 4), it is often assumed that in the positive branch of polarization $\Delta[p(g)]$ is controlled only by Fresnel reflection from the surfaces of the particles and that P_x occurs at $2\vartheta_B$. Let us investigate these assumptions quantitatively.

We saw in Chapter 6 that the light transmitted through irregular, rough-surfaced particles retains some negative polarization. However, assume for the moment that the volume-scattered light is unpolarized. For simplicity, also assume that the refracted light is emitted isotropically, although this is not essential to the argument. Then the reflection from the surface should be described by the Fresnel reflection coefficients $R_\perp(g/2)$ and $R_\parallel(g/2)$, and the volume scattering by the unpolarized remainder $(w - S_e)$, so that

$$w[p(g)]_\perp = R_\perp(g/2) + (w - S_e)/2,$$

$$w[p(g)]_\parallel = R_P(g/2) + (w - S_e)/2.$$

From equation (13.8) the polarization of the medium would then be given by

$$P(i, e, g) = \frac{1}{2} \frac{\Delta R(g)}{R_\perp(g/2) + R_\parallel(g/2) + w - S_e + w[H(\mu_{0e}/K)H(\mu_e/K) - 1]}, \tag{13.12}$$

where

$$\Delta R(g) = R_\perp(g/2) - R_\parallel(g/2) \tag{13.13}$$

is the difference between the perpendicular and parallel Fresnel reflection coefficients.

Equation (13.12) is plotted in Figure 13.3 as a function of g for a macroscopically smooth area viewed at $e = 60°$ in the principal plane of a medium of particles whose real refractive index is 1.50 for several values of w and $K = 1$. Note that the maximum of P occurs close to $2\vartheta_B$ only when $w = S_e$, that is, when the particle is completely opaque. As w increases, P_p decreases, and g_p shifts toward longer phase angles. The reason for this shift is that the numerator of equation (13.12) is

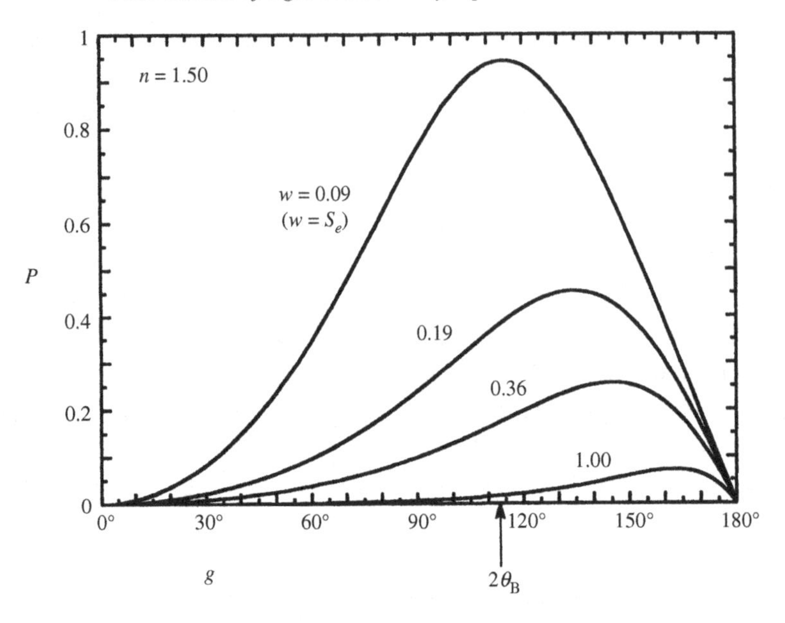

Figure 13.3 Theoretical positive-polarization curves of light scattered by a particulate medium of particles of refractive index $n = 1.5$ if the only polarized component is the light specularly reflected from the particle surfaces. Curves for several different values of the single-scattering albedo are shown. The angle of emergence is $e = 60°$.

ΔR, which does not peak at $2\vartheta_B$, but at much larger phase angles near 160°. $\Delta R(g)$ is plotted versus g for several values of the refractive index in Figure 13.4.

Thus one or more of the assumptions on which equation (13.12) is based must be incorrect. Let us recall these assumptions: (1) multiply scattered radiance is unpolarized; (2) $p(g)$ is approximately isotropic; (3) radiance transmitted through the particle is unpolarized. These assumptions were tested experimentally by measuring the bidirectional-radiance factors $\pi r(i, e, g)$ of several size fractions of olivine basalt powders in the principal plane for parallel and perpendicular directions of polarization. Unpolarized light was incident, while i was varied with $e = 60°$.

Assuming that the surface is macroscopically smooth, equation (13.7) can be written in the form

$$P(i, e, g) = \frac{\frac{1}{2}(1/4\pi)[\mu_0/(\mu_0 + \mu)]w\Delta[p(g)]}{r(i, e, g)},$$

which may be solved for $w\Delta[p(g)]$,

$$w\Delta[p(g)] = 8\frac{\mu_0 + \mu}{\mu_0}\pi r(i, e, g)P(i, e, g). \tag{13.14}$$

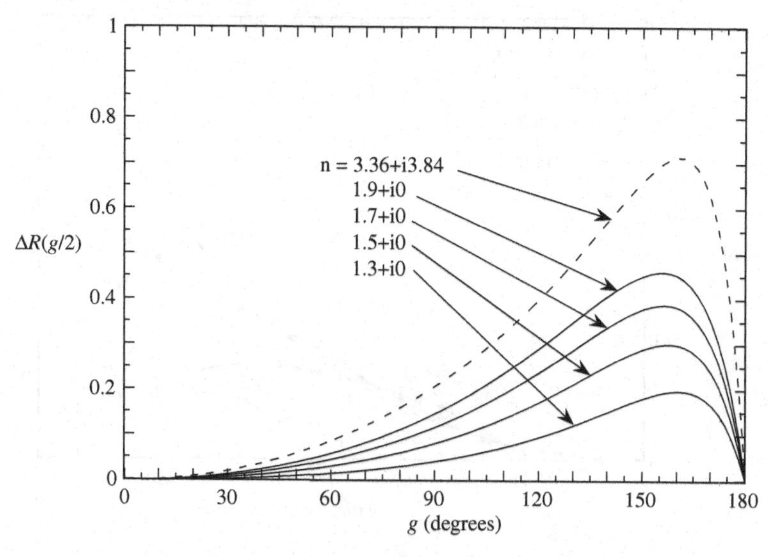

Figure 13.4 Difference between the perpendicular and parallel components of the Fresnel reflection coefficient as a function of phase angle for several values of the refractive index. The dashed curve is for the refractive index of metallic iron at $\lambda = 0.55\,\mu$m. Calculated from the values of Yolken and Kruger (1965).

The quantity $w\Delta[p(g)]$ was calculated from (13.14) using the measured values of r and P. The result is plotted in the top of Figure 13.5 along with the Fresnel difference function $\Delta R(g)$ calculated for $n_r = 1.7$, which is representative of the refractive indices of the minerals in the olivine basalt. If all of the above assumptions are valid the data for $w\Delta[p(g)]$ should be close to the curve of $\Delta R(g)$. This is indeed the case when $g \lesssim 80°$. However, as the phase angle increases, the points of $w\Delta[p(g)]$ fall below the $\Delta R(g)$ curve. Shown in the bottom of Figure 13.5 is a similar analysis for lunar soil, which is seen to exhibit the same behavior as the fine basalt powder. Thus $w\Delta[p(g)] \neq \Delta R(g)$.

If the doubly scattered radiance is not randomly polarized its polarization is highly likely to be positive, because that of the singly scattered radiance is positive. This would have to be added to $\Delta R(g)$, which would increase the discrepancy. Making $p(g)$ highly forward scattering would move g_p toward smaller angles; however, this would increase $r(i, e, g)$ at large phases, which is a measured quantity and cannot be altered. The only remaining way in which g_p can be reduced is if the forward-transmitted component of the light is not randomly polarized, but retains some negative polarization, which partially cancels the positively polarized reflection from the surface. This was seen to be the case for large translucent particles studied in the laboratory by McGuire and Hapke (1995) and also for the olivine particle of Figure 6.7b.

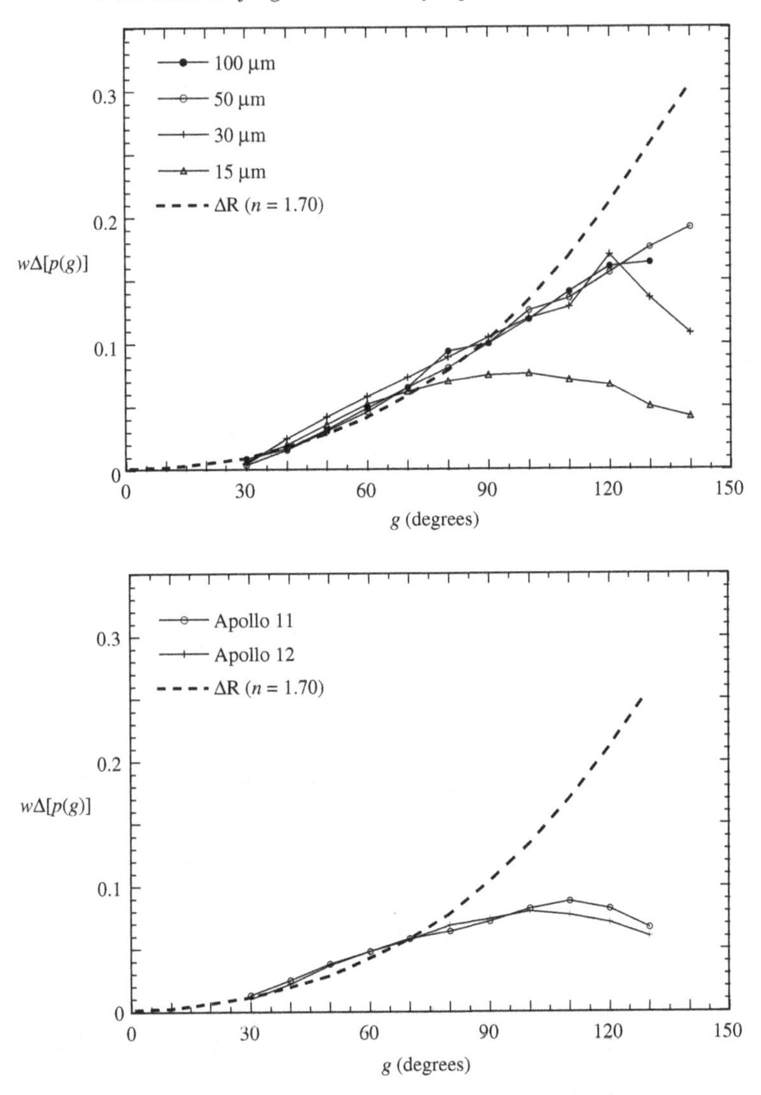

Figure 13.5 Difference between the the perpendicular and parallel components of polarization vs. phase angle measured for several different powders and compared with the difference between the Fresnel reflection coefficients. (Top) Powdered terrestrial olivine basalt of several sizes. (Bottom) *Apollo 11* and *12* soil samples. The dashed curves give $\Delta R(g)$ calculated for $n = 1.7$.

Additional support for this conclusion comes from observations of the Moon. According to Dollfus and Bowell (1971), g_p and P_p increase as albedo decreases on the lunar surface. This behavior is consistent with the hypothesis that the transmitted light is partially negatively polarized, because brighter areas are more transparent and hence have smaller P_p and g_p.

Vanderbilt *et al.* (1985) have published data on the intensity and polarization of light of several wavelengths reflected from wheat. The spectrum of the reflected radiance exhibits the strong absorption bands of chlorophyll at 0.48 and 0.66 μm, but the difference spectrum $I_\perp - I_\parallel$ does not. The authors argue correctly from this observation that only specular reflection contributed to $I_\perp - I_\parallel$. However, the largest phase angles in their measurement were about 80°, but the transmitted light is appreciable only at larger phase angles. Alternatively, this vegetation may have too large an optical thickness for the refracted light to have appreciable negative polarization. By contrast, Woessner and Hapke (1987) have measured the polarization of light scattered by clover and concluded that the light transmitted through this vegetation affects $I_\perp - I_\parallel$ in a manner similar to that for the silicate materials of Figure 13.5.

It may be concluded that for a medium of nonopaque particles the amplitude and angle of the maximum of the positive branch of polarization is not determined by the Brewster angle, but by the negatively polarized light that is transmitted through the particles in the forward direction, which becomes important for $g \gtrsim 2\vartheta_C$. The polarization curves have maxima around 100° because the refracted light is an important component of the scattering for larger phase angles. This component is controlled by particle size and by the internal absorption and scattering coefficients. However, for a completely opaque mineral, only surface reflection contributes to the particle scattering, so that P_p and g_p should be close to the values predicted from the Fresnel equations. The differences between the Fresnel coefficients of materials with large imaginary components of refractive index, such as metals, also peak at very large phase angles (Figure 13.4). Thus, a value of g_p significantly less than 160° is evidence that the regolith particles are translucent, rather than opaque.

13.3.2 The polarization–albedo relation (the Umov effect)

Returning to equation (13.8), for any given values of i, e, and g the denominator increases monotonically with w. In addition, if nonopaque minerals are present, any increase in w due to a decrease in internal absorption will be accompanied by a decrease in the numerator at large phase angles because of the increased, negatively polarized, transmitted light. Both effects will cause the amplitude of the positive branch to decrease as w increases. Hence, there is an inverse relation between the amplitude of the positive branch of the polarization curve and the reflectance. This relation is known as the *Umov effect,* after the Russian scientist who first observed and explained it (Umov, 1905). The Umov effect is one of those phenomena that keep getting rediscovered from time to time.

The Umov effect is illustrated in Figure 13.1. Before full Moon, the eastern hemisphere, which is dominated by brighter highlands, is illuminated and has a lower P_p, whereas after full Moon, the western hemisphere, whose surface has a large number of dark maria, is illuminated and has a higher P_p.

Because of the Umov effect, a form of spectroscopy can be carried out by measuring $P(g)$ as a function of λ, which has the advantage of requiring the measurement of only a ratio, rather than of an absolute intensity.

Two measures of reflectance are the normal albedo A_n of a resolved surface area and the physical or geometric albedo A_p of the integrated radiance from a body, which is the weighted average of A_n. Thus, the Umov effect implies that there should be an inverse relation between P_p and A_n or A_p, and this is indeed found to be the case.

Unfortunately, although the Umov effect describes a general trend, there is no unique relation that is valid for all materials. There are several reasons for this. First, equation (13.8) shows that P_p is determined by w, $\Delta[p(g)]$, $p(g)$, and the H functions at large phase angles, whereas A_n and A_p depend on these quantities at $g = 0$. In general, these functions change in different ways in different materials as w changes. Second, the negative polarization in the transmitted component can be decreased by increasing the amount of internal scattering in the particles, which randomizes the polarization, but may not affect w appreciably. Third, w may also be increased either by decreasing the absorption of the transmitted light, which decreases both the numerator and the denominator of (13.8), or by increasing the real part of the refractive index, which primarily increases the numerator.

However, homogeneous classes of materials may possess a quasi-unique Umov relation. An example is the lunar regolith, which is relatively homogeneous because it consists of mafic silicate minerals and glasses and is the product of meteorite impacts. Dollfus and Bowell (1971) found that areas on the Moon observed telescopically from Earth obey the empirical relation

$$\log A_n = -c_1 \log P_p - c_2, \tag{13.15}$$

where $c_1 \simeq 0.724$ and $c_2 \simeq 1.81$. This is illustrated in Figure 13.6. Other materials, such as certain terrestrial silicate rock powders and pulverized meteorites, obey similar rules, except that the constants have different numerical values. Geake and Dollfus (1986) have suggested that there is a correlation between c_2 and particle size, although the generality of such a relation is doubtful.

13.3.3 The slope–albedo relation

The inverse relation between the amplitude of the positive branch and the albedo also manifests itself in other properties of $P(g)$ in the positive region. In particular, KenKnight *et al.* (1967) and Widorn (1967) independently pointed out that the slope $h_i = dP/dg$ of the polarization at the inversion angle g_i is inversely related to

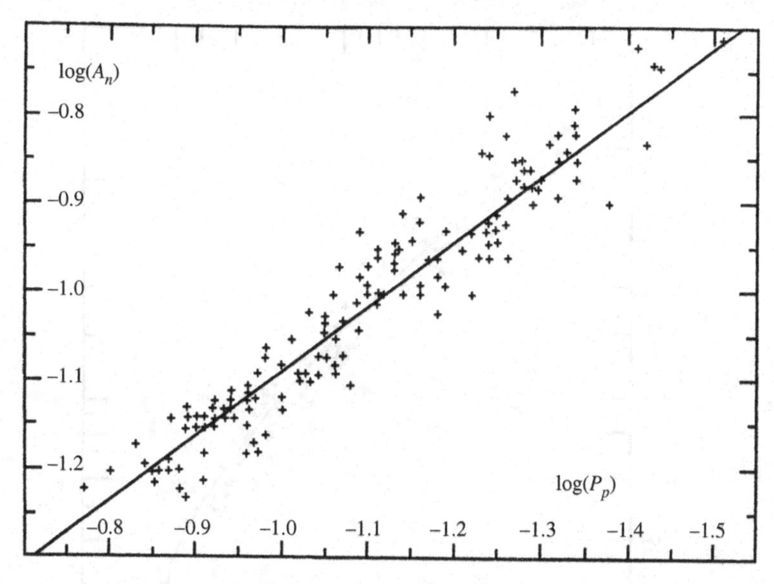

Figure 13.6 Plot of the normal albedo against the maximum polarization for 144 areas on the Moon. The line is the empirical function $\log A_p = -0.724 \log P_p - 1.81$. (Reproduced fom Dollfus and Bowell [1971], copyright 1971 with permission of Springer-Verlag.)

albedo. This relation is shown in Figure 13.7. Over a fairly large range of albedos, h_i and A_n obey the empirical rule

$$\log A_n = -c_3 \log h_i - c_4, \tag{13.16}$$

where $c_3 \simeq 1.00$ and $c_4 \simeq 3.77$ (Bowell *et al.*, 1973). The physical albedo A_p may be substituted for A_n in (13.16). However, the curve saturates for $A_n < 0.04$, so that there is not a unique relation between albedo and slope for very dark materials.

The slope–albedo relation appears to be unique for a wider range of materials than the maximum polarization–albedo relation. Probably, the reason is that the numerator of (13.8) is less affected by the transmitted, negatively polarized light at phase angles significantly smaller than g_p, and also the quantities in the denominator of (13.8) are closer to their zero phase values at g_i than at g_p.

This relation has important applications in the determination of asteroid albedos and diameters. For many asteroids, g_i is small enough that h_i can be determined from the Earth, whereas the measurement of P_p at g_p requires a polarimeter on board a spacecraft in the outer solar system. If the albedo of an object can be found from h_i, its size can be calculated from a measurement of absolute integral brightness, even though the object is too small to be resolved (Zellner *et al.*, 1974; Zellner and Gradie, 1976).

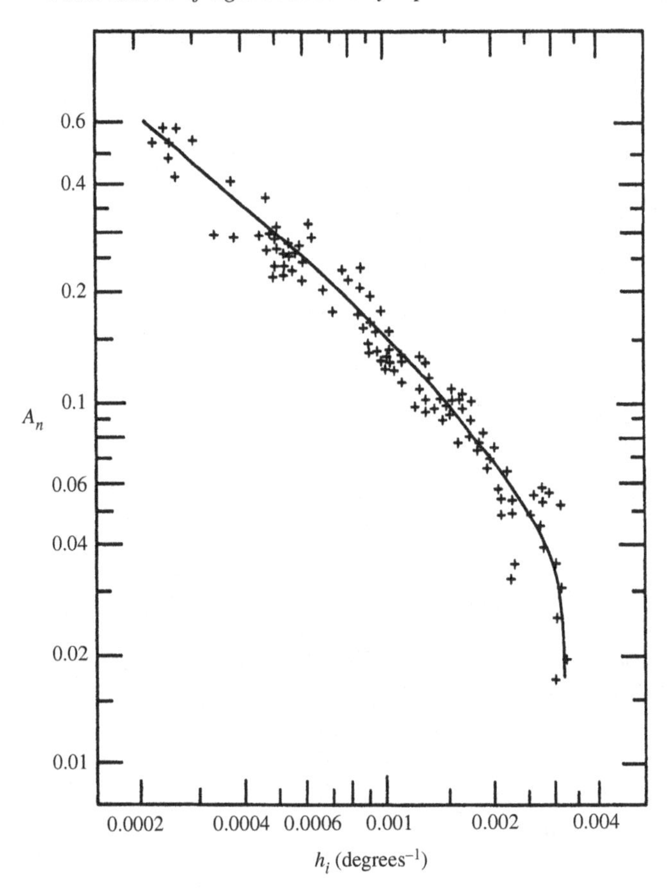

Figure 13.7 Log–log plot of normal albedo against the slope of the polarization phase curve at the inversion angle for 95 samples of lunar soil, pulverized terrestrial rocks, and pulverized meteorites. (Reproduced from Geake and Dollfus [1986], copyright 1986 with permission of the Royal Astronomical Society.)

13.4 The negative branch of polarization

13.4.1 Introduction

The negative branch of the polarization phase curve is one of the enigmas of planetary remote sensing. In spite of the fact that it has been known since the 1920s, it has defied repeated attempts to account for it quantitatively. Yet virtually all pulverized materials display negative polarization at small phase angles, and the strength of the negative branch increases as the particle size decreases. Indeed, the observation by Lyot of a well-developed negative branch of polarization in light reflected from the Moon was one of the earliest indications that the lunar surface was covered with a fine-grained regolith.

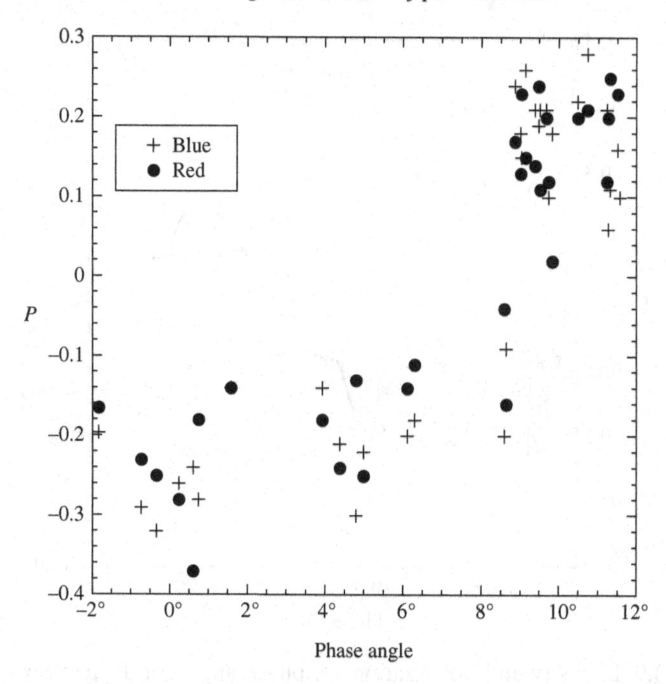

Figure 13.8 Polarization phase curve of Europa in red (crosses) and blue (circles) wavelengths. (Data from Rosenbush *et al.* [1997] and Rosenbush and Kiselev [2005].)

The negative branch is bimodal in some materials, with a narrow negative peak at a phase angle around 1° and only 1° or 2° wide, called the *polarization opposition effect* (POE), and a *broad negative polarization* (BNP) peak centered at a larger phase angle and 10° or 20° wide. Lyot (1929) first observed what is, in retrospect, believed to be a POE peak in MgO smoke deposited on a plate. The negative polarization in most of the materials studied by Lyot were BNP peaks. However, he did not distinguish between the two. The first unambiguous observations of a bimodal negative branch were by Rosenbush *et al.* (1997) in the Galilean satellites of Jupiter (Figure 13.8). The two peaks are illustrated in Figure 13.9 for 17-μm SiC abrasive powder.

It is not clear whether or not all negative polarization branches are, in fact, bimodal. The POE peak may actually be present in material for which it appears to be missing, but be so weak that it is hidden in the broad peak, or it may be missed because of low angular resolution of the measurements. The latter appears to be the case for the Moon which, paradoxically, is one of the most intensely studied objects in the universe. A separate POE peak has never been reported in light reflected from the Moon. However, high-angular-resolution laboratory measurements of lunar soil appear to show two distinct peaks (Figure 13.10). There are several reasons why a

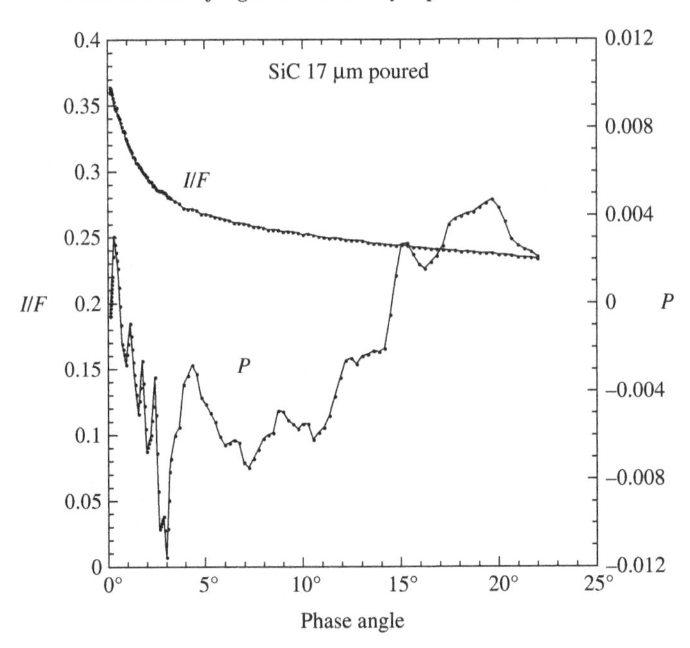

Figure 13.9 Intensity and polarization vs. phase angle for 17-μm SiC abrasive powder showing the coherent backscatter opposition effect, the narrow polarization opposition effect, and the broad negative branch of polarization. The linear decrease in intensity at larger phase angles is probably part of the unresolved shadow-hiding peak.

lunar POE might have been missed. The source of illumination of the lunar surface is the Sun, which has an angular diameter of $1/2°$ at the Moon; thus all lunar phase curves are averaged over $1/2°$. Most full Moons occur at phase angles larger than $1°$, and the Moon enters the Earth's shadow at phase angles smaller thatn $1°$. Hence, the Moon only enters the POE region infrequently, and when it does most of the peak is obscured. Finally, observations of lunar polarizations typically are spaced at about $1°$ intervals.

13.4.2 The polarization opposition effect (POE)

The POE peak is caused by the same phenomenon as the coherent backscatter opposition effect and can be qualitatively explained as a coherent effect. The mechanism was first proposed independently by Shkuratov (1989) and Muinonen (1990). The coherent-backscatter phenomenon is discussed in detail in Chapter 9. Portions of the incident wave that are multiply reflected between scatterers over the same path, but in opposite directions, combine coherently upon emerging from the medium to produce a peak of increased intensity by constructive interference.

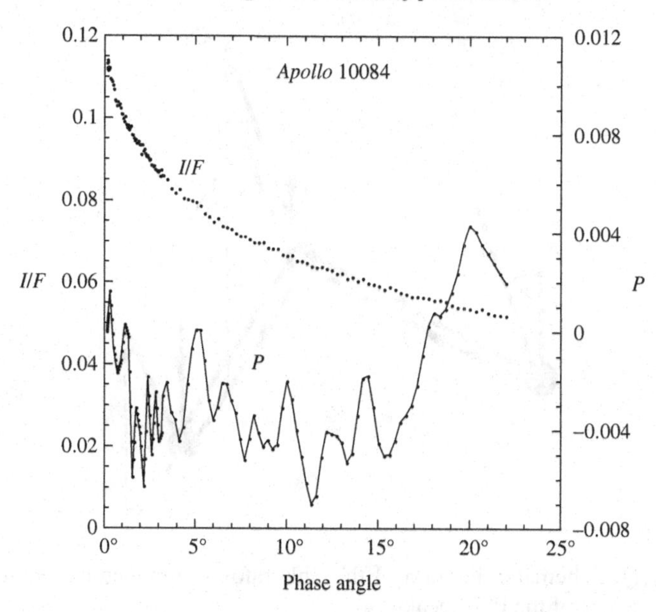

Figure 13.10 Intensity and polarization versus phase angle for *Apollo* lunar soil sample 10084 showing the POE, BNP, CBOE, and probably part of the SHOE.

Shkuratov (1989) and Muinonen (1990) pointed out that this phenomenon is inherently anisotropic in such a way as to emphasize the portions of the wave scattered transversely, perpendicular to the scattering plane. Muinonen (1990) calculated the polarization from double scatterings between Rayleigh particles (dipoles) of varying separations and position. Shkuratov (1989) calculated the polarization from double scatterings between particles larger than the wavelength, so that most of the light scattered into the CBOE was by Fresnel reflection from the particle surfaces. Mishchenko *et al.* (2000) calculated the variation of polarization with phase angle of a medium of Rayleigh scatterers using an exact vector solution of the equation of radiative transfer and was able to fit Lyot's measurements of MgO powder.

The process is illustrated schematically in Figure 13.11. Plane C–2–3–D is the scattering plane. The portions of the incident wave that are transversely scattered perpendicular to the scattering plane along paths A–1–2–A′ and B–2–1–B′ are in phase for any value of g. As discussed in Chapter 9, the intermediate scatterings that contribute the most to the CBOE take place around angles not too different from 90°. At these angles the coefficients for both Rayleigh scattering and Fresnel reflection for the waves with their electric vectors perpendicular to the plane A–1–2–B are larger than for the electric vectors parallel to this plane. Portions of the wave longitudinally scattered parallel to the scattering plane along paths C–2–3–C′

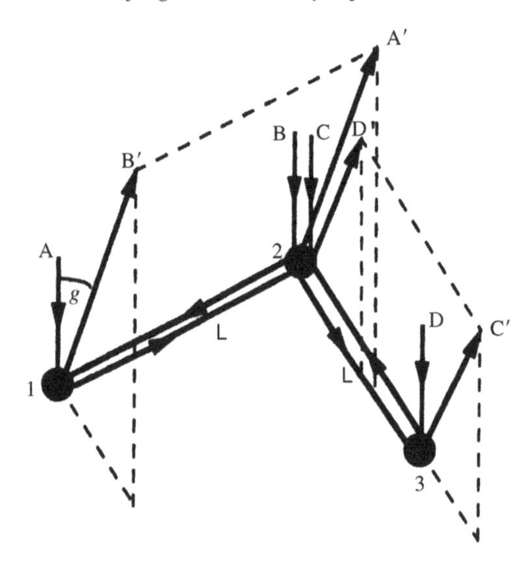

Figure 13.11 Schematic diagram of the Shkuratov–Muinonen coherent double-scattering model of the POE peak.

and D–3–2–D′ are in phase only at $g = 0$ and will be out of phase if $g \gtrsim \lambda/2\pi L$, where L is the separation of the scatterers. However, the reflection coefficients for the waves with their electric vectors perpendicular to the plane C–2–3–D are larger than for the electric vectors parallel to this plane. Hence, the intensity of the negatively polarized, transverse, doubly scattered wave is roughly twice that of the positively polarized, longitudinal, doubly scattered wave, except at phase angles close to zero. This produces a negative POE peak close to zero phase with a similar angular width as the CBOE.

While there is no doubt that the POE peak is a coherent phenomenon, present theories suffer from the same problems as the CBOE. These predict that the POE peak should be wide enough to be observable only when the particles of the medium are around a wavelength in size, They also predict a strong dependence on wavelength, particle size, and spacing. None of these been consistently reported (e.g., Figure 13.8; see also Shkuratov *et al.*, 2002).

13.4.3 The broad negative polarization (BNP) branch

By contrast with the POE peak, which is at least qualitatively understood, there is no agreement on the cause of the BNP peak. After the POE model was proposed it was thought that this might account for the entire negative branch. However, with the discovery of a separate POE peak this explanation became inadequate. The BNP peak is in the same general location as the shadow-hiding opposition effect.

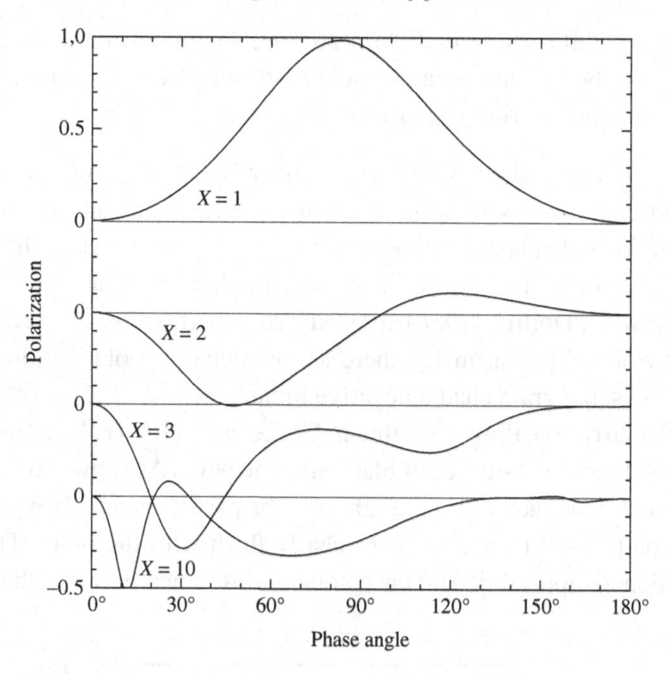

Figure 13.12 Theoretical polarization phase curves for different sized spheres with $n = 1.50 + i0$, showing a BNP-like negative polarization feature.

Whether or not this is a coincidence is unknown, It may be that there are several causes, with the nature of the media determining which cause dominates.

It has long been known that spherical particles somewhat larger than the wavelength have a negative polarization peak at small phase angles (Figure 13.12) whose angular width decreases with increasing size relative to the wavelength. Muinonen *et al.* (2007) studied scattering by spherical and nonspherical particles using the DDA method (Chapter 6) and concluded that this peak is caused by coherent interference between different parts of the same particle. They also found that negative polarization can be caused by wavelength-scale roughness on the surface of a larger particle. They suggested that this is the cause of the BNP peak.

This hypothesis is attractive because it explains why the negative branch is particularly well developed for media of small particles. The difficulty with it is that, since the peak is an interference phenomenon, its width is strongly dependent on wavelength and particle size. However, Dollfus and Bowell (1971) measured the polarization of a large number of regions on the Moon at several wavelengths between 0.33 and 1.05 μm. Over this interval, in which the wavelength varied by a factor of 3, there was no change in the negative branch. Although media of small particles generally have stronger BNP peaks, media of large particles also display them.

Furthermore, while single-particle scattering may contribute to the negative branch, it cannot be the entire cause. Several observations indicate that multiple scattering is a major contributor to the negative branch:

(1) Dollfus (1956; see also Bowell and Zellner, 1974 and Dollfus *et al.*, 1989) described experiments in which carbon particles rising in the smoke above a flame did not display negative polarization. However, when those particles were collected into a thick coating on a plate, they had a negative branch.
(2) Similarly, when Dollfus (1956) allowed a stream of well-separated sand grains to fall in front of a polarimeter, there was no negative polarization, but a thick layer of the same grains had a negative branch.
(3) In a particularly revealing experiment, Geake *et al.* (1984) placed a single layer of glass particles on a surface of black silicone putty. As shown in Figure 13.13, the layer exhibited negative polarization. The particles were then pushed down into the putty until their tops were nearly flush with the putty. The negative polarization disappeared. The only notable difference between the two media

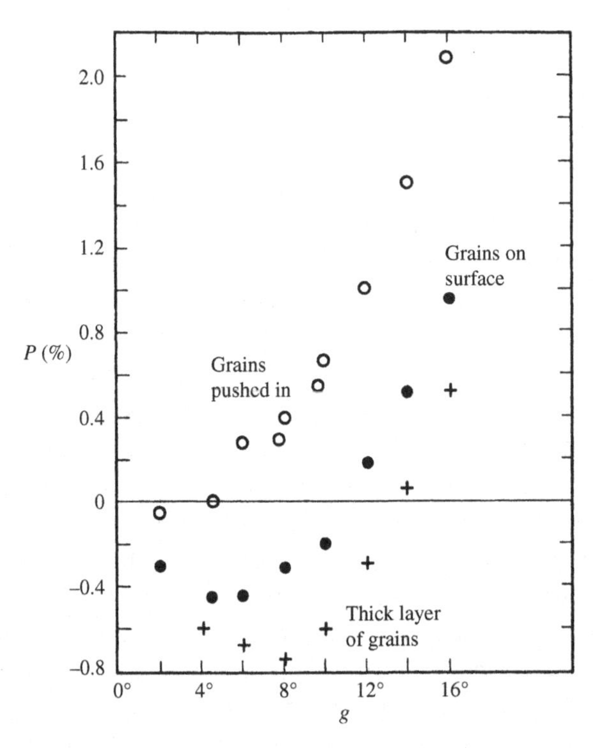

Figure 13.13 Polarization phase curves for powdered glass on silicone putty. Open circles, glass grains resting on surface of putty; filled circles, grains pushed into surface; crosses, thick layer of glass powder. (Reproduced from Geake *et al.* [1984], copyright 1984 with permission of the Royal Astronomical Society.)

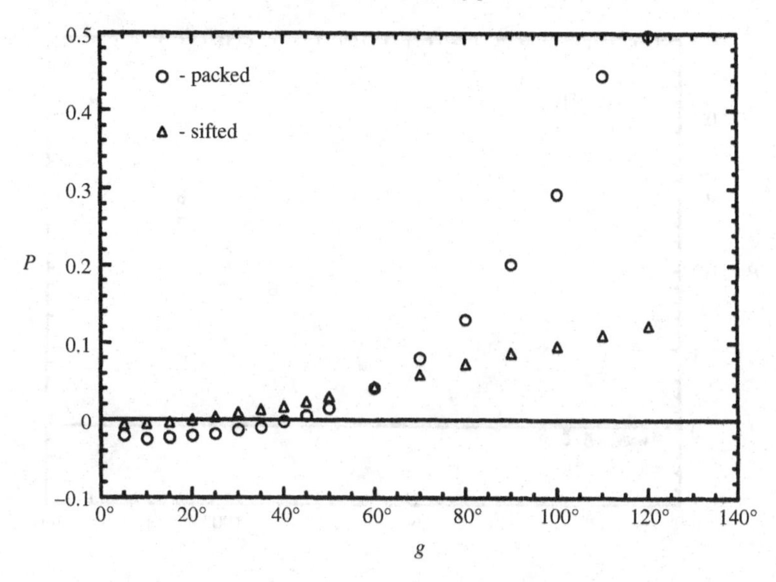

Figure 13.14 Polarization phase curves of SiC powder about 15 μm in size, showing the effect of porosity.

was that pushing the particles into the clay had the effect of separating them with an opaque layer and preventing double scatterings.

(4) The amplitude of the negative branch depends strongly on the filling factor in a particulate medium, in the sense that a more closely packed powder has a stronger negative polarization. This is illustrated in Figure 13.14. A commercial silicon carbide abrasive powder with particles about 15 μm in size was sedimented in acetone, which was then allowed to evaporate. The sediment had a strong negative branch. Microscopic examination showed that the grains in the sediment were closely packed. However, when the particles were sifted into a low-density deposit, the negative polarization was much weaker. Shkuratov *et al.* (2002) reported similar results.

The experiment with the silicon carbide abrasive indicated not only that the efficacy of the mechanism that produces the negative polarization is increased by placing the particles closer together but also that close packing increases the positive polarization dramatically, probably because the light transmitted through the particle is more readily blocked as the filling factor increases.

A possible mechanism for producing the broad negative branch is that the negatively polarized light transmitted through one particle in the forward direction is reflected off another in the backward direction to produce negative polarization at small phase angles. However, although the light forward-refracted through a

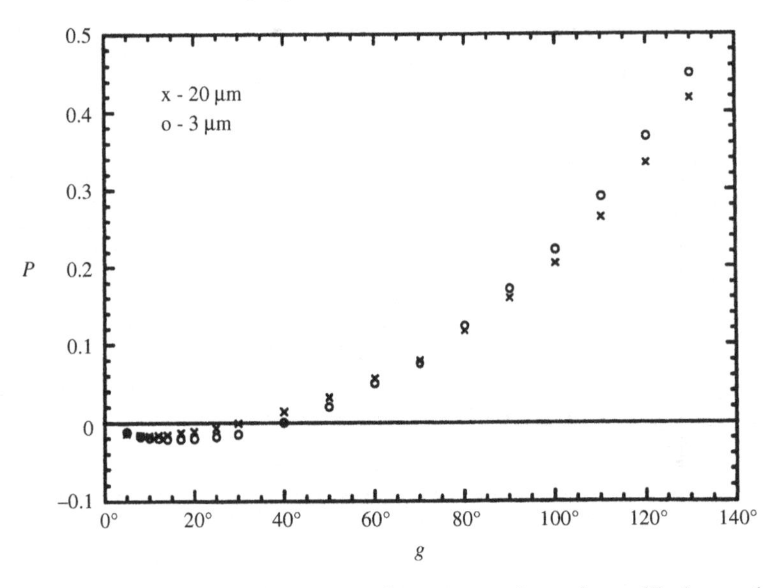

Figure 13.15 Polarization phase curve for two powders of metallic iron spheres about 20 μm (crosses) and 3 μm (circles) in size. Note that the smaller particles have larger negative polarization.

transparent particle is negatively polarized, transparency is not required. Powders of highly opaque materials, such as metals, display negative polarization (Figure 13.15; see also Shkuratov *et al.*, 2002).

One of the earliest explanations offered for the negative polarization, and the one that underlies many of the models proposed to explain it, is due to Ohman (1955). Ohman pointed out that double specular reflection, in which the intermediate path is perpendicular to the scattering plane, would generate negative polarization. This is illustrated schematically in Figure 13.16. The Fresnel coefficient for ray S-1-A, which is reflected from the surface of particle 1, is larger for the component with its electric vector perpendicular to the scattering plane, and so is positively polarized. However, for the doubly reflected ray S-1-2-B, which reflects at nearly right angles from particles 1 and 2, there is an intermediate scattering plane that is perpendicular to the main scattering plane. Hence, for the doubly scattered ray there are two Fresnel reflections, each of which produces light that is preferentially polarized parallel to the main scattering plane.

That the Ohman mechanism can produce negative polarization is verified by Figure 13.17, which shows the polarization in the light scattered by two large copper spheres just touching each other and resting on black velvet. When the line between the centers of the spheres is parallel to the scattering plane there is only a small amount of negative polarization. Strong negative polarization was observed when

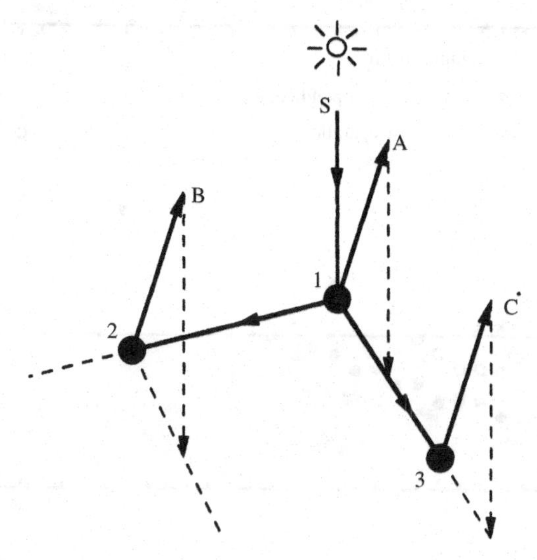

Figure 13.16 Schematic diagram of the Ohman incoherent double-scattering model of the BNP peak.

the line between the centers was perpendicular to the plane and sideways reflections could occur. A single sphere produced no negative polarization. (However, it is of interest to note that even the parallel configuration produced weak negative polarization.)

The difficulty with applying the Ohman mechanism to particulate media concerns the statistical azimuthal symmetry of the positions of the particles. Longitudinal scatterings such as S-1–3-C in Figure 13.16, in which all rays are parallel to the scattering plane, and for which the net polarization is positive, should be just as frequent as transverse scatterings, such as S-1–2-B. Quantitative calculations show that the polarizations caused by the longitudinal and transverse scatterings cancel each other to a high degree, leaving only the small net positive polarization due to single scatterings.

Most Ohman-type models to produce the BNP peak rely on some hypothesized geometric property of the surface to block the light from one of the longitudinal scatterings. Because of the coincidence in location with the SHOE intensity peak many persons have investigated shadows. Alternately, the surface may assumed to be covered with vertical-walled pits or cracks lined with particles (e.g., Wolff, 1975, 1980, 1981; Steigman, 1978; Bandermann *et al.*, 1972; Shkuratov 1982; Kolokolova, 1985, 1990). When such a pit is viewed from any direction other than vertical, the side closest to the detector is not visible, so that the light scattered from that surface element does not contribute to the polarization.

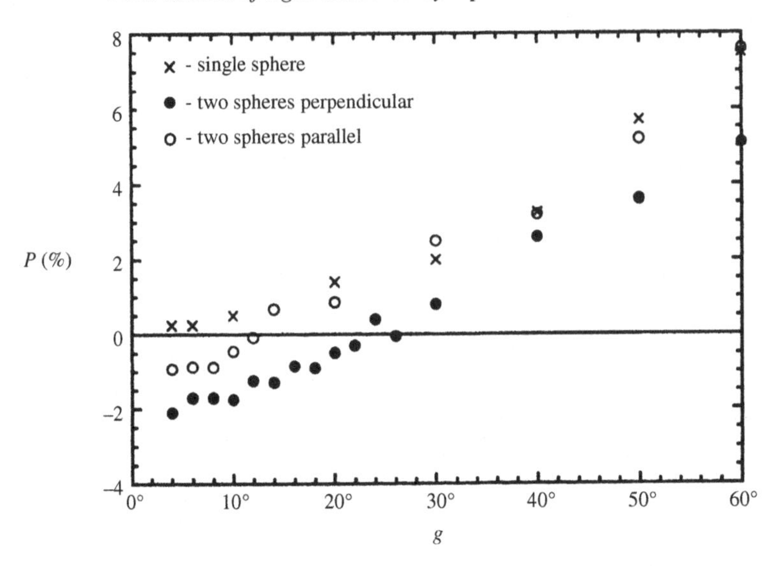

Figure 13.17 Polarization phase curves for a single copper sphere approximately 5 mm in diameter (crosses), for two copper spheres with the line perpendicular to the scattering plane (filled circles), and for two copper spheres with the line joining their centers parallel to the scattering plane (open circles).

The Wolff model is a detailed example of these types of models and is a numerical calculation in the form of a FORTRAN computer program. It consists of semi-empirical functions containing several parameters that, if properly chosen, can indeed describe the intensity and polarization quite well. The surface is assumed to be covered with pits, and scattering from each wall of a pit is assigned an empirical blocking function. One of the longitudinal scatterings is assigned a larger blocking function than any of the transverse scatterings, thus producing negative polarization. However, the basis for the choice of the values of the various parameters is obscure, and the Wolff model has not been widely accepted.

The difficulty with models involving pits is that pits do not seem to be necessary to cause negative polarization. Microscopic examination showed that the surface of the sedimented silicon carbide powder shown in Figure 13.14 was quite smooth. Geake *et al.* (1984) measured the polarization from pits in black silicone putty that had been covered with a layer of glass particles. To be sure, the glass-lined pits produced negative polarization, but so did a simple layer without pits.

Some authors have assumed that the interstices between particles would act as pits. However, this is unconvincing. Negative polarization is observed from layers of particles that are equant, convex, and gently rounded in shape. There is a vast difference between the side of such a particle and the sharp vertical wall of a pit. No models that involve pits appear to be adequate to account for the negative polarization from general particulate media. However, negative polarization is observed in light scattered from some samples of volcanic foam and scoria, and the pit models may be appropriate for this type of material.

A different type of mechanism was proposed by Hopfield (1966), who pointed out that the light diffracted past a straight edge is partially polarized with the electric vector parallel to the edge. Hopfield considered a particle of square cross section above a diffusely scattering substrate. If viewed from any angle other than zero phase, a shadow is visible under that edge which is nearest the detector and oriented perpendicular to the scattering plane. The light diffracted from this edge is reduced in intensity, because it comes from the shadow. However, Zellner (Geake *et al.*, 1984) quoted some unpublished experiments he conducted with K. Lumme showing that the negative polarization produced by this mechanism was far too weak to explain the negative branch.

Videen *et al.* (2003) found that coherence is able to provide the necessary asymmetry for the Ohman mechanism. They carried out second-order numerical modeling of a cloud of specular reflectors. When the coherent interference between two parts of the incident wave that traverses the same path in opposite directions

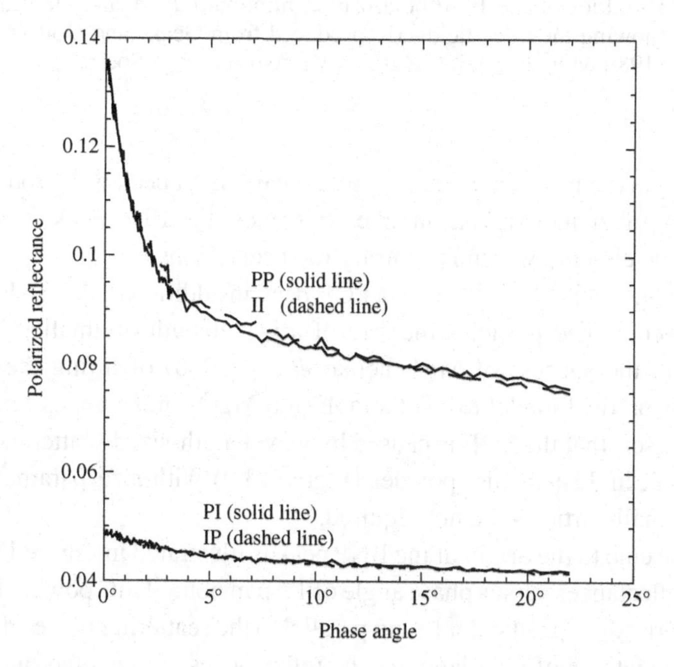

Figure 13.18 Individual components of the reflectance of poured 17-μm SiC powder when linearly polarized light is incident and measured. See text for definitions of the labels on the curves.

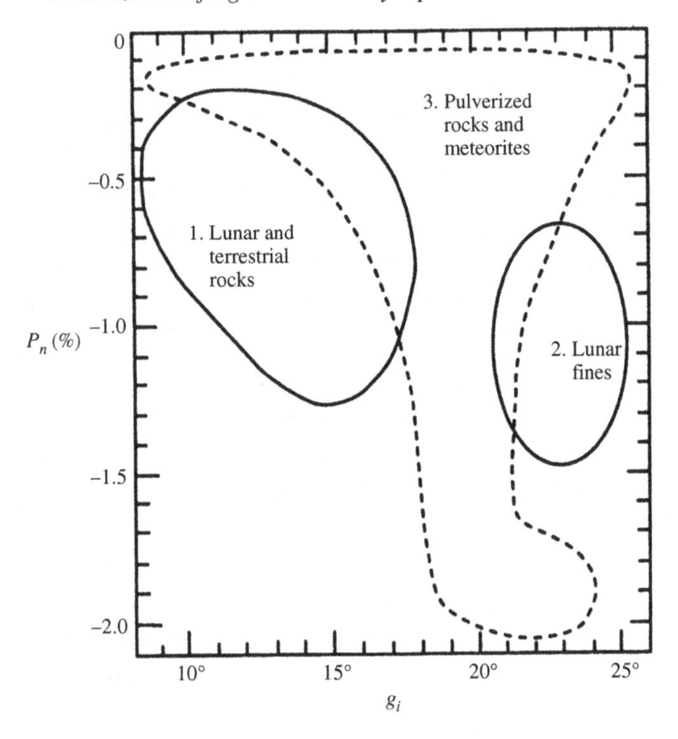

Figure 13.19 Plot of the BNP polarization minimum P_n against the inversion angle g_i showing the three fields. (Reproduced from Geake and Dollfus [1986], copyright 1986 with permission of the Royal Astronomical Society.)

was included in the model a broad negative polarization peak was produced, However, as with other models that invoke coherence, the BNP peak is predicted to depend on wavelength, which is contrary to observations.

Videen *et al.* were able to produce a POE peak in addition to the BNP peak when their media contained particles the size of a wavelength or smaller. This result, together with the success of Mishchenko *et al.* (2000) of fitting their Rayleigh scattering theoretical model to Lyot's POE of a MgO smoke deposit has led most workers to assert that the POE is caused by wavelength-sized scatterers. However, the POE peak in 17-μm SiC powder (Figure 13.9) with size parameter $X = 85$ shows that small particles are not required.

A possible clue to the origin of the BNP peak is illustrated in Figure 13.18, which shows the reflectances versus phase angle of 17-μm poured SiC powder illuminated by light polarized perpendicular to and parallel to the scattering plane and measured in the two directions of polarization. The reflectances are denoted by two letters: either P (perpendicular to the scattering plane) and I (in or parallel to the scattering plane). The first letter is the direction of polarization of the incident light and the second is that in which the reflectance was measured. Then

$$\mathrm{P} = \frac{\mathrm{PP} + \mathrm{IP} - \mathrm{II} - \mathrm{PI}}{\mathrm{PP} + \mathrm{IP} + \mathrm{II} + \mathrm{PI}}.$$

Figure 13.18 shows that the cross-polarized components PI and IP are indistinguishable.

Hence, effectively,

$$\mathrm{P} = \frac{\mathrm{PP} - \mathrm{II}}{\mathrm{PP} + \mathrm{IP} + \mathrm{II} + \mathrm{PI}}$$

However, the copolarized reflectance II when the incident and scattered light are polarized in the scattering plane is slightly larger than the copolarized reflectance PP when the incident and scattered light are perpendicular to the scattering plane. Why this difference arises is unclear.

In spite of the lack of theoretical understanding, Dollfus and his co-workers (Geake and Dollfus, 1986; Dollfus *et al.*, 1989) have discovered a number of empirical relations between the size of the broad negative branch of polarization and the properties of various particulate media. These relations are summarized in Figure 13.19, which is a plot of the amplitude of the polarization minimum P_n versus g_i, the inversion angle. There appear to be three regions, which are labeled 1, 2, and 3 in Figure 13.19. Coarse chunks of terrestrial and lunar rocks fall in region 1, and finely pulverized terrestrial volcanic rocks and meteorites lie in region 3. Lunar soil occupies region 2. Most asteroids fall in region 3, implying that these bodies are covered with a fine-grained regolith. However, until both peaks of the negative branch of the polarization phase curve are better understood, the general applicability of these types of empirical relations remains uncertain.

13.5 Summary

The following is a summary of our current understanding of polarization and the implications for the interpretation of remote polarization measurements of planetary regoliths.

- The positive branch is reasonably well understood. For materials that are sufficiently dark that multiple scattering makes only a minor contribution, the polarization in the positive branch is similar to the volume-averaged polarization of a single particle of the medium. The maximum polarization decreases as the albedo increases (Umov effect) and is a crude measure of albedo. The phase angle of the maximum does not occur at the Brewster angle and cannot be used to measure n_r.
- The slope of the polarization at which the inversion occurs is linearly proportional to the albedo and can be used to estimate the albedo.

- The polarization opposition effect is qualitatively understood. It apparently is caused by coherent backscattering and is predicted by theories of the CBOE. However, like the CBOE, the predicted dependence on wavelength and porosity is not observed, either in the laboratory or in solar system objects, implying that our understanding of the phenomenon is incomplete.
- The broad negative branch of polarization is not understood at all, and a wide variety of explanations have been proposed, none of which are generally accepted. It is in the same angular region as the SHOE, but attempts to model it as a shadow-hiding phenomenon have not met with success. It arises from a small difference between the two copolarized components of reflectance, but what causes this difference is unknown. However, the BNP is wider and deeper when the particles of the medium are not too much larger than the wavelength.

14
Reflectance spectroscopy

14.1 Introduction

One of the objectives of studying a planet by reflectance is to infer various properties of the surface by inverting a set of remote measurements. In the laboratory, the objective of a reflectance measurement is usually to determine the spectral absorption coefficient of the material, or at least some quantity proportional to it, by inversion of the spectral reflectance.

For several reasons reflectance spectroscopy is a powerful technique for measuring the characteristic absorption spectrum of a particulate material. The dynamic range of the measurement is extremely large. Multiple scattering amplifies the contrast within weak absorption bands in the light transmitted through the particles, while strong bands can be detected by anomalous dispersion in the radiation reflected from the particle surfaces. Hence, the measurement of a single spectrum can give information on the spectral absorption coefficient over a range of several orders of magnitude in α. This is especially true for the range $n_i = 10^{-3} - 10^{-1}$, where both transmission and reflection techniques are difficult. Another advantage of reflectance methods is that sample preparation is convenient and simply requires grinding the material to the desired degree of fineness and sieving it to constrain the particle size. By contrast, if $\alpha(\lambda)$ is measured by transmission, the sample must be sliced into a thin section that must then be polished on both sides, and the range by which $\alpha(\lambda)$ can vary is not much more than about one order of magnitude.

The characteristics of a medium that can be measured directly by reflectance are those that occur in the expression for the bidirectional reflectance, equations (9.47) and (12.63), and are of two types: those that describe the average scattering properties of single particles, w and $p(g)$; and those that describe the physical structure of the surface K, $\overline{\theta}$, and the opposition-effect parameters. In this chapter we will discuss these quantities and the methods for inverting the reflectance data to estimate them.

Because the reflectance equations are nonlinear the relation between the absorption coefficient and the reflectance is not trivial. This relation will be discussed in detail, especially the shapes of absorption bands when seen in reflectance. The effects of layers on band contrast will also be considered. Spectral information is carried primarily by the volume-average single-scattering albedo. Hence, the first step in finding the spectral absorption coefficient α is to retrieve w from the reflectance. The next step is to use an appropriate relation to find α from w, and ways of doing this will be discussed. A few alternate methods, including the widely used Kubelka–Munk model, will also be discussed, along with some special cases of interest in planetary spectroscopy.

14.2 Measurement of reflectances

In this section we shall briefly describe the methods by which the various types of reflectances and albedos described in Chapters 8–11 can be measured. In order to find the reflectance, both the scattered radiance and the incident intensity must be measured. This can be done by measuring the radiance scattered from the surface and then rotating the detector to measure the incident irradiance. However, under many experimental conditions this is inconvenient or even impossible. Hence, a common procedure is to calibrate the incident light by measuring the intensity reflected from a standard material of known scattering properties. Then the reflectance is given by

$$r(\text{sample}) = \frac{I(\text{sample})}{I(\text{standard})} r(\text{standard}). \tag{14.1}$$

Usually the standard is measured at i, e, and g close to 0°, but outside the opposition surge.

Desirable properties of a standard are that it be highly reflecting with no absorption bands in the spectral region of interest, and stable with time. Standards in wide use in the near-UV/visible/near-IR region of the spectrum include finely ground MgO (Middleton and Sanders, 1951), $BaSO_4$ (Grum and Luckey, 1968), and polytetrafluoroethylene (PTFE) (Weidner and Hsia, 1981; Weidner *et al.*, 1985; Fairchild and Daoust, 1988). The latter compound is available under the trade name Halon (Allied Chemical Co.). One of the most widely used standards consists of pressed PTFE and is sold under the trade name Spectralon, available from the Labsphere Co. It has a calibrated hemispherical albedo of 0.992 in the near-UV/visible/near-IR wavelength region. Depending on the wavelength, materials used as standards in the IR include sulfur, gold, and KBr (Nash, 1986; Salisbury *et al.*, 1987).

The *bidirectional reflectance* can be measured by illuminating a material with light from a source of small angular aperture, as seen from the surface, and observing the scattered radiance with a detector that also subtends a small angle from the surface. To adequately constrain the scattering parameters, it is desirable that the reflectance be measured at many values of i, e, and g. However, frequently it is measured at only one fixed set of angles, in which case the values of the parameters cannot be uniquely determined.

The *hemispherical reflectance* (*directional–hemispherical reflectance*) is measured using an integrating sphere. This device consists of a hollow cavity whose inner walls are covered with a highly reflective, diffusely scattering paint, perforated by two small openings, or ports, one for admitting the incident irradiance and another for observing the radiance inside the sphere. The material whose reflectance is being measured is placed inside the sphere, where it is illuminated by collimated light. The scattered radiance is sampled by a detector that views the inside of the sphere, but not the material directly. For accurate values, the measurements must be corrected for losses out of the ports (Hisdal, 1965; Kortum, 1969).

The *hemispherical–directional reflectance* of a material can be measured using an integrating sphere operating in the reverse direction from that used to measure the directional–hemispherical reflectance. That is, light from the source does not illuminate the material directly, but only after multiple diffuse scatterings inside the sphere. The radiance scattered into a specific direction by the sample is viewed directly by a detector of small angular aperture.

The *bihemispherical reflectance* or *spherical reflectance* of a material can, in principle, be measured by covering an opaque sphere (actually, a hemisphere is sufficient) with an optically thick coating of the sample and placing it at the center of an integrating sphere. One side of the sphere is illuminated by collimated light, and the radiance that is scattered in all directions is measured by a detector that does not view the target directly. It was shown in Section 11.C.7 that the bihemispherical reflectance of a flat surface illuminated by a uniform flux from an entire hemisphere is equal to the spherical albedo of a sphere illuminated by collimated irradiance from one direction. I am not aware of any published values of bihemispherical reflectances that have actually been measured in this way in the laboratory. However, Bond or spherical albedos are routinely determined for planets, satellites, and other bodies of the solar system.

The *diffusive reflectance* should not be confused with the bihemispherical reflectance. The diffusive reflectance is strictly a useful mathematical artifice with no direct physical meaning and thus cannot be measured. The Kubelka–Munk reflectance (Section 14.8) is a special case of the diffusive reflectance and similarly cannot be measured.

The *physical albedo* or *geometric albedo* of a body of the solar system can, in principle, be determined by measuring its brightness at zero phase angle. Usually the intensity is calibrated by observing comparison stars of known brightness. Using the intensity of sunlight, which is also assumed to be known, the radiance that would be detected if the body were replaced by a perfectly diffusing (Lambert) disk of the same radius can be calculated. The physical albedo is the ratio of the brightness of the body to that of the disk.

Unfortunately, orbits and trajectories of both natural bodies and artificial spacecraft are such that few bodies are ever observed exactly at zero phase angle. Hence, the brightness of an object must be measured at other phase angles and extrapolated to zero phase. The opposition effect complicates the extrapolation, and several conventions for carrying it out are in use. One convention is based on the fact that if the brightness I of a body as a function of phase angle is plotted on a logarithmic scale, such as astronomical magnitude m,

$$m = -2.5 \log_{10}(I/I_r),$$

where I_r is some reference brightness, the data very nearly fall on a straight line over a limited range of phase angles. The opposition brightness is determined by extrapolating to zero phase. Some workers ignore the opposition effect, whereas others include it by using either a linear extrapolation of small-phase-angle data on a semilogarithmic plot or fitting a theoretical equation to the data. When it is necessary to use a physical albedo, the convention by which it was calculated should be ascertained.

The *spherical albedo* or *Bond albedo* is determined by measuring the integrated brightness of a body at a large number of phase angles and numerically calculating the phase integral and physical albedo. The spherical albedo is then calculated using equation (11.48).

The *normal albedo* of an area on the surface of a body is determined in the same way as the physical albedo. The brightness is measured at several phase angles to yield the photometric function of the surface, which is then extrapolated to zero phase angle, and the normal albedo is calculated.

14.3 Inverting the reflectance to find the scattering parameters

14.3.1 Numerical inversion methods

If a large enough data set is available, the reflectance can, in principle, be inverted to find all of the parameters in the reflectance function. A major difficulty is that the parameters enter the function nonlinearly. Frequently a data set will consist of measurements of the bidirectional reflectance of a surface or the integral phase function of a planet at a much larger number of angles than the total number of parameters in the relevant expression for the reflectance. In this case the problem of finding the parameters is overdetermined. The general theory of finding parameters in an overdetermined problem is called *regression analysis* and is an important area in applied mathematics.

The most common method of parameter fitting is to compare the calculated and measured reflectances while varying the parameters according to some scheme until the best agreement is obtained. The usual criterion for finding the best fit of a theoretical reflectance model to an overdetermined data set is to minimize the root-mean-square (rms) residual between the calculated and measured data points

$$\varepsilon_{\mathrm{rms}} = \left[\frac{\sum_{j=1}^{\mathcal{N}} \left(I_{dj} - I_{cj} \right)^2}{\mathcal{N} - 1} \right]^{1/2}, \tag{14.2a}$$

where I_{dj} is the intensity of the jth measured data point, I_{cj} is the intensity calculated for that geometry using the appropriate equation, and $\mathcal{N}$ is the number of points. (The denominator is usually taken as $\mathcal{N} - 1$ rather than $\mathcal{N}$ because if there is only one data point then $\varepsilon_{\mathrm{rms}}$ is meaningless.) However, certain parameters are most strongly influenced by the angles at which the reflectance is small. Hence, in finding the best fit, it is important that points of low intensity be given the same weight as points of high intensity. This can be done by using either relative or logarithmic residuals, rather than linear residuals. That is, the preferred procedure is to define $\varepsilon_{\mathrm{rms}}$ either as

$$\varepsilon_{\mathrm{rms}} = \left[\frac{\sum_{j=1}^{\mathcal{N}} \frac{\left(I_{dj} - I_{cj} \right)^2}{I_{dj}}}{N - 1} \right]^{1/2} \tag{14.2b}$$

or

$$\varepsilon_{\mathrm{rms}} = \left[\frac{\sum_{j=1}^{N} (\log I_{cj} - \log I_{dj})^2}{N - 1} \right]^{1/2} \tag{14.2c}$$

In planetary photometry the intensity is frequently given in astronomical magnitudes, which is a logarithmic scale, so that equation (14.2c) is an appropriate form.

When the single-particle angular-scattering function $p(g)$ is found by fitting equation (9.47), or one of the expressions derived from it, to the data, an important caveat must be kept in mind. In equation (9.47) all of the effects of particle anisotropy have been placed in the single-scattering term, whereas in a rigorous formulation the multiple-scattering term also exhibits some of the effects. Hence, when a data set is inverted, the calculated $p(g)$ overestimates the anisotropy (Figure 8.9). A few inversion methods will now be discussed.

The trial-and-error method This method can be used on a personal computer with a plotting program. The data points and the appropriate equation with trial parameters are entered into the plotting routine, and a trial curve is generated. Of all the parameters in the equation of radiative transfer, the reflectance is dominated by the single-scattering albedo $w(\lambda)$ or, equivalently, the albedo factor $\gamma(\lambda)$ (Hapke *et al.*, 1981; Helfenstein and Veverka, 1989). This parameter determines both the amplitude of the reflectance and its general variation with angle, while $p(g)$ essentially fine-tunes the angular dependence of the reflectance. The opposition effect is important only at small phase angles, the macroscopic roughness parameter has the greatest effect at large phase angles. Hence, with a little practice, the effect of changing each parameter becomes intuitive, and satisfactory fits can be found rapidly. This method was used by Hapke and Wells (1981), by Domingue and Hapke (1989), and by Domingue *et al.* (1991).

Automated computer fitting routines The rms residual can be regarded as a surface in a multidimensional space in which $\varepsilon_{\rm rms}$ is one coordinate and the reflectance parameters are the other coordinates. Mathematically, the objective of finding the best fit is that of finding the lowest minimum on this surface. There are several techniques for minimizing the residuals in such types of problems, as discussed in standard textbooks (e.g., Bevington, 1969). Computer programs are available in several places, such as the Numerical Algorithms Group (NAG) library. Many plotting programs have a subroutine for parameter fitting. One method is to set up a grid in the parameter space and search all points on the grid. The degree of fineness of such a search is limited by the available computer time and memory. Other methods involve varying the parameters according to some scheme to successively lower the residual until a minimum is located. Helfenstein (helfenst@astro.cornell.edu) has developed a program that combines grid search and parameter variation to fit equation (9.47) to observational data.

A problem with automated methods is that most of them require trial guesses of the initial values of the parameters. If a trial value is far from the actual value the computer may find a local minimum in the residual, rather than the deepest global minimum. Hence, any additional information that can constrain the parameters should be used. In addition, the solution should be checked for uniqueness (Domingue and Hapke, 1989).

The trial-value problem can be made more manageable by realizing that most of the parameters influence the curve only over restricted ranges of angles. For example, the opposition effect is important only at small phase angles. In this range the continuum reflectance is fairly constant. Hence, the amplitude and width parameters can be estimated by assuming that within this range the reflectance behaves like an opposition effect superimposed on a constant continuum. Roughness affects the reflectance mainly where $70° < e < 90°$, and can be ignored to a first approximation elsewhere. Using the reflectance data outside these angles reduces the problem to finding only w and $p(g)$. If the estimated solution is then extrapolated to the limb the brightness difference between the extrapolated and measured values can be used to estimate $\overline{\theta}$ by asking through what angle an area at the limb must be tilted to bring the two into agreement. Values found in this way can then be used as initial guesses in an automated fitting program.

The exception to this discussion is the porosity parameter K. Unfortunately, the angular distribution of reflectance is extremely insensitive to K so that it is impossible to determine this parameter remotely. Hence, in the absence of other information, K is usually set equal to 1.00. However, this means that the value of w retrieved from the reflectance may be too high.

Other simplifications may be necessary. It is seldom the case that the range of angles at which the observations are made is sufficient to allow the retrieval of all photometric parameters. For example, most particles have a forward-scattering lobe, but unless the reflectance is measured at several phase angles considerably greater than 90° the strength and width of this lobe cannot be determined. In that case the analyst must decide whether to guess at these parameters or solve for a $p(g)$ with only a backscattering lobe. If only one lobe is assumed, it must be realized that normalization of $p(g)$ will cause the retrieved height of the backward lobe to be too high and, consequently, the retrieved value of w will be too low.

If the material has a narrow particle size distribution the angular width of the SHOE will be much wider than that of the coherent backscatter peak, so that the two maybe disentangled. However, if the particle size distribution is wide, as is the case with the lunar regolith, the SHOE will be almost as narrow as the CBOE, making unique retrieval impossible. In that case the best procedure may be to simply set the amplitude of the SHOE equal to zero and solve only for CBOE parameters. This will cause the retrieved amplitude to be too high, possibly even >1.

14.3.2 Analytic solution for the single-scattering albedo

Frequently the reflectances of both the sample and standard are measured in only one geometric configuration. This is usually the situation in laboratory spectrometers and may also be true in astronomical observations. In that case, all the parameters except w must be guessed at and then w found from the single observation. The simplest assumption that can be made concerning $p(g)$ is that the particles scatter isotropically, so that we may set $p(g) = 1$. If the measurement is of the bidirectional reflectance, the angles may be outside the opposition peak, and if the measurement is of an integral reflectance, the opposition effect has little influence; hence, the opposition effect amplitude may be set equal to 0. In the laboratory and in planetary observations not too close to the limb or terminator, roughness may be assumed to be unimportant, so $\overline{\theta}$ may be set equal to 0. Unfortunately, as we have seen, it is difficult to constrain the porosity parameter K unless the filling factor is known independently, so that K may be taken to be equal to 1.00.

If these assumptions are made w can be found analytically. The equations are easier to solve for γ than for w directly. Once γ is found, the single-scattering albedo can be calculated from

$$w = 1 - \gamma^2. \tag{14.3}$$

Analytic solutions for several types of reflectances in which $p(g) = 1$ and $\overline{\theta} = B_{C0} = B_{S0} = 0$ are given below.

The diffusive reflectance Although r_0 has no direct physical meaning, the diffusive reflectance equations are useful for making rapid semiquantitative estimates of γ or w. The inversion formulas were derived previously (equations 8.27 and 8.28. Repeating them here, for completeness,

$$\gamma = \frac{1 - r_0}{1 + r_0}, \tag{14.4a}$$

$$w = \frac{4r_0}{(1 + r_0)^2}. \tag{14.4b}$$

The relative reflectance The relative reflectance Γ is the bidirectional reflectance relative to a medium of nonabsorbing, isotropic scatterers, illuminated and observed at the same set of angles. From equation (10.7),

$$\Gamma(\mu_0, \mu, g) = \frac{1 - \gamma^2}{(1 + 2\gamma\mu_0)(1 + 2\gamma\mu)}. \tag{14.5a}$$

Solving this for γ gives

$$\gamma(\Gamma) = \frac{[(\mu_0 + \mu)^2\Gamma^2 + (1 + 4\mu_0\mu\Gamma)(1 - \Gamma)]^{1/2} - (\mu_0 + \mu)\Gamma}{1 + 4\mu_0\mu\Gamma} \tag{14.5b}$$

The radiance factor The bidirectional reflectances of surfaces or objects in the planetary literature are commonly given as the radiance factor I/F, the reflectance relative to a perfect Lambert surface observed and illuminated normally,

$$\frac{I}{F} = \frac{w}{4}\frac{\mu_0}{\mu_0 + \mu}\frac{1 + 2\mu_0}{1 + 2\gamma\mu_0}\frac{1 + 2\mu}{1 + 2\gamma\mu} = \frac{1}{4}\frac{\mu_0}{\mu_0 + \mu}(1 + 2\mu_0)(1 + 2\mu)\Gamma(\gamma).$$

Let

$$\Gamma_L = 4\frac{\mu_0+\mu}{\mu_0}\frac{1}{(1+2\mu_0)(1+2\mu)}\frac{I}{F}.$$

Then

$$\gamma(\Gamma_L) = \frac{\left[(\mu_0+\mu)^2\Gamma_L{}^2 + (1+4\mu_0\mu\Gamma_L)(1-\Gamma_L)\right]^{1/2} - (\mu_0+\mu)\Gamma_L}{1+4\mu_0\mu\Gamma_L}. \quad (14.6a)$$

In the laboratory and in the field the usual way of measuring a reflectance is to detemine the intensities of light scattered from the sample and from a calibrated standard at $i = e = 0$, and calculate the ratio of the brightness of the sample to that of the standard. It is usually assumed that the standard scatters like a perfect diffuse surface, so that this ratio is interpreted as the radiance factor I/F. However, this is usually not quite true. Piatek *et al.* (2004) found that the Spectralon standard mentioned above has the angular distribution of the bidirectional reflectance of a particulate medium with $w = 1$ and $p(g) = 1$, so that its reflectance illuminated and observed normally is $r_s = 9/8\pi$. It is highly likely that other standards have similar reflectance distributions. If the reflectance relative to such an isotropic standard is denoted by I/F_I, then

$$\frac{I}{F_I} = \frac{2}{9}\frac{\mu_0}{\mu_0+\mu}(1+2\mu_0)(1+2\mu)\Gamma(\gamma).$$

Then I/F_I may be solved for γ to give

$$\gamma(\Gamma_I) = \frac{\left[(\mu_0+\mu)^2\Gamma_I{}^2 + (1+4\mu_0\mu\Gamma_I)(1-\Gamma_I)\right]^{1/2} - (\mu_0+\mu)\Gamma_I}{1+4\mu_0\mu\Gamma_I}, \quad (14.6b)$$

where

$$\Gamma_I = \frac{9}{2}\frac{\mu_0+\mu}{\mu_0}\frac{1}{(1+2\mu_0)(1+2\mu)}\frac{I}{F_I}$$

when $\mu_0 = \mu = 1$, then $\Gamma_I = I/F_I$.

The hemispherical reflectance (directional–hemispherical reflectance) The relative hemispherical reflectance of a medium of isotropically scattering particles is equal to the hemispherical reflectance, equation (11.3),

$$r_h = \frac{1-\gamma}{1+2\gamma\mu_0}, \quad (14.7)$$

which can be solved for γ to give

$$\gamma(r_h) = \frac{1-r_h}{1+2\mu_0 r_h}. \quad (14.8)$$

The hemispherical–directional reflectance From the symmetry between r_{hd} and r_h we may write, by inspection,

$$\gamma(r_{hd}) = \frac{1 - r_{hd}}{1 + 2\mu r_{hd}}. \tag{14.9}$$

The spherical reflectance (bihemispherical reflectance) The relative spherical reflectance is equal to the spherical reflectance, equation (11.19),

$$r_s = \frac{1-\gamma}{1-\gamma}\left(1 - \frac{1}{3}\frac{\gamma}{1+\gamma}\right), \tag{14.10}$$

which can be solved for γ to give

$$\gamma(r_s) = \frac{(24r_s + 25)^{1/2} - (6r_s + 1)}{2(3r_s + 2)}. \tag{14.11}$$

14.4 Absorption bands in reflectance

14.4.1 Relation between the reflectance and the absorption coefficient

The wavelength at which a solid-state absorption band occurs is the major diagnostic characteristic for determining composition by remote sensing. Similarly, the depths of the absorption bands in the spectra of minerals can provide information on relative abundance. However, because the process of diffuse reflectance is nonlinear, both the band center and shape seen in reflectance differ from those measured by transmission. In the past this has caused some workers to unjustly reject reflectance spectroscopy as being unreliable. Thus, it is important that anyone who desires to use reflectance as a spectroscopic tool be aware of these effects, and to understand how the absorption coefficient affects the various parameters in the single-scattering albedo and the reflectance.

It will be sufficient to discuss a medium that consists of only one type of particle that is large compared to the wavelength. In that case,

$$w = N\sigma Q_S / N\sigma Q_E = Q_S / Q_E = Q_s, \tag{14.12}$$

since $Q_E = 1$ and $Q_S = Q_s$ for large particles in a medium in which the particles are in contact. Also,

$$\gamma^2 = 1 - w = (N\sigma Q_E - N\sigma Q_s)/N\sigma Q_E = Q_A / Q_E = Q_A. \tag{14.13}$$

It will be assumed that the scattering efficiency of an equant particle can be described to a sufficient approximation by the equivalent-slab model, equation (6.20):

$$w = Q_s = S_e + (1 - S_e)\frac{(1 - S_i)\Theta}{1 - S_i\Theta}, \tag{14.14}$$

where S_e and S_i are the average Fresnel reflection coefficients for externally and internally incident light, respectively (equations 6.49 and 6.50), and Θ is the internal-transmission factor. From Chapter 6

$$\Theta = \frac{r_i + \exp\left[-\sqrt{\alpha(\alpha+s)}\langle D\rangle\right]}{1 + r_i \exp\left[-\sqrt{\alpha(\alpha+s)}\langle D\rangle\right]}, \tag{14.15a}$$

where s is the internal scattering coefficient inside the particle,

$$r_i = \frac{1 - \sqrt{\alpha/(\alpha+s)}}{1 + \sqrt{\alpha/(\alpha+s)}}, \tag{14.15b}$$

and $\langle D\rangle$ is the effective particle size, defined as the average distance traveled by rays that traverse the particle once without being internally scattered.

In Chapter 5 it was shown that for a perfect sphere, $\langle D\rangle \simeq 0.9\,D$. For a distribution of irregular particles, $\langle D\rangle$ is expected to be of the same magnitude as, but somewhat smaller than, the size. There are several reasons for this. First, as discussed in Chapter 6, almost any departure from sphericity decreases $\langle D\rangle$. Second, most natural media consist of a distribution of particle sizes. Reflectance measurements emphasize the brightest particles, which are the smaller particles if they are not opaque. Third, the mixing formulas typically assign the highest weights to the smallest particles. Hence, when $\langle D\rangle$ is determined by reflectance, it will usually correspond to the smallest particles in the distribution, rather than to some mean size. This was seen in Figures 6.16 and 6.19.

In order to understand the dependence of the reflectance on α and n_i it will be sufficient to use the diffusive reflectance $r_0 = (1-\gamma)/(1+\gamma)$, because its behavior is representative of all the types of reflectances. (If desired, r_0 may be interpreted as $\Gamma(60°, 60°, g \gg h)$ or as $r_h(60°)$. See equations [11.20].)

Figures 14.1 and 14.2 illustrate the dependence of the single-scattering albedo w and the reflectance r_0 on the absorption. To facilitate understanding, the curves are plotted in two ways: by using as the independent variable the imaginary part of the refractive index n_i (Figures 14.1a and 14.2a) and the effective absorption thickness $\alpha\langle D\rangle$ (Figures 14.1b and 14.2b). In these figures it has been assumed that $X = \pi\langle D\rangle/\lambda$, so that $\alpha\langle D\rangle = 4n_i X$. For purposes of illustration, the various parameters have been taken to have the following values: $\lambda = 0.5\,\mu\text{m}$, $\langle D\rangle = 50\,\mu\text{m}$ (corresponding to $X = 300$) and $\langle D\rangle = 500\,\mu\text{m}$ (corresponding to $X = 3000$), $s = 00.06\,\mu\text{m}^{-1}$, and $n_r = 1.50$.

In general, three different reflectance regimes may be distinguished. These will be referred to as the *volume-scattering region*, the *weak surface-scattering* region, and the *strong surface-scattering* region. These are illustrated in Figure 14.1, which plots w against n_i (top) and $\alpha\langle D\rangle$ (bottom), and Figure 14.2, which plots r_0 against the same quantities.

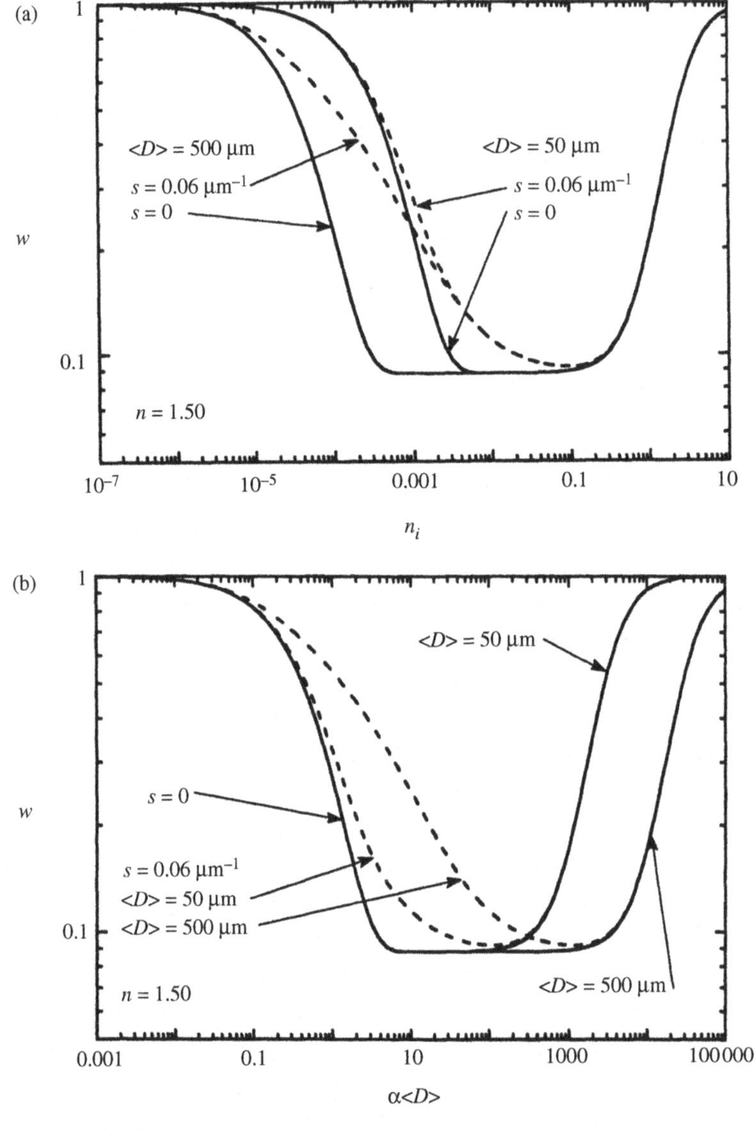

Figure 14.1 Single-scattering albedo $w = Q_s$ for particles of sizes and internal scattering coefficients indicated. The real part of the refractive index is $n_r = 1.50$; (a) w vs. k; (b) w vs. $\alpha\langle D\rangle$.

The volume-scattering region When $\alpha\langle D\rangle \ll 1$, Figures 14.1 and 14.2 show that w and r_0 are both close to 1. Note, however, that even for $n_i \sim 10^{-7}$, which for the 500 μm particles corresponds to $\alpha\langle D\rangle \sim 10^{-3}$, there is an easily measurable difference between r_0 and 1.0, so that the magnitude of the reflectance is sensitive to n_i even for such small values of α. When $\alpha\langle D\rangle \lesssim 3$, the reflectance is dominated

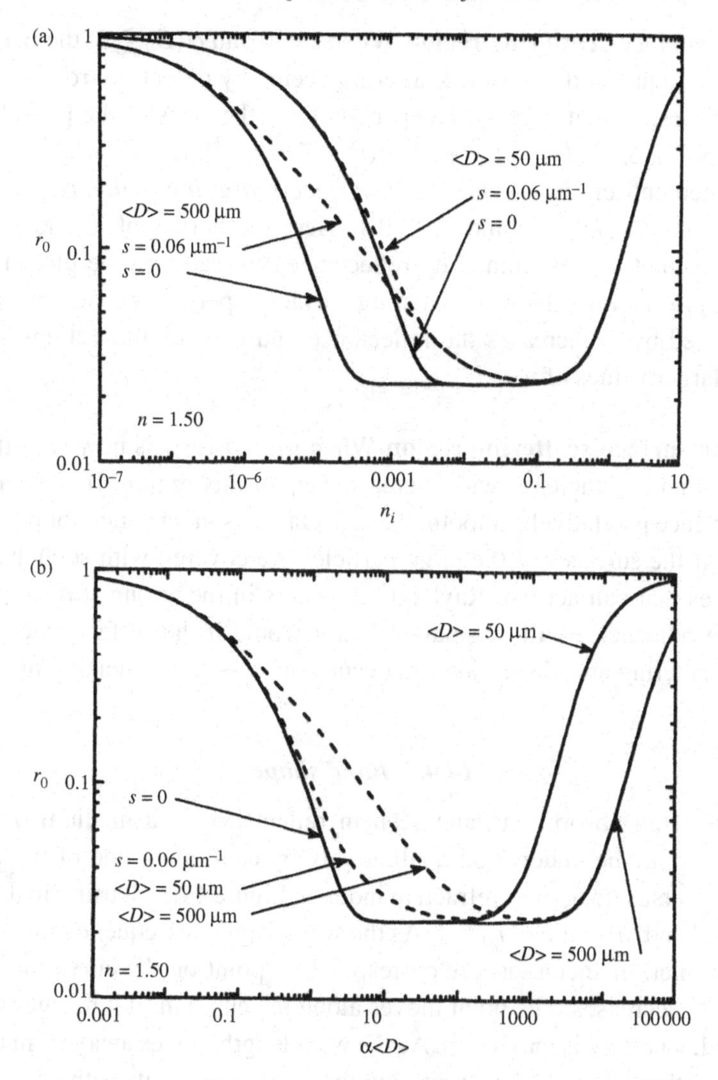

Figure 14.2 Diffusive reflectance r_0 for media composed of the particles in Figure 14.1; (a) r_0 vs. n_i; (b) r_0 vs. $\alpha\langle D\rangle$.

by light that has been refracted, transmitted, and scattered within the volume of the particle. Hence, this part of the curve is called the *volume-scattering region*. In the part of this region where $\alpha\langle D\rangle \lesssim 0.1$, w and r_0 decrease slowly as n_i increases. The rate of decrease becomes more rapid when $0.1 \lesssim \alpha\langle D\rangle \lesssim 3$, so that the reflectance is highly sensitive to n_i there. The reflectance is equally sensitive to particle size: as $\langle D\rangle$ increases, the part of the r_0-vs.-n_i curve in the volume-scattering region shifts to the left, and the reflectance decreases. This dependence of reflectance on particle size has been noted by many authors (e.g., Adams and Felice, 1967).

The weak surface-scattering region When $s = 0$ and $\alpha\langle D\rangle \gtrsim 3$, the particles are essentially opaque, and all of the scattering occurs by reflection from the surfaces of the particles, so that $w \simeq S_e$, independently of the sizes of the particles. From equation (5.37), $S_e \propto [(n_r - 1)^2 + n_i^2]/[(n_r + 1)^2 + n_i^2]$. Hence, when $n_i \lesssim 0.1$, S_e is determined entirely by n_r. The *weak surface-scattering region* occurs between the place where $\alpha\langle D\rangle \gtrsim 3$ and $n_i \lesssim 0.1$. Here the curves of w and r_0 are flat, so that n_i cannot be determined by reflectance, but can only be placed between upper and lower limits. However, adding surface asperities or internal scatterers, parameterized by s, increases the reflectance and extends the volume-scattering region to larger values of n_i.

The strong surface-scattering region When $n_i \gtrsim 0.1$, S_e is now sensitive to n_i, and both w and r_0 increase with increasing n_i in this region. If $X \gg 1$ and the particle surface is relatively smooth, the reflectance is independent of particle size. However, if the surfaces of the large particles are covered with scratches, edges, or asperities that can act like Rayleigh absorbers in the strong surface-scattering region, the reflectance will be smaller than it would if the surfaces were smooth. Unusual scattering and absorption can occur if $n^2 \simeq -2$, as discussed in Chapter 5.

14.4.2 Band shape

The shape of an absorption band seen in reflectance in a medium of particles with $X \gg 1$ may be understood qualitatively by examining one of the curves of reflectance versus imaginary refractive index in Figure 14.3. Starting in the wing of an isolated band, $n_i = 1$ and $r_0 \simeq 1$. As the wavelength or frequency moves toward the band center, n_i increases, the corresponding point on the curve moves to the right, and r_0 decreases. The point moves along the curve until it reaches the center of the band, where n_i is maximum. As the wavelength moves away from the center toward the other wing, n_i decreases, and the point retraces its path back along the reflectance curve to very small n_i.

The shape of a band, including the slope of the spectrum at any wavelength, is determined by the reflectance regime in which the center of the band occurs, which depends on both α and $\langle D\rangle$. This is illustrated in Figure 14.3 for a Lorentz absorption band (Chapter 3) shown in Figure 14.3a.

If $\alpha\langle D\rangle$ is small enough that all points of the band fall in the volume-scattering region, then the reflectance spectrum of the band is similar to a transmission spectrum in shape. This case is illustrated in Figure 14.3b. The reflectance decreases from wing to center and then increases back to wing again, and the band center is at the same wavelength as in transmission. If the reflectance of a monomineralic powder is transformed to an espat curve (Section 6.5.5), the espat $W(\lambda)$ will be

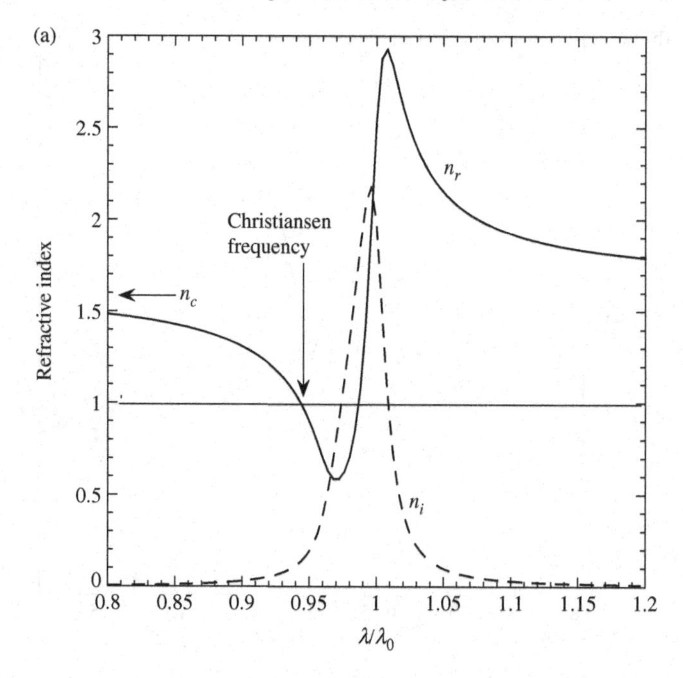

Figure 14.3 (a) Refractive index of an absorption band calculated from the Lorentz model (Chapter 3) with the following values of the parameters: continuum refractive index $n_c = 1.6$; band width parameter $\Xi/\nu_0 = 0.02$; band strength parameter $\nu_p/\nu_0 = 0.02$. The center of the band is at λ_0. (b) Bidirectional reflectance $I/F = \pi r(i, e, g)$ spectrum at $i = e = g = 0$ of a particulate material with the absorption band of Figure 14.3a, except that the band is sufficiently weak that it is entirely within the volume-scattering region. The opposition effect has been ignored. The curve was calculated from the equivalent-slab model (Chapter 6) for w, and the two-stream expression for the reflectance (Chapter 8) of a medium of isotropic scatterers. The following values for the parameters were used: particle size $D/\lambda_0 = 10$; band strength parameter $\nu_p/\nu_0 = 1 \times 10^{-5}$. (c) Reflectance spectrum of a particulate material with an absorption band whose center lies in the weak surface-scattering region. All parameters same as in Figures 14.3a and 14.3b, except $\nu_p/\nu_0 = 0.01$. (d) Reflectance spectrum of a particulate material with an absorption band whose center lies in the strong surface-scattering region. All parameters same as in Figure 14.3a and 14.3b, except $\nu_p/\nu_0 = 0.2$.

directly proportional to the absorption coefficient. Thus, if the band $a(\lambda)$ has a Gaussian shape, $W(\lambda)$ will also be Gaussian.

If the value of $\alpha\langle D\rangle$ at the center of the band is strong enough to be in the weak surface-scattering region, then the reflectance saturates as the wavelength enters that region. This case is illustrated in Figure 14.3c. A further increase in n_i or α does not cause a corresponding decrease in reflectance. Instead, the reflectance remains constant until $\alpha\langle D\rangle$ moves out of the weak surface region back into the volume-scattering region, and the reflectance increases. In this case the bottom of

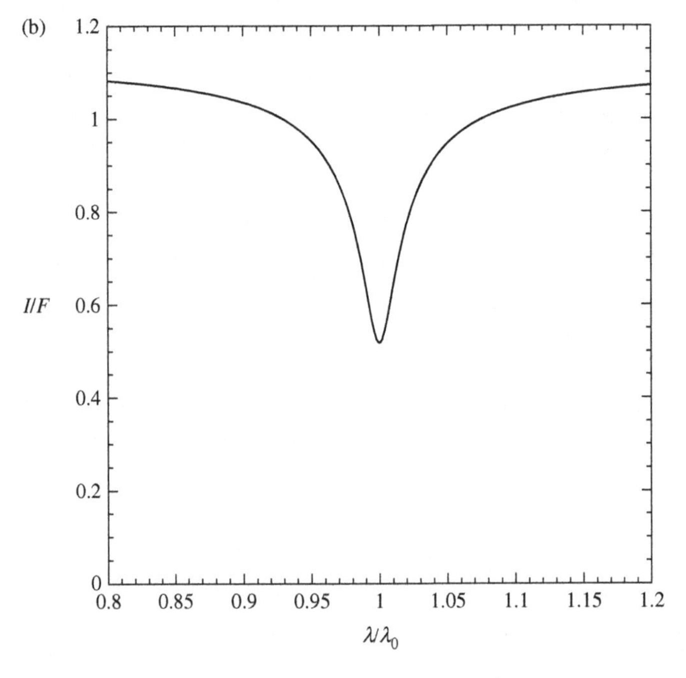

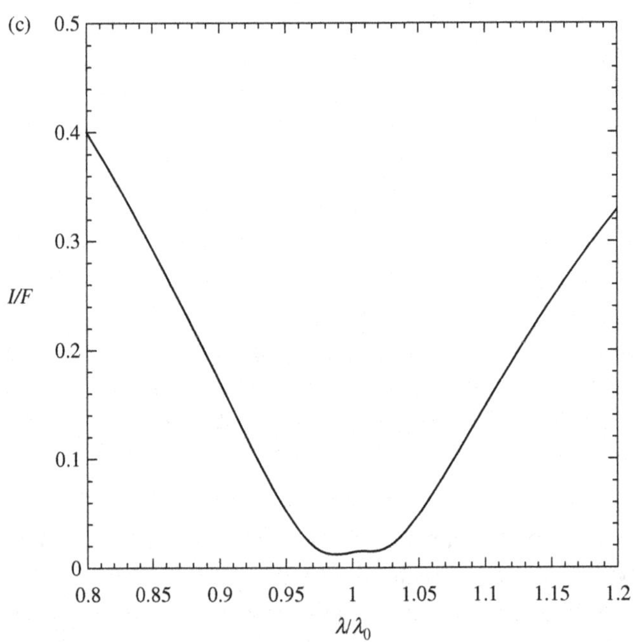

Figure 14.3 (*cont.*)

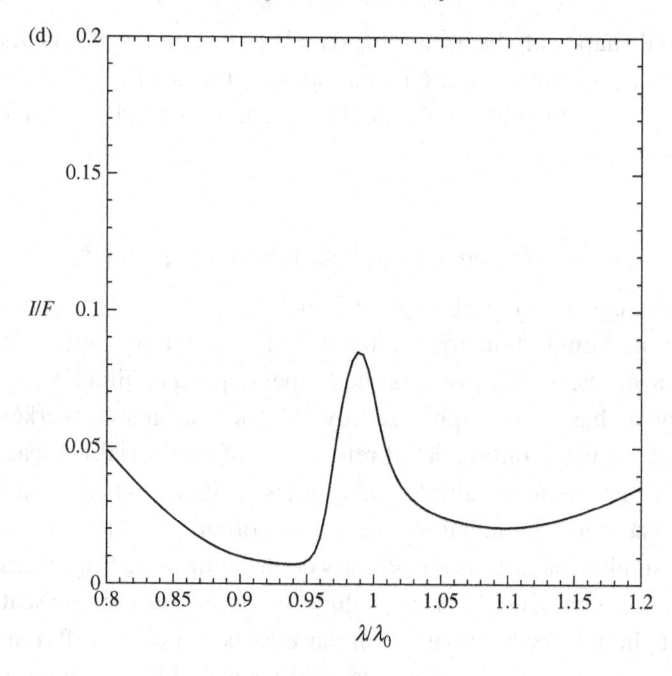

Figure 14.3 *(cont.)*

the band is cut off and replaced by a flat line. Also, the heights of the wings are increased by multiple scattering effects, relative to the wings seen in transmission. A naive observer might easily interpret the band as two unresolved, overlapping bands.

If $n_i > 0.1$ at the band center then the reflectance there will lie in the strong surface-scattering regime. This case is illustrated in Figure 14.3d. As n_i increases, the reflectance first decreases through the volume-scattering region, but then increases as it enters the strong surface region. The reflectance goes through a maximum near the center of the band. However, as seen in Chapter 3, anomalous dispersion effects cause n_r to be a function of frequency, so that the position of the maximum of the reflectance curve is displaced toward the higher-frequency (shorter-wavelength) side of the actual band center. As n_i decreases again, the reflectance decreases through a second minimum and then increases into the wing.

Thus, a strong band has two minima on either side of the maximum corresponding to the band center. This type of minimum will be called a *transition minimum*, because it occurs in the weak surface-scattering transition region between the volume-scattering and strong surface-scattering regions. The anomalous dispersion behavior of n_r causes the shorter-wavelength transition minimum to be deeper than the longer-wavelength one. A naive observer could easily mistake this band for two overlapping, partially resolved bands.

The unusual shapes of absorption bands when the band center is dominated by surface scattering means that automated methods for band identification, such as the one proposed by Huguenin and Jones (1986), must be used with great care when strong bands are present.

14.4.3 Dependence of band depth on geometry

Thus far, the shapes of the absorption bands have been discussed in terms of r_0, which in its simplest interpretation is independent of angle. However, for physically real cases, the band depths and shapes depend on illumination and viewing geometry, as has been emphasized by Veverka and his co-workers (Veverka *et al.*, 1978a,b,c). If the bidirectional reflectance of a material is measured with i and e close to normal, then multiple scattering will significantly increase the wings of the band, where w is high, but will be less important near the band center, where w is low and single scattering is the primary contributor to the brightness. However, if the reflectance is measured at large values of i and e, only single scattering contributes throughout the whole band. Similar effects can cause differences in band shapes between spectra of the same material measured bidirectionally and using an integrating sphere to measure the directional–hemispherical reflectance (Gradie and Veverka, 1982). The dependence of the relative band shape on angle and type of reflectance is illustrated in Figure 14.4.

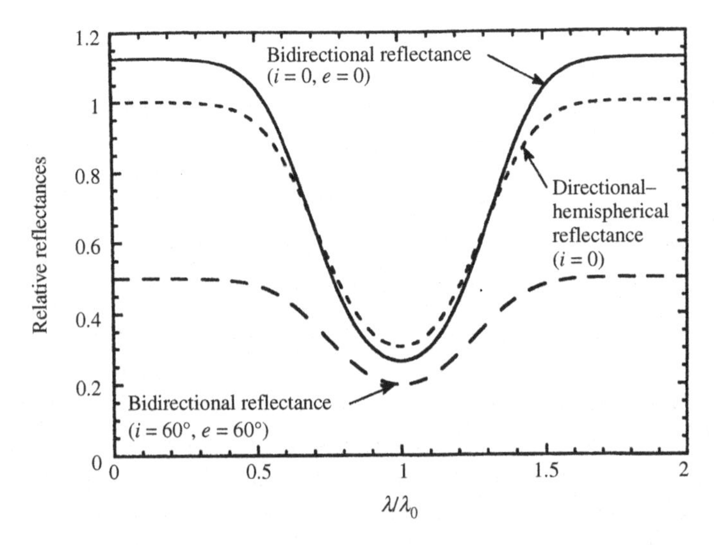

Figure 14.4 Reflectance spectra of a particulate medium of isotropic scatters with a Gaussian absorption band, illustrating the effects of illuminating and viewing geometry on the band depth.

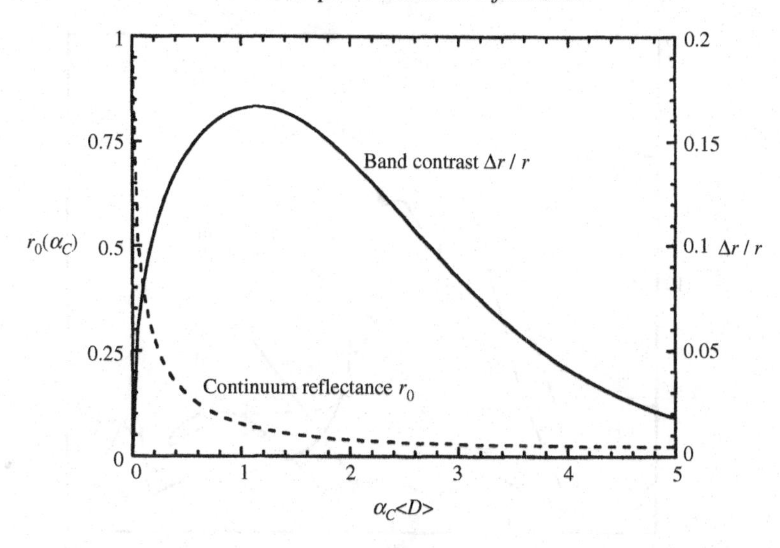

Figure 14.5 Relation between band contrast and particle size. Solid line, relative band contrast; dashed line, continuum reflectance. The relative contrast in absorbance is 20%.

14.4.4 Dependence of band contrast on particle size

A question of interest is the relation between the depth of an absorption band seen in reflectance and the particle size of the scattering medium. It may be addressed by using equations (14.14) and (14.15) for w to calculate the diffusive reflectance r_0. This has been done in Figure 14.5, where for purposes of illustration we have taken $n_r = 1.50$ and $s = 0$. It was assumed that the absorption coefficient at the center of the band was $\alpha_B = 1.20\alpha_C$, where α_C is the value in the continuum. Figure 14.5 shows the relative band contrast in reflectance $\Delta r/r = [r_0(\alpha_C) - r_0(\alpha_B)]/r_0(\alpha_C)$ as $\alpha_C\langle D\rangle$ is varied. The continuum reflectance $r_0(\alpha_C < D >)$ is also shown.

When $\alpha_C\langle D\rangle$ is very small and the particles are optically thin, the band contrast is small also. As the particle size $\langle D\rangle$ increases, $\Delta r/r$ increases to a maximum value roughly equal to the relative band contrast in absorbance, 20% at $\alpha_C\langle D\rangle \simeq 1$. The reflectance then decreases monotonically. As the particle size continues to increase, the band contrast now decreases and becomes small as the particles become optically thick, and the reflectance saturates in the weak surface-scattering region. Note that the band contrast is not a monotonic function of the reflectance or of the particle size.

When the optical thickness of the particle is large the reflectance is dominated by surface scattering. Hence, the shape and depth of the bands are virtually independent of the particle size.

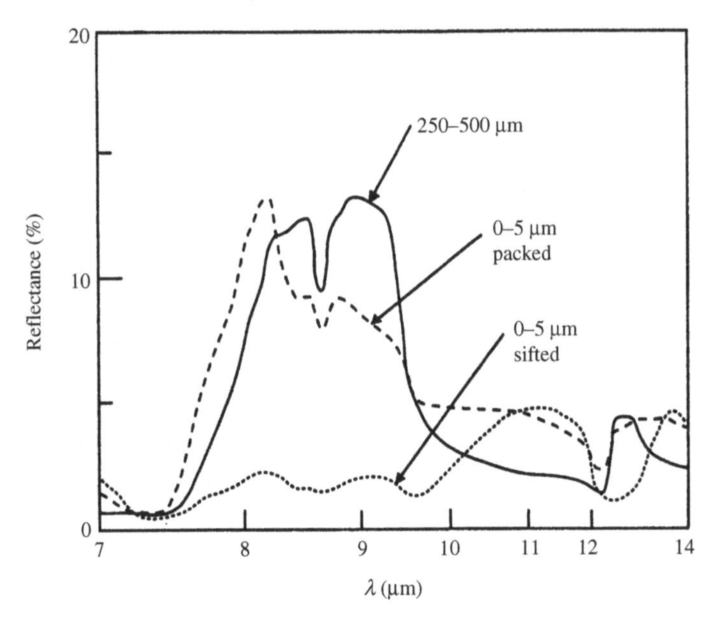

Figure 14.6 Reflectance spectra in the restrahlen region of two different size fractions of quartz particles showing the effect of packing on the spectrum of the smaller size particles. The spectrum of the coarser particles had little dependence on packing. (Reproduced from Salisbury and Eastes [1985], copyright 1985 with permission courtesy of Elsevier.)

14.4.5 Effects of porosity

Salisbury and Eastes (1985) measured the reflectance of quartz particles of sizes less than 5 μm at wavelengths between 7 and 14 μm (Figure 14.6). The powder had a very low reflectance when sifted to form a medium of high porosity. However when the powder was packed, the reflectance increased by a factor that exceeded 5 in the quartz *restrahlen* band, where the single scattering albedo was low. However, the reflectance changed little outside the band, where the single-scattering albedo was higher. This behavior is qualitatively consistent with the predictions of the effects of porosity in Chapter 9, although the observed change is larger than expected inside the band.

14.5 The reflectance spectra of intimate mixtures

Because the reflectance is a nonlinear function of the single-scattering albedo, the dependence of reflectance of mixtures will, in general, not be a linear function of the spectra of the end members (Nash and Conel, 1974; Clark, 1983). This is especially true if the albedos are high. However, the reflectance of an intimate mixture can be calculated from the reflectances of the pure end members by using the methods of Section 10.7. First, w_j and $p_j(g)$ are found by inverting the appropriate equation for the reflectance r_j of the jth end member. Then w and $p(g)$ for the mixture are calculated using the mixing formulas and are inserted into the reflectance equation to calculate the reflectance of the mixture.

Conversely, if the identities and spectra of the individual members of a mixture are known, the weights $N_j \sigma_j Q_{Ej} \propto M_j Q_{Ej}/\rho_j D_j$ in the mixture can be found by trial and error by fitting calculated spectra to the measured spectrum of the mixture.

Deconvolutions of mixtures to find the fractions of the end members have been done in laboratory investigations by Smith *et al.* (1985) and Mustard and Pieters (1987,1989). Jenkins *et al.* (1985) applied these concepts to the analysis of lunar reflectance spectra.

Figure 14.7, from the paper by Mustard and Pieters (1989), compares the fractions of the end members in binary and ternary mixtures calculated by deconvolution of the reflectance spectra with the actual values. They deconvolved the data in two ways, one assuming that all particles scatter isotropically, and the other allowing for anisotropic scattering. As might be expected, the anisotropic fit gave smaller residuals. However, making the simplifying assumption that the scatterers are isotropic still allows the abundances to be estimated to better than 7%, except for opaque minerals, for which the errors are somewhat larger.

The following example illustrates the mixing equations. Suppose we have two powders consisting of isotropic scatterers larger than the wavelength. Suppose that the hemispherical reflectances of the two materials, measured at $i = 60°$, are $r_{hl} = 0.05$ and $r_{h2} = 0.90$. What will the reflectances of various mixtures of the two powders be? Using equation (14.8), the single-scattering albedos corresponding to these reflectances are calculated to be $w_1 = 0.181$ and $w_2 = 0.997$. For a binary mixture of particles, equation (10.46) is

$$w = \frac{w_1 + C w_2}{1 + C},$$

where C is the weighting factor

$$C = \frac{\mathcal{M}_1}{\mathcal{M}_2} \frac{\rho_2}{\rho_1} \frac{D_2}{D_1}$$

$\mathcal{M}_j$, is the bulk density of material of type j, ρ_j is its solid density, and D_j is its size. Note that the weighting factor C depends on the ratio of particle sizes in addition to the relative amounts of the two materials.

Figure 14.8 illustrates the dependence of the reflectance of the mixture on the mass mixing ratio $\mathcal{M}_2/(\mathcal{M}_1 + \mathcal{M}_2)$ for three different size ratios, $D_2/D_1 = 0.01$, 1.0, and 100. For simplicity the figure assumes that $\rho_1 = \rho_2$. Figure 14.8 makes the important point that when there is large disparity in the sizes of the components of

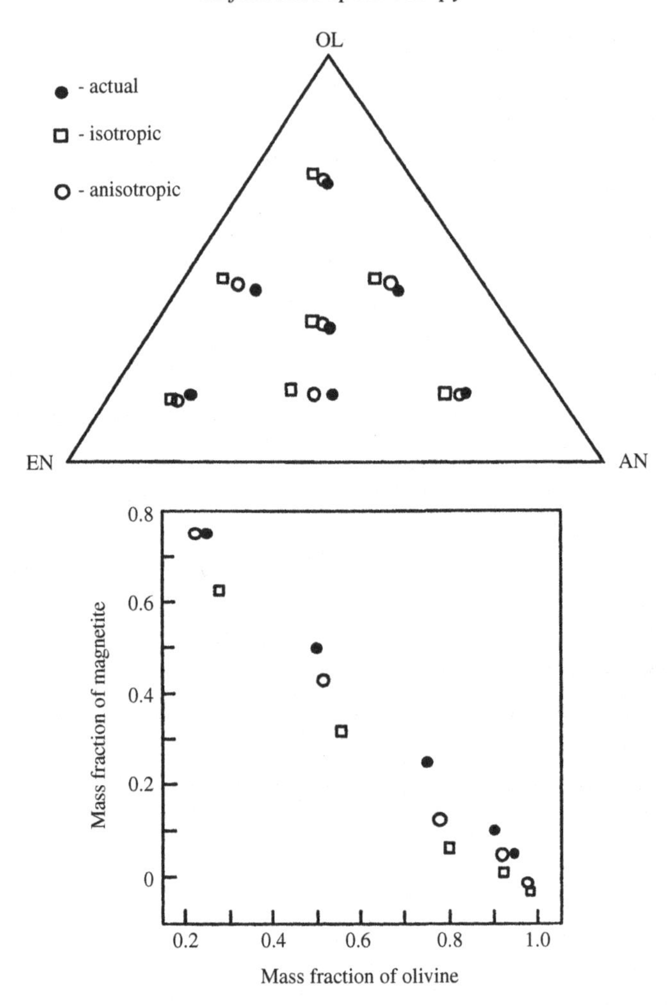

Figure 14.7 Determination of mass fractions in intimate mixtures by deconvolution of reflectance spectra. Top, ternary mixtures of olivine, enstatite, and anorthite; bottom, binary mixtures of olivine and magnetite. The filled circles are the actual mass fractions; the open squares are the results of deconvolution assuming isotropic scattering; the open circles are the results of deconvolution assuming nonisotropic scattering. (Reproduced from Mustard and Pieters [1989], copyright 1989 by the American Geophysical Union.)

an intimate mixture, the fine particles can have an effect all out of proportion to their mass fraction. For example, Clark and his co-workers (Clark and Lucey, 1984; Clark and Roush, 1984) found that the addition of a small amount of finely divided carbon black to coarse particles of ice drastically reduced the reflectance and the contrast in the absorption bands of the ice. Note that if the bright material in the example in

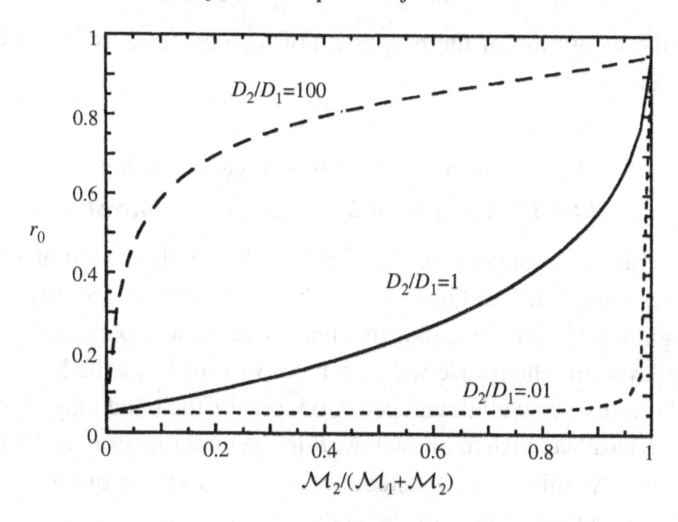

Figure 14.8 Diffusive reflectances of intimate mixtures of bright and dark particles as a function of the mass ratio for three different particle size ratios. Note that when the low-albedo particles are much smaller than the high-albedo particles, the reflectance of the mixture is almost independent of the amount or the reflectance of the bright material.

Figure 14.8 possessed any absorption bands, they would be almost totally masked by only a small amount of the dark material.

The same effect also accounts for the low albedo of the Moon. Lunar regolith consists of pulverized rocks and glasses of the same composition as the rocks. If a lunar rock or a glass made by melting the rock in vacuum are finely ground, the resulting powders are found to have a much higher albedo than the soil (Wells and Hapke, 1977). However, in the regolith many of the rock and glass fragments are welded together into particles called agglutinates, which are quite dark because they also contain submicroscopic particles of metallic iron. The soil particles are coated with deposits of vaporized rock that also contain metallic iron grains. About 0.5% of the soil consists of this submicroscopic metallic iron, an amount that is sufficient to lower the reflectance of the mixture to the observed value (Hapke *et al.*, 1975; Hapke, 2001).

It is frequently stated in the literature that in a mixture of large and small particles, the small particles "coat" the large particles and prevent light from reaching them, so that the large particles cannot influence the reflectance. This is a physically incorrect explanation, because the effect would occur even if the particles were so far apart they never touch. Small particles have a large influence because of the combined effects of the nonlinear dependence of reflectance on single-scattering albedo plus the weighting of the properties of the components by cross-sectional area rather than volume.

14.6 Absorption bands in layered media

14.6.1 The effect of layers on band contrast

Often in the both the laboratory and the field we deal with layered media. Thus, an important question of practical interest is how layers affect our ability to detect and measure diagnostic absorption bands that may be present. A band may be displayed by the upper layer, in which case we wish to know how thick the layer must be for the band to be visible or well developed. Conversely, the band may be in the lower layer, in which case we wish to know how thin the upper layer must be in order not to hide the band. As might be expected, the answers to these questions depend on the scattering properties of both of the layers.

To be rigorous, these questions should be addressed using the two-layer bi-directional equations developed in Section 10.6. However, for a semiquantitative discussion, the two-layer diffusive model, equation (10.23), may be used. Define the band contrast of a layered medium as

$$C(\tau_0) = (r_w - r_c)/r_w,$$

where r_w is the reflectance in the wing of the band, and r_c is the reflectance at the center. This contrast will be a function of the optical thickness τ_0 of the upper layer. Define the relative band contrast $\Delta C(\tau_0)$ as

$$\Delta C(\tau_0) = \frac{[(r_w(\tau_0) - r_c(\tau_0)]/r_w(\tau_0)}{[(r_w(\infty) - r_c(\infty)]/r_w(\infty)}. \tag{14.16}$$

That is, ΔC is the ratio of the band contrast observed in a layered medium to the intrinsic contrast the band would have if the medium exhibiting it were infinitely thick and not covered by any other material.

To illustrate the effects of layering on band contrast, we will consider four examples, in which it is assumed that one of the layers has a band with an intrinsic contrast of 20%.

Example 1: band in top layer, top layer dark, bottom layer bright

Suppose the bottom layer has a reflectance $r_L = 0.90$, and that in the wings of the band the top layer has $r_U = 0.090$, with the corresponding $\gamma = 0.83$. In the center of the band, $r_U = 0.072$, with the corresponding $\gamma = 0.87$. Using equation (10.23), the reflectance r_0 and relative band contrast ΔC may be calculated as functions of the optical thickness τ_0 of the upper layer. The curve of reflectance is plotted in Figure 14.9a, and that of relative contrast in Figure 14.9b. As τ_0 increases, r_0 decreases, while ΔC increases rapidly. Both reach approximately their thick-layer

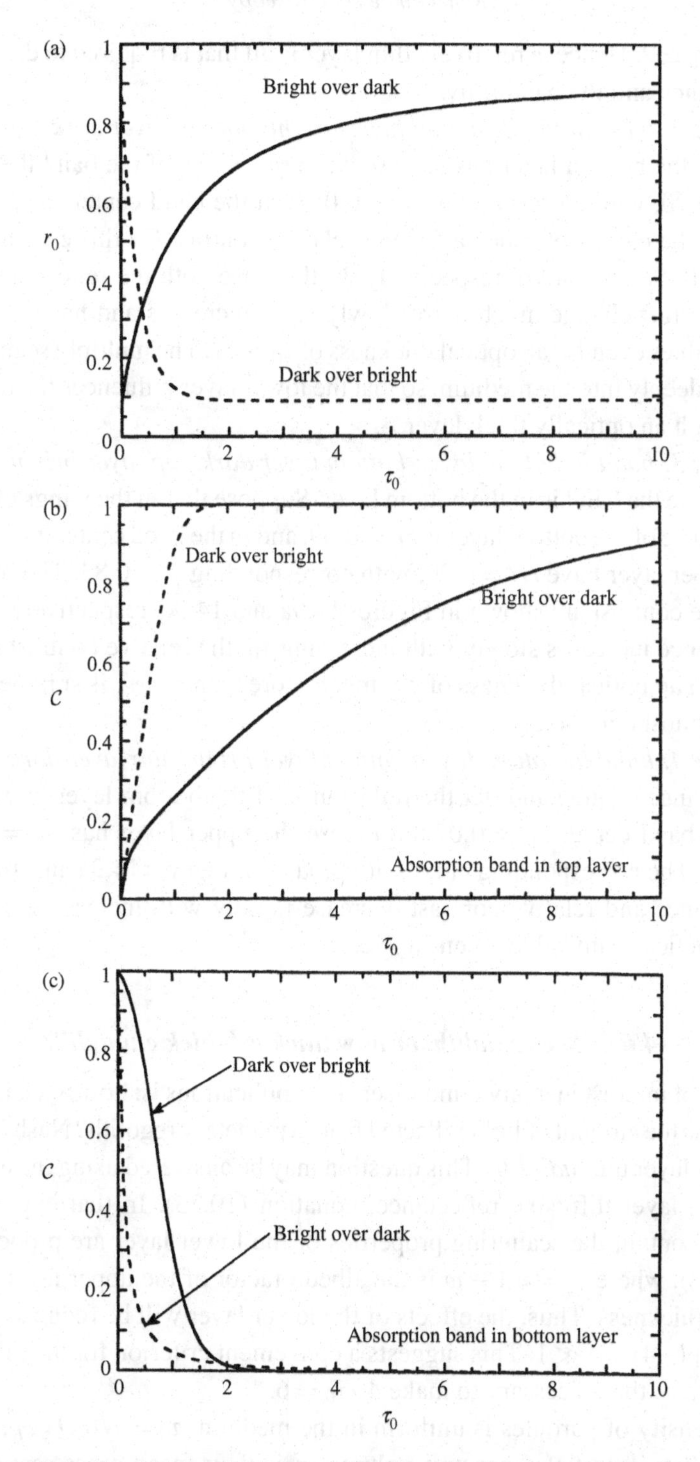

Figure 14.9 Reflectance and relative absorption-band contrast in two-layer media showing the effects of the reflectances of the top and bottom layers and the layer in which the band is located. See text for details. (a) Diffusive reflectance. (b) Band contrast, band in upper layer, $(\Delta r/r)_U = 20\%$. (c) Band contrast, band in lower layer, $(\Delta r/r)_L = 20\%$.

values by $\tau_0 \simeq 2$. Hence, a relatively thin layer is all that is required to develop both the reflectance and the band fully.

Example 2: band in top layer, top layer bright, bottom layer dark

Suppose the bottom layer has $r_L = 0.09$; in the wings of the band the top layer has $r_U = 0.90$, with corresponding $\gamma = 0.053$; in the band center, $r_U = 0.72$ and $\gamma = 0.16$. The curves of reflectance and relative contrast for this case are plotted in Figures 14.9a and 14.9b, respectively. In this case both the reflectance and the relative contrast change much more slowly as τ_0 increases and have not reached their full values even for an optical thickness of $\tau_0 > 11$. The multiply scattered light penetrates deeply into the medium, so that the lower layer influences the reflectance even through an optically thick layer.

Example 3: band in bottom layer, bottom layer dark, top layer bright

In this case the band is in the bottom layer. Suppose that in the wings of the band the reflectance of the bottom layer is $r_L = 0.09$, and in the band center it is $r_L = 0.72$. Let the upper layer have $r_U = 0.90$, with corresponding $\gamma = 0.83$. The reflectance and relative contrast are shown in Figures 14.9a and 14.9c, respectively. Although the reflectance increases slowly with increasing τ_0, the relative contrast decreases rapidly, and an optical thickness of not much more than $\tau_0 \sim 1$ is sufficient to hide the band almost completely.

Example 4: band in bottom layer, bottom layer bright, top layer dark

In the wings of the band, let the reflectance of the bottom layer be $r_L = 0.90$, and at the band center $r_L = 0.072$. Suppose the upper layer has $r_U = 0.09$ and $\gamma = 0.053$. The corresponding curves are plotted in Figures 14.9a and 14.9c. Both the reflectance and relative contrast decrease rapidly with increasing τ_0, and the band is practically invisible when $\tau_0 \gtrsim 2$.

14.6.2 The radialith, or how thick is "thick enough"?

A question of interest in many remote-sensing applications is: how thick is the layer that controls the amount of light reflected from a planetary regolith? Nash (1983) has termed this layer the *radialith.* This question may be answered using the expression for the two-layer diffusive reflectance, equation (10.23). In that expression the terms that contain the scattering properties of the lower layer are proportional to $\exp(-4\gamma\tau_0)$, where $\gamma = \sqrt{1-w}$ is the albedo factor of the upper layer, and τ_0 is its optical thickness. Thus, the effects of the lower layer will be reduced to a small value if $\exp(-4\gamma\tau_0) \ll 1$. This suggests a convenient criterion for the thickness of the radialith as that necessary to make $4\gamma\tau_0 = 6$.

If the density of particles is uniform in the medium, $\tau_0 = N\sigma Q_E z_R$, where N is the number of particles per unit volume, σ is their mean cross-sectional area, Q_E is their extinction efficiency, and z_R is the thickness of the radialith. For large particles, $Q_E = 1$, and we may write, approximately,

$$\tau_0 \simeq N\frac{1}{4}D^2 z_L = \frac{3}{2}\left(\frac{4}{3}N\frac{1}{8}D^3\right)\frac{z_L}{D} = \frac{3}{2}\phi\frac{z_L}{D} \tag{14.17}$$

where ϕ is the filling factor, and D is the mean particle size. Hence, the criterion that $4\gamma\tau_0 = 6$ is equivalent to

$$z_L/D = 1/\phi\gamma = 1/\phi\sqrt{1-w}. \tag{14.18}$$

If w is small, then γ is not too different from 1, and if the particles are close together, so that $\phi \sim \frac{1}{2}$, then z_R is only a few particle layers thick. However, if the material is only weakly absorbing so that $w \sim 1$, then the espat approximation may be used to calculate w, so that $\gamma = \sqrt{1-w} \approx \sqrt{(1-w)/w} = \sqrt{W} \simeq \sqrt{\alpha D_e}$, and expression (14.25) is

$$z_R \simeq \frac{D}{\phi\sqrt{\alpha D_e}} \simeq \frac{1}{\phi}\sqrt{\frac{D}{2\alpha}}. \tag{14.19}$$

If the particles are very weakly absorbing, and if, in addition, the porosity of the layer is high, the radialith can be very thick indeed.

The same criterion may be used to estimate the thickness required for laboratory samples in order that the substrate does not influence the reflectance. If the particles are absorbing over the wavelength range of interest, then only a few monolayers are required. If the absorbance is very small, then equation (10.25) may be used to estimate the necessary thickness,

$$r_0 = [r_L + (1-r_L)\tau_0]/[1+(1-r_L)\tau_0].$$

Suppose our criterion is that r_0 change by less than 1% no matter what the reflectance of the lower substrate r_L. This requires $\tau_0 > 99$, which from (14.17) means that the layer must be more than $66/\phi$ particles thick. If the filling factor is 50%, the sample must contain more than 130 monolayers. For example, if the grain size is of the order of 80 μm, the sample must be at least 1 cm thick.

14.7 Retrieving the absorption coefficient from the single-scattering albedo

14.7.1 Introduction

We are now in a position to solve for the absorption coefficient α, or, equivalently, the imaginary part of the index of refraction n_i, from the reflectance, and to understand the regimes in which the various expressions that will be derived are valid. Recall that the single-scattering albedo w must first be found using one of the methods described previously. If the material is a mixture, its reflectance must be deconvolved to determine the single-scattering albedos of the components. Next, a suitable model, such as those described in Chapters 5 and 6, must be chosen to relate the single-scattering albedo to the fundamental properties of the scatterers, which are the refractive index $n = n_r + in_i$ and the particle size D and shape.

In laboratory studies it is often possible to determine, or at least estimate, all of the particle parameters except n_i, which may then be calculated. The size and shape can be estimated from microscopic examination of the particles, using either an optical or scanning electron microscope. The real part of the index of refraction may be available in handbooks if the identity of the material is known, or it can be determined by a number of standard methods, such as measurement of the Brewster angle (Chapter 4), or the Becke-line method (Bloss, 1961). Even in remote-sensing measurements there may be clues, such as the distinctive wavelengths of the absorption bands, to some of these parameters.

14.7.2 Solving for α directly

Solving (14.14) for Θ gives

$$\Theta = \frac{w - S_e}{1 - S_e - S_i + S_i w}. \tag{14.20}$$

If $s = 0$, then $\Theta = \exp(-\alpha\langle D\rangle)$ and equation (14.27) contains three parameters: n_r, n_i or α, and $\langle D\rangle$. If the particle size distribution is known, then $\langle D\rangle$ can be estimated by assuming that it is equal to the smallest particles in the distribution. If n_r is known and the reflectance is in the volume-scattering region, then S_e and S_i can be calculated from n_r using the equations given in Chapter 6. Then α is given by

$$\alpha = -\frac{1}{\langle D\rangle} \ln\left(\frac{w - S_e}{1 - S_e - S_i + S_i w}\right). \tag{14.21}$$

If s is not small, then α and s can be found by measuring the reflectances of powders of the same material but of different particle sizes such that the reflectances are in the volume-scattering region. From the measured values of w Θ can be calculated for large and small particles. Then one has two different equations of the form of (14.15) for two unknowns, from which α and s can be found by iteration.

Although the reflectance is sensitive to α over the entire volume-scattering region, the part of the curve where $\alpha\langle D\rangle \ll 1$ is especially subject to systematic errors. Special problems in this region include trace impurities and the fact that the reflectance of the standard must be known precisely. To minimize such

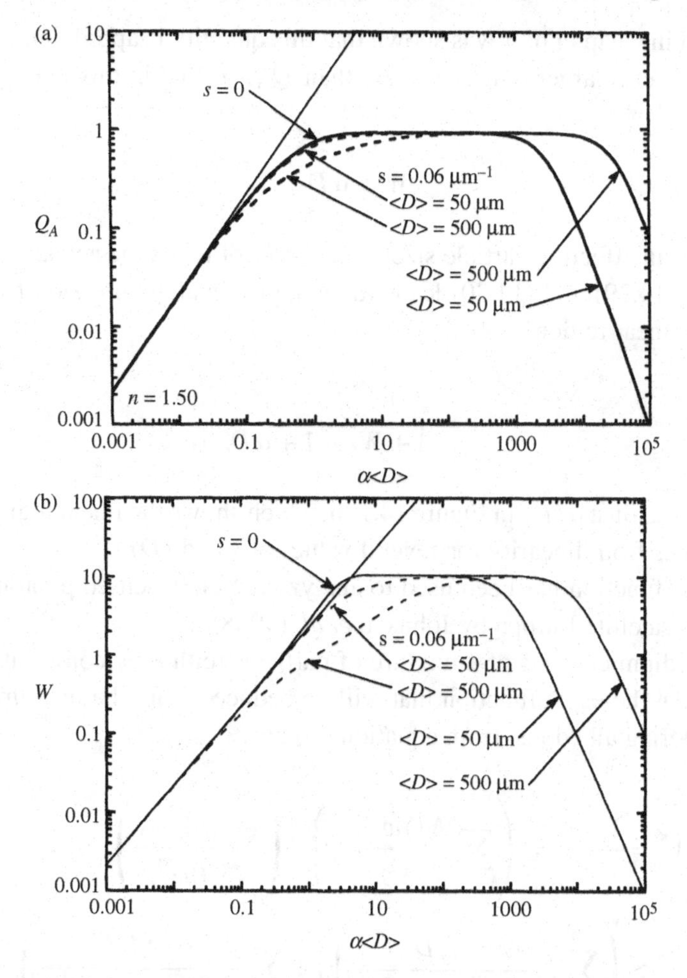

Figure 14.10 (a) Absorption efficiency Q_A vs. $\alpha\langle D\rangle$ for the particles in Figure 14.1. The straight line has unit slope. (b) Espat function W vs. $\alpha\langle D\rangle$ for the particles in Figure 14.1. The straight line has unit slope.

errors when measuring α by reflectance, a particle size should be chosen to make $0.1 \lesssim \alpha\langle D\rangle \lesssim 1$, if possible.

14.7.3 The espat function

In Figure 14.10a the absorption efficiency Q_A is plotted against $\alpha\langle D\rangle$ for the same particles as in Figures 14.1 and 14.2, and it is seen to be proportional to α only for $\alpha\langle D\rangle \lesssim 0.07$. The espat (effective single-particle absorption thickness) function is

$$W = Q_A/Q_S = (1-w)/w \tag{14.22}$$

was defined in Chapter 6. It was shown that this quantity is approximately proportional to α over a larger range of $\alpha\langle D\rangle$ than Q_A, so that in this linear region we may write

$$W \simeq \alpha D_e, \tag{14.23}$$

where D_e is an effective particle size of the order of twice the actual particle size. Combining (14.29) and (14.30) leads to an approximate expression for w that is valid in the linear region,

$$w = \frac{1}{1+W} \simeq \frac{1}{1+\alpha D_e}. \tag{14.24}$$

W is plotted against $\alpha\langle D\rangle$ in Figure 14.10b, which shows the linear region and also the departures from linearity for several values of s and $\langle D\rangle$.

The espat function has been used to analyze remotely sensed photometric data on Jupiter's satellite Europa by Johnson *et al.* (1988).

If the medium consists of a mixture of particles (either by composition or size or both), then the espat function that will be deduced from the measured volume single-scattering albedo is, from equation (10.56),

$$W = \frac{1-w}{w} = \left(\sum_j \frac{\mathcal{M}_j w_j}{\rho_j D_j} W_j\right) \Big/ \left(\sum_j \frac{\mathcal{M}_j w_j}{\rho_j D_j}\right)$$
$$\simeq \left(\sum_j \frac{\mathcal{M}_j \alpha_j D_{ej}}{\rho_j D_j (1+\alpha_j D_{ej})}\right) \Big/ \left(\sum_j \frac{\mathcal{M}_j}{\rho_j D_j (1+\alpha_j D_{ej})}\right), \tag{14.25}$$

where the subscript j denotes the property of the jth type of particle.

For a monomineralic medium with a small particle size distribution the weighting functions in (14.25) are equal, and $W \simeq \alpha D_e$. If $\alpha(\lambda)$ is known at some wavelength, then D_e of a powder can be found by measuring $W(\lambda)$ of the powder at the same wavelengths. Once D_e is known $\alpha(\lambda)$ can be found from the measured reflectance over the entire range of wavelengths in which the particles are volume scatterers.

Two important limitations of the espat function must be emphasized. First, Figure 14.10b shows that W is linearly proportional to α only when neither $\alpha\langle D\rangle$ nor $s\langle D\rangle$ is large; that is, the particles must not be optically thick. Second, if a medium is not monominerallic or if it has a wide particle size distribution, then equation (14.25) shows that the volume-average espat function is proportional to the weighted sum of the absorption coefficients of the individual components of a mixture *only* if $\alpha D_e \ll 1$ for *each* component.

14.7.4 An example of retrieving n_i from the reflectance

To illustrate these concepts, let us turn to a material whose properties are well known and which is of practical interest in remote sensing: water ice. The measured complex refractive index of H_2O ice as a function of wavelength is shown as the solid and dashed lines in Figure 14.11a. The reflectance spectrum of an H_2O frost, measured over the same wavelength range, is shown as the solid line in Figure 14.11b. The reflectance and refractive index spectra were measured independently of each other. The absorption spectrum shows a strong fundamental vibrational band at 3.08 μm, plus a number of overtone bands whose strength decreases with decreasing wavelength in the near infrared. It also shows part of a strong electron-excitation band in the vacuum ultraviolet. In reflectance, the infrared fundamental band is expressed as a maximum at 3.15 μm, with two associated transition minima at 2.85 and 3.45 μm. The 2.85 minimum is lower than the 3.45 minimum. The weaker remaining bands are all in the volume-scattering regime and are expressed as minima.

As a test of the ability of the models to correctly predict the reflectance of a material, the refractive index of H_2O ice was inserted into equations (14.14) and (14.15) for w, assuming $s = 0$, and the diffusive-reflectance spectrum $r_0(\lambda) = (1-\gamma)/(1+\gamma)$ was calculated. The diffusive reflectance was used because angular information on portions of the reflectance was not available. The particle size of the sample in Figure 14.11b was not known, hence D_e was taken to be an unknown parameter. The value that gave the best fit was $D_e = 125$ μm. The calculated reflectance spectrum is shown as the points in Figure 14.11b.

As a test of the ability of the method to retrieve the imaginary refractive index, the measured reflectance spectrum of ice in Figure 14.11b was inverted. The reflectance was assumed to be given to sufficient accuracy by $r_0(\lambda)$ and equation (14.4) used to calculate $(w\lambda)$ from the reflectance. The real part of the index of refraction was approximated as constant at $n_r = 1.3$, independent of wavelength; D_e was assumed to be 125 μm, and equation (14.21) was used to calculate α. From the dispersion relation, n_i was then found. The result is shown as the points in Figure 14.11a. The agreement is again seen to be excellent, except at two wavelengths: the strong fundamental band, and in the visible. The reason for the less satisfactory agreement at 3 μm is because n_r was assumed to be constant, so that the anomalous dispersion of n_r was neglected. The recovery of n_i in this region could be improved by using the Kramers–Kronig relations (Section 4.4). The reasons for poor agreement in the visible are the large systematic measurement errors when the reflectance is close to 1, as discussed in Section 14.7.2.

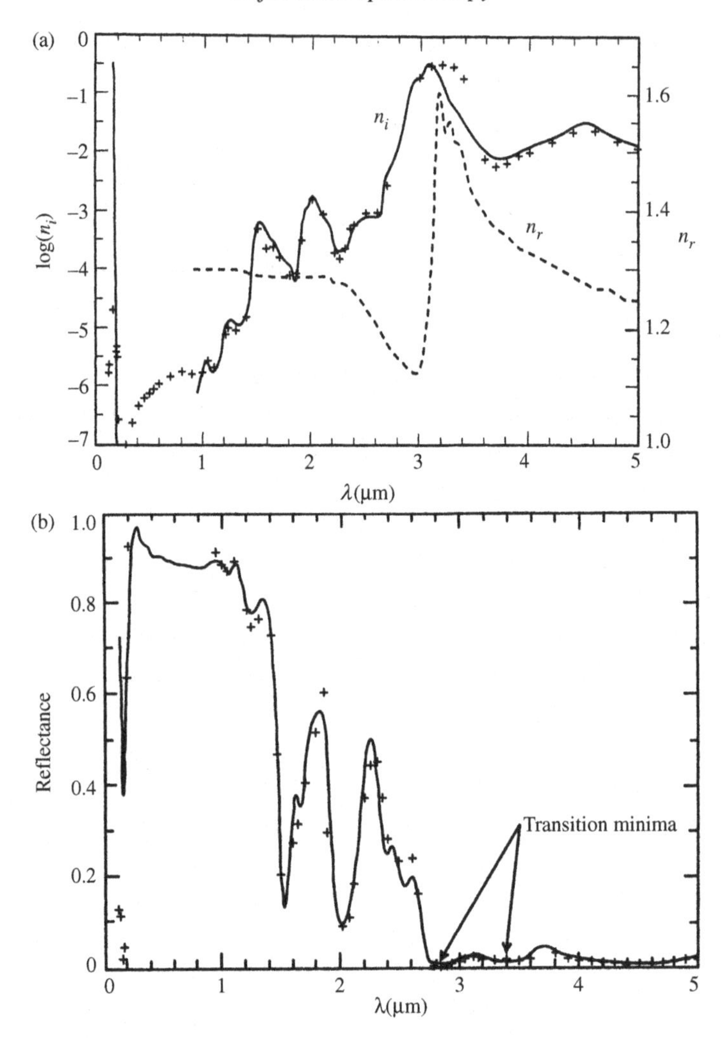

Figure 14.11 (a) Complex refractive index of water ice. Dashed line, measured n_r; solid line, measured n_i; crosses, n_i calculated from the reflectance of frost assuming n_r is constant. Data from Irvine and Pollack (1968) and Browell and Anderson (1975). (b) Spectral reflectance of water frost. Solid line, measured reflectance; crosses, diffusive reflectance calculated from the measured refractive index shown in Figure 11.10a, assuming $\langle D \rangle = 125\,\mu$m. The arrows show the two transition minima on either side of the band center. Data from Smythe (1975) and Hapke *et al.* (1981).

14.8 Other methodologies

14.8.1 Kubelka–Munk theory

Several other methodologies have been suggested for retrieving the absorption coefficient from the reflectance. The oldest and most widely used is the pioneering model by Kubelka and Munk (1931); see also Wendtland and Hecht (1966) and Kortum (1969). The KM model of is actually a form of the diffusive reflectance and is subject to the same inherent limitations.

The KM model is a form of the two-stream solution to the radiative-transfer equation. The radiances traveling into the upward and downward directions are denoted by I_1 and I_2, respectively. Then their divergences are $\frac{1}{2}(dI_1/dz)$ and $-\frac{1}{2}(dI_2/dz)$, respectively. The factors $\frac{1}{2}$ and $-\frac{1}{2}$ arise from averaging the cosine of the direction of propagation over the upward and downward hemispheres (see Section 8.7). The model contains two parameters, the volume absorption coefficient A_{KM}, and the volume scattering coefficient S_{KM}. Then the extinction coefficient is $(S_{KM}+A_{KM})$. In KM theory, A_{KM} is assumed to be equal to $\langle\alpha\rangle$, the true absorption coefficient α inside the particles of the medium reduced by averaging over a volume large compared with that of a single particle, and S_{KM} is assumed to be caused by undefined, uniformly distributed scattering centers that are entirely independent of the absorption. The radiance is assumed to be scattered only if its direction of propagation has been changed from the hemisphere into which it was moving to the oppositely-going hemisphere. Light that is scattered into the same hemisphere is interpreted as unscattered. As in the diffusive reflectance, the volume source term $\mathcal{F}$ is zero. The only source is the incident irradiance converted as it passes through the upper surface into a uniform diffuse radiance emerging into the downward-going hemisphere, which constitutes the boundary condition at the surface.

With these assumptions, the equations governing the radiance consist of two coupled differential equations,

$$\frac{1}{2}\frac{dI_1}{dz} = -(A_{KM}+S_{KM})I_1 + S_{KM}I_2, \tag{14.26a}$$

$$-\frac{1}{2}\frac{dI_1}{dz} = -(A_{KM}+S_{KM})I_2 + S_{KM}I_1. \tag{14.26b}$$

Let

$$d\tau_{KM} = -(A_{KM}+2S_{KM})dz, \tag{14.27a}$$

$$w_{KM} = \frac{2S_{KM}}{A_{KM}+2S_{KM}}, \tag{14.27b}$$

and

$$\gamma_{KM} = \sqrt{1-w_{KM}}. \tag{14.27c}$$

Then (14.26) can be put into the form

$$-\frac{1}{2}\frac{dI_1}{d\tau_{KM}} = -I_1 + \frac{w_{KM}}{2}(I_1+I_2), \tag{14.28a}$$

$$\frac{1}{2}\frac{dI_1}{d\tau_{KM}} = -I_2 + \frac{w_{KM}}{2}(I_2+I_1). \tag{14.28b}$$

Comparing equations (14.28) with equations (8.18) for the diffusive reflectance we see that they have exactly the same form. Because the boundary conditions are also the same, we may write down the expression for the *Kubelka–Munk reflectance* by comparison with (8.25),

$$r_{KM} = \frac{1-\gamma_{KM}}{1+\gamma_{KM}} = \frac{1-\sqrt{1-2S_{KM}/(A_{KM}+2S_{KM})}}{1+\sqrt{1-2S_{KM}/(A_{KM}+2S_{KM})}} = \frac{1-\sqrt{A_{KM}/(A_{KM}+2S_{KM})}}{1+\sqrt{A_{KM}/(A_{KM}+2S_{KM})}}. \tag{14.29}$$

Solving r_{KM} for A_{KM}/S_{KM} gives the so-called *Kubelka–Munk remission function*, $f(r_{KM})$,

$$f(r_{KM}) = \frac{A_{KM}}{S_{KM}} = \frac{(1-r_{KM})^2}{2r_{KM}}. \tag{14.30}$$

By assumption, A_{KM} is identified with the volume-averaged absorption coefficient $\langle\alpha\rangle$, and S_{KM} is independent of $\langle\alpha\rangle$; thus, $f(r_{KM})$ should be proportional to $\langle\alpha\rangle$, with the constant of proportionality equal to $1/S_{KM}$. In particular, the true absorption coefficient α of a monominerallic medium should be proportional to $f(r_{KM})$. Experimentally, it is found that if $f(r_{KM})$ is calculated from the measured directional–hemispherical reflectance, while α is independently measured by transmission, the two quantities are indeed proportional to each other for small values of $f(r_{KM})$.

However, with increasing absorption, the slope of the remission function decreases, and the curve saturates. There has been a great deal of discussion in the literature as to the reason for this failure at larger absorptions. The usual explanation is that the scattering coefficient S_{KM} is somehow "wavelength-dependent."

Because the proportionality between α and $f(r_{KM})$ was found experimentally to be valid only for small values of the remission function, a technique called the *dilution method* that attempts to measure larger values of α by reflectance was developed. A small number of the strongly absorbing particles are mixed with a sufficient quantity of weakly absorbing particles of some other material to bring the mixture into the linear region. However, although this somewhat extends the accuracy, in practice the amount of improvement is found to be minor.

The principal difficulty with KM theory is that the parameters A_{KM} and S_{KM} are completely misinterpreted. This is the source of the linearity failure in the remission function and lack of improvement provided by the dilution method. The problem is that the absorption and scattering are not distributed evenly throughout the medium, as assumed by the theory, but are localized into particles. If the KM differential equations are compared with the diffusive reflectance equations in Chapter 8 for a particulate medium, then A_{KM} is seen to be equivalent to A, and S_{KM} to $S/2$.

Thus, A_{KM} is *not* equal to the average internal absorption coefficient of the particles, but is the volume absorption coefficient $A = \sum_j N_j \sigma_j Q_{Aj}$, which is an entirely different quantity. Similarly, $S_{KM} = S/2 = \sum_j N_j \sigma_j Q_{Sj}/2$. Furthermore, because for large particles $Q_{Sj} + Q_{Aj} = Q_{Ej} = 1$, S_{KM} is *not* independent of A_{KM}, but is coupled to it, $S_{KM} = \sum_j N_j \sigma_j (1 - Q_{Aj})/2 = (\sum_j N_j \sigma_j - A_{KM})/2$. For large particles, the reason that S_{KM} is not constant has nothing to do with wavelength-dependent scattering, but occurs because S_{KM} and A_{KM} *both* depend on the absorption.

From equation (14.4b), the single-scattering albedo is seen to be related to the diffusive reflectance by $w = 4r_0/(1+r_0)^2$. Hence, for the diffusive reflectance the espat function is given by

$$W = \frac{1-w}{w} = \frac{(1-r_0)^2}{4r_0}. \tag{14.31}$$

Comparing (14.31) with (14.30) shows that the remission function is equal to twice the espat function, the factor of 2 arising from the differing definitions of S and S_{KM}. We can now understand why the remission function behaves as it does. It is equal to twice the espat function and hence is subject to the same limitations. It is proportional to the internal absorption coefficient of a monomineralic material if the reflectance is in the volume-scattering region, but it saturates as the reflectance enters the weak surface-scattering region. Even though Kubelka–Munk theory misinterprets the physical nature of the remission function, $f(r_{KM})$ turns out to be proportional to the true particle absorption coefficient in the volume-scattering region because of the fortuitous mathematical behaviors of Q_A and Q_S. This is why the theory is useful.

We can also understand why the dilution method does not appreciably improve the linearity. It was shown in Section 14.4 that the weighting factors of α_j in the espat of a mixture are independent of α_j only if $\alpha_j D_{ej} = 1$ for all components. Suppose a small amount of strongly absorbing material, with properties denoted by subscript 2, is mixed with a large amount of material with properties denoted by subscript 1. Material 1 has a high albedo with no aborption bands in the wavelength range of interest. Then the espat function of the mixture is, from equation (14.25),

$$W = \frac{N_1\sigma_1(1-w_1) + N_2\sigma_2(1-w_2)}{N_1\sigma_1 w_1 + N_2\sigma_2 w_2} = \frac{W_1 + CW_2}{1+C} \approx (1-C)W_1 + CW_2 \tag{14.32}$$

to first order in C, where $C = N_2\sigma_2 w_2/N_1\sigma_1 w_1 << 1$. Thus, W is proportioinal to W_2, which is proportional to α_2 if $\alpha_2 D_2$ is small enough to be in the volume-scattering region. However, w_2 and W_2 saturate if the absorbance of material 2 is in the weak surface-scattering region, irrespective of the presence of the dilutant. The only way to prevent saturation is to grind both materials so finely that they fall in the volume-scattering region. But a dilutant is not required to accomplish this, so the whole dilution method appears to be superfluous.

Thus, the dilution method offers no substantial advantage in extending the linear region of the remission function. Instead, the particles should be ground to a small enough size that the espat function will be in the volume-scattering region; that is, $\alpha D \lesssim 3$ for the largest anticipated value of α. If $n_i \geq 0.1$, this means that $D \lesssim 2.5\lambda$. Larger values of n_i will place the reflectance in the strong surface-scattering region and allow the retrieval of n_i by Kramers–Kronig analysis (Section 4.4).

14.8.2 The Shkuratov albedo model

The Shkuratov albedo model (Shkuratov *et al.*, 1999b) is an analytical model for the reflectance of a particulate material. Angular properties are unspecified. The medium is assumed to consist of parallel layers consisting of discrete particles and the spaces between them. Light transmitted and absorbed by individual layers are summed to give the reflectance.

Let R_b and R_f be the fraction of externally incident light specularly reflected from the particle surfaces in the backward and forward directions, respectively, and R_i is the fraction of internally incident light reflected. Shkuratov gives the following empirical approximations:

$$R_b = (0.28n_r - 0.20)R_e,$$
$$R_e = R_0 + 0.05,$$
$$R_f = R_e - R_b,$$
$$R_i = 1.04 - 1/n_r^2,$$

where $R_0 = (n_r - 1)^2/(n_r + 1)^2$. Let

$$T_e = 1 - R_e,$$
$$T_i = 1 - R_i.$$

Let r_b and r_f be the fractions of light scattered by an average particle into the backward and forward directions, respectively. Shkuratov shows that

$$r_b = R_b + \frac{1}{2}T_e T_i R_i \exp(-2\tau)/(1 - R_i \exp(-\tau)),$$
$$r_f = R_f + T_e T_i \exp(-\tau) + \frac{1}{2}T_e T_i R_i \exp(-2\tau)/(1 - R_i \exp(-\tau)),$$

where $\tau = \alpha\langle D\rangle$, and $\langle D\rangle$ is the average path length through the particle. Next, let ρ_b and ρ_f be the fraction of light scattered into the backward and forward directions by a layer. Shkuratov gives

$$\rho_b = \phi r_b,$$
$$\rho_f = \phi r_f + 1 - \phi.$$

where ϕ is the filling factor. Then the albedo is

$$A = \frac{1-\rho_b^2-\rho_f^2}{2\rho_b} - \sqrt{\frac{1-\rho_b^2-\rho_f^2}{2\rho_b} - 1}.$$

The albedo may be inverted to find the absorption coefficient,

$$\alpha = -\frac{1}{<D>} \ln\left[\frac{b}{a} + \sqrt{\left(\frac{b}{a}\right)^2 - \frac{c}{a}}\right],$$

where

$$\begin{aligned}
a &= T_e T_i (yR_i + \phi T_e),\\
b &= yR_b R_i + \frac{\phi}{2} T_e^2 (1+T_i) - T_e(1-\phi R_b),\\
c &= 2yR_b - 2T_e(1-\phi R_b) + \phi T_e^2,\\
y &= (1-A)^2/2A.
\end{aligned}$$

A difficulty with the Shkuratov albedo model is that the reflectance is predicted to be relatively insensitive to porosity (Shkuratov *et al.*, 1999b), in contrast with the experimental data which exhibit strong porosity-dependence. Also, the equations are considerably more algebraically complex than the comparably accurate diffusive reflectance model.

14.8.3 The mean optical path length (MOPL)

Many authors have noted the similarity between reflectance in the volume-scattering region and transmittance. In both cases, absorption affects the intensity by transmission through the transparent material. The amount of light that penetrates through a thickness x of material is given by the transmittance $t = \exp(-\alpha x)$, so that $-\ln t \propto \alpha$. The quantity $\alpha x = -\ln t$ in a transmission measurement is called the *absorbance*, and $x = -(1/\alpha)\ln t$ is the *optical path*. For reflectance, Kortum (1969) defines a quantity analogous to the absorbance, which he calls the *effective absorbance* $A_e = -\ln r$, where r is a reflectance. Similarly, Clark and Roush (1984) define the mean optical path length as

$$\text{MOPL} = \frac{A_e}{\alpha} = -\frac{1}{\alpha} \ln r. \tag{14.33}$$

If $\alpha \propto -\ln r$, then the MOPL should be a constant for a given material as α changes. Let us see if the MOPL is approximately constant anywhere in the volume-scattering region. If so, then $-\ln r$ should be approximately proportional to α, which would indeed be a useful relation. We shall again use the diffusive reflectance r_0 because

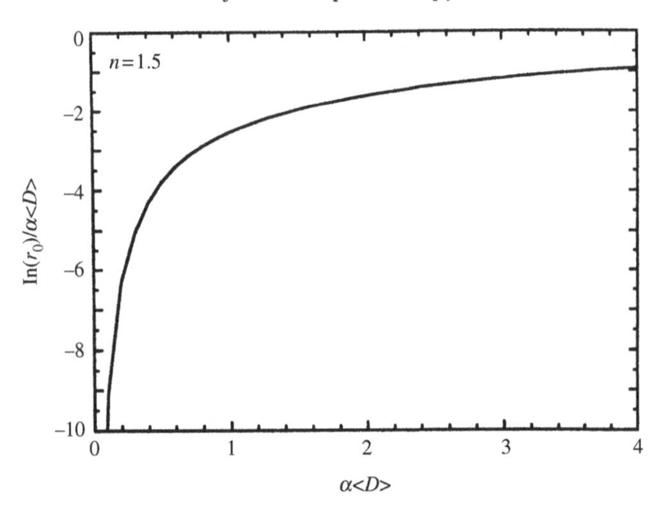

Figure 14.12 Mean optical path length $\ln r_0$ divided by $\alpha\langle D\rangle$ plotted against $\alpha\langle D\rangle$.

its behavior is representative of most types of reflectances, and it is mathematically simple.

In the volume-scattering region, $w \simeq 1/(1+\alpha D_e)$. If $\alpha D_e \ll 1$, $\gamma = \sqrt{1-w} \simeq \sqrt{\alpha D_e}$. Hence, $r_0 \simeq (1-\gamma)(1+\gamma)$; $1-2\gamma \simeq e^{-2r}$, and MOPL $\simeq 2(D_e/\alpha)^{1/2}$. At the opposite extreme, if αD_e is large enough that only single scattering contributes appreciably, then $r_0 \simeq w/4$, and MOPL $\simeq [\ln 4 + \ln(1+\alpha D_e)]/\alpha$. In neither case is the MOPL even approximately constant. However, it does have one virtue, as pointed out by Clark and Roush (1984): if an absorption band has a Gaussian shape, the effective absorbance of that band will also be approximately Gaussian.

Figure 14.12 shows a plot of $(\ln r_0)/\alpha\langle D\rangle$ vs. $\alpha\langle D\rangle$ for a powder in which $s=0$. There does not seem to be any region in which the MOPL is even roughly constant. Hence, neither the effective absorbance nor the MOPL seems to be a particularly useful quantity especially when contrasted with the espat function.

14.9 Particulate media with $X \ll 1$

The case of a medium of closely packed solid particles for which the mean $X \ll 1$ is not well understood. Many workers have attempted to apply effective-medium theories to this type of material. However, it was emphasized in Section 7.1 that effective-medium theories are inadequate because they do not take scattering into account. The limited experimental evidence discussed in Section 7.5.1 suggests that the effective sizes of the scatterers in media of very small particles are agglomerates of the order of the wavelength. It was also suggested, although this needs to be verified experimentally, that an effective-medium theory be used to calculate the refractive index n, which may then be inserted into Mie theory to calculate the w and $p(g)$ of the agglomerates. The latter quantities may then be used in the appropriate expression for the reflectance.

This discussion suggests that if $n_i \ll 1$, then the espat function W should be proportional to α, even for small particles. However, the value of the constant of proportionality D_e is not clear; it may be of the order of λ.

The situation is even more confusing when n_i is not small. Under certain conditions, $w \ll 1$, as evidenced by the fact that fine metallic powders, such as gold black, are among the darkest materials known. This is consistant with predictions of Mie theory and suggests that each particle scatters and absorbs quasi-independently.

14.10 Planetary applications

14.10.1 The near-IR Fe^{2+} *bands in silicates*

The absorption band in the vicinity of 1000 nm in silicates is one of the most important spectral features for the compositional remote sensing of bodies of the solar system whose surfaces can be seen from space. It is a weak, forbidden band caused by the excitation of electrons in the Fe^{2+} ion coordinated by six O^{-2} ions in pyroxene, olivine, and feldspar (Burns, 1993). In pyroxenes the minimum of the band is sensitive to the amount of Ca^{2+}, being at 900 nm in low-Ca orthopyroxene and at 950 nm in high-Ca clinopyroxene. Pyroxene has a second band near 2000 nm. The 950-nm band was first detected in the reflectance spectrum of the Moon by McCord and Adams (Adams, 1968) and was one of the early indications that the crust of the Moon was mafic in composition. The band in olivine is a composite of three closely spaced bands with the minimum lying slightly longward of 1000 nm. The band in anorthositic feldspar is at 1250 nm, where it is caused by ferrous impurities, and so is extremely weak and difficult to detect remotely.

Anorthite, clinopyroxene, and olivine are three of the major minerals in the regolith of the Moon, the other important constituents being ilmenite and impact-melted glass. Also present in the regolith at the 0.5% level are tiny grains of submicroscopic metallic Fe (SMFe).which range in size from a few nanometers to about a micrometer. (The SMFe is sometimes also denoted by $npFe^0$, standing for nano-phase metallic iron. This notation is rather curious, since "phase" usually refers to the structure of a material – e.g., solid, liquid, or gas, not size.) Although the SMFe is only a minor component by mass, it has a major effect on the spectrum.

14.10.2 The modified Gaussian method

We have seen that the theoretical form of the absorption coefficient, expressed in frequency, of a pure absorption band is Lorentzian. However, in practice, the bands often have a Gaussian-like shape. Although each individual oscillator has a Lorentzian shape, their band centers and widths are shifted randomly by lattice distortions, thermal vibrations, and impurities, so that the aggregate of them resembles a Gaussian. Other distortions of shape occur during the translation of the bands from absorption coefficient of individual grains to reflectance of a powder.

Sunshine and her colleagues (Sunshine *et al.*, 1990; Sunshine and Pieters, 1993) studied the systematics of the 1000-nm band in mixtures of pulverized pyroxene and olivine. They found that Gaussians describe the bands poorly in frequency, but were a good fit when the spectra were expressed as functions of wavelength. In the modified Gaussian method the measured spectrum is fitted in wavelength space by a series of Gaussian bands superposed on a continuum consisting of a series of straight lines. This is facilitated by a computer program that finds the band centers and widths, as well as the intercepts and slopes of the continuum lines, that give the best fit. The bands of an observed spectrum of a planetary regolith resolved in this way can then be compared to those of pure candidate materials to find the composition.

14.10.3 Space weathering

The space environment is often thought to be inert. In fact, the surface of a body without an atmosphere is subject to a variety of processes that alter its physical state and reflectance spectrum. These processes are known collectively as *space weathering,* and consist of meteoritic impact comminution, vitrification and vaporization, sputtering by the solar wind, deposition of the vapors generated by impact and sputtering, and irradiation by energetic nuclei of solar and cosmic origin (see review by Hapke, 2001).

Hypervelocity meteorite impacts grind up a pristine solid rock surface, which increases the albedo and alters the depths of absorption bands as described in this chapter. Impact-melted rocks have a higher albedo and bands that are wider and deeper than those of the parent rocks. Many of the regolith particles are agglutinates, which are agglomerates of smaller particles held together by melt glass. Most of the vapor generated by impacts and sputtering does not leave the Moon, but coats the particles of the regolith. The vapor deposition process reduces some of the Fe^{2+} to Fe^0 and produces the SMFe particles (Hapke *et al.*, 1975; Hapke, 2001). According to the results of Chapter 5, the absorption efficiency of particles smaller than the wavelength is proportional to $1/\lambda$. Hence, the coatings absorb more strongly at shorter wavelengths and cause the reflectance spectrum of the Moon to be redder than the incident sunlight. They also obscure the absorption bands of the mineral and glass particles they coat.

Over time the optical effects of space weathering have caused the regoliths of the Moon and Mercury to become dark and slightly reddish and are slowly darkening bright craters and their rays. The weathering is countered by the addition of fresh unaltered material brought up from below the surface by impacts. When the two processes are in equilibrium the regolith is said to be *mature*. In the asteroid belt the velocities of most of the impacts are thought to be too low to cause appreciable melting or vaporization. However, the solar wind still causes sputter vaporization and deposition, although at a smaller rate than on the Moon. Thus, it takes much longer for the regoliths of asteroids to mature. Space weathering by comminution and sputter deposition can alter the spectrum of a material with the composition of chondritic meteorites, the most common type of meteorite, so that it is similar to that of S-asteroids, the most common type of asteroid (Hapke, 2001; Chapman, 2004).

Now, the scattering efficiency of the SMFe particles is proportional to $(D/\lambda)^4$, so that they have a negligible effect on the scattering coefficient of the medium. Their effect on the absorption coefficient can be calculated (Hapke, 2001) using the Maxwell-Garnet effective-medium model (equation 7.5) for the dielectric constant K_e of a material containing SMFe,

$$K_e = K_{eh} + \frac{3\phi_{Fe} K_{eh} [(K_{eFe} - K_{eh}) / (K_{eFe} + 2K_{eh})]}{1 - \phi_{Fe} [(K_{eFe} - K_{eh}) / (K_{eFe} + 2K_{eh})]},$$

where ϕ_{Fe} is the fraction of SMFe by volume, subscript h refers to the host material and subscript Fe to the SMFe. The complex refractive index is $n = n_r + in_i = \sqrt{K_e}$, and the absorption coefficient is $\alpha = 4\pi n_i/\lambda$. Since $\phi_{Fe} \ll 1$, it is sufficiently accurate to keep only terms to first order in this quantity. Then the calculation of the absorption coefficient is straightforward and gives

$$\alpha = \alpha_h + \frac{36\pi}{\lambda}\phi_{Fe}\frac{n_{rh}^3 n_{rFe} n_{iFe}}{(n_{rFe}^2 - n_{iFe}^2 + 2n_{rh}^2)^2 + (2n_{rFe}n_{iFe})^2}.$$

This quantity turns out to be independent of the size distribution of the SMFe particles so long as they are $< \lambda$.

14.10.4 The spectral-ratio–albedo diagram

The spectral-ratio–albedo diagram was developed by P. Lucey and his colleagues (Lucey *et al.*, 1995, 1998, 2000), and is a convenient and powerful, empirical method for estimating the amount of FeO and degree of maturity of a regolith from its visible and near-IR spectrum. An example of the diagram is shown in Figure 14.13. It plots the ratio $r(\mathrm{IR}))/r(\mathrm{V})$ against $r(\mathrm{V})$, where $r(\mathrm{V})$ is the reflectance in visible light at the wavelength ($\sim 750\,\mathrm{nm}$) where the reflectance is

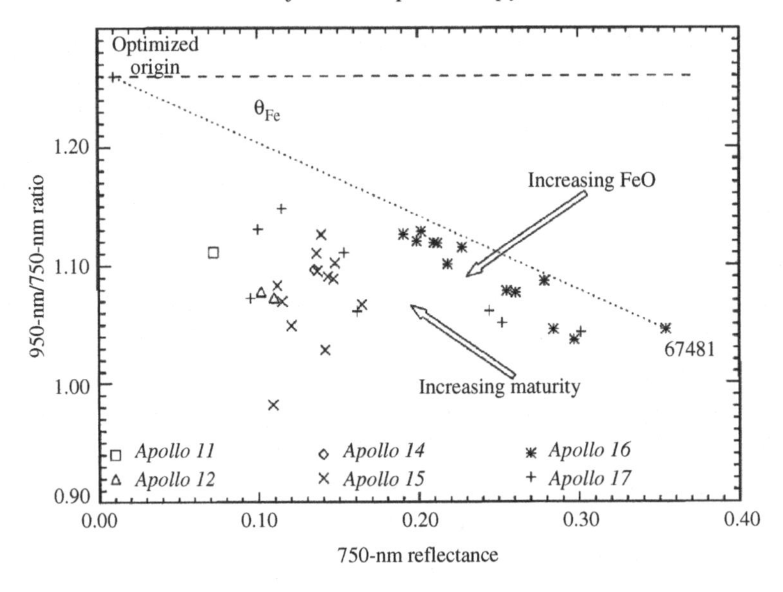

Figure 14.13 The spectral-ratio–albedo diagram for *Apollo* lunar samples. (Reproduced from Blewett *et al.* [1997], copyright 1997 by the American Geophysical Union.)

maximum just before it begins to decrease into the 1000-nm ferrous band, and $r(\mathrm{IR})$ is the reflectance at the center of the band ($\sim 950\,\mathrm{nm}$).

Lucey *et al.* analyzed the spectra of a large number of terrestrial, lunar, and meteoritic minerals and rocks and discovered that materials with the same FeO content all fall approximately on a line radial to a certain point, called the "optimized origin," on the diagram. Increasing the FeO increases the band depth, which decreases the spectral ratio, and decreases the albedo. This moves the line downward and to the left in such a manner that the points on it remain radial to the apparent origin point. That is, the line rotates clockwise as the FeO increases. On the other hand, adding SMFe by space weathering reduces the visible reflectance on the short-wavelength wing of the band more than at the band center, so that so that the band depth decreases, causing the point corresponding to a material of given composition to move upward and to the left. The material remains approximately on its original, constant-FeO line as the weathering increases. The advantage of the spectral-ratio–albedo diagram is that spectral differences due to FeO composition are approximately orthogonal to changes caused by space weathering, so that the two effects may be separated.

Lucey *et al.* define two parameters on the diagram: the spectral Fe parameter θ_{Fe}, which is the angle between a horizontal line through the apparent origin and a constant FeO line, and the optical maturity parameter OMAT, which is the distance along a constant FeO line and the apparent origin,

$$\mathrm{OMAT} = \left\{ [r(IR)/r(V) - y_0]^2 + [r(V) - x_0]^2 \right\}^{1/2},$$

where x_0 and y_0 are the coordinates of the apparent origin. The θ_{Fe} and OMAT parameters are found to be linearly correlated with FeO content and maturity, respectively. The correlation coefficients and the location of the apparent origin must be calibrated empirically and are specific to each data set. For the *Clementine* observations of the Moon the calibration was done by comparing the spacecraft observations of *Apollo* landing sites with laboratory measurements of samples from those same sites.

15

Thermal emission and emittance spectroscopy

15.1 Introduction

The region of the electromagnetic spectrum in the vicinity of 10-μm wavelength is referred to as the *mid-infrared*, but is also called the *thermal infrared* because objects at the temperature of the Earth's surface emit radiation strongly there. It is important because many materials have strong vibrational absorption bands at these wavelengths (Chapter 3). In most remote-sensing measurements these bands can be detected only through their effects on the radiation that is thermally emitted by the planetary surface being studied. Many substances have overtone or combinations of these bands at shorter wavelengths; although they can be observed in reflected light, their depths and shapes may be affected by emitted thermal radiation. It will be seen that there are complementary relations, known as Kirchhoff's laws, between reflectance and emissivity at the same wavelength. Hence, much of the preceding discussions of reflectance can also be applied to emissivity.

Figure 15.1 shows the spectrum of sunlight reflected from a surface with a diffusive reflectance of 10%, compared with the spectrum of thermal emission from a black body in radiative equilibrium with the sunlight, at various distances from the Sun. Clearly, thermal emission can be ignored at short wavelengths, and reflected sunlight at long, but at intermediate wavelengths the radiance received by a detector viewing the surface includes both sources.

In this chapter, expressions will be derived for the radiant power received by a detector viewing a particulate medium, such as a powder in the laboratory or a planetary regolith, when either or both reflected sunlight and thermally emitted radiation are present. The reflectance models developed in previous chapters will be extended to include the effects of thermal radiation. One of the effects of large-scale surface roughness is its ability to cause a thermal shadow-hiding opposition effect, which will be discussed near the end of the chapter. However, a general treatment of macroscopic surface roughness is beyond the scope of this book.

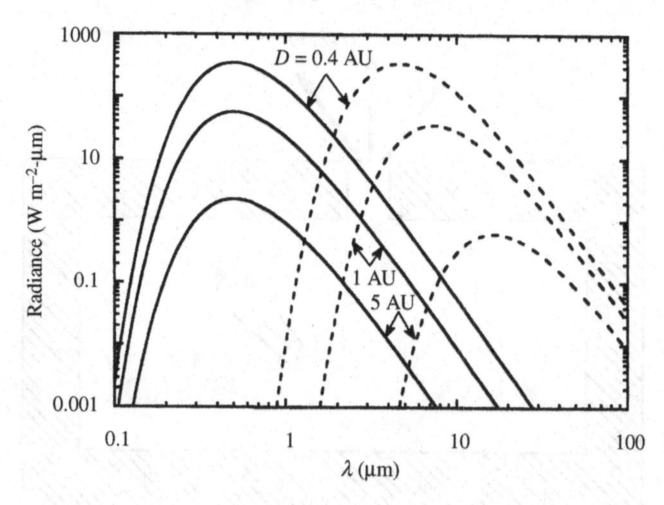

Figure 15.1 Comparison of sunlight reflected (solid lines) from a surface with a visual albedo of 0.1 with the radiation thermally emitted (dashed lines) from the surface with an IR emissivity of 1.00, for three different distances from the Sun.

There is a large body of literature that treats thermal emission and radiation transfer in planetary and stellar atmospheres (e.g., Chandrasekhar, 1960; Goody, 1964; Sobolev, 1975; Van de Hulst, 1980). Theoretical treatments of emittance from a particulate medium have been published by several authors, including Vincent and Hunt (1968), Conel (1969), Emslie and Aronson (1973), and Aronson *et al.* (1979). However, virtually all of the published theories of emittance by powders derive only the hemispherically integrated radiance, whereas it is the directional emittance that is measured remotely. The situation in which both reflectance and emittance contribute to the radiance has hardly been discussed at all.

15.2 Black-body thermal radiation

Suppose a hollow cavity is surrounded by material that is optically thick at all wavelengths and is heated to a uniform absolute temperature T (Figure 15.2). Then it is found that the spectral radiance in the cavity is given by

$$I(\lambda, T) = \frac{1}{\pi} U(\lambda, T), \tag{15.1a}$$

where $U(\lambda, T)$ is the Planck function,

$$U(\lambda, T) = \frac{2\pi h_0 c_0^2}{\lambda^5} \frac{1}{e^{hc_0/\lambda k_0 T} - 1}, \tag{15.1b}$$

h_0 is Planck's constant ($h_0 = 6.626 \times 10^{-34}$ J sec), c_0 is the speed of light, and k_0 is Boltzmann's constant ($k_0 = 1.381 \times 10^{-23}$ J K^{-1}). The quantities $c_1 = 2\pi h_0 c_0^2 =$

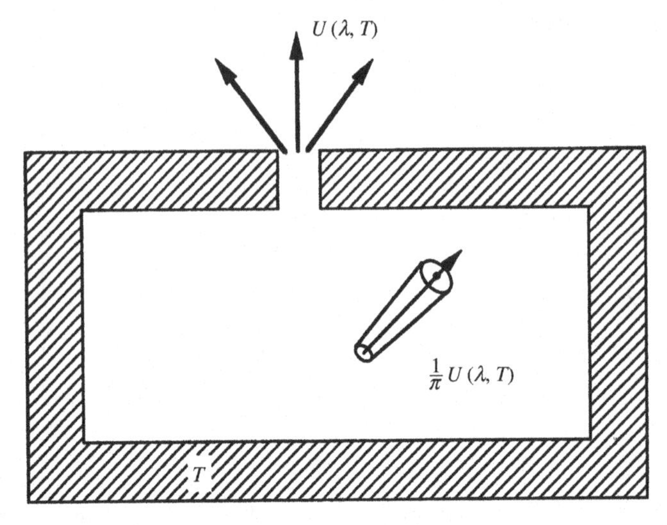

Figure 15.2 Thermal radiance inside a black-body enclosure whose walls are at temperature T and the power per unit area emerging from a small hole in the wall.

$3.742 \times 10^{-16}\,\mathrm{W\,m^2}$ and $c_2 = h_0 c_0 / k_0 = 0.01439\,\mathrm{m\,K}$ are called, respectively, the *first* and *second radiation constants*. The unit of $U(\lambda, T)$ is power per unit area per unit wavelength interval. The radiance in the cavity is found to be independent of direction, position, shape of the cavity, and composition of its walls.

If a small hole is drilled through one of the walls of the cavity (Figure 15.2), the radiant power per unit area emerging into the hemisphere above the hole is

$$P_{\mathrm{em}}(\lambda, T) = \int_0^{\pi/2} I(\lambda, T) \cos 2\pi \sin e\, d\, e = U(\lambda, T). \tag{15.2}$$

Such a hole approximates an ideal surface that does not reflect any light incident on it and that emits thermal radiation with 100% efficiency, and so is called a *black body*. The radiation described by $U(\lambda, T)$ is *black-body thermal radiation,* and the cavity plus its walls is a *black-body enclosure* or *isothermal enclosure.*

At short wavelengths the exponential in the denominator of (15.1b) becomes large, and

$$U(\lambda, T) \simeq \frac{2\pi h_0 c_0^2}{\lambda^5} e^{-h_0 c_0/\lambda k_0 T}. \tag{15.3a}$$

At long wavelengths this exponential may be expanded as $e^x \simeq 1 + x$, to give

$$U(\lambda, T) \simeq \frac{2\pi c_0 k_0 T}{\lambda^4}, \tag{15.3b}$$

which is known as the *Rayleigh–Jeans law.*

Setting the derivative $\partial U(\lambda, T)/\partial\lambda = 0$, shows that the Planck function has a maximum at

$$\lambda = 2898/T \ \mu\text{m}, \tag{15.3c}$$

where T is degrees Kelvin. This relation, which is known as the *Wien displacement law*, shows that the wavelength of the maximum shifts toward shorter values as the temperature increases.

The total power per unit area $V(T)$ emitted from the surface of a black body can be found by integrating $U(\lambda, T)$ over all wavelengths:

$$V(T) = \int_0^\infty U(\lambda, T)d\lambda.$$

This integral may be evaluated by making the substitution $x = h_0c_0/\lambda k_0T$, giving

$$V(T) = \frac{2\pi k_0^4 T^4}{h_0^3 c_0^2}\int_0^\infty x^3(e^x - 1)^{-1}dx.$$

But

$$\begin{aligned} x^3(e^x - 1)^{-1} &= x^3e^{-x}(1 - e^{-x})^{-1} \\ &= x^3e^{-x}(1 + e^{-x} + e^{-2x} + \cdots) \\ &= x^3(e^{-x} + e^{-2x} + e^{-3x} + \cdots). \end{aligned}$$

Hence,

$$\int_0^\infty x^3(e^x - 1)^{-1}dx = \sum_{j=1}^\infty \int_0^\infty x^3e^{-jx}dx = \sum_{j=1}^\infty \frac{3!}{j^4} = \frac{\pi^4}{15},$$

which gives

$$V(T) = \sigma_0 T^4, \tag{15.4}$$

where $\sigma_0 = 2\pi^5k_0^4/15h_0^3c_0^2 = 5.671 \times 10^{-8}\,\text{W}\,\text{m}^{-2}\,\text{K}^{-4}$ is the *Stefan–Boltzmann constant.* Equation (15.4) is the *Stefan–Boltzmann law.* The unit of $V(T)$ is power per unit area. The corresponding integrated radiance inside an isothermal enclosure is

$$I(T) = V(T)/\pi.$$

15.3 Emissivity

15.3.1 Emissivity and emittance

The power thermally emitted by a surface is called the *emittance.* If the emittance of the surface of an optically thick sample of real material is measured, it is found that the spectrum generally is similar to the Planck function, but usually is smaller by an amount that may vary with wavelength. The ratio of the actual power $U_a(\lambda, T)$ to that emitted by an ideal black surface is the *spectral emissivity:*

$$\varepsilon(\lambda) = \frac{U_a(\lambda, T)}{U(\lambda, T)}. \tag{15.5}$$

If ε is independent of wavelength, the surface is called a *gray body.*

The *integrated emissivity* $\bar{\varepsilon}$ is the average emissivity weighted by the thermal spectrum:

$$\bar{\varepsilon} = \frac{1}{V(T)} \int_0^\infty \varepsilon(\lambda) U(\lambda, T) d\lambda. \tag{15.6}$$

Just as there are several kinds of reflectances that are distinguished by the degree of collimation of the source and detector, several kinds of emissivities may be defined, depending on the degree of collimation of the detector. These are the directional, conical, and hemispherical emissivities, denoted, respectively, by ε_d, ε_c, and ε_h. Yet another kind of emissivity refers to the radiation emitted by a particle. In this book, ε with no subscript will denote the emissivity of a single particle.

15.3.2 The emissivity of a solid, smooth surface

It is found experimentally that the radiance inside an isothermal enclosure is independent of the nature of composition of the walls. Suppose the walls are smooth and polished. Then the radiant power incident per unit area on a wall inside such an enclosure per unit wavelength is $U(\lambda, T)$. A fraction $S_e(\lambda)$ per unit wavelength of this power is reflected, where $S_e(\lambda)$ is the hemispherically integrated Fresnel reflection coefficient of the wall (Chapter 5), and a fraction $1 - S_e(\lambda)$ is absorbed. Thus, if the radiance inside the cavity is to be independent of the composition of the wall, the power absorbed must be exactly balanced by the power emitted by the wall. That is, for a specularly reflecting surface, $\varepsilon_h(\lambda)U(\lambda, T) = [1 - S_e(\lambda)]U(\lambda, T)$, so that

$$\varepsilon_h(\lambda) = 1 - S_e(\lambda). \tag{15.7}$$

Equation (15.7) is one form of a relationship known as *Kirchhoff's law.* This law will be discussed in more detail in Section 15.4. Although there are many kinds of emissivities, the one defined by equation (15.7) is usually the quantity that is discussed in most physics textbooks.

15.3.3 Emissivity and emissivity factor of a particle

Because we are interested in the thermal emission from particulate media, it is necessary to know the emissivity of a particle. This quantity can be found by considering the energy balance of a particle located inside an isothermal enclosure. As usual throughout this book, except where explicitly stated otherwise, it is assumed that we are dealing with ensembles of randomly oriented particles. As discussed in Chapter 6, the geometric cross sections and efficiencies of particles are taken to be equivalent to their rotationally averaged values, and so are independent of the direction of the incident radiance.

Suppose an enclosure contains a particle with a rotationally averaged geometric cross-sectional area σ, extinction, scattering, and absorption efficiencies $Q_E(\lambda)$, $Q_S(\lambda)$, and $Q_A(\lambda)$, respectively, and emissivity $\varepsilon(\lambda)$. Then the total power per unit wavelength that is intercepted by the particle is

$$\int_{4\pi} \sigma Q_E(\lambda)(1/\pi)U(\lambda, T)d\Omega = 4\sigma Q_E(\lambda)U(\lambda, T),$$

of which a fraction $Q_A(\lambda)/Q_E(\lambda)$ is absorbed. The power thermally emitted by the particle is

$$\int_{4\pi} \varepsilon(\lambda)\sigma(1/\pi)U(\lambda, T)d\Omega = 4\varepsilon(\lambda)\sigma U(\lambda, T).$$

Hence, if the radiance inside the enclosure is not to be altered by the particle, we must have

$$\varepsilon(\lambda) = Q_A(\lambda). \tag{15.8}$$

An alternate proof of this equality for homogeneous spheres has been given by Kattawar and Eisner (1970).

Define the *particle emissivity factor* $\mathcal{E}(\lambda)$ as

$$\mathcal{E}(\lambda) = \frac{\varepsilon(\lambda)}{Q_E(\lambda)}. \tag{15.9a}$$

Then

$$\mathcal{E}(\lambda) = \frac{\varepsilon(\lambda)}{Q_E(\lambda)} = \frac{Q_A(\lambda)}{Q_E(\lambda)} = \frac{Q_E(\lambda) - Q_S(\lambda)}{Q_E(\lambda)} = 1 - \varpi(\lambda), \tag{15.9b}$$

where $\varpi(\lambda)$ is the particle single-scattering albedo. For large particles close together, $Q_E = 1$ and $\mathcal{E} = \varepsilon$. Equation (15.9b) is Kirchhoff's law for single particles.

15.3.4 The thermal source function

Suppose a small volume element $\Delta\upsilon$ contains a number of different types of particles. Then the power emitted per unit wavelength per unit solid angle from the volume is

$$\Delta P_e = \sum_j N_j \Delta\upsilon \sigma_j \varepsilon_j(\lambda)(1/\pi)U(\lambda, T),$$

where N is the number of particles per unit volume, and the subscript j denotes the type of particle. It was shown in Chapter 7 that one of the quantities appearing in the radiative-transfer equation is the power thermally emitted per unit volume into unit solid angle. This quantity is known as the *thermal volume emission coefficient* and is denoted by $\mathcal{F}_T$. Hence,

$$\mathcal{F}_T = \frac{\Delta P_e}{\Delta \upsilon} = \sum_j N_j \sigma_j \varepsilon_j(\lambda) \frac{1}{\pi} U(\lambda, T)$$
$$= \sum_j N_j \sigma_j Q_{Aj}(\lambda) \frac{1}{\pi} U(\lambda, T) = \frac{A(\lambda)}{\pi} U(\lambda, T), \quad (15.10a)$$

where $A(\lambda)$ is the volume absorption coefficient of a particulate medium (equation [7.37]). Because of the assumption of random particle orientation, $\mathcal{F}_T$ is independent of direction, although it may depend on position.

Similarly, the *volume thermal source function* $F_T = \mathcal{F}_T / E$ was defined in Chapter 7, where E is the volume extinction coefficient. Equation (15.10a) shows that for a collection of particles the volume thermal source function is

$$F_T = \frac{A(\lambda)}{E} \frac{U(\lambda, T)}{\pi}. \quad (15.10b)$$

Define the *volume emissivity factor* of a particulate medium as

$$\mathcal{E}_V(\lambda) = \frac{\pi}{U(\lambda, T)} F_T = \frac{1}{E(\lambda)} \sum_j N_j \sigma_j \varepsilon_j(\lambda) = \frac{A(\lambda)}{E(\lambda)} = \frac{E(\lambda) - S(\lambda)}{E(\lambda)}$$
$$= 1 - w(\lambda) = \gamma^2(\lambda), \quad (15.11)$$

where S is the volume scattering coefficient (equation [7.36]), w is the volume single-scattering albedo (equation [7.21a]), and γ is the volume albedo factor.

Thus, the thermal source function in a particulate medium can be written

$$F_T = \frac{\mathcal{E}_V(\lambda)}{\pi} U(\lambda, T) = \frac{\gamma^2(\lambda)}{\pi} U(\lambda, T). \quad (15.12)$$

15.3.5 The directional emissivity of a particulate medium

The directional emissivity $\varepsilon_d(e, \lambda)$ is the ratio of the thermal radiance $I(e, \lambda, T)$ emerging from the surface of a particulate medium at a uniform temperature T into a given direction making an angle e with the zenith to the thermal radiance $U(\lambda, T)/\pi$ emerging from a black body at the same temperature:

$$\varepsilon_d(e, \lambda) = \pi \frac{I(e, \lambda, T)}{U(\lambda, T)}. \quad (15.13)$$

In this section, the directional emissivity of an infinitely thick particulate medium of isotropic scatterers will be calculated using the method of invariance. This method, which was used previously in Chapter 8 to calculate the bidirectional reflectance, is based on the principle that if an optically thin layer of particles is added to the top of an infinitely thick medium consisting of the same type of particles, neither the emittance nor the reflectance will be changed. We will calculate the first-order changes caused by the addition of such a layer and then require that the sum of all these changes must equal zero. For convenience and economy of notation, the explicit dependences of the various quantities on λ and T will be dropped. For simplicity we will also assume that the particles are all of one type; the extension to media of mixtures is straightforward.

Consider a semi-infinite medium consisting of N particles per unit volume, with bidirectional reflectance $r(i, e, g)$. Because the particles emit and scatter isotropically, $r(i, e, g)$ and $\varepsilon_d(e)$ are independent of azimuth or phase angle and may be written $r(\mu_0, \mu)$ and $\varepsilon_d(\mu)$, respectively. Let a layer of thickness Δz and optical thickness $\Delta\tau = N\sigma Q_E \Delta z = E\Delta z \ll 1$ be added to the top of the medium. Assume that $\Delta\tau$ is so small that interactions of light with the layer involving powers of $\Delta\tau$ greater than 1 can be ignored. Then the layer will cause five separate changes proportional to $\Delta\tau$ in the emitted radiation. These changes are shown schematically in Figure 15.3.

(1) Radiance $\varepsilon_d(e)(U/\pi)$ emitted by the lower medium into a direction Ω_e making an angle e with the vertical is attenuated by extinction in the added layer (Figure 15.3a). The radiance emerging from the upper layer is $I = \varepsilon_d(e)(U/\pi)\exp(-\Delta\tau/\mu) \simeq \varepsilon_d(e)(U/\pi)(1 - \Delta\tau/\mu)$, to first order in $\Delta\tau$. Hence, the change due to this effect is

$$\Delta I_1 = -\varepsilon_d(e)\frac{U}{\pi}\frac{\Delta\tau}{\mu}. \tag{15.14a}$$

(2) The added layer emits an additional amount of light into the direction Ω_e (Figure 15.3b). Consider a cylindrical volume coaxial with the emitted ray with cross-sectional area σ and length $\Delta z/\mu$, where $\mu = \cos e$. Then the radiance emitted toward Ω_e by the particles in this volume is

$$\Delta I_2 = N\sigma\varepsilon\frac{\Delta z}{\mu}\frac{U}{\pi} = \frac{\mathcal{E}}{\pi}U\frac{\Delta\tau}{\mu} = \frac{\mathcal{E}_V}{\pi}U\frac{\Delta\tau}{\mu}, \tag{15.14b}$$

since for media of a single type of particle $\mathcal{E} = \mathcal{E}_V$.

(3) The added layer emits an amount of light in the downward direction. This light is scattered by the lower medium into the direction Ω_e (Figure 15.3c). Consider a cylinder of area σ and length $\Delta z/\mu_0$ coaxial with a direction Ω_i making an

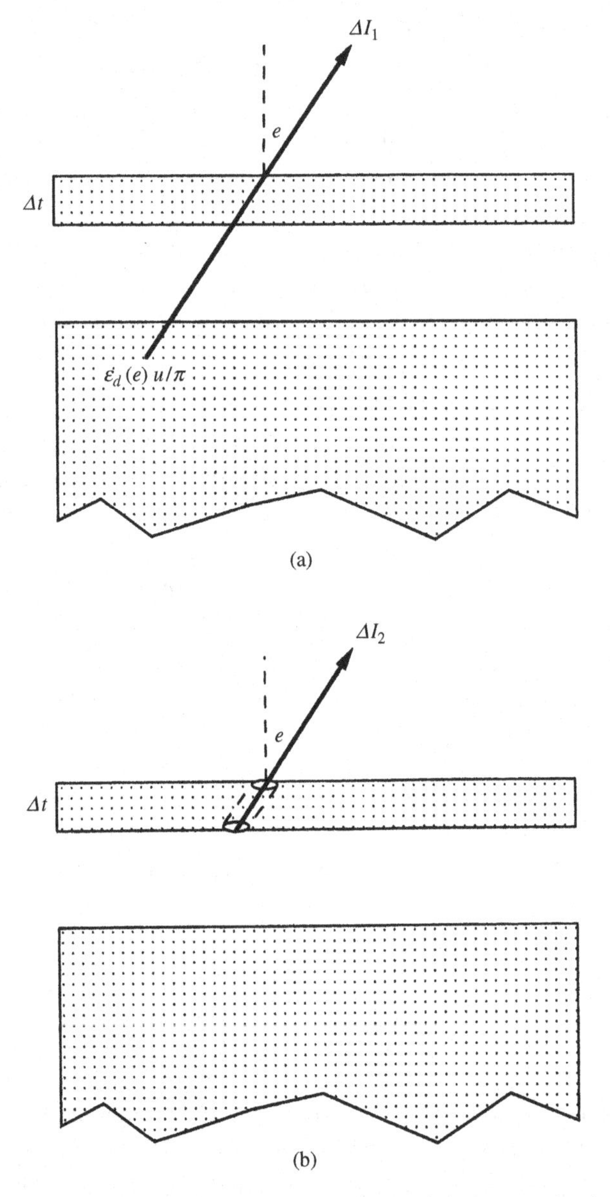

Figure 15.3 Schematic diagram of the five first-order changes in the thermally emitted radiance caused by adding a thin layer of optical thickness $\Delta\tau$ to the top of an infinitely thick medium.

angle i with the normal to the layer, where $\mu_0 = \cos i$. This cylinder emits a radiance $(\mathcal{E}_V/\pi)U(\Delta\tau/\mu_0)$ into this direction toward the lower medium. The medium scatters a fraction $r(\mu_0, \mu)$ of this radiance into the direction Ω_e. The

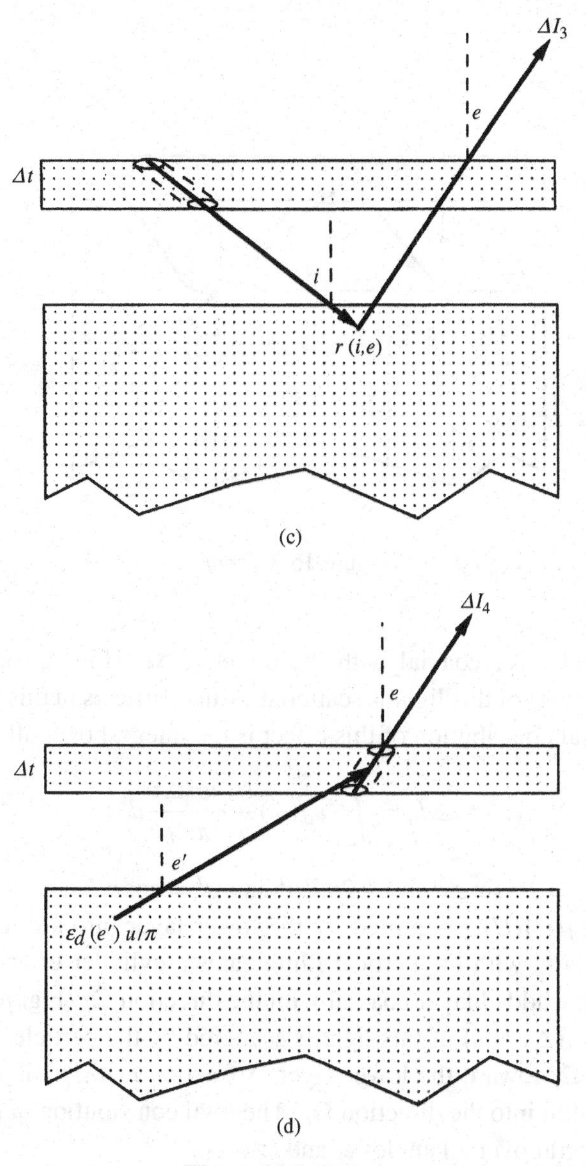

Figure 15.3 (*cont.*)

total contribution of this effect is the integral over all angles i:

$$\Delta I_3 = \int_{2\pi} \frac{\mathcal{E}_V}{\pi} U \frac{\Delta\tau}{\mu_0} r(\mu_0, \mu) d\Omega_i, \tag{15.14c}$$

where $d\Omega_i = 2\pi \sin i di = -2\pi d\mu_0$.

(4) Radiance $\varepsilon_d(e')(U/\pi)$ emitted by the lower layer into a direction $\Omega_{e'}$ making an angle e' with the vertical illuminates a cylinder of area σ and length $\Delta z/\mu$

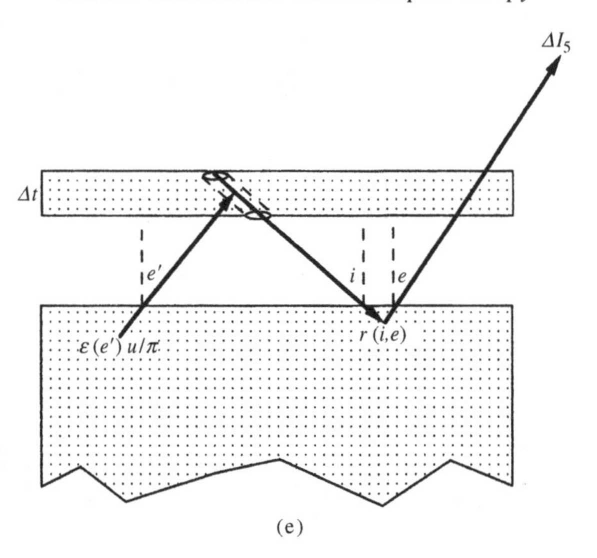

Figure 15.3 *(cont.)*

in the added layer coaxial with the direction Ω_e (Figure 15.3d). A fraction $(w/4\pi)(\Delta\tau/\mu)$ of this light is scattered by the particles in this cylinder toward Ω_e. The total contribution of this effect is the integral over all angles e':

$$\Delta I_4 = \int_{2\pi} \varepsilon_d(e')\frac{U}{\pi}\frac{w}{4\pi}\frac{\Delta\tau}{\mu}d\Omega_{e'}, \tag{15.14d}$$

where $d\Omega_{e'} = 2\pi \sin e' de' = -2\pi d\mu'$ and $\mu' = \cos e'$.

(5) Radiance $\varepsilon_d(e')(U/\pi)$ emitted by the lower layer into a direction $\Omega_{e'}$ making an angle e' with the vertical illuminates a cylinder of area σ and length $\Delta z/\mu_0$ in the added layer coaxial with the direction Ω_i (Figure 15.3e). A fraction $(w/4\pi)(\Delta\tau/\mu_0)$ of this light is scattered by the particles in this cylinder parallel to Ω_i toward the lower layer. A fraction $r(\mu_0, \mu)$ is scattered by the lower medium into the direction Ω_e. The total contribution of this effect is the double integral over all angles e' and i:

$$\Delta I_5 = \int_{2\pi}\int_{2\pi} \varepsilon_d(e')\frac{U}{\pi}\frac{w}{4\pi}\frac{\Delta\tau}{\mu_0}r(\mu_0, \mu)d\Omega_i d\Omega_{e'}. \tag{15.14e}$$

The sum of all the changes ΔI_1 through ΔI_5 must be zero. Hence, after dividing through by $U\Delta\tau/\pi\mu$, we obtain

$$\varepsilon_d(e) = \mathcal{E}_V + \mathcal{E}_V\mu \int_{2\pi} \frac{r(\mu_0, \mu)}{\mu_0}d\Omega_i + \frac{w}{4\pi}\int_{2\pi} \varepsilon_d(e')d\Omega_{e'}$$

$$+\frac{w}{4\pi}\left[\int_{2\pi}\frac{r(\mu_0,\mu)}{\mu_0}d\Omega_i\right]\left[\int_{2\pi}\varepsilon_d(e')d\Omega_{e'}\right]. \tag{15.15}$$

Now, it was shown in Section 8.7.3.2 that a medium of isotropically scattering particles has a bidirectional reflectance

$$r(\mu_0,\mu)=\frac{w}{4\pi}\frac{\mu_0}{\mu_0+\mu}H(\mu_0)H(\mu),$$

where $H(\mu)$ satisfies the integral equation

$$H(\mu)=1+\frac{w}{2}\mu H(\mu)\int_0^1\frac{H(\mu_0)}{\mu_0+\mu}d\mu_0.$$

Hence,

$$\mu\int_{2\pi}\frac{r(\mu_o,\mu)}{\mu_0}d\Omega_i=\frac{w}{4\pi}\mu H(\mu)\int_0^1\frac{H(\mu_0)}{\mu_0+\mu}2\pi d\mu_0=H(\mu)-1,$$

and (15.14) becomes

$$\begin{aligned}\varepsilon_d(e)&=\mathcal{E}_V+\mathcal{E}_V[H(\mu)-1]+\frac{w}{2}\int_0^1\varepsilon_d(\mu')d\mu'\\&\quad+\frac{w}{2}[H(\mu)-1]\int_0^1\varepsilon_d(\mu')d\mu'\\&=H(\mu)\left[\mathcal{E}_v+\frac{w}{2}\int_0^1\varepsilon_d(\mu')d\mu'\right].\end{aligned} \tag{15.16}$$

Integrating (15.16) with respect to μ,

$$\int_0^1\varepsilon_d(\mu)d\mu=\left[\int_0^1H(\mu)d\mu\right]\left[\mathcal{E}_V+\frac{w}{2}\int_0^1\varepsilon_d(\mu)d\mu\right],$$

and solving for the integral of the emissivity gives

$$\int_0^1\varepsilon_d(\mu)d\mu=\mathcal{E}_V\left[\int_0^1H(\mu)d\mu\right]\left[1-\frac{w}{2}\int_0^1H(\mu)d\mu\right]^{-1}. \tag{15.17}$$

But according to (8.51),

$$\int_0^1H(\mu)d\mu=H_0=\frac{2}{1+\gamma}.$$

Inserting this into (15.17) gives

$$\int_0^1\varepsilon_d(\mu)d\mu=\frac{\mathcal{E}_V}{\gamma}\frac{2}{1+\gamma}. \tag{15.18}$$

Substituting this result into (15.16) gives

$$\varepsilon_d(e) = \frac{\mathcal{E}_v}{\gamma} H(\mu) = \gamma H(\mu). \tag{15.19}$$

Note that, like the equation for the reflectance, which was derived in the same manner, (15.19) is an exact, general solution for ε_d and makes no assumptions about the medium, other than that it is composed of particles that emit and scatter isotropically and that the added layer can be made optically thin. If approximation (8.53) is used for $H(\mu)$, (15.19) becomes

$$\varepsilon_d(e) \simeq \gamma \frac{1+2\mu}{1+2\gamma\mu}. \tag{15.20}$$

Thus, the radiance emerging from the surface of an optically thick particulate medium at uniform temperature T is

$$I(e) = \frac{\gamma(\lambda)}{\pi} H(\lambda, \mu) U(\lambda, T). \tag{15.21}$$

The directional emissivity ε_d is plotted versus e for several values of w in Figure 15.4.

When w is small, $\varepsilon_d \simeq 1$ at all angles, and the surface emits like a black body. Hence, for low-albedo materials, the assumption that the emissivity is independent of angle is a good approximation.

As w increases, ε_d decreases, and also ε_d decreases as e increases. The reason for the dependence on e is that the multiply scattered flux is important when the

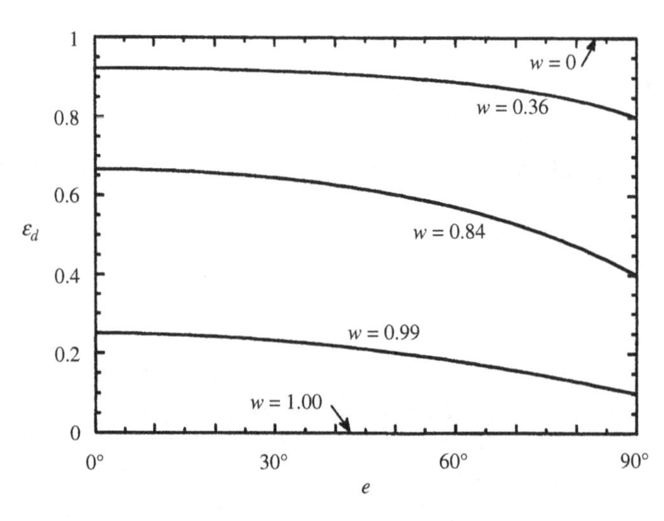

Figure 15.4 Directional emissivity of a particulate medium as a function of the angle of emergence for several values of the single-scattering albedo.

albedo is large. This flux decreases toward the surface because of leakage from the surface. As e increases, the field of view includes a greater contribution from the smaller flux closer to the surface. Only a few measurements of the directional (as opposed to hemispherical) emissivity have been published. However, Jakowsky *et al.* (1990) have measured the directional emissivities of natural sand and playa surfaces, and they seem to be in qualitative agreement with the theoretical model of this chapter.

15.3.6 The hemispherical emissivity of a particulate medium

The hemispherical emissivity $\varepsilon_h(\lambda)$ of a particulate medium can be found by integrating the upward component of the emitted radiance, equation (15.21), over the upward hemisphere. The power emitted per unit area is

$$P_{em} = \varepsilon_h U = \int_{2\pi} I(\mu)\mu d\Omega_e = \int_0^1 \frac{\gamma}{\pi} H(\mu) U \mu 2\pi d\mu = 2\gamma H_1(\gamma) U, \quad (15.22)$$

where $H_1 = \int_0^1 H(\mu)\mu d\mu$ is the first moment of the H function. Hence, the hemispherical emissivity is

$$\varepsilon_h = 2\gamma H_1. \quad (15.23)$$

Using approximation (8.58b), $H_1 \simeq [1/(1+\gamma)](1+\frac{1}{6}r_0)$, where $r_0 = (1-\gamma)/(1+\gamma)$ is the diffusive reflectance, gives

$$\varepsilon_h \simeq \frac{2\gamma}{1+\gamma}\left(1+\frac{1}{6}r_0\right). \quad (15.24a)$$

The hemispherical emissivity ε_h is plotted versus w in Figure 15.5. Note that when w and r_0 are small,

$$\varepsilon_h \simeq \frac{2\gamma}{1+\gamma}. \quad (15.24b)$$

15.4 Kirchhoff's law

Kirchhoff's law is an extremely powerful and useful rule which states that there is a complementary relationship between emissivity and reflectance. Kirchhoff's law allows the emissivity to be calculated from the reflectance, which often is more convenient to measure in the laboratory. However, as we have seen, there are several different kinds of reflectances and emissivities, and it is not always obvious which ones form the complementary pairs. From the derivations of this chapter, the following quantities obey Kirchhoff's law.

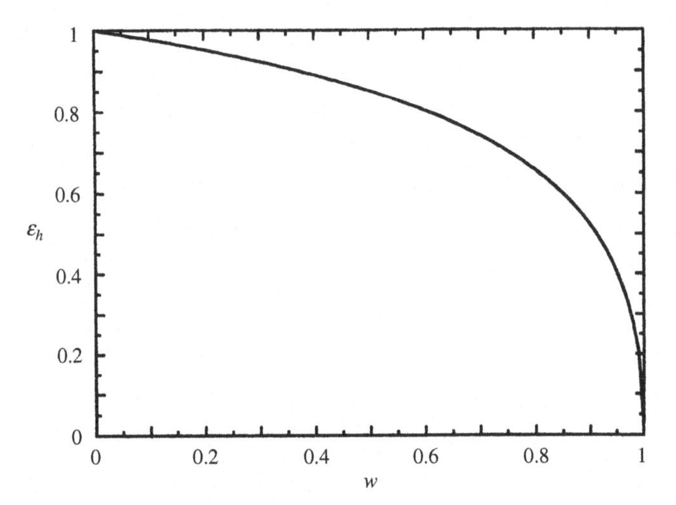

Figure 15.5 Hemispherical emissivity of a particulate medium vs. the single-scattering albedo.

In Section 15.3.1 it was seen that the hemispherical emissivity of a smooth surface is

$$\varepsilon_h = 1 - S_e, \tag{15.25a}$$

where S_e is the integral of the Fresnel reflection coefficients over a hemisphere. From Kirchhoff's law it follows immediately that the directional emissivity of a smooth surface is

$$\varepsilon_d(e) = 1 - R(e), \tag{15.25b}$$

where $R(e)$ is the average of the Fresnel reflection coefficients over the two directions of polarization.

From equations (15.8) and (5.6), the emissivity of a single particle is

$$\varepsilon = Q_A = Q_E - Q_S, \tag{15.26}$$

where Q_E and Q_S are the volume-average particle extinction and scattering efficiencies, respectively.

From equations (15.11) and (7.17), the volume emissivity factor of a particulate medium is

$$\mathcal{E}_V = 1 - w, \tag{15.27}$$

where w is the volume single-scattering albedo in the medium.

From equations (15.19) and (11.16), the directional emissivity of a particulate medium of isotropic scatterers is

$$\varepsilon_d(e) = 1 - r_{hd}(e), \tag{15.28}$$

where $r_{hd}(e)$ is the hemispherical–directional reflectance of the medium. Hence, the directional emissivity may be calculated from a measurement of the hemispherical–directional reflectance. However, because $r_{hd}(e)$ has the same functional dependence on e as the directional–hemispherical reflectance $r_h(i)$ has on i, the latter quantity may also be used to calculate $\varepsilon_d(e)$.

From (15.23) and (11.18), the hemispherical emissivity of a particulate medium of isotropic scatterers is

$$\varepsilon_h = 1 - r_s, \tag{15.29}$$

where r_s is the spherical or bihemispherical reflectance. Hence, if it is desired to calculate the hemispherical emissivity from reflectance, the spherical or Bond albedo is the quantity that must be measured. As discussed in Chapter 11, this is inconvenient to do in the laboratory. However, according to (11.19), an approximate expression for the spherical reflectance is $r_s \simeq r_0\{1 - \frac{1}{3}[\gamma/(1+\gamma)]\}$. Substituting this into (15.29) gives the following approximate expression for the hemispherical emissivity:

$$\varepsilon_h \simeq (1 - r_0)\left(1 + \frac{r_0}{6}\right). \tag{15.30}$$

If w and r_0 are small,

$$\varepsilon_h \simeq 1 - r_0.$$

For media in which the scatterers are approximately isotropic, several different kinds of reflectances reduce to the diffusive reflectance at certain angles. Hence, the spherical reflectance and hemispherical emissivity may also be calculated from a measurement of one of those reflectances at the appropriate angle.

15.5 Combined reflectance and emittance

In between the near and thermal infrared regions the radiance from a body may consist of a mixture of reflected sunlight and thermally emitted radiance. This topic will be treated in more detail in Chapter 16. However, if the temperature near the surface of a medium is constant the combined radiance emitted in a given direction is

$$I(i, e, g) = Jr(i, e, g) + \varepsilon_d(e)U_0(T), \tag{15.31}$$

where $r(i, e, g)$ is the bidirectional reflectance, and the hemispherically integrated power emitted by the surface is

$$P_{em}(i) = J\mu_0 r_h(i) + \varepsilon_h U_0(T), \tag{15.32}$$

where $r_h(i)$ is the hemispherical reflectance.

15.6 Emittance spectroscopy

15.6.1 Brightness temperature

In applications of emittance spectroscopy to remote sensing, the radiance in the thermal infrared part of the spectrum emitted from the surface of a planet into a given direction is measured. The usual procedure is to fit a Planck function, which is assumed to be representative of the temperature of the surface, to the overall data, with the constraint that the Planck function can nowhere exceed the measured radiance. Dividing the measured spectrum by the fitted Planck function gives the directional emissivity spectrum.

An alternative procedure is to use the measured radiance to calculate a quantity called the *brightness temperature*. The brightness temperature is a useful concept for characterizing the radiance coming from an object. It is defined as the temperature that a perfect black body of the same size and distance would have to have in order to emit the measured radiance at a given wavelength. It is not necessarily related to the actual temperature of the surface, especially if the radiance is emitted by a nonthermal process. Thus, if $I(\lambda)$ is the radiance emerging from a surface, then the brightness temperature T_b is given by

$$I(\lambda) = \frac{1}{\pi} U(\lambda, T_b),$$

which can be solved for T_b to give

$$T_b(\lambda) = \frac{h_0 c_0}{\lambda k_0} \left[\ln \left(1 + \frac{2 h_0 c_0^2}{\lambda^5 I(\lambda)} \right) \right]^{-1}. \qquad (15.33)$$

If $I(\lambda) = \varepsilon_d U(\lambda, T)/\pi$, the spectrum of $T_b(\lambda)$ is similar to that of $\varepsilon_d(\lambda)$, although the relationship obviously is not linear. The advantage of using the brightness temperature is that this procedure avoids possible errors caused by the subjective judgments involved in fitting the Planck function to the observed radiance.

15.6.2 Absorption bands in emissivity

After the spectrum of either $\varepsilon_d(\lambda)$ or $T_b(\lambda)$ is obtained, it is then inspected for absorption bands that may be diagnostic of composition. An excellent discussion of the interpretation of emissivity spectra may be found in the work of Salisbury (1993). Christensen and his colleagues have made extensive spectral measurements of the surface of Mars in the thermal IR (e.g., Christensen *et al.*, 2008a, b).

Because of the complementary relation between the emissivity and reflectance, exactly the same principles as discussed in Chapter 14 apply to the emissivity.

In particular, both the reflectance and the emissivity of a particulate medium are primarily controlled by the single-scattering albedo w (or, equivalently, the albedo factor γ), which in turn depends on the complex refractive index $n = n_r + in_i$, the effective particle size $\langle D \rangle$, and the near-surface internal scattering coefficient s of the particles.

As with reflectance, the relation between emissivity and absorption coefficient or imaginary refractive index may be divided up into three regions: volume scattering, weak surface scattering, and strong surface scattering. The behavior of absorption bands seen in emissivity is the inverse of those seen in reflectance. Illustrations of a weak, intermediate, and strong band seen as features in the bidirectional reflectance of a particulate medium were given in Figure 14.7. The same bands seen as directional emissivity features are shown in Figure 15.6.

In the volume-scattering region an absorption band sufficiently weak that it is entirely in the volume-scattering region is expressed as a peak in the emissivity spectrum (Figure 15.6a), located at the band center. A band whose center extends into the weak surface-scattering region is a positive emissivity feature with a flattened peak (Figure 15.6b). A strong band whose center extends into the strong surface-scattering region is expressed as two peaks on either side of a dip (Figure 15.6c). The depth of the minimum relative to the peaks increases as the band strength increases. The minimum is not at the band center, but is displaced toward shorter wavelengths.

15.6.3 The Christiansen and transparency features

The peak on the short-wavelength side of the band in the spectrum of a particulate medium is called the *Christiansen feature* because it is fortuitously located close to the Christiansen wavelength where the real part of the refractive index is equal to 1.00. However, in general, the maximum is not exactly at λ_C. It was originally thought that the peak occurred at λ_C because reduced particle surface scattering would make the powder transparent there, allowing radiation from the hotter interior to more readily escape. However, even though the surface scattering is small the particles are not transparent at λ_C. (In the example of Figure 15.6c, $\alpha(\lambda_C)D \sim 14$.) The Christiansen feature is identical with the transition minimum where the reflectance of a strong band is changing from being dominated by volume scattering to surface scattering as n_i increases. Similarly, the maximum on the long-wavelength side of the emissivity band is identical with the second transition minimum in a strong band, where the reflectance is changing from strong-surface to volume scattering as n_i decreases.

Silicates have very strong restrahlen bands between 8.5 and 12.0 μm, depending on composition, caused by the stretching of Si—O bonds. They also have slightly

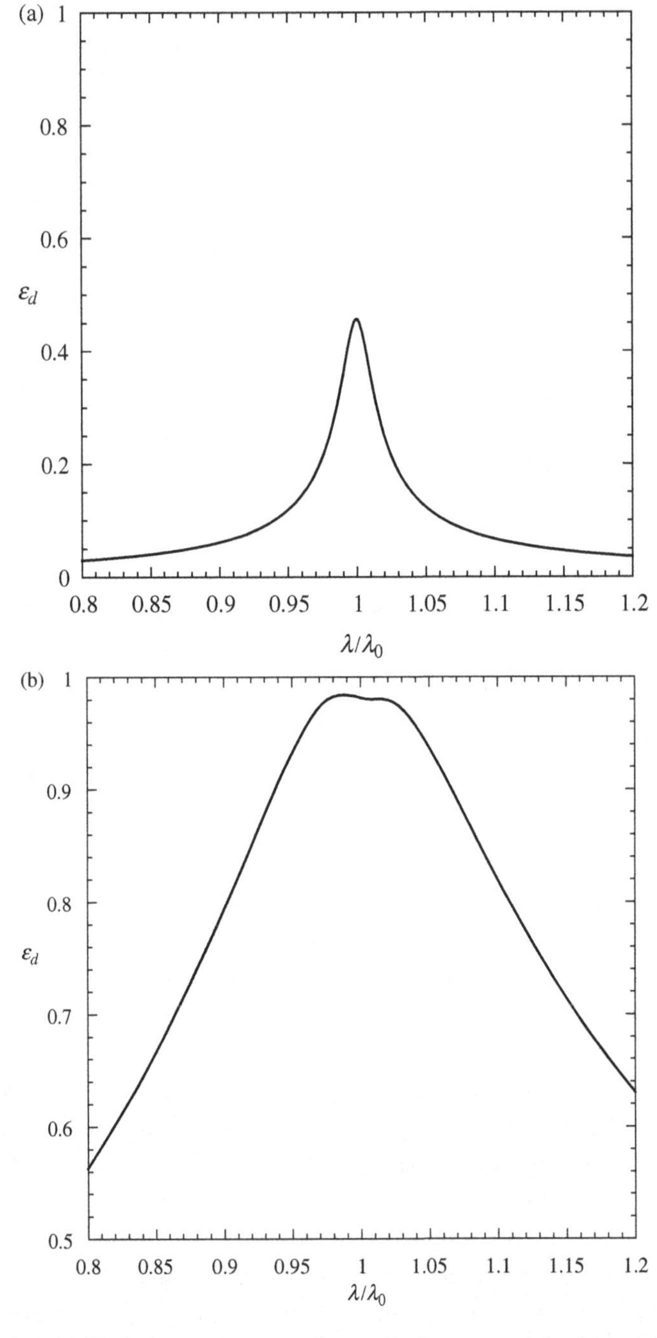

Figure 15.6 (a) Emittance spectrum of a particulate material with an absorption band of Lorentzian shape that is sufficiently weak that it is entirely within the volume scattering region. The center of the band is located at λ_0. (b) Same as for Figure 15.6a except that the band center is in the weak surface-scattering region. (c) Same as for Figure 15.6a except that the band center is in the strong surface-scattering region.

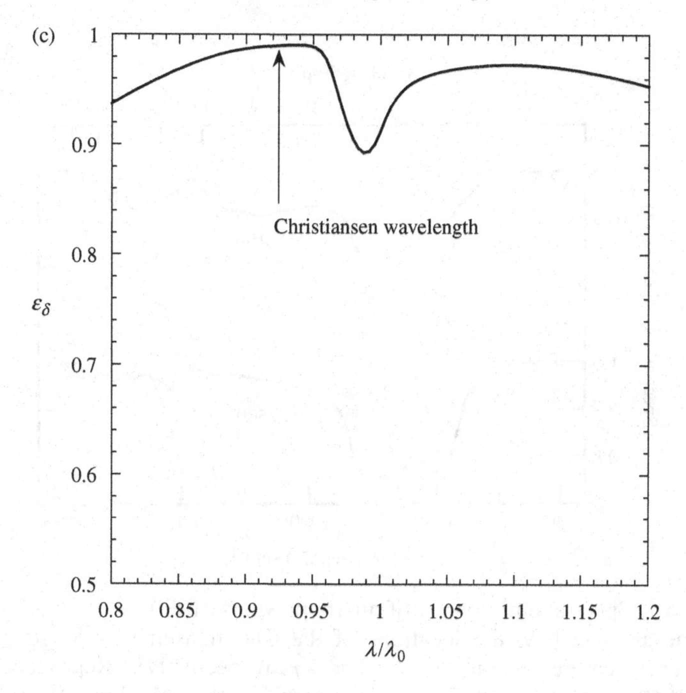

Figure 15.6 (*cont.*)

less intense bands between 16.5 and 25 μm associated with Si—O—Si bending vibrations. In between the two bands the particles are in the volume-scattering region, which produces a maximum in reflectance and a minimum in emissivity. This emissivity minimum is called the *transparency feature*. The Christiansen and transparency features in quartz are illustrated in Figure 15.7.

Logan *et al.* (1973), Salisbury and Walter (1989), and Walter and Salisbury (1989) have shown that the wavelengths of the Christiansen and transparency features of a substance are diagnostic of composition, and they have emphasized the potential of this technique for remote sensing in the thermal infrared. Thus, the Christiansen and transparency features are correlated with each other and are diagnostic of composition. Figure 15.8 shows this correlation.

15.6.4 Deconvolution of emissivity spectra of mixtures

It was emphasized in Chapters 10 and 14 that the reflectance of a mixture is a nonlinear function of the abundances of the end members because of the nonlinear dependence of the reflectance on single-scattering albedo. However, this nonlinearity is caused by the multiply scattered component, so that for low-albedo materials the reflectance is almost linear. In spectral regions where a strong absorption band

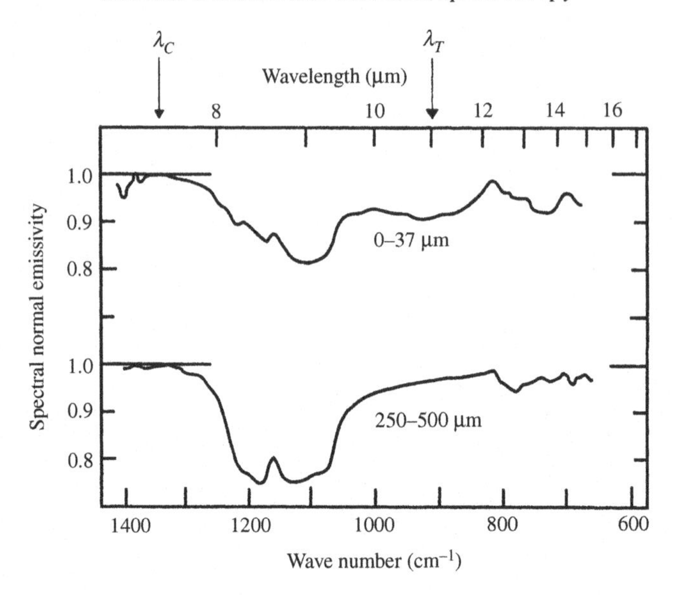

Figure 15.7 Spectral emissivities of quartz powders of different sizes and packings. The arrows show the locations of the Christiansen wavelength and the transparency feature (denoted by λ_C and λ_T, respectively). (Reproduced from Conel [1969], copyright 1969 by the American Geophysical Union.)

is displayed as an emissivity minimum the single-scattering albedo is inherently small. Thus, to a good approximation the emissivity of a mixture is a linear function of its end members in those regions. This allows the identification of the individual minerals in planetary regoliths from measurements of thermal IR emissivity using linear deconvolution and comparison with spectra of candidate minerals (Ramsey and Christensen, 1998). Many commercial software packages contain linear deconvolution algorithms.

15.6.5 Contrast in emissivity bands

In general, the contrasts in absorption bands observed in emissivity spectra are much smaller than the contrasts observed in reflectance, particularly in specular reflection from a polished surface. This point has been strongly emphasized by Conel (1969). Thus, there is an unfortunate paradox that often limits the usefulness of emittance spectroscopy for compositional remote sensing. In spite of the fact that the compositionally diagnostic restrahlen absorption bands are strong in the thermal infrared, the emissivity contrast that can be observed in a particulate medium may be very small. There are several reasons for the loss of contrast.

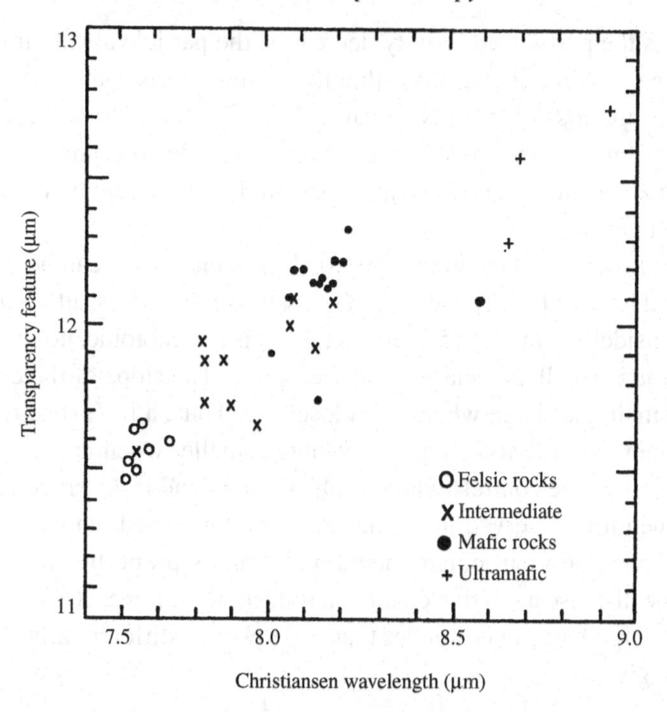

Figure 15.8 Correlation between the Christiansen wavelength and the transparency feature for silicates of different compositions. (Reproduced fom Salisbury and Walter [1989], copyright 1989 by the American Geophysical Union.)

First, as the particle size in a medium decreases, the contrast in the vicinity of the very strong *Reststrahlen* bands may decrease also (Hunt and Vincent, 1968; Salisbury, 1993) as the bands move from the strong into the weak surface-scattering region. If grinding moves the band all the way into the volume-scattering region, a restrahlen band may be expressed as a dip in large particles and as a peak in fine ones (Arnold and Wagner, 1988). If the medium consists of a mixture of coarse and fine particles, such effects may cause the bands to be almost unobservable.

Further complicating this discussion is the fact that many media have particles that are smaller than the wavelength in the thermal IR. The particle efficiencies and single-scattering albedos are reasonably well understood when their sizes are large compared with λ. As was discussed in Chapter 7, when $\langle D \rangle \ll \lambda$, it is not clear what effective particle size and refractive index should be used for the calculation of w in the radiative-transfer equation. However, the spectra seem to behave qualitatively as predicted from the assumption that the scattering from each particle is incoherent and quasi-independent of that from other particles. As seen in Chapter 5, absorbing particles smaller than the wavelength are almost perfect absorbers, which is equivalent to being almost perfect emitters, independent of wavelength.

Second, as the particle emissivity decreases, the particle albedo increases, thus increasing the multiple scattering within the medium. It is sometimes stated in the literature that the loss of contrast is caused by void spaces in the medium acting like thermodynamic enclosures. However, this is a misleading physical picture. The loss of contrast occurs because the increased multiple scattering partly cancels the decreased emissivity.

Third, the loss of contrast in the *Reststrahlen*-band region can be seen from the mathematical form of the dependence of emissivity on single-scattering albedo. For example, consider ε_h. Figure 15.5 shows that ε_h is a monotonic, nonlinear function of w. When w is small, ε_h is large, and vice versa. The slope of the curve is small when w is small, and large when w is close to 1. Thus, a high spectral contrast is observed when $w \sim 1$, but the contrast is much smaller when $w \ll 1$.

Let us compare the contrast when an absorption band is observed in reflectance with that when it is observed in emittance. For a low-albedo material the bidirectional reflectance of a particulate medium is directly proportional to w, so that a change Δw will cause a relative change in reflectance $\Delta r/r \simeq \Delta w/w$. This contrast can be quite large since w is smaller than 1. However, differentiating (15.20) with respect to w gives

$$\frac{1}{\varepsilon_d}\frac{d\varepsilon_d}{dw} = -\frac{1}{2\gamma^2(1+2\gamma\mu)}.$$

Thus,

$$\frac{\Delta\varepsilon_d}{\varepsilon_d} = -\frac{\Delta w}{2[1-w][1+2\sqrt{1-w}\mu]}.$$

If $w \ll 1$ and $\mu \simeq 1$, $\Delta\varepsilon_d/\varepsilon_d \simeq \Delta w/6$, which can be quite small.

If both reflectance and emittance are important, thermal emission will partially fill in an absorption band being observed in reflectance. The following is an example of the magnitude of this effect, calculated from equation (15.32) for a medium of isotropic scatterers. Suppose an asteroid at 3 AU from the Sun has a surface temperature of 200 K. The asteroid is illuminated by sunlight, which has the spectrum of a black body at a temperature of 5770 K. Suppose that the regolith has a single-scattering albedo $w = 0.50$, except in an absorption band, where $w = 0.40$ at the center of the band.

Figure 15.9 plots the band contrast [I(continuum) – I (band center)]/[I (continuum)] of the radiance emerging from the surface versus the wavelength of the band. If the band is shorter than about 4 μm, thermal emission is negligible compared with scattered sunlight, and the band is seen with its full contrast. As the wavelength of the band increases, thermal emission fills in the band and decreases the contrast, until the band becomes unobservable at 5.8 μm. For wavelengths longer than about 6 μm the contrast is negative; that is, the radiance is dominated by thermal emission, and the band manifests itself as a weak maximum rather than a strong minimum.

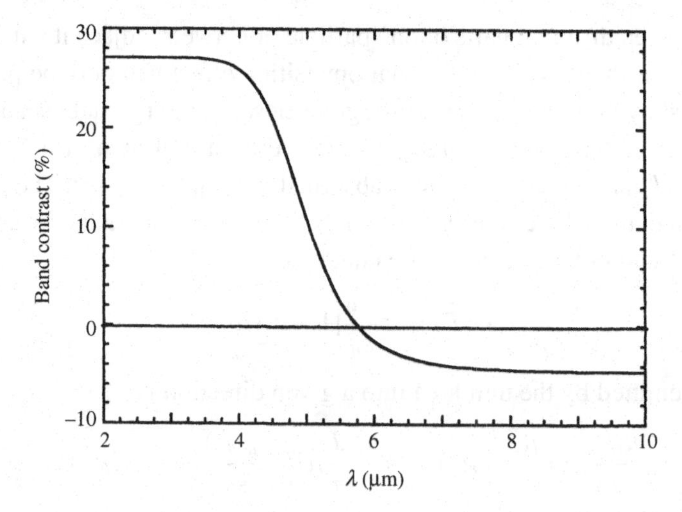

Figure 15.9 Change in the contrast of an absorption band as the band center moves from a wavelength where the emitted radiance is dominated by reflectance to domination by thermal emission.

15.7 The thermal shadow-hiding opposition effect: thermal beaming

15.7.1 The thermal SHOE

Both the shadow-hiding and coherent-backscatter mechanisms for an opposition effect require a preferred direction. For radiation produced by reflectance the incident irradiance provides that direction. However, there is no preferred direction for thermally emited radiance so that pure thermal radiation cannot have an opposition effect. However, a type of SHOE can occur if the temperature of a medium is maintained by the incident irradiance, usually sunlight, and one part of the medium can cast a shadow on another part. Then at any phase angle except zero a thermal radiation detector sees a mixture of hot illuminated surfaces and cold shadows, but at zero phase angle it sees no shadows.

In order for a medium to have an appreciable thermal opposition effect two conditions must be met. First, the portions of the medium that radiate to space must be able to change temperature sufficiently rapidly that they can stay approximately in phase with the illumination. That is, the medium must have a small thermal inertia (see Chapter 16). Second, the shadows must be so cold that thermal radiation from them is small compared with the illuminated regions.

The second condition implies that the illuminated and shadowed regions of the medium must be sufficiently well insulated from one another that large temperature differences can be maintained. For bodies with fine-grained regoliths, like the Moon, thermal conduction prevents appreciable temperature differences from existing between different parts of one particle or between adjacent particles of the soil. Hence, for these bodies a thermal opposition effect can only be produced by shadows cast by large-scale surface irregularities, not interparticle shadowing.

The power of the incident sunlight reflected by a unit area of a surface in all directions is $J\mu_0 r(\mu_0)$ and the power absorbed is $J\mu_0[1-r(\mu_0)]$. The power thermally radiated in all directions is $\varepsilon_h V(T)$. If the surface is in radiative equilibrium, the absorbed and radiated powers are equal so that

$$V(T) = J\frac{\mu_0}{\varepsilon_h}[1 - r_h(\mu_0)]. \tag{15.34}$$

The power emitted by the unit area into a given direction is

$$\frac{dP_{em}}{dA} = \frac{\mu}{\pi}\varepsilon_d(\mu)V(T) = \frac{J}{\pi}\mu\mu_0\frac{\varepsilon_d(\mu)}{\varepsilon_h}[1 - r_h(\mu_0)], \tag{15.35}$$

so that the emitted radiance is

$$I(i, e, g) = \frac{dP_{em}}{d\Omega} = \frac{J}{\pi}\mu_0\frac{\varepsilon_d(\mu)}{\varepsilon_h}[1 - r_h(\mu_0)]. \tag{15.36}$$

This constitutes a kind of visual–thermal bidirectional "reflectance"

$$r_{vis-th}(i, e, g) = \frac{1}{J}\frac{dP_{em}}{d\Omega} = \frac{\mu_0}{\pi}\frac{\varepsilon_d(\mu)}{\varepsilon_h}[1 - r_h(\mu_0)]. \tag{15.37}$$

For a mediuim of isotropic scatterers, $\varepsilon_d = \gamma H(\mu), \varepsilon_h = 2\gamma H_1,$ and $r_h = 1 - \gamma H(\mu_0)$, so

$$r_{vis-th}(i, e, g) = \frac{\gamma}{2\pi}\mu_0 H(\mu_0)H(\mu)/H_1. \tag{15.38}$$

It was shown in Chapter 12 that the reflectance of a medium with a macroscopically smooth surface can be corrected for large-scale roughness by replacing the cosines by effective cosines (indicated by subscript *e*) and multiplying the smooth-surface reflectance by the shadowing function $\mathcal{S}(\mu_0, \mu, \psi)$ where these quantities are derived in Chapter 12. Thus,

$$r_{vis-th}(i, e, g) = \frac{\mu_{0e}}{\pi}\frac{\varepsilon_d(\mu_e)}{\varepsilon_h}[1 - r_h(\mu_{0e})]\mathcal{S}(\mu_0, \mu, \psi). \tag{15.39}$$

This equation describes the thermal SHOE of a rough surface.

Expressions for the visual–thermal "physical albedo" and "phase integral" of a spherical body can be developed in a similar manner. If the surface is rough the phase integral will include the factor $\mathcal{K}(g, \overline{\theta})$ (Chapter 12), which thus describes the thermal SHOE of the integrated IR radiance from the body.

An interesting example of a thermal SHOE peak was observed in the rings of Saturn by a detector aboard the *Cassini* spacecraft (Altobelli *et al.*, 2009). This peak is about 40° wide (Figure 15.10) and is believed to be caused by shadows cast by

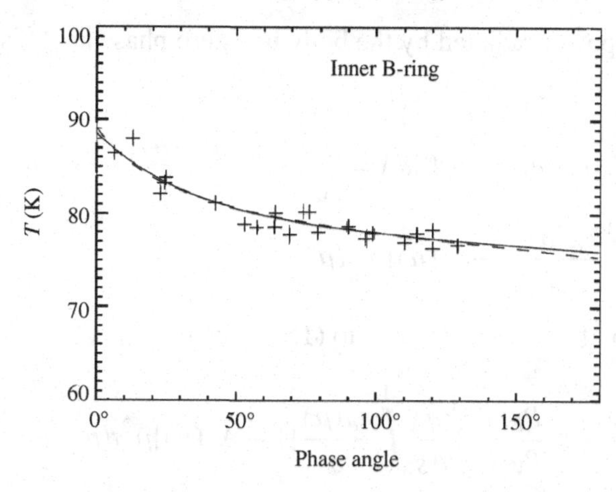

Figure 15.10 The thermal opposition effect in Saturn's B-ring observed by *Cassini*. The crosses are the observations; the solid line is a simple linear plus exponential function; the dashed line is from the SHOE model of Chapter 9. (Reproduced from Altobelli *et al.* [2009], copyright 2009 by the American Geophysical Union.)

one ring particle falling on another. It is different from the narrow visual opposition effect of Figures 9.3 and 9.15, which is believed to be due to CBOEs in the surface structures of the individual ring particles. Thus, this thermal opposition surge is an example of a medium in which the particles are not in contact, and so are well insulated from one another. In this case the thermal SHOE can be described by the expressions for the SHOE in the reflectance of a layer of particles, equations (9.19) and (9.20).

15.7.2 Thermal beaming

The radiometric technique (Morrison and Lebofsky, 1979) is a major method for determining the diameters and physical albedos of objects that are too small to be resolved. It is especially important in the study of asteroids and the smaller moons of the outer planets. This method involves comparing measured radiances in the visible and thermal IR from the object at small phase angle.

Assume that a body of radius R rotates sufficiently slowly that the surface is in radiative equilibrium with incident sunlight. The power of the sunlight scattered into zero phase is, by definition of the physical albedo,

$$P_V = J\pi R^2 \frac{1}{\pi} A_p = JR^2 A_p = JR^2 \frac{A_S}{q} \tag{15.40}$$

where A_p is the visual physical albedo, $q = A_p/A_S$ is the visual phase integral, and A_S is the visual spherical or Bond albedo.

The thermal power radiated by the body into zero phase angle is from equation (15.35)

$$P_T = \int_A \frac{dP_{em}}{dA}(i = e, e, g = 0)dA = \int_0^{\pi/2} \frac{J}{\pi}\mu^2 \frac{\varepsilon_d(\mu)}{\varepsilon_h}[1 - r_h(\mu)]2\pi R^2 \sin e de$$

$$= 2JR^2 \int_0^1 \frac{\varepsilon_d(\mu)}{\varepsilon_h}[1 - r_h(\mu)]\mu^2 d\mu. \quad (15.41)$$

Taking the ratio of equations (15.41) to (15.40) gives

$$\frac{P_T}{P_V} = 2\frac{q}{A_S} \int_0^1 \frac{\varepsilon_d(\mu)}{\varepsilon_h}[1 - A_h(\mu)]\mu^2 d\mu. \quad (15.42)$$

The so-called *standard model* assumes that $\varepsilon_d(\mu)$ is constant and equal to ε_h, and $A_h(\mu_0)$ is constant and equal to A_S. In that case the integral in equations (15.41) and (15.42) can be evaluated to give

$$P_T = \frac{2}{3}R^2 J(1 - A_S), \quad (15.43)$$

and

$$\frac{P_T}{P_V} = \frac{2}{3}q\frac{1 - A_S}{A_S}. \quad (15.44)$$

A value for q is assumed (usually the lunar value) and A_S found from the measured values of P_T and P_V using equation (15.44). Then R is found from equation (15.43) and A_p from A_S/q.

However, when applied to bodies whose properties are known the standard model yields radii that are systematically too large and albedos that are too small. To correct for departures from the assumptions of the standard model a concept called *thermal beaming* is introduced. This assumes that a body possesses an opposition effect in the infrared similar to that in the visible. The directional emissivity at zero phase in equation (15.42) is multiplied by an empirical parameter, the *thermal beaming factor* η, while keeping $\varepsilon_d = \varepsilon_h$ and $r_h = A_S$. Then equations (15.43) and (15.44) become

$$P_T = \frac{2}{3\eta}R^2 J(1 - A_S) \quad (15.45)$$

and

$$\frac{P_T}{P_V} = \frac{2}{3}q\frac{1 - A_S}{\eta A_S}. \quad (15.46)$$

The factors η and q are found by fitting these equations to observations of the Moon, the Galilean satellites, and asteroids whose sizes and albedos are known (Lebofsky *et al.*, 1986) and assuming that they are similar for other bodies. Then the values of R, A_S, and A_p can be found from measurements of P_V and P_T as before.

Spencer (1990) has shown that the thermal beaming factor can readily be accounted for by a thermal SHOE due to unresolved surface roughness (see equation [15.39]).

16

Simultaneous transport of energy by radiation and thermal conduction

16.1 Introduction

The temperature distribution in the regolith of a body of the solar system is important for several reasons. First, the mean temperature is one of the fundamental properties of an object; second, the regolith is the part of the planet studied by remote-sensing methods; and third, the temperature distribution in the upper layers constitutes the boundary condition for thermal models of the deeper interior. The temperatures in the regolith are governed by three energy-transport processes that interact within the medium: visible and near-IR solar radiation, thermal radiation, and heat conduction. However, most models treat the energy as being carried entirely by thermal conduction inside the medium, the interaction with radiation being treated only as a boundary condition. In this chapter all three processes occurring in the medium are treated simultaneously. In doing so it will be seen that radiative conductivity of heat and the solid-state greenhouse effect appear intrinsically without any ad hoc assumptions.

In Section 16.2 the time-dependent, one-dimensional radiative and heat conduction equations are introduced. The time-independent equations are solved analytically for two important cases: a layer of isotropically scattering particles heated from below and radiating into a vacuum, simulating conditions often used in laboratory IR and thermal measurements, and a semi-infinite medium of isotropic scatterers heated by incident visible light and radiating into a vacuum, simulating a planetary regolith in equilibrium with sunlight. In the final section some time-dependent problems are discussed briefly.

16.2 Equations

16.2.1 Basic equations

The system that will be considered in this chapter is a semi-infinite, horizontally stratified, particulate medium in a vacuum. As in the rest of this book, it is assumed that the radiance can be adequately described by the equation of radiative transfer. Let a subscript λ on a quantity explicitly denote that the quantity is dependent on wavelength. Then the radiative transfer equation for radiance $I_\lambda(z,\Omega,t)$ of wavelength λ at time t and depth z below the surface, traveling in direction Ω making an angle ϑ with the normal to the surface is

$$\cos\vartheta\frac{\partial I_\lambda(z,\Omega,t)}{\partial z} = -E_\lambda(z)I_\lambda(z,\Omega,t) + \frac{1}{4\pi}\int_{4\pi} I_\lambda(z,\Omega',t)G_\lambda(z,\Omega',\Omega)d\Omega'$$
$$+\frac{J_\lambda F(t)}{4\pi}G_\lambda(z,\Omega_i,\Omega)\exp\left[-\frac{1}{\mu_0(t)}\int_z^\infty E_\lambda(z')dz'\right] + \frac{A_\lambda}{\pi}U_\lambda(T), \quad (16.1)$$

where E_λ is the extinction coefficient, $G(z,\Omega,\Omega)$ is the volume angular scattering coefficient, J_λ is the incident irradiance, $F(t)$ describes the time-variation of the irradiance, $\mu_0 = \cos i$, A_λ is the volume-average absorption coefficient, and $U_\lambda(T)$ is the Planck function, which radiates isotropically.

The temperature T is governed by the heat equation,

$$\rho C\frac{\partial T(z,t)}{\partial t} = \frac{\partial}{\partial z}\left[k\frac{\partial T(z,t)}{\partial z}\right] + \int_\lambda A_\lambda(z)J_\lambda F(t)\exp\left[-\frac{1}{\mu_0(t)}\int_z^\infty E_\lambda(z')dz'\right]d\lambda$$
$$+\int_\lambda A_\lambda(z)\left[\int_{4\pi} I_\lambda(z,\Omega,t)d\Omega\right]d\lambda$$
$$-\int_\lambda A_\lambda(z)\left[\int_{4\pi}\frac{1}{\pi}U_\lambda(z,t)d\Omega\right]d\lambda, \quad (16.2)$$

where ρ is the bulk density, C is the specific heat per unit mass, and k is the solid-state thermal conductivity. The left-hand side of equation (16.2) describes the rate of change of heat content per unit volume in a layer of thickness dz at depth z. The first term on the right-hand side is the difference between the heat conducted into the layer and the heat conducted out. The second term on the right describes the absorption of the incident collimated irradiance that has penetrated to depth z and been absorbed in the layer. The third term is the diffuse radiance absorbed by the layer. The fourth term describes the loss of heat by thermal radiation from the particles in the layer at a temperature $T(z,t)$.

Note that the heat equation is coupled to the radiative-transfer equation only through integrals of the radiance over wavelength, not the spectral radiance. Therefore, solving this equation requires the calculation of the wavelength-integrated radiance. In some situations the only radiation is in the thermal IR region and the wavelength distribution approximates a Planck function. However, in a medium illuminated by a source of visible light, such as the Sun or an incandescent lamp, the radiance is appreciable in two spectral regions, the visual, including the near-UV/visible/near-IR, and the thermal IR. Then the spectrum of the radiance is bimodal, consisting of two separate wavelength regions: one with a spectrum similar to that of the incident irradiance J_λ and one with a spectrum similar to $U(T_C)$ where $U(T_C)$ is the Planck function of a black body at a temperature T_C that characterizes the medium. If the two spectra overlap they can still be separated by suitable extrapolation from the nonoverlapping regions. Then the radiative transfer equation can be separated into two discrete equations by integration and/or averaging over these spectral regions.

Let the subscript V denote integration or averaging over wavelength over the visual portion of the spectrum and T over the thermal IR. Define the following quantities,

$$J(t) = \int_{visual\,\lambda} J_\lambda(t)d\lambda,$$

$$I_V(z,\Omega,t) = \int_{visual\,\lambda} I_\lambda(z,\Omega,t)d\lambda,$$

$$I_T(z,\Omega,t) = \int_{thermal\,\lambda} I_\lambda(z,\Omega,t)d\lambda,$$

$$V(T) = \int_{thermal\,\lambda} U_\lambda(T)d\lambda = \sigma_0 T^4(z,t),$$

where *visual* λ denotes the visual wavelength region where I_λ is appreciable, and similarly for *thermal* λ. Define the visual value of any parameter X_λ by

$$X_V(z) = \int_{visual\,\lambda} X_\lambda(z)I_\lambda(z,\Omega,t)d\lambda / I_V(z,\Omega,t).$$

Assuming that the spectrum of I_λ is similar to that of J_λ we can calculate X_V to a good approximation by

$$X_V(z) \approx \int_0^\infty X_\lambda(z)J_\lambda(t)d\lambda / J(t).$$

Similarly, for the thermal region define

$$X_T(z) = \int_{thermal\,\lambda} X_\lambda(z)I_\lambda(z,\Omega,t)d\lambda / I_T(z,\Omega,t),$$

and to a good approximation

$$X_T(z) \approx \int_0^\infty X_\lambda(z)U_\lambda(T_C)d\lambda / V(T_C).$$

Integrating over each spectral region, the radiative-transfer equation can be separated into a wavelength-integrated equation for visual radiance

$$\cos\vartheta \frac{\partial I_V(z,\Omega,t)}{\partial z} = -E_V(z)I_V(z,\Omega,t) + \frac{1}{4\pi}\int_{4\pi} I_V(z,\Omega',t)G_V(z,\Omega',\Omega)d\Omega' + \frac{JF(t)}{4\pi}G_V(z,\Omega_i,\Omega)\exp\left[-\frac{1}{\mu_0(t)}\int_z^\infty E_V(z')dz'\right], \quad (16.3)$$

and an equation for thermal IR radiance

$$\cos\vartheta \frac{\partial I_T(z,\Omega,t)}{\partial z} = -E_T(z)I_T(z,\Omega,t) + \frac{1}{4\pi}\int_{4\pi} I_T(z,\Omega',t)G_T(z,\Omega',\Omega)d\Omega' + \frac{A_T}{\pi}V(T). \quad (16.4)$$

Then the heat equation becomes

$$\rho C\frac{\partial T(z,t)}{\partial t} = \frac{\partial}{\partial z}\left[k\frac{\partial T(z,t)}{\partial z}\right] + A_T(z)\left[\int_{4\pi} I_T(z,\Omega,t)d\Omega\right] - 4A_T V(T) + A_V(z)JF(t)\exp\left[-\frac{1}{\mu_0(t)}\int_z^\infty E_V(z')dz'\right] + A_V(z)\left[\int_{4\pi} I_V(z,\Omega,t)d\Omega\right]. \quad (16.5)$$

Equation (16.4) assumes that the incident irradiance is negligible in the thermal IR region. If it is not negligible then an additional term similar to the third term on the right of equation (16.3) must be added to (16.4).

Now

$$\tau_V = \int_z^\infty E_V dz, \text{ or } d\tau_V = -E_V dz,$$

$$\tau_T = \int_z^\infty E_T dz, \text{ or } d\tau_T = -E_T dz,$$

$$G_V(z,\Omega',\Omega) = S_v(z)p_V(z,\Omega',\Omega),$$

$$G_T(z,\Omega',\Omega) = S_T(z)p_T(z,\Omega',\Omega),$$

$$w_V = S_V/E_V, w_T = S_T/E_T,$$

where the Ss are the volume-scattering coefficients and the ws are the volume particle single-scattering albedos. It was shown in Chapter 15 that

$$A_\lambda = \varepsilon_T E_\lambda = \gamma_T^2 E_\lambda$$

where ε_T is the volume particle emissivity and $\gamma_T = \sqrt{1 - w_T}$. Then the separated equations become

$$-\cos\vartheta \frac{\partial I_V(\tau_V, \Omega, t)}{\partial \tau_V} = -I_V(\tau_V, \Omega, t) + \frac{w_V(\tau_V)}{4\pi} \int_{4\pi} I_V(\tau_V, .\Omega', t) p_V(z, \Omega', \Omega) d\Omega' + JF(t) \frac{w_V(\tau_V)}{4\pi} \exp\left[-\frac{\tau_V}{\mu_0(t)}\right], \tag{16.6}$$

$$-\cos\vartheta \frac{\partial I_T(\tau_T, \Omega, t)}{\partial \tau_T} = -I_T(\tau_T, \Omega, t) + \frac{w_T(\tau_T)}{4\pi} \int_{4\pi} I_T(\tau_T, \Omega', t) p_T(\tau_T, \Omega', \Omega) d\Omega' + \frac{\varepsilon_T(\tau_T)}{\pi} V(T), \tag{16.7}$$

$$\rho C \frac{\partial T(\tau_T, t)}{\partial t} = E_T(\tau_T) \frac{\partial}{\partial \tau_T}\left[E_T(\tau_T) k \frac{\partial T(\tau_T, t)}{\partial \tau_T}\right] + A_T(\tau_T) \int_{4\pi} I_T(\tau_T, \Omega, t) d\Omega - 4A_T(\tau_T) V(T) + A_V(\tau_T) JF(t) \exp\left[-\frac{\eta \tau_T}{\mu_0(t)}\right] + A_V(\tau_T) \int_{4\pi} I_V(\tau_T, \Omega, t) d\Omega, \tag{16.8}$$

where

$$\eta = \tau_V / \tau_T.$$

In order to solve these equations one temporal and six spatial boundary conditions are required, which are specific to a given situation. In time-dependent problems a condition on T at some given time must be specified. Three of the spatial boundary conditions are that the radiances and temperature must be finite everywhere, or if the medium is not infinitely thick, then values at the lower boundary must be specified. Two more result from the fact that at the upper boundary there cannot be any downward-going sources of visual or thermal radiance (the incident irradiance J is included as a source term in the visual equation and is not a boundary condition). The remaining boundary condition is on T and comes from the requirement that the flux of heat across the upper boundary can only be carried by thermal radiation and none by conduction. However, the conducted heat flux $k\partial T/\partial z$ must be continuous across any horizontal plane, including the upper surface of the medium. Since $k = 0$

above the surface but not below it, this requires

$$\frac{\partial T}{\partial z}(0,t) = 0. \tag{16.9}$$

16.2.2 *Equations for a uniform medium of isotropically scattering particles in the two-stream approximation*

16.2.2.1 *Equations*

In this section we will employ these equations to media of isotropic scatterers with properties independent of depth using the two-stream approximation. The solution for aniosotropically scattering particles has been published in Hapke (1996a). (However, anyone who consults that paper should be cautioned that it contains a large number of typographical errors which the author was not able to correct because of an editorial mixup.) Since the particles scatter equally in all directions, $p_V(z,\Omega',\Omega) = p_T(z,\Omega',\Omega) = 1$. In a uniform medium all parameters in the equations are independent of z. Then the equations become

$$-\cos\vartheta\frac{\partial I_V(\tau_V,\Omega,t)}{\partial\tau_V} = -I_V(\tau_V,\Omega,t) + \frac{w_V}{4\pi}\int_{4\pi} I_V(\tau_V,\Omega,t)d\Omega + JF(t)\frac{w_V}{4\pi}\exp\left[-\frac{\tau_V}{\mu_0(t)}\right], \tag{16.10}$$

$$-\cos\vartheta\frac{\partial I_T(\tau_T,\Omega,t)}{\partial\tau_T} = -I_T(\tau_T,\Omega,t) + \frac{w_T}{4\pi}\int_{4\pi} I_T(\tau_T,\Omega,t)d\Omega + \frac{\gamma_T^2}{\pi}V(T), \tag{16.11}$$

$$\rho C\frac{\partial T(\tau_T,t)}{\partial t} = kE_T^2\frac{\partial^2 T}{\partial\tau_T^2} + E_T\gamma_T^2\int_{4\pi} I_T(\tau_T,\Omega,t)d\Omega - 4E_T\gamma_T^2 V(T) + E_V\gamma_V^2 JF(t)\exp\left[-\frac{\eta\tau_T}{\mu_0(t)}\right] + E_V\gamma_V^2\int_{4\pi} I_V(\tau_T,\Omega,t)d\Omega. \tag{16.12}$$

In the two-stream approximation (see Chapters 7 and 8) the radiances in the equations are replaced by their averages over the upward-going and downward-going hemispheres. Let I_{1V} and I_{2V} be the average visual radiances going respectively in the upward and downward directions, and let I_{1T} and I_{2T} be similar thermal radiances,

$$I_{1V}(\tau_V, t) = \frac{1}{2\pi} \int_{\text{upper hemisphere}} I_V(\tau_V, \Omega, t) d\Omega, I_{2V}(\tau_V, t)$$

$$I_{2V}(\tau_V, t) = \frac{1}{2\pi} \int_{\text{lower hemisphere}} I_V(\tau_V, \Omega, t) d\Omega,$$

$$I_{1T}(\tau_V, t) = \frac{1}{2\pi} \int_{\text{upper hemisphere}} I_T(\tau_T, \Omega, t) d\Omega, I_{2T}(\tau_T, t)$$

$$I_{2T}(\tau_T, t) = \frac{1}{2\pi} \int_{\text{lower hemisphere}} I_T(\tau_T, \Omega, t) d\Omega.$$

Then the radiances averaged over all directions are

$$\varphi_V(z, t) = (I_{1V} + I_{2V})/2 = \frac{1}{4\pi} \int_{4\pi} I_V(z, \Omega, t) d\Omega,$$

$$\varphi_T(z, t) = (I_{1T} + I_{2T})/2 = \frac{1}{4\pi} \int_{4\pi} I_T(z, \Omega, t) d\Omega,$$

and the average hemispherical radiance differences are

$$\Delta\varphi_V = (I_{1V} - I_{2V})/2,$$

$$\Delta\varphi_T = (I_{1T} - I_{2T})/2.$$

After making these substitutions and combining the resulting expressions (see Chapter 8 for details), the equations for the visual radiance become

$$\frac{1}{4} \frac{\partial^2 \varphi_V(\tau_V, t)}{\partial \tau_V^2} = \gamma_V^2 \varphi_V(\tau_V, t) - JF(t) \frac{w_V}{4\pi} \exp\left[-\frac{\tau_V}{\mu_0(t)}\right], \tag{16.13a}$$

$$\Delta\varphi_V(\tau_V, t) = \frac{1}{2} \frac{\partial \varphi_V(\tau_V, t)}{\partial \tau_V}, \tag{16.13b}$$

with the condition at the surface that $I_{2V}(0, t) = 0$, which requires

$$\Delta\varphi_V(0, t) = \varphi_V(0, t) = \frac{1}{2} \frac{\partial \varphi_V}{\partial \tau_V}(0, t), \tag{16.13c}$$

and the equations for the thermal radiance are

$$\frac{1}{4\gamma_T^2} \frac{\partial^2 \varphi_T(\tau_T, t)}{\partial \tau_T^2} = \varphi_T(\tau_V, t) - \frac{1}{\pi} V(T), \tag{16.14a}$$

$$\Delta\varphi_T(\tau_T, t) = \frac{1}{2} \frac{\partial \varphi_T(\tau_T, t)}{\partial \tau_T}, \tag{16.14b}$$

with the condition at the surface that $I_{2T}(0,t)=0$, so that

$$\Delta\varphi_V(0,t)=\varphi_V(0,t)=\frac{1}{2}\frac{\partial\varphi_V}{\partial\tau_V}(0,t). \tag{16.14c}$$

The heat equation becomes

$$\begin{aligned}\rho C\frac{\partial T(\tau_T,t)}{\partial t}&=kE_T^2\frac{\partial^2 T(\tau_T,t)}{\partial\tau_T^2}+4\pi E_T\gamma_T^2\left\{\varphi_T(\tau_T,t)-\frac{1}{\pi}V(T)\right\}\\&+E_V\gamma_V^2\left\{JF(t)\exp\left[-\frac{\eta\tau_T}{\mu_0(t)}\right]+4\pi\varphi_V(\tau_Tt)\right\},\end{aligned} \tag{16.15a}$$

where $\eta=\tau_V/\tau_T=E_V/E_T$. An alternative form of the heat equation may be obtained by substituting equations (16.14a) into (16.15a), and after dividing by kE_T^2, gives

$$\begin{aligned}\frac{\rho C}{kE_T^2}\frac{\partial T(\tau_T,t)}{\partial t}&=\frac{\partial^2 T(\tau_T,t)}{\partial\tau_T^2}+\pi q\frac{\partial^2\varphi_T(\tau_T,t)}{\partial\tau_T^2}\\&+\eta\gamma_V^2 q\{J_V F(t)\exp[-\eta\tau_T/\mu_0(t)]+4\pi\varphi_V(\tau_T,t)\},\end{aligned} \tag{16.15b}$$

where

$$q=1/kE_T, \tag{16.15c}$$

with the condition at the surface for either form of the equation,

$$\frac{\partial T}{\partial\tau_T}(0,t)=0. \tag{16.15d}$$

It must be emphasized that all of these surface boundary conditions apply only to situations where there is no atmosphere and the surface sees cold space. If the medium is at the bottom of an atmosphere then at the surface the heat flux conducted through the medium must be continuous with the flux conducted through the air above the surface. Also, the downward-going radiances I_{2V} and I_{2T} at the surface must be equal to the visual and IR radiances scattered and radiated downward onto the surface by the atmosphere.

A further check on the validity of any solution is that deep in the interior the thermal radiance must be in thermodynamic equilibrium with the particle temperature. This requires

$$\varphi_T(\tau_T\gg 1,t)=\frac{1}{\pi}V(T). \tag{16.16}$$

16.2.2.2 The boundary-layer approximation

Because the equations for T and φ_T are nonlinear, exact solutions are difficult to obtain except numerically by computer, which is often inconvenient. A useful, approximate relation can be derived by realizing that because the temperature gradient is zero at the surface there is a small region near the surface in which the temperature is almost constant. Within this boundary layer equation (16.14a) is approximately

$$\frac{1}{4\gamma_T^2}\frac{\partial^2\varphi_T(\tau_T,t)}{\partial\tau_T^2} = \varphi_T(\tau_V,t) - \frac{1}{\pi}V(T_S),$$

where $T_S(t) = T(0,t)$ is the surface temperature. The solution to this equation that also satisfies the boundary conditions on φ_T is

$$\varphi_T(\tau_v,t) = \frac{V(T_S)}{\pi} - \frac{1}{1+\gamma_T}\frac{V(T_S)}{\pi}\exp(-2\gamma_T\tau_T). \tag{16.17}$$

At the surface this solution becomes

$$\varphi_S = \frac{\gamma_T}{1+\gamma_T}\frac{V(T_S)}{\pi} = \frac{\gamma_T}{1+\gamma_T}\frac{\sigma_0}{\pi}T_S^4, \tag{16.18}$$

where $\varphi_S(t) = \varphi(0,t)$. Equation(16.18) provides an alternative boundary condition at the surface.

Furthermore, the requirement that $\varphi_T \to V(T)/\pi$ as τ_T increases away from the surface suggests that a good approximate solution for φ_T can be obtained by replacing $V(T_S)$ by $V(T)$ in the first term of equation(16.17),

$$\begin{aligned}\varphi_T(\tau_v,t) &\approx \frac{V(T)}{\pi} - \frac{1}{1+\gamma_T}\frac{V(T_S)}{\pi}\exp(-2\gamma_T\tau_T)\\ &= \frac{\sigma_0}{\pi}T^4(\tau_T,t) - \frac{1}{1+\gamma_T}\frac{\sigma_0}{\pi}T_S^4\exp(-2\gamma_T\tau_T).\end{aligned} \tag{16.19}$$

Equation(16.19) is the boundary-layer approximation for φ_T. It is exact for both large and small τ_T. Comparisons with exact computer solutions of the radiative and heat equations and solutions using the boundary-layer approximation will be given below. It will be seen that in all cases the solutions differ by only a few percent.

16.2.2.3 The equations in reduced form

Frequently it is useful to put the equations in nondimensional form in such as way as to reduce the number of parameters to a minimum. Let t_R be an interval of time that characterizes the system of interest, and T_R be a temperature that is characteristic of the system. Define

$$\begin{aligned}\varphi_V{}^* &= \varphi_V/J,\\ \varphi_T{}^* &= \pi\varphi_T/\sigma_0 T_R^4,\\ T^* &= T/T_R\\ t^* &= t/t_R.\end{aligned}$$

Then equations (16.13)–(16.15) become

$$\frac{1}{4}\frac{\partial^2\varphi_V^*(\tau_V,t^*)}{\partial\tau_V^2}=\gamma_V^2\varphi_V^*(\tau_V,t^*)-F(t^*)\frac{w_V}{4\pi}\exp\left[-\frac{\tau_V}{\mu_0(t^*)}\right], \quad (16.20a)$$

$$\varphi_V^*(0,t^*)=\frac{1}{2}\frac{\partial\varphi_V^*}{\partial\tau_V}(0,t^*), \quad (16.20b)$$

$$\frac{1}{4\gamma_T^2}\frac{\partial^2\varphi_T^*(\tau_T,t^*)}{\partial\tau_T^2}=\varphi_T^*(\tau_V,t^*)-T^{*4}, \quad (16.21a)$$

$$\varphi_V^*(0,t^*)=\frac{1}{2}\frac{\partial\varphi_V^*}{\partial\tau_V}(0,t^*), \quad (16.21b)$$

$$\frac{t_D}{t_R}\frac{\partial T^*(\tau_T,t^*)}{\partial t^*}=\frac{\partial^2T^*(\tau_T,t^*)}{\partial\tau_T^2}+q_T\frac{\partial^2\varphi_T^*(\tau_T,t^*)}{\partial\tau_T^2}$$
$$+F(t^*)q_V\left\{\exp\left[-\eta\tau_T/\mu_0(t^*)\right]+4\pi\varphi_V^*(\tau_T,t^*)\right\}, \quad (16.22a)$$

$$\frac{\partial T}{\partial\tau_T}(0,t^*)=0, \quad (16.22b)$$

where

$$t_D=\rho C/kE_T^2 \quad (16.23a)$$

$$q_T=\sigma_0T_R^3/kE_T, \quad (16.23b)$$

$$q_V=J\eta\gamma_v^2/kE_TT_R, \quad (16.23c)$$

$$\eta=E_V/E_T. \quad (16.23d)$$

16.3 Some time-independent applications of the equations

16.3.1 An optically thick layer heated from below

In this section we will address the problem of a particulate medium of thickness $Z\gg 1/E_T$ heated from below by a plate at constant temperature T_P and radiating into cold space from the top surface. This might represent the situation in a laboratory measurement of emissivity where the surface of the medium sees only cold surfaces close to absolute zero.

For this problem all quantities are independent of t and the visual radiance is negligible, so $F(t)=1$, and $\partial T/dt=J_V=\varphi_V=0$. Then equation (16.15b) becomes

$$\frac{d^2T(\tau_T)}{d\tau_T^2} + \pi q \frac{d^2\varphi_T(\tau_T)}{d\tau_T^2} = 0. \tag{16.24}$$

Integrating from 0 to τ_T gives

$$\frac{dT(\tau_T)}{d\tau_T} - \frac{dT(0)}{d\tau_T} + \pi q \frac{d\varphi_T(\tau_T)}{d\tau_T} - \pi q \frac{d\varphi_T(0)}{d\tau_T} = 0.$$

Applying boundary conditions (16.14c) and (16.15c) this is

$$\frac{dT(\tau_T)}{d\tau_T} + \pi q \frac{d\varphi_T(\tau_T)}{d\tau_T} = 2\pi q \varphi_S. \tag{16.25}$$

Integrating again we obtain

$$T(\tau_T) + \pi q \varphi_T(\tau_T) = T_S + \pi q \varphi_S + 2\pi q \varphi_S \tau_T. \tag{16.26}$$

This equation shows that the quantity $(T + \pi q \varphi_T)$ is linearly proportional to τ_T.

At the bottom of the layer $T = T_P$, $\varphi_T = \varphi_P$, and $\tau_T = \tau_P = E_T Z$, and equation (16.26) is

$$T_P + \pi q \varphi_P = T_S + \pi q \varphi_S + 2\pi q \varphi_S \tau_P. \tag{16.27}$$

Since the layer is optically thick, the radiation is in thermodynamic equilibrium with the temperature of the grains at the plate; hence $\varphi_P = V(T_P)/\pi$. Inserting this, together with equation (16.18), into equation (16.27) gives

$$T_P + q\sigma_0 T_P^4 = T_S + q\frac{\gamma_T}{1+\gamma_T}\sigma_0 T_S^4(1+2\tau_P), \tag{16.28}$$

where T_P is the known plate temperature. Equation (16.28) is a transcendental equation for T_S that can be rapidly solved by iteration. Once T_S is obtained it can be inserted into equation (16.18) to get φ_S.

Inserting equation (16.19) into (16.26), we obtain

$$T(\tau_T) + q\left[\sigma_0 T^4(\tau_T) - \frac{1}{1+\gamma_T}\sigma_0 T_S^4 \exp(-2\gamma_T \tau_T)\right]$$
$$= T_S + q\frac{\gamma_T}{1+\gamma_T}\sigma_0 T_S^4(1+2\tau_T), \tag{16.29}$$

which can be solved by iteration. The values for $T(\tau_T)$ are then inserted into equation (16.19) to give the solution for $\varphi_T(\tau_T)$. This completes the solution of the problem using the boundary layer approximation. An example of the temperature and thermal radiance calculated from equations (16.19) and (16.29) is given in Figure 16.1. For comparison, the figure also shows the exact numerical solution. The temperature shown is relative to that of the heating plate T_P. The IR radiance is relative to the thermal equilibrium radiance at the plate $\sigma_0 T_P^4/\pi$.

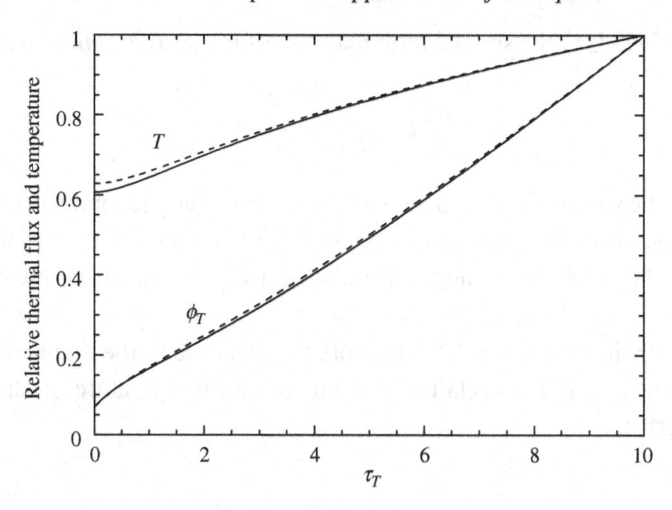

Figure 16.1 Reduced temperature and thermal radiance as a function of thermal optical depth in a layer of powder in a vacuum lying on a hot plate and radiating into a cold environment from its upper surface. The heater plate is located at an optical depth of 10. Other parameters are $w_T = 0.36$ and $q = 0.26$. Solid line is the numerical solution; dashed line is the approximate solution using the boundary-layer approximation. (Reproduced from Hapke [1996], copyright 1996 by the American Geophysical Union.)

16.3.2 Radiative conductivity

The power radiated from the upper surface of the slab is

$$P_{rad} = \int_0^{\pi/2} I_{1T}(0) 2\pi \cos e \sin e de = \pi I_{1T}(0) = 2\pi \varphi_S.$$

Since there are no sources or sinks of energy in the slab the heat flux $2\pi\phi_S$ must be constant through the slab. From equations(16.19) and(16.25)

$$\begin{aligned} P_{rad} &= 2\pi\varphi_S = kE_T \frac{dT(z)}{d\tau_T} + \pi \frac{d\varphi_T(z)}{d\tau_T} = k\frac{dT(z)}{dz} + \frac{\pi}{E_T}\frac{d\varphi_T(z)}{dz} \\ &= k\frac{dT(z)}{dz} + \frac{\pi}{E_T}\frac{d}{dz}\left[\frac{V(T)}{\pi} - \frac{1}{1+\gamma_T}\frac{V(T_S)}{\pi}\exp(-2\gamma_T E_T z)\right] \\ &= k\frac{dT(z)}{dz} + \frac{4}{E_T}\sigma_0 T^3(z)\frac{dT(z)}{dz} + \frac{2\gamma_T}{1+\gamma_T}\sigma_0 T_S^4 \exp(-2\gamma_T E_T z). \quad (16.30) \end{aligned}$$

The first term on the right in equation (16.30) is the heat carried by ordinary conduction, where the coefficient of dT/dz is the solid-state conductivity k. The last two terms are the flux of energy carried by radiation. Far from the surface where the exponential is negligible the second term is the heat carried by radiation. Thus, the coefficient of dT/dz in the second term may be defined as the *radiative conductivity*

$$k_{rad} = \frac{4}{E_T}\sigma_0 T^3(z). \tag{16.31}$$

However, as the surface is approached within the boundary layer the first and second terms decrease while the third term increases until at the surface all of the heat flux is carried by the third term, where it is radiated away with an effective temperature T_S and hemispherical emissivity $\varepsilon_h = 2\gamma_T(1+\gamma_T)$. Compare this expression for the emissivity with equations (15.24). Note that because of the boundary condition that $dT/dz(0) = 0$, P_{rad} is relatively insensitive to temperature gradients deeper under the surface.

16.3.3 Planetary regolith in equilibrium with sunlight

The second system to be considered is a semi-infinite medium of isotropically scattering particles illuminated at angle of incidence i by a constant source of visible light and radiating into a vacuum. This simulates the surface of a solar-system body tidally locked so the same side always faces the Sun. It also simulates conditions sometimes used in laboratory thermal measurements.

In the classical treatment of the steady-state problem all absorption and emission of radiation is assumed to take place at the surface and constitutes the boundary conditions. In that case the heat equation is simply $d^2T/dz^2 = 0$. The only solution to this equation that satisfies the boundary conditions is $T(z) =$ constant. Then the visual light absorbed $(1-A_h)J\mu_0$ must be equal to the IR radiated from the surface $\varepsilon_h\sigma_0 T^4$, so that

$$T = [(1 - A_h)\, J\mu_0/\varepsilon_h\sigma_0]^{1/4}.$$

For a perfect black body, $A_h = 0$ and $\varepsilon_h = 1$, so that at $\mu_0 = 1$, $T = T_{BB}$, where T_{BB} is the *black-body radiative equilibrium temperature*

$$T_{BB} = (J_V/\sigma_0)^{1/4}. \tag{16.32}$$

Although no real object is a perfect black body, T_{BB} is a useful quantity with which to compare actual temperatures on bodies of the solar system. In this section we will derive more realistic expressions for the temperature and radiation within the medium, using the two-stream and boundary-layer approximations.

Equation (16.13a) for the directionally averaged visual radiance is independent of the thermal radiative and heat equations and can be solved separately. The solution with $F(t) = 1$ that satisfies the condition that the radiance is finite everywhere is given by equations (8.39), (8.42), and (8.43). Inserting these solutions into equation

(16.15b) gives

$$\frac{d^2T(\tau_T)}{d\tau_T^2} + \pi q \frac{d^2\phi_T}{d\tau_T^2} = -\eta q \gamma_v^2 \left[A\exp(-\eta\tau_T/\mu_0) + B\exp(-2\gamma_V\eta\tau_T)\right], \tag{16.33a}$$

where

$$A = J\frac{(1-4\mu_0^2)}{1-4\gamma_V^2\mu_0^2}, \tag{16.33b}$$

$$B = J\frac{2\mu_0(1-\gamma_V)(1+2\mu_0)}{1-4\gamma_V^2\mu_0^2}, \tag{16.33c}$$

and $q = 1/kE_T$.

Integrating equation (16.33a) from 0 to τ_T and applying the boundary conditions gives

$$\frac{dT(\tau_T)}{d\tau_T} + \pi q\left[\frac{d\varphi_T(\tau_T)}{d\tau_T} - 2\varphi_S\right]$$
$$= \gamma_V^2 q\left\{A\mu_0[\exp(-\eta\tau_T/\mu_0) - 1] + \frac{B}{2\gamma_V}[\exp(-2\gamma_V\eta\tau_T) - 1]\right\}$$

where $\varphi_S = \varphi_T(0)$. Integrating again, we obtain

$$T(\tau_T) - T_S + \pi q\left[\varphi_T(\tau_T) - \varphi_S - 2\varphi_S\tau_T\right]$$
$$= -\gamma_V^2 q\left(A\mu_0\left\{\frac{\mu_0}{\eta}[\exp(-\eta\tau_T/\mu_0) - 1] + \tau_T\right\}\right.$$
$$\left. + \frac{B}{2\gamma_V}\left\{\frac{1}{2\eta\gamma_V}[\exp(-2\gamma_V\eta\tau_T) - 1] + \tau_T\right\}\right), \tag{16.34}$$

where $T_S = T(0)$. Now, $T(\tau_T)$ must remain finite as $\tau_T \to \infty$. The only way this can be true is if the sum of the coefficients of τ_T vanishes. Thus,

$$2\varphi_S = \gamma_V^2(A\mu_0 + B/2\gamma_V)/\pi.$$

After substituting for A and B this becomes

$$\varphi_S = J\frac{\gamma_V\mu_0}{2\pi}\frac{1+2\mu_0}{1+2\gamma_V\mu_0}, \tag{16.35}$$

which becomes the boundary condition on φ_T. Then equation (16.34) is

$$T(\tau_T) - T_S + \pi q\left[\varphi_T(\tau_T) - \varphi_S\right]$$
$$= \frac{q}{\eta}\left\{A\gamma_V^2\mu_0^2[1 - \exp(-\eta\tau_T/\mu_0)] + \frac{B}{4}[1 - \exp(-2\gamma_V\eta\tau_T)]\right\}. \tag{16.36}$$

Substituting the boundary-layer approximation, equation (16.19), and putting $\tau_T = E_T|z|$, we obtain

$$T(z) - T_S + \frac{1}{kE_T}\left[\sigma_0 T^4(z) - \frac{1}{1+\gamma_T}\sigma_0 T_S^4 \exp(-2\gamma_T E_T|z|) - \pi\phi_S\right]$$
$$= \frac{1}{kE_V}\left\{A\gamma_V^2\mu_0^2[1-\exp(-E_V|z|/\mu_0)] + \frac{B}{4}[1-\exp(-2\gamma_V E_V|z|)]\right\}, \tag{16.37}$$

where, from equations (16.18) and (16.30),

$$T_S = \left(\frac{1+\gamma_T}{\gamma_T}\varphi_S\right)^{1/4} = \left[J\frac{\mu_0(1+\gamma_V)}{2\sigma_0}\frac{1+2\mu_0}{1+2\gamma_V\mu_0}\right]^{1/4}. \tag{16.38}$$

$T(z)$ can rapidly be found from equation (16.37) by iteration, and $\varphi_T(z)$ from the boundary-layer approximation,

$$\varphi_T(z) = \frac{\sigma_0}{\pi}T^4(z) - \frac{1}{1+\gamma_V}\frac{\sigma_0}{\pi}T_S^4\exp(-2\gamma_T E_T|z|). \tag{16.39}$$

This completes the solution for the distribution of temperature and radiances in the medium.

Figures 16.2–16.5 show the temperature and thermal radiance as a function of depth for several values of the parameters in these equations. For comparison, the exact numerical solutions are also shown. Since the particles are assumed to be large compared to the IR wavelength, the extinction lengths should not vary strongly with wavelength. Hence, for these figures it was assumed that $E_V = E_T$, so that $\eta = 1$ and $\tau_T = \tau_V$. The low value of the visual single-scattering albedo of $w_V = 0.36$ might be representative of a relatively low-albedo body, such as the Moon or an asteroid. The high value of $w_V = 0.91$ might be representative of somewhat dirty ice on the surface of a rocky–icy satellite, while $w_V = 0.9996$ is representative of clean, fresh snow. In the thermal IR $w_T = 0.36$ is probably representative of most materials. A value of $w_T = 0.91$ probably is unrealistically high because most materials of interest for planetary remote sensing have restrahlen bands at IR wavelengths. However, it is included to show the effect of a high IR albedo, such as that of an alkali halide. The thermal radiances shown are relative to J/π, which is the IR radiance from the surface of a nonconducting black body illuminated by visual irradiance J. The temperatures are relative to the black-body temperature $(\pi J/\sigma_0)^{1/4}$ of the surface of the body.

In all cases, as the optical depth increases φ_T rises rapidly from a value of $\varphi_T = \varphi_S$ at the surface and then levels off to either a constant value or to a more gentle rate of increase. The rapid rise is a direct result of leakage of thermal radiation from the surface. Thermodynamic equilibrium is approached when $\tau_T > \sim 3$.

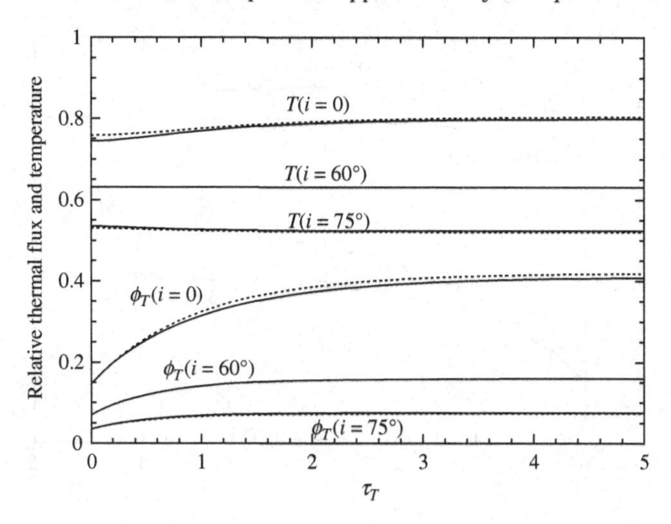

Figure 16.2 Reduced temperature and thermal radiance as a function of thermal optical depth in a regolith in vacuum in equilibrium with sunlight, showing the effect of varying the angle of incidence i. Other parameters are $w_V = 0.36$, $w_T = 0.36$, $q = 0.26$. Solid line is the numerical solution; dashed line is the approximate solution using the boundary-layer approximation. (Reproduced from Hapke [1996], copyright 1996 by the American Geophysical Union.)

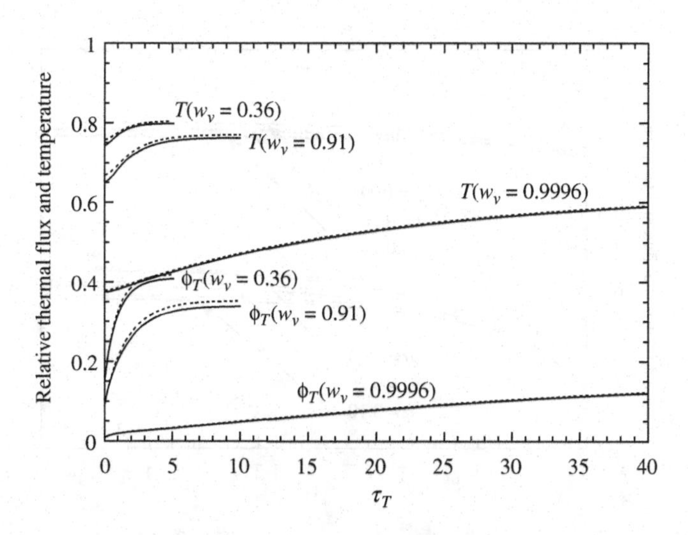

Figure 16.3 Reduced temperature and thermal radiance as a function of thermal optical depth in a regolith in vacuum in equilibrium with sunlight, showing the effect of varying the visual single-scattering albedo w_V. Other parameters are $w_T = 0.36$, $q = 0.26$, $i = 0$. Solid line is the numerical solution; dashed line is the approximate solution using the boundary-layer approximation. (Reproduced from Hapke [1996], copyright 1996 by the American Geophysical Union.)

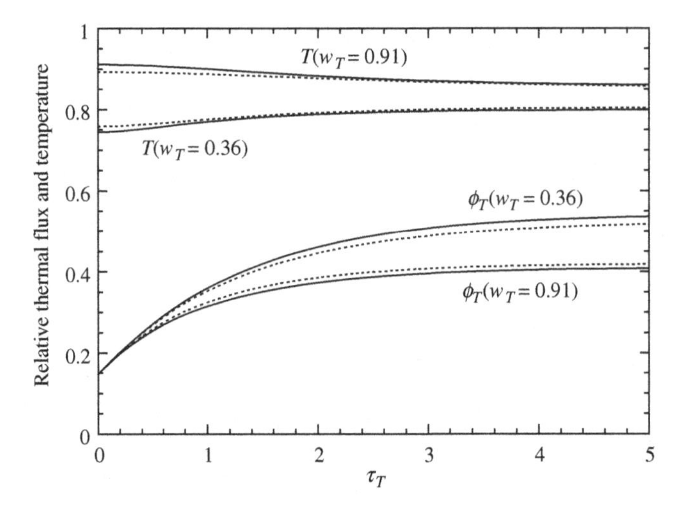

Figure 16.4 Reduced temperature and thermal radiance as a function of thermal optical depth in a regolith in vacuum in equilibrium with sunlight, showing the effect of varying the thermal single-scattering albedo w_T. Other parameters are $w_V = 0.36$, $q = 0.26$, $i = 0$. Solid line is the numerical solution; dashed line is the approximate solution using the boundary-layer approximation. (Reproduced from Hapke [1996], copyright 1996 by the American Geophysical Union.)

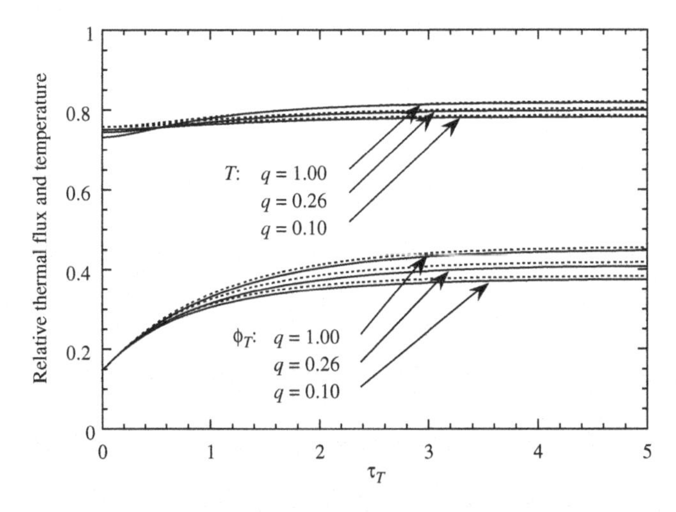

Figure 16.5 Reduced temperature and thermal radiance as a function of thermal optical depth in a regolith in vacuum in equilibrium with sunlight, showing the effect of varying the parameter $q = 1/kE_T$. Other parameters are $w_V = 0.36$, $w_T = 0.36$, $i = 0$.. Solid line is the numerical solution; dashed line is the approximate solution using the boundary-layer approximation. (Reproduced from Hapke [1996], copyright 1996 by the American Geophysical Union.)

The behavior of T is more complex. The slope of T is zero at the surface, as required by the boundary condition, but as the optical depth increases T may exhibit a positive or negative slope, depending on the value of γ_T compared with γ_V and μ_0. If these quantities are comparable T has a positive slope. However, the slope of T may be small or even negative if γ_T is small. Heating by unscattered visible irradiance occurs over an optical depth $\sim 1/\mu_0$, and by multiply scattered visible radiance over a distance $\sim 1/\gamma_V$, but the IR flux is radiated from the surface from optical depths $\sim 1/\gamma_T$. If $1/\gamma_T > 1/\gamma_V$ or $1/\mu_0$ then heat must be conducted from the surface to supply power to the deeper IR radiative sink. This requires a negative temperature gradient. It should also be emphasized that the widths of the gradients in actual depth z, rather than optical depth τ_T, will depend on particle size and filling factor.

16.3.4 The solid-state greenhouse effect

An atmospheric greenhouse effect can occur on a planet whose atmosphere is transparent to visible light but opaque to thermal radiation. The energy of visible sunlight is absorbed at the ground level but radiated back into space as thermal radiation from the upper layers of the atmosphere. This requires a temperature gradient to transport heat from the surface to the radiating layer. On Venus this effect produces a surface temperature that is nearly triple the black-body temperature. Brown and Matson (1987) and Matson and Brown (1989) have pointed out that an analogous solid-state greenhouse effect could occur in a planetary regolith. Visible sunlight is deposited over a depth $\sim 1/E_V\sqrt{1-w_V}$, but thermally radiated from a depth $\sim 1/E_T\sqrt{1-w_T}$. If the regolith is weakly absorbing in the visible and strongly absorbing in the IR so that $w_V \sim 1$ but $w_T \ll 1$ this could cause a considerable increase in temperature from the surface to the interior.

The increase in temperature from the surface to the deep interior in a regolith irradiated by sunlight can be calculated from equation (16.37) by letting $z \to \infty$,

$$T(\infty) = T_S + \frac{1}{kE_T}\left[\pi\varphi_S - \sigma_0 T^4(\infty)\right] + \frac{1}{kE_V}\left[A\gamma_V^2\mu_0^2 + \frac{B}{4}\right].$$

After substituting for A, B, and φ_S from equations (16.33) and (16.35) this equation can be considerably simplified. The final result is

$$T(\infty) = T_S - \frac{\sigma_0}{kE_T}T^4(\infty) + \frac{J}{2kE_T}\mu_0(1+2\mu_0), \tag{16.40}$$

where T_S is given by equation (16.38).

The surprising result is that equation (16.40) is independent of both w_V and w_T. The reason is that although a small absorption coefficient allows the visible radiance to be deposited deep in the regolith, it also causes a high visual albedo so

Table 16.1. *The solid-state greenhouse effect on a nonrotating body*

Location	R_e(AU)	kE_T(W m^{-2} K^{-1})	T_S(K)	$T(\infty)$(K)	ΔT_{Gr}(K)
Moon	1	10	397	422	25
Jovian satellites	5	1	176	188	12
	5	10	176	180	4
	5	100	176	178	2

that less sunlight is absorbed by the medium. The two effects exactly cancel each other when the particles are isotropic scatterers so that only a relatively small rise in temperature occurs.

Let us estimate the greenhouse effect for a few locations in the solar system. The incident irradiance $J = S_\odot/R_\odot^2$, where $S_\odot = 1360$ W m^{-2} is the irradiance of sunlight at the earth and $R_\odot$ is the distance of the body from the Sun in astronomical units. Assuming that the particles are equant and larger than the wavelength, $Q_E = 1$, the filling factor is $\phi = \pi ND^3/6$, and the extinction coefficient is $E_T = E_V = N\sigma Q_E = (6\phi/\pi D^3)(\pi D^{2-}/4)Q_E = 3\phi/2D$. Thus $kE_T = 3k\phi/2D$. The solid-state thermal conductivity is the bulk conductivity of the medium and not just the conductivity of the material of which the particles are made. It includes the fact that the particles are in good thermal contact with each other in only a few small places on their surfaces. Hence the bulk thermal conductivity is primarily determined by the sizes of the contact areas and is relatively insensitive to the composition of the material. Fountain and West (1970) measured the radiative and solid-state thermal conductivity of lunar soil in the laboratory and found $k \approx 0.001$ W m^{-1}K^{-1}. Since the atmosphereless, icy bodies of the solar system have regoliths that, like the Moon's, are generated by meteoritic impacts, Matson and Brown (1989) argued that this value if k is probably also representative of the icy bodies. For lunar soil, $\phi \sim 0.30$ (Carrier *et al.*, 1973) and $D \sim 50\,\mu$m (McKay *et al.*, 1974), so that $kE_T \sim 10$ W m^{-2}K^{-1}.

The time-independent greenhouse rise in temperature $\Delta T_{Gr} = T(\infty) - T_S$ is calculated for $\mu_0 = 1$ in Table 16.1 for bodies at the location of the Moon and the Jovian satellites for several values of kE_T. It is seen to be only a few 10's of degrees. A somewhat larger greenhouse effect can occur if the particles in the medium scatter anisotropically (Hapke, 1996b). If the particles are forward-scattering in the visual, but backscattering in the IR, and if the bodies are allowed to rotate, temperature rises as large as 40 K are possible in the regoliths of the Jovian icy satellites.

A solid-state greenhouse effect is intrinsically different from an atmospheric greenhouse effect. In the latter the optical depth at which visual energy is deposited is essentially independent of the bolometric albedo, whereas in the former the two are tightly coupled. In addition, radiation can contribute to the thermal conductivity even at the low temperatures of the Jovian satellites and reduce the temperature differences within the medium.

16.3.5 Physical meaning of the parameters in the reduced equations

The quantity $\rho C/k$ that appears in equations (16.23) is known as the *thermal diffusivity*. The related quantity $t_D = \rho C/kE_T^2$ has the units of time and is called the *thermal diffusion time*. The physical meaning of the diffusion time may be seen by writing it in the form

$$t_D = \frac{\rho C \Delta T l_{eT}}{k\frac{\Delta T}{l_{eT}}}$$

where ΔT is a small rise in temperature and $l_{eT} = 1/E_T$, the extinction length for thermal radiation Now, $\rho C \Delta T l_{eT}$ is the amount of heat necessary to increase the temperature of a layer of thickness l_{eT} by ΔT, and $k\Delta T/l_{eT}$ is the flux of heat conducted through the layer. Thus, t_D is a measure of the time required to change the temperature distribution of a system by solid-state conduction when some parameter the system is changed. For typical soils in a vacuum the thermal diffusion time is of the order of seconds.

The diffusion time may also be written

$$t_D = \frac{k\rho C}{k^2 E_T^2} = (q\Gamma)^2$$

where $q = 1/kE_T$ and

$$\Gamma = \sqrt{k\rho C} \tag{16.41}$$

is the *thermal inertia.* If the thermal inertia is high, relatively large amounts of heat are required to change the temperature of the medium appreciably, and heat is efficiently conducted between the surface and the interior of the medium, so that only small temperature fluctuations result from changing J. Conversely, a small value of the thermal inertia means that changing J causes large temperature fluctuations.

From equation (16.23b) the parameter q_T is

$$q_T = \frac{\sigma_0 T_R^3}{kE_T} = \frac{1}{4}\frac{k_{rad}}{k}.$$

Thus q_T is a measure of the flux of heat carried by thermal radiation relative to that carried by conduction.

Assuming $E_V \approx E_T$ in equation (16.23c) the parameter $q_V = J(1-w_V)/kE_T T_R$. If we take $T_R = T_{BB} = (J/\sigma_0)^{1/4}$ this becomes

$$q_V = \frac{\sigma_0 T_{BB}^3 (1-w_V)}{kE_T} = \frac{1}{4}\frac{k_{rad}}{k}(1-w_V).$$

Now, $(1-w_V)$ is the fraction of J absorbed per unit volume. Hence, q_V is a measure of the amount of absorbed visible light carried away by thermal radiation relative to conduction.

16.4 Time-dependent radiative and conductive models

Thus far, only problems that are constant in time have been considered because they allow analytic solutions. However, in order to realistically model the thermal regime in a planetary regolith, time-dependent models must be considered. Unfortunately this renders the problem analytically intractable and solutions must be obtained numerically.

One of the early efforts to do this was the classical paper of Wesselink (1948). He assumed a constant thermal conductivity and that the visual and thermal radiation interacted with the medium only at the surface. Then the heat equation is simply

$$\rho C \frac{\partial T(z,t)}{\partial t} = k \frac{\partial^2 T(z,t)}{\partial z^2},$$

with the boundary condition that the conductive heat flux under the surface must be equal to the thermal and visual radiative fluxes above the surface,

$$JF(t)\mu_0(t) = \varepsilon_h \sigma_0 T^4(0,t) + k\frac{\partial T(z,t)}{\partial z},$$

where $F(t)$ is a binary function equal to 1 during the day and 0 at night. To reduce these equations to dimensionless form, let $T\# = T/T_R$, where $T_R = (J_V/\varepsilon_h \sigma_0)^{1/4}$, $t\# = t/P$, where P is the Moon's period of rotation, and $z\# = z/z_R$, where $z_R = (kP/\rho C)^{1/2}$. Making these substitutions, the differential equation becomes

$$\frac{\partial T^\#(z^\#, t^\#)}{\partial t^\#} = \frac{\partial^2 T^\#(z^\#, t^\#)}{\partial z^{\#2}},$$

and the boundary condition is

$$F(t^\#)\mu_0(t^\#) = T^{\#^4} - \frac{\Gamma}{\varepsilon_h \sigma_0 T_R^3 \sqrt{P}} \frac{\partial T^\#(0,t^\#)}{\partial z^\#}.$$

Now, the values of the quantities σ_0, T_R, and P are known, and to a good approximation $\varepsilon_h \approx 1$, so Γ is the only unknown. By comparing solutions of these equations to measured data Γ can be found. The best value of the thermal inertia found by

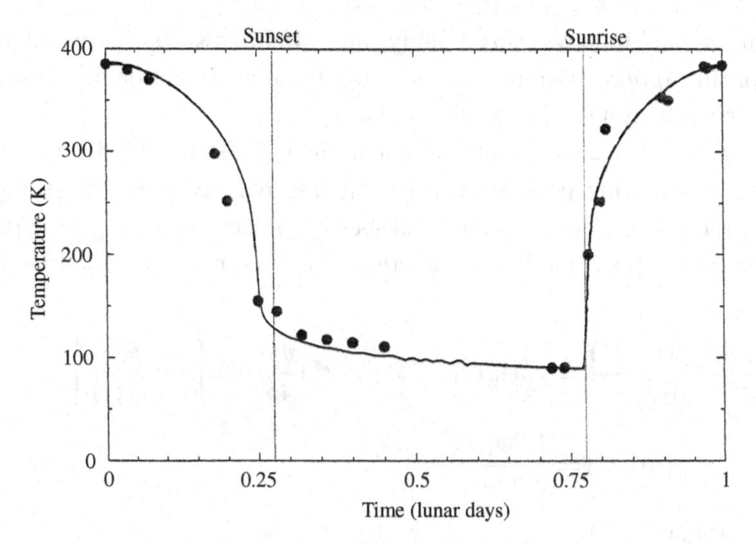

Figure 16.6 Brightness temperature of the lunar surface versus time for an area on the equator. Line is the model with $\Gamma = 43$ J m^{-2} K^{-1} s$^{-1/2}$, $kE_T = 18.4$ J s^{-1} m^{-2} s^{-1}. Dots are observations of Shorthill (1972). (Reproduced from Hale and Hapke [2002], copyright 2002 with permission of Elsevier.)

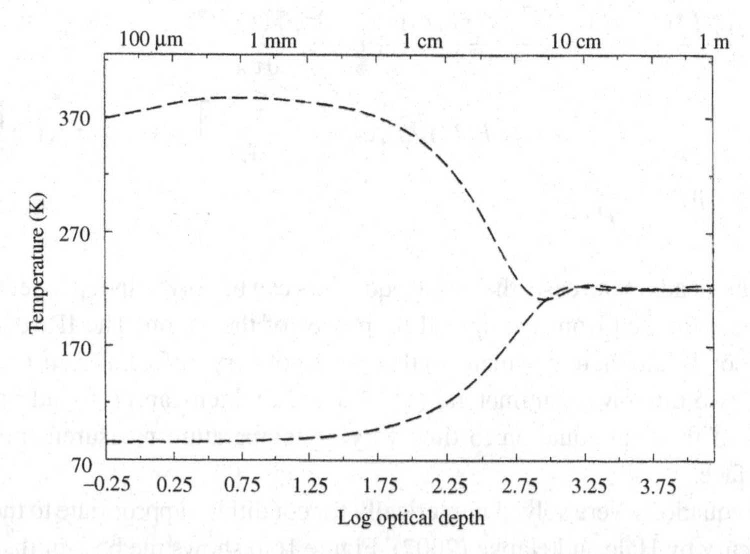

Figure 16.7 Calculated temperature versus thermal optical depth in the lunar regolith for an area on the lunar equator for the same model parameters as in Figure 16.6. The two curves show the temperatures at noon and just before dawn. (Reproduced from Hale and Hapke [2002], copyright 2002 with permission of Elsevier.)

Wesselink for the Moon was surprisingly small, 43 Jm^{-2}K^{-1}s$^{-1/2}$, which implied, well before the *Apollo* missions, that the regolith was an exceptionally good thermal insulator, consistent with fine powder in vacuum.

The more realistic time-dependent equations (16.13)–(16.15) for a regolith of isotropically scattering particles can be put into reduced form by letting $T_R = T_{BB}$, where $T_{BB} = (J/\sigma_0)^{1/4}$ is the black-body temperature, $t_R = P$, the period of revolution, and $F(t)$ is the binary function. Then the reduced equations take the form

$$\frac{1}{4}\frac{\partial^2\varphi_V^*(\tau_V, t^*)}{\partial\tau_V^2} = \gamma_V^2\varphi_V^*(\tau_V, t^*) - F(t^*)\frac{w_V}{4\pi}\exp\left[-\frac{\tau_V}{\mu_0(t^*)}\right],$$

$$\varphi_V^*(0, t^*) = \frac{1}{2}\frac{\partial\varphi_V^*}{\partial\tau_V}(0, t^*),$$

$$\frac{1}{4\gamma_T^2}\frac{\partial^2\varphi_T^*(\tau_T, t^*)}{\partial\tau_T^2} = \varphi_T^*(\tau_V, t^*) - T^{*4},$$

$$\varphi_T^*(0, t^*) = \frac{1}{2}\frac{\partial\varphi_T^*}{\partial\tau_V}(0, t^*),$$

$$\frac{t_D}{P}\frac{\partial T^*(\tau_T, t^*)}{\partial t^*} = \frac{\partial^2 T^*(\tau_T, t^*)}{\partial\tau_T^2} + q_T\frac{\partial^2\varphi_T^*(\tau_T, t^*)}{\partial\tau_T^2}$$
$$+ q_V F(t^*)\gamma_V^2\left\{\exp\left[-\frac{\tau_T}{\mu_0(t^*)}\right] + 4\pi\varphi_V^*(\tau_T, t^*)\right\},$$

$$\frac{\partial T^*}{\partial\tau_T}(0, t^*) = 0.$$

As with the steady-state case, the visual equations can be solved independently, and, γ_V can be estimated from the optical properties of the Moon. The IR reflectance of the Moon is known to be small, so that γ_T is not very different from 1.00. Thus there are two unknown parameters, (kE_T) and Γ, which can be found by fitting solutions of the heat equation to time-varying temperature measurements of the lunar surface.

These equations were solved numerically for conditions appropriate to the Moon and Mercury by Hale and Hapke (2002). Figure 16.6 shows the best-fit theoretical solutions to thermal measurements of an area on the lunar equator. The values found by these authors were $\Gamma = 43\,\mathrm{Jm^{-2}K^{-1}s^{-1/2}}$, which is the same value obtained by Wesselink, and $kE_T = 18.4\,\mathrm{J\,s^{-1}\,m^{-2}\,K^{-1}}$. Figure 16.7 shows the calculated distribution of temperature with depth corresponding to these values at noon and just before dawn. Note that the diurnal temperature fluctuations are confined to a layer a few centimeters thick.

Appendix A

A brief review of vector calculus

Vectors are simply a kind of shorthand notation in which complicated quantities can be expressed succinctly. In this appendix a number of useful relations in vector calculus are given without proof. More detailed discussions can be found in standard texts, such as those by Lass (1950), Chorlton (1976), or Arfken and Weber (2005).

Consider two vectors $\mathbf{A}(x, y, z)$ and $\mathbf{B}(x, y, z)$ differing in direction by angle θ (see Figure A.1):

$$\mathbf{A} = \mathbf{u}_x A_x(x, y, z) + \mathbf{u}_y A_y(x, y, z) + \mathbf{u}_z A_z(x, y, z) \tag{A.1}$$

$$\mathbf{B} = \mathbf{u}_x B_x(x, y, z) + \mathbf{u}_y B_y(x, y, z) + \mathbf{u}_z B_z(x, y, z) \tag{A.2}$$

where u_x, u_y, and u_z are unit vectors in the x, y, and z directions, respectively. Then **A** and **B** can be multiplied together in two ways: as a scalar product (dot product)

$$\mathbf{A} \cdot \mathbf{B} = |A||B| \cos\theta = A_x B_x + A_y B_y + A_z B_z, \tag{A.3}$$

or as a vector product (cross-product)

$$\begin{aligned} \mathbf{A} \times \mathbf{B} &= \mathbf{n}|A||B|\sin\theta \qquad \text{(A.4)} \\ &= \mathbf{u}_x(A_y B_z - A_z B_y) + \mathbf{u}_y(A_z B_x - A_x B_z) + \mathbf{u}_z(A_x B_y - A_y B_x), \end{aligned}$$

where **n** is a unit vector perpendicular to the plane of **A** and **B** with direction that a right-handed screw would move if rotated from **A** to **B** (Figure A.1).

Two useful identities, which can easily be proved by direct substitution, are

$$\mathbf{A} \times (\mathbf{B} \times \mathbf{C}) = (\mathbf{A} \cdot \mathbf{C})\mathbf{B} - (\mathbf{A} \cdot \mathbf{B})\mathbf{C} \tag{A.5}$$

and

$$\mathbf{A} \cdot [\mathbf{B} \times (\mathbf{C} \times \mathbf{D})] = (\mathbf{A} \times \mathbf{B}) \cdot (\mathbf{C} \times \mathbf{D}). \tag{A.6}$$

Both scalar and vector derivatives can be formed by using the differential vector operator "nabla" or "del," which in rectangular coordinates is

$$\nabla = \mathbf{u}_x \frac{\partial}{\partial x} + \mathbf{u}_y \frac{\partial}{\partial y} + \mathbf{u}_z \frac{\partial}{\partial z}, \tag{A.7}$$

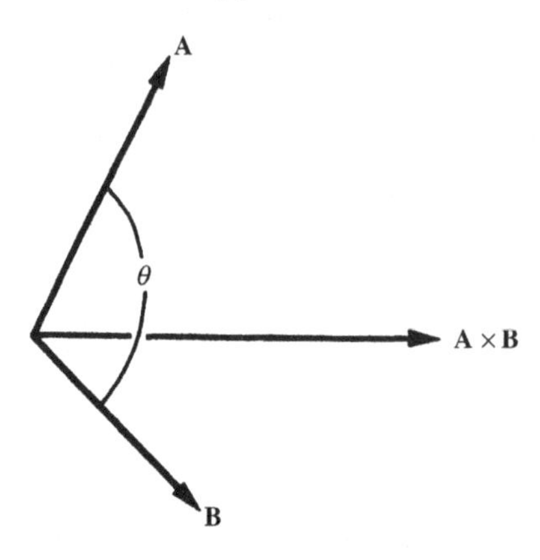

Figure A.1 Relation between vectors in a cross-product.

and treating it symbolically like a vector quantity, as follows. Let $F(x, y, z)$ be a scalar field. Then the *gradient* of the scalar is a vector, defined as

$$\text{grad}\, F = \nabla F = \mathbf{u}_x \frac{\partial F}{\partial x} + \mathbf{u}_y \frac{\partial F}{\partial y} + \mathbf{u}_z \frac{\partial F}{\partial z}. \tag{A.8}$$

For example, the gradient of the electric potential is the electric field.

Let **G(x,y,z)** be a vector field, $\mathbf{G} = \mathbf{u}_x G_x(x, y, z) + \mathbf{u}_y G_y(x, y, z) + \mathbf{u}_z G_z(x, y, z)$. The *divergence* of a vector is analogous to the scalar product of ∇ and **G** and is a scalar:

$$\text{div}\, \mathbf{G} = \nabla \cdot \mathbf{G} = \frac{\partial G_x}{\partial x} + \frac{\partial G_y}{\partial y} + \frac{\partial G_z}{\partial z}. \tag{A.9}$$

The divergence is a measure of the way fields change as they diverge from a source – for example, electric-field lines around a charge.

The *curl* of a vector is analogous to vector product of ∇ and **G** and is a vector:

$$\text{curl}\, \mathbf{G} = \nabla \text{x}\, \mathbf{G} = \mathbf{u}_x \left(\frac{\partial G_z}{\partial y} - \frac{\partial G_y}{\partial z} \right) + \mathbf{u}_y \left(\frac{\partial G_x}{\partial z} - \frac{\partial G_z}{\partial x} \right) + \mathbf{u}_z \left(\frac{\partial G_y}{\partial x} - \frac{\partial G_x}{\partial y} \right). \tag{A.10}$$

Curls are associated with rotational or looping behavior – for example, the vortex flow of fluids or magnetic-field lines around a line of current. The direction of the curl vector is perpendicular to **G** and parallel to the way the thumb of your right hand would point if you curled your fingers along **G**, as shown in Figure A.2.

An operator frequently encountered in physical problems is the *Laplacian*, div · grad, or "del squared." In rectangular coordinates (x, y, z),

$$\text{div} \cdot \text{grad} = \nabla^2 = \frac{\partial^2}{\partial x^2} + \frac{\partial^2}{\partial y^2} + \frac{\partial^2}{\partial z^2}; \tag{A.11}$$

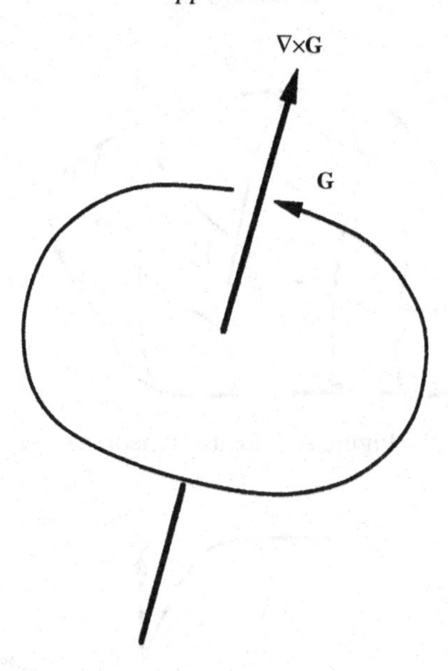

Figure A.2 Curl of a vector.

in cylindrical coordinates (r, ϕ, z),

$$\nabla^2 = \frac{1}{r}\frac{\partial}{\partial r}\left(r\frac{\partial}{\partial r}\right) + \frac{1}{r^2}\frac{\partial^2}{\partial \varphi^2} + \frac{\partial^2}{\partial z^2}; \tag{A.12}$$

and in spherical coordinates (r, θ, ϕ)

$$\nabla^2 = \frac{1}{r^2}\frac{\partial}{\partial r}\left(r^2\frac{\partial}{\partial r}\right) + \frac{1}{r^2 \sin\theta}\frac{\partial}{\partial \theta}\left(\sin\theta\frac{\partial}{\partial \theta}\right) + \frac{1}{r^2 \sin^2\theta}\frac{\partial^2}{\partial \varphi^2}. \tag{A.13}$$

The Laplacian can operate on either a vector or a scalar.

Two useful identities, which can be proved by direct substitution, are

$$\left.\begin{array}{l} \mathrm{curl}(\mathrm{curl}\,\mathbf{G}) = \mathrm{grad}(\mathrm{div}\,\mathbf{G}) - (\mathrm{div}\cdot\mathrm{grad})\mathbf{G}, \\ \text{or} \\ \nabla \times (\nabla \times \mathbf{G}) = \nabla(\nabla\cdot\mathbf{G}) - \nabla^2\mathbf{G}, \end{array}\right\} \tag{A.14}$$

and

$$\left.\begin{array}{l} \mathrm{div}(\mathbf{A}\times\mathbf{B}) = \mathbf{A}\cdot\mathrm{curl}\,\mathbf{B} - \mathbf{B}\cdot\mathrm{curl}\,\mathbf{A}, \\ \text{or} \\ \nabla\cdot(\mathbf{A}\times\mathbf{B}) = \mathbf{A}\cdot\nabla\times\mathbf{B} - \mathbf{B}\cdot\nabla\times\mathbf{A}. \end{array}\right\} \tag{A.15}$$

A powerful integral relation, which holds if a vector $\mathbf{G}$ is defined over a volume V bounded by a closed surface A, is Gauss's theorem,

$$\int_V div\mathbf{G}dV = \oint_A \mathbf{G}\cdot\mathbf{n}_A dA, \tag{A.16}$$

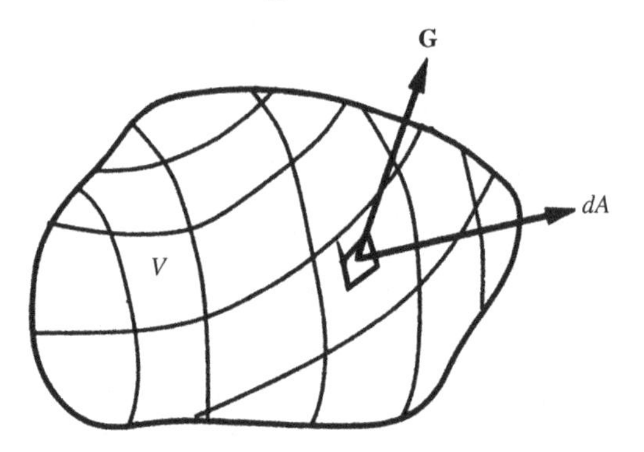

Figure A.3 Gauss's theorem.

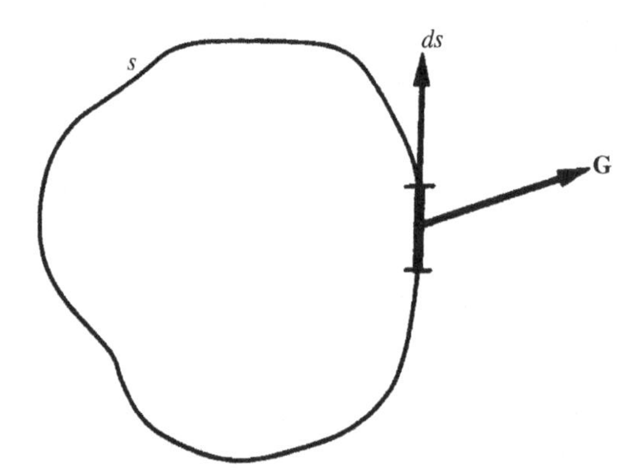

Figure A.4 Stokes's theorem.

where $\mathbf{n}_A$ is a unit vector perpendicular to the increment of surface area dA. Gauss's theorem states that the integral of the divergence of **G** over the volume V is equal to the closed integral of the component of **G** normal to dA over the surface (Figure A.3).

A second integral relation is Stokes's theorem, which holds if **G** is defined over a surface A bounded by a closed line s:

$$\int_{A} \text{curl } \mathbf{G} \cdot \mathbf{n}_A dA = \oint_{s} \mathbf{G} \cdot \mathbf{n}_S ds. \tag{A.17}$$

where $\mathbf{n}_S$ is a unit vector parallel to the line element ds. Stokes's theorem states that the integral of the normal component of the curl of **G** over the surface is equal to the closed integral of the component of **G** tangential to ds around the curve (Figure A.4).

Appendix B

Functions of a complex variable

In this appendix the properties of a function of a complex variable are reviewed briefly. For rigorous proofs and applications of the formulas given here, the reader is referred to standard textbooks (e.g., Paliouras, 1975; Brown and Churchill, 1996; Arfken and Weber, 2005).

Let z be a complex number,

$$z = x + iy = \rho e^{i\phi} \tag{B.1}$$

where $i = \sqrt{-1}$,

$$\rho = \left(x^2 + y^2\right)^{1/2}, \tag{B.2}$$

$$e^{i\phi} = \cos\phi + i\sin\phi, \tag{B.3}$$

so that $x = \rho\cos\phi$ and $y = \rho\sin\phi$. The complex conjugate z^* of z is found by replacing i by $-i$:

$$z^* = x - iy = \rho e^{-i\phi}. \tag{B.4}$$

Let $f(z)$ be a function of $z = x + iy$, with real and imaginary parts u and v, respectively, $f(z) = u(x, y) + iv(x, y)$, and with complex conjugate $f^*(z) = u(x, y) - iv(x, y)$. A function $f(z)$ is *analytic* at a point if the derivative df/dz exists and is independent of the direction in the complex plane in which the derivative is taken. The conditions for this to be true may be found as follows.

Suppose the derivative of $f(z)$ is taken along a direction parallel to the x-axis; then $df/dz = \partial u/\partial x + i\partial v/\partial x$. On the other hand, suppose the derivative is taken parallel to the y-axis; then $df/dz = (1/i)\partial u/\partial y + \partial v/\partial y$. In order that the two derivatives be equal, it must be true that

$$\partial u/\partial x = \partial v/\partial y \text{ and } \partial v/\partial x = -\partial u/\partial y. \tag{B.5}$$

These are the *Cauchy–Riemann conditions.*

The *Cauchy–Goursat theorem* states that if $f(z)$ is analytic in some region of the complex plane, then

$$\oint_C f(z)dz = 0, \tag{B.6}$$

where C is any closed curve in the region. A consequence of this theorem is that the integral of $f(z)$ along any path joining the points z_1 and z_2

$$F(z_1, z_2) = \int_{z_1}^{z_2} f(z)dz, \tag{B.7}$$

z_2 is independent of the specific path as long as it is entirely in the region where $f(z)$ is analytic. Furthermore, the value of f at a point z is determined by the values along any closed curve C surrounding z in the analytic region by *Cauchy's integral formula*,

$$f(z) = \frac{1}{2\pi i}\oint_C \frac{f(z')}{z'-z}dz'. \tag{B.8}$$

However, if the point z lies outside the closed curve C then

$$\oint_C \frac{f(z')}{z'-z}dz' = 0. \tag{B.9}$$

If $f(z) = u + iv$ is analytic throughout the upper half-plane, including the real axis, and also has the property that $f(z) \to 0$ as $|z| \to \infty$, then (B.9) can be transformed into an integral along the real axis, as follows. Choose the real axis to pass through the point z and evaluate the integral around the closed curve C shown in Figure B.1. Then let $R \to \infty$; then the last integral on the right vanishes. Next let $r \to 0$; then the second integral on the right becomes $-i\pi f(x)$. Hence,

$$\begin{aligned}\oint_C \frac{f(z')}{z'-z}dz' &= 0 \\ &= \int_{-R}^{z-r} \frac{f(x')}{x'-x}dx' + \int_0^{\pi} \frac{f(z')}{re^{i\phi}} re^{i\varphi} i d\phi + \int_{z+r}^{R} \frac{f(x')}{x'-x}dx' \\ &\quad + \int_{\pi}^{0} \frac{f(z')}{Re^{i\theta}-z} Re^{i\theta} i d\theta.\end{aligned} \tag{B.10}$$

Let $R \to \infty$; then the last integral on the right vanishes. Next let $r \to 0$; then the second integral on the right becomes $-i\pi f(x)$. Hence,

$$f(x) = u + iv = \frac{1}{i\pi}\int_{\infty}^{-\infty} \frac{f(x')}{x'-x}dx'. \tag{B.11}$$

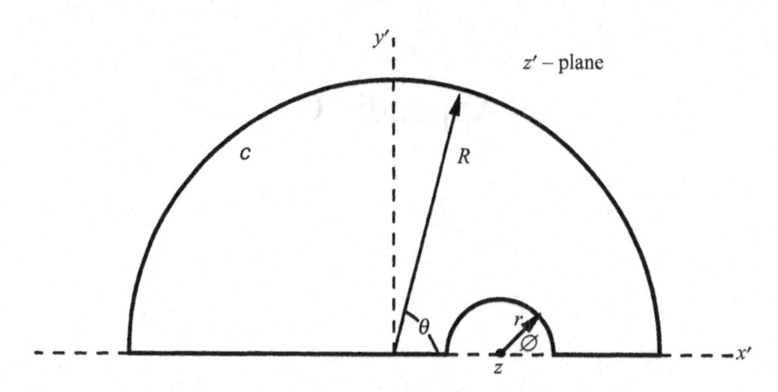

Figure B.1

where

$$u(x) = \frac{1}{\pi}\int_{\infty}^{-\infty} \frac{v(x')}{x'-x}dx' \tag{B.12}$$

$$v(x) = -\frac{1}{\pi}\int_{\infty}^{-\infty} \frac{u(x')}{x'-x}dx'. \tag{B.13}$$

Thus, if $f(z)$ has the properties stated earlier, its real and imaginary parts are not independent, but are related through (B.12) and (B.13).

Now, if u and v represent physical quantities, then it is highly likely that $f(z)$ is analytic and also possesses the property of *crossing symmetry*; that is, $f(-z) = f^*(z)$, so that $u(-z) = u(z)$ and $v(-z) = -v(z)$. In that case equations (B.12) and (B.13) can be put in a particularly useful form by breaking the integrals up into separate integrals from $-\infty$ to 0 and 0 to $+\infty$ and using the symmetry relations:

$$\begin{aligned} u(x) &= \frac{1}{\pi}\int_{-\infty}^{0} \frac{v(x')}{x'-x}dx' + \frac{1}{\pi}\int_{0}^{\infty} \frac{v(x')}{x'-x}dx' \\ &= \frac{1}{\pi}\int_{0}^{\infty} \frac{v(x)}{x'+x}dx' + \frac{1}{\pi}\int_{-\infty}^{0} \frac{v(x)}{x'-x}dx' = \frac{2}{\pi}\int_{0}^{\infty} \frac{x'v(x')}{x'^2-x^2}dx', \end{aligned} \tag{B.14}$$

$$\begin{aligned} v(x) &= -\frac{1}{\pi}\int_{-\infty}^{0} \frac{u(x')}{x'-x'}dx' - \frac{1}{\pi}\int_{0}^{\infty} \frac{u(x')}{x'-x}dx' \\ &= \frac{1}{\pi}\int_{0}^{\infty} \frac{u(x')}{x'+x}dx' - \frac{1}{\pi}\int_{-\infty}^{\infty} \frac{u(x')}{x'-x}dx' = -\frac{2x}{\pi}\int_{0}^{\infty} \frac{u(x')}{x'^2-x^2}dx'. \end{aligned} \tag{B.15}$$

Equations (B.14) and (B.15) are the Kramers–Kronig relations.

Appendix C

The wave equation in spherical coordinates

C.1 Solution of the wave equation by separation of variables

In this appendix the general solutions to the wave equation in spherical coordinates using the separation-of-variables method is reviewed. Because these equations are encountered over and over again in physics, planetary science, and engineering applications their solutions have been extensively studied and tabulated. Thus, we will encounter and begin to understand the properties of such often-used functions as Legendre, Bessel, and Hankel functions and spherical harmonics. For further information, the many excellent textbooks on mathematical methods in physics should be consulted, such as Morse and Feshbach (1953), Margenau and Murphy (1956), or Arfken and Weber (2005).

The wave equation (Chapter 2) is

$$\nabla^2 F - (1/\upsilon^2)\partial^2 F/\partial t^2 = 0 \tag{C.1}$$

where F is the field quantity, υ is the propagation velocity, t is the time, and ∇^2 is the Laplacian operator. In spherical coordinates

$$\nabla^2 F = \frac{1}{r^2}\frac{\partial}{\partial r}\left(r^2\frac{\partial F}{\partial r}\right) + \frac{1}{r^2 \sin\theta}\frac{\partial}{\partial \theta}\left(\sin\vartheta\frac{\partial F}{\partial \theta}\right) + \frac{1}{r^2 \sin^2\theta}\frac{\partial^2 F}{\partial \psi^2}, \tag{C.2}$$

where r is the radial coordinate, ϑ is the polar angle, and ψ is the azimuth angle (Figure C.1). In a large class of problems it is possible to write the solution to the wave equation as the product of separate functions of the independent variables:

$$F(r, \vartheta, \psi, t) = R(r)\Theta(\theta)\Phi(\psi)T(t). \tag{C.3}$$

(The reader should be cautioned that this does not always work. Not infrequently one will obtain an apparent solution to a differential equation, only to find that it is not possible to satisfy the boundary conditions.) It is almost always found that more than one solution of the form (C.3) exists (usually an infinite number of them), so that the complete solution to (C.1) consists of a sum of terms of the form of (C.3).

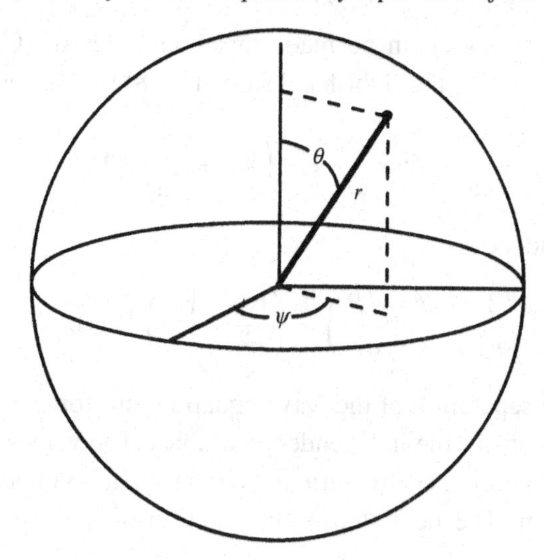

Figure C.1 Spherical coordinates.

To solve (C.1) using the separation-of-variables method, insert (C.3) into (C.1) and multiply both sides by υ^2/F to obtain

$$\upsilon^2\left[\frac{1}{Rr^2}\frac{d}{d\mathbf{r}}\left(r^2\frac{dR}{d\mathbf{r}}\right)+\frac{1}{\Theta r^2\sin\theta}\frac{d}{d\theta}\left(\sin\theta\frac{d\Theta}{d\theta}\right)+\frac{1}{\Psi r^2\sin^2\theta}\frac{d^2\Psi}{d\psi^2}\right]=\frac{1}{T}\frac{d^2T}{dt^2}. \tag{C.4}$$

Now, the left-hand side of (C.4) is solely a function of the spatial variables r, ϑ, and ψ, while the right-hand side is a function of time t only. Because these variables are independent, the only way the equality can hold is if both sides are equal to some constant. Call this separation constant K_1. Equating the right-hand side of (C.4) to K_1 gives

$$d^2T/dt^2 - \mathsf{K}_1 T = 0. \tag{C.5}$$

Equating the left-hand side of (C.4) to K_t and rearranging gives

$$\frac{\mathsf{K}_1}{\upsilon^2}r^2\sin^2\theta+\frac{\sin^2\theta}{R}\frac{d}{dr}\left(r^2\frac{dR}{dr}\right)+\frac{\sin\theta}{\Theta}\frac{d}{d\vartheta}\left(\sin\theta\frac{d\Theta}{d\theta}\right)=-\frac{1}{\Psi}\frac{d^2\Psi}{d\psi^2}. \tag{C.6}$$

Because each side of (C.6) is a function of different independent variables, both sides must equal a constant. Call this second separation constant K_2. Equating the right-hand side to K_2 gives

$$\frac{d^2\Psi}{d\psi^2}+\mathsf{K}_2\Psi=0. \tag{C.7}$$

Setting the left-hand side of (C.6) equal to K_2 gives

$$-\frac{\mathsf{K}_1}{\upsilon^2}r^2+\frac{1}{R}\frac{d}{dr}\left(r^2\frac{dR}{dr}\right)=\frac{\mathsf{K}_2}{\sin^2\theta}-\frac{1}{\Theta\sin\theta}\frac{d}{d\theta}\left(\sin\vartheta\frac{d\Theta}{d\theta}\right). \tag{C.8}$$

The same argument may again be made that both sides of (C.8) must equal a constant, say K_3. Equating the right-hand side of (C.8) to this constant gives

$$\frac{1}{\sin\theta}\frac{d}{d\theta}\left(\sin\theta\frac{d\Theta}{d\theta}\right)+\left(\mathsf{K}_3-\frac{\mathsf{K}_2}{\sin^2\theta}\right)\Theta=0, \tag{C.9}$$

and for the left-hand side,

$$\frac{1}{r^2}\frac{d}{dr}\left(r^2\frac{dR}{dr}\right)-\left(\frac{\mathsf{K}_1}{\upsilon^2}+\frac{\mathsf{K}_3}{r^2}\right)R=0. \tag{C.10}$$

This completes the separation of the wave equation into four separate differential equations for functions of the independent variables. The values of the separation constants must be determined by various boundary and symmetry conditions of the specific problem. The next step is the solution of the separated differential equations.

C.2 Solution for $T(t)$

The solutions of (C.5) are of the form $\exp(\pm\sqrt{\mathsf{K}_1}t)$, as may be verified by direct substitution. Most applications are concerned with waves that vary periodically with time, so that

$$T(t)=T_0e^{\pm 2\pi i\nu t}, \tag{C.11}$$

where ν is the frequency and T_0 is a constant that is determined by conditions at $t=0$. Hence,

$$\mathsf{K}_1=-4\pi^2\upsilon^2. \tag{C.12}$$

Note that we could have obtained the same result simply by assuming that the wave equation has periodic solutions of the form $F(r,\theta,\psi,t)=\mathcal{F}(r,\theta,\psi)\exp(\pm 2\pi\nu it)$ and substituting this into (C.1) to obtain

$$\nabla^2F+\mathsf{K}^2F=0, \tag{C.13}$$

where

$$\mathsf{K}=2\pi\nu/\upsilon=2\pi/\lambda \tag{C.14}$$

is the wavenumber and λ is the wavelength. If $\mathsf{K}=0$, equation (C.13) is called Laplace's equation, and if $\mathsf{K}\neq 0$, it is known as Helmholtz's equation. The parameter K may be a constant or a function of r, ϑ, or ψ, but may not be a function of F. The Helmholtz equation is widely encountered in many physical problems, including the wave equation, the heat equation, the diffusion equation, and Schrödinger's equation of quantum mechanics.

C.3 Solution for $\Psi(\psi)$

Equation (C.7) has solutions of the form $\sin\sqrt{\mathsf{K}_2}\psi$ and $\cos\sqrt{\mathsf{K}_2}\psi$, or, equivalently, $\exp(\pm i\sqrt{\mathsf{K}_2}\psi)$. We will choose the trigonometric form of the solutions, although which form to use is strictly a matter of taste. To evaluate K_2, note that if ψ is increased by 2π, Φ must return to the same value. Thus, $\sqrt{\mathsf{K}_2}$ must be an integer. Furthermore, because decreasing ψ by 2π gives the same effect as increasing ψ by 2π, only positive integers need be considered. Hence, there are an infinite number of solutions to (C.7) of the form

$$\Psi(\psi) = A_m \cos m\psi + B_m \sin m\psi, \tag{C.15}$$

where

$$m = \sqrt{\mathsf{K}_2} = \text{integer} \geq 0, \tag{C.16}$$

and A_m and B_m are constants that must be determined by the boundary conditions.

C.4 Solution for $\Theta(\theta)$; Legendre functions

The general solutions to equation (C.9) for Θ are the set of functions known as confluent hypergeometric functions. They have the property that they are infinite at $\theta = 0$ and π unless K_3 has the special value

$$\mathsf{K}_3 = l(l+1), \tag{C.17}$$

where l is an integer $\geq m$. Because most problems of interest include 0 and π, the solutions with singularities are not physically acceptable, and the separation constant must be chosen to have the value given by (C.17).

The solutions of (C.9) with $\mathsf{K}_2 = m^2$ and $\mathsf{K}_3 = l(l+1)$ are the associated Legendre functions, $P_{lm}(\theta)$, so that

$$\Theta(\theta) = E_{lm} P_{lm}(\theta), \tag{C.18}$$

where E_{lm} is a constant. These functions are tabulated in many places (e.g., Jahnke and Emde, 1945; Abramowitz and Stegun, 1972). If the second index on P_{lm} is omitted, the value of m is understood to be 0 (i.e., $P_l = P_{l0}$), and the functions are called simply Legendre polynomials. The associated Legendre functions can be expressed in terms of sines and cosines. The first few are as follows:

$$P_{00} = P_0 = 1 \quad P_{10} = P_1 = \cos\theta \quad P_{20} = P_2 = \tfrac{1}{2}(3\cos^2\theta - 1) \quad P_{30} = P_3 = \tfrac{1}{2}(5\cos^3\theta - 3\cos\theta)$$

$$P_{11} = \sin\theta \quad P_{21} = 3\cos\theta\sin\theta \quad P_{31} = \tfrac{3}{8}(\sin\theta + 5\sin 3\theta)$$

$$P_{22} = 3\sin^2\theta \quad P_{32} = 15\cos\theta\sin^2\theta$$

$$P_{33} = 15\sin^3\theta$$

The four lowest Legendre polynomials are plotted in Figure C.2.

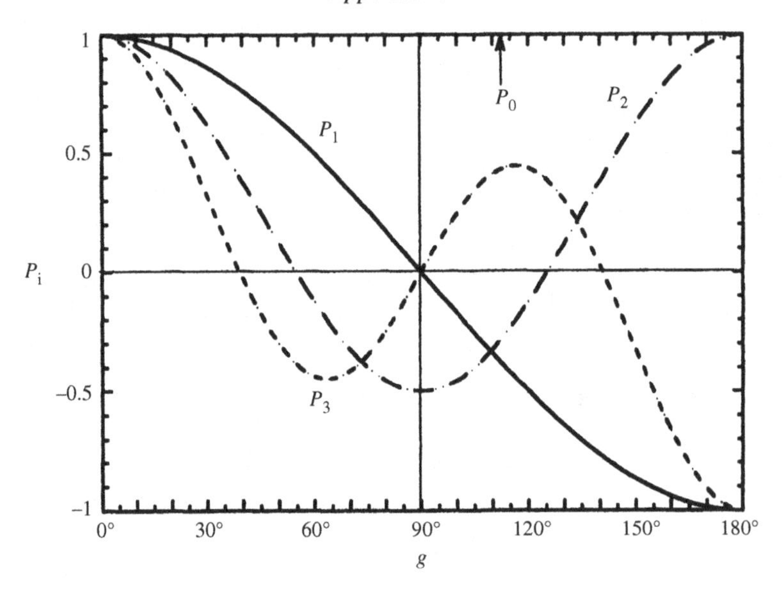

Figure C.2 The first four Legendre polynomials.

The Legendre polynomials obey the addition theorem, which states that if angles g, i, and e are related by the law of cosines, such that

$$\cos g = \cos i \cos e + \sin i \sin e \cos \psi,$$

then

$$P_n(\cos g) = P_n(\cos i) P_n(\cos e) + 2 \sum_{m=1}^{\infty} \frac{(n-m)!}{(n+m)!} P_{nm}(\cos i) P_{nm}(\cos e) \cos m\psi. \tag{C.19}$$

C.5 Solution for $R(r)$; Bessel and related functions

If $\mathsf{K}_1 = 0$, the solution to (C.10) is of the form

$$R(r) = C_l r^l + D_l r^{-l(l+1)}, \tag{C.20}$$

where C_l and D_l are constants that must be determined from the boundary conditions. However, if $\mathsf{K}_1 = 0$, then $\nu = 0$, so that this solution for $R(r)$ applies only to static situations. Because such problems cannot involve propagating waves, they will not be considered further here.

The solutions to (C.10) with and K_1 and $\mathsf{K} \neq 0$ and $\mathsf{K}_2 = l(l+1)$ are

$$R(r) = \left[C_l J_{(l+\frac{1}{2})}(\mathsf{K}r) + D_l Y_{(l+\frac{1}{2})}(\mathsf{K}r) \right] / \sqrt{r}, \tag{C.21}$$

where $J_p(z)$ is a Bessel function of the first kind of argument z and order p, and $Y_p(z)$ is a Bessel function of the second kind of argument z and order p. Bessel functions of the second kind are also called Neumann functions. As usual, the constants C_l and D_l must be determined by the boundary conditions.

The functions that result when Bessel functions of half-integer order are divided by $\sqrt{z}$ are called spherical Bessel functions because they occur in the solutions of Helmholtz's equation in spherical coordinates. Thus, the spherical Bessel functions are defined as

$$\begin{aligned} j_p(z) &= \sqrt{\frac{\pi}{2z}} J_{(p+\frac{1}{2})}(z) \\ y_p(z) &= \sqrt{\frac{\pi}{2z}} Y_{(p+\frac{1}{2})}(z). \end{aligned} \tag{C.22}$$

The Bessel functions are tabulated and discussed in many places (e.g., Watson, 1958; Jahnke and Emde, 1945; Abramowitz and Stegun, 1972). Jahnke and Emde are particularly helpful because the Legendre functions and Bessel functions are displayed graphically in several ways, which allows the student to rapidly grasp their properties. The spherical Bessel functions are oscillatory functions whose amplitude generally decreases with increasing argument. In particular, $j_0(z) = (\sin z)/z$ and $y_0(z) = -(\cos z)/z$.

Just as solutions to various problems often are expedited by combining trigonometric functions into complex functions, such as $(\cos\vartheta \pm i \sin\vartheta)$, it is frequently convenient to do the same with the Bessel functions. Thus, Bessel functions of the third kind, or Hankel functions, are defined as

$$\begin{aligned} H_p^{(1)}(z) &= J_p(z) + iY_p(z) \\ H_p^{(2)}(z) &= J_p(z) - iY_p(z). \end{aligned} \tag{C.23}$$

Similarly, the spherical Hankel functions are defined as

$$\begin{aligned} h_p^{(1)}(z) &= \sqrt{\frac{\pi}{2z}} H^{(1)}_{(p+\frac{1}{2})}(z) \\ h_p^{(2)}(z) &= \sqrt{\frac{\pi}{2z}} H^{(2)}_{(p+\frac{1}{2})}(z). \end{aligned} \tag{C.24}$$

For any given value of l there will be $l+1$ possible values of m (m goes from 0 to l). Because l can be any integer ≥ 0, the general solution to the wave equation that is periodic in time will be a sum of the various solutions for $R(r)\Theta(\vartheta)\Psi(\psi)$ for all possible values of l and m. Hence (absorbing T_0 and E_{lm} into the other constants), the general solution to the wave equation is

$$\begin{aligned} &F(r,\theta,\psi,t) \\ &= \sum_{l=0}^{\infty}\sum_{m=0}^{l} [C_l J_l(\mathsf{K}r) + D_l y_l(\mathsf{K}r)] P_{lm}(\theta)[A_{lm}\cos m\psi + B_{lm}\sin m\psi]e^{-2\pi i\nu t}. \end{aligned} \tag{C.25}$$

Because the associated Legendre functions and the trigonometric functions always appear multiplied together in the solutions to the wave equation in spherical coordinates, their products are called the even and odd spherical harmonic functions:

$$\begin{aligned}\Phi^{(e)}_{lm}(\vartheta,\psi) &= P_{lm}(\cos\vartheta)\cos m\psi \\ \Phi^{(o)}_{lm}(\vartheta,\psi) &= P_{lm}(\cos\vartheta)\sin m\psi.\end{aligned} \tag{C.26}$$

Thus, the general solution to the wave equation may be written in a form equivalent to (C.24):

$$\begin{aligned}&F(r,\theta,\psi,t)\\ &= \sum_{l=0}^{\infty}\sum_{m=0}^{l}[C_l h_l^{(1)}(\mathsf{K}r) + D_l h_l^{(2)}(\mathsf{K}r)][A_{lm}\Phi^{(e)}_{lm}(\theta,\psi) + B_{lm}\Phi^{(o)}_{lm}(\theta,\psi)]e^{-2\pi i\nu t}.\end{aligned} \tag{C.27}$$

The physical meaning of writing the solution in the form of (C.27) may be seen from the asymptotic properties of the spherical Hankel functions. When $\mathsf{K}r \gg l^2$,

$$\begin{aligned}h_l^{(1)}(\mathsf{K}r)e^{-2\pi i\nu t} &= (-i)^{l+1}e^{2\pi i(r/\lambda-\nu t)}/\mathsf{K}r \\ h_l^{(2)}(\mathsf{K}r)e^{-2\pi i\nu t} &= i^{l+1}e^{-2\pi i(r/\lambda+\nu t)}/\mathsf{K}r.\end{aligned} \tag{C.28}$$

The first expression represents a wave propagating radially outward whose amplitude falls off as $1/r$, that is, whose power falls off as $1/r^2$. Similarly, the second represents a wave propagating radially inward. The part of (C.27) proportional to the spherical harmonics describes how the strengths of these spherical waves vary with direction.

C.6 Complete orthogonal sets

The spherical Bessel functions and spherical harmonic functions are members of a large class of functions known as *complete orthogonal sets*. Let $f_p(z)$ be a member of a set of functions of argument z and order p, defined over the interval $a \le z \le b$. Then the set is *complete* if any well-behaved mathematical function $g(z)$ can be represented over that interval as a sum of the set:

$$g(z) = \sum_p A_p f_p(z), \tag{C.29}$$

where the A_ps are constants. The set is *orthogonal* if

$$\begin{aligned}\int_a^b f_P(z) f_{p'}(z)dz &= 0 \text{ if } p \neq p', \\ &= \text{constant} \neq 0 \text{ if } p = p'.\end{aligned} \tag{C.30}$$

Thus, to find the expansion coefficients of $g(z)$ in terms of the $f_p(z)$s, multiply (C.29) by $f_p(z)$ and integrate:

$$\int_a^b g(z) f_{p'}(z) dz = \sum_p A_p \int_a^b f_p(z) f_{p'}(z) dz = A_{p'} \int_a^b f_{p'}(z) f_{p'}(z) dz, \quad \text{(C.31)}$$

so

$$A_{p'} = \left[\int_a^b g(z) f_{p'}(z) dz \right] / \left[\int_a^b f_{p'}(z) f_{p'}(z) dz \right]. \quad \text{(C.32)}$$

A typical example is the familiar set of trigonometric functions. The ability of a sum of sine and cosine functions to represent any periodic function is well known and forms the basis for the technique of Fourier analysis and synthesis. That these functions are orthogonal can be illustrated by the following. Let $f_p(z) = \cos pz$, $a = 0$, and $b = 2\pi$, and let p and p' be integers. Then

$$\begin{aligned} &\int_0^{2\pi} \cos pz \cos p'z dz \\ &= \int_0^{2\pi} \frac{1}{2} [\cos(p + p')z + \cos(p - p')z] \, dz = 0 \text{ if } p \neq p', \\ &= \pi \text{ if } p = p'. \end{aligned} \quad \text{(C.33)}$$

Appendix D

Fraunhofer diffraction by a circular hole

In this appendix we consider the problem of the diffraction of light by a circular hole in an opaque wall. The situation is shown in Figure D.1. An electromagnetic wave of irradiance $\boldsymbol{J}$ and wavelength λ is incident on a hole of radius $a >> \lambda$ in an opaque wall of infinite extent oriented perpendicular to the direction of propagation. We wish to know the pattern of the electric fields and the resultant intensity at a large distance on the other side of the wall. This can be done by using Huygens's principle, in which each point on a wave front is considered to be a source of wavelets that travel radially outward and combine coherently with wavelets from other points to produce new wave fronts. The Fraunhofer approximation to calculating the diffraction assumes that the *Fresnel condition*

$$\frac{a}{\lambda}\left(\frac{a}{L_{SW}}+\frac{a}{L_{WP}}\right) << 1 \tag{D.1}$$

is satisfied, where L_{SW} is the distance from the source to the wall and L_{WP} is the distance from the wall to the point at which the diffraction pattern is to be calculated.

Choose a coordinate system so that the wall lies in the x–y plane with $z = 0$ and the origin at the center of the hole. Let the electric field at the hole be E_{e0}, which is related to the amplitude of the incident irradiance $\boldsymbol{J}$ by equation (2.24), $J = \sqrt{\varepsilon_{e0}/\mu_{m0}}E_{e0}^2/2$. The intensity of a Huygens wavelet generated by the incident field at a given point in the hole falls off inversely with the square of the distance from that point, so that the field strength of the wavelet falls off as the reciprocal of the first power of the distance. Thus the wavelet field generated by an incremental area $dA_1 = \zeta\, d\zeta\, d\phi$, located at point P_1 at position $(x_1, y_1, z_1) = (\zeta\cos\varphi, \zeta\sin\varphi, 0)$ in the hole, at a point P_2 a distance l away, is $(CE_{e0}/l)\, e^{2\pi^i(l/\lambda - \nu t)} dA_1$, where C is a quantity to be determined. Because of cylindrical symmetry we may choose the x–z plane so that it contains P_2 without loss of generality, so that the coordinates of P_2 are $(x_2, y_2, z_2) = (r\sin\theta, 0, r\cos\theta)$, where r is the distance of P_2 from the center of the hole and θ is the scattering angle relative to the center of the hole.

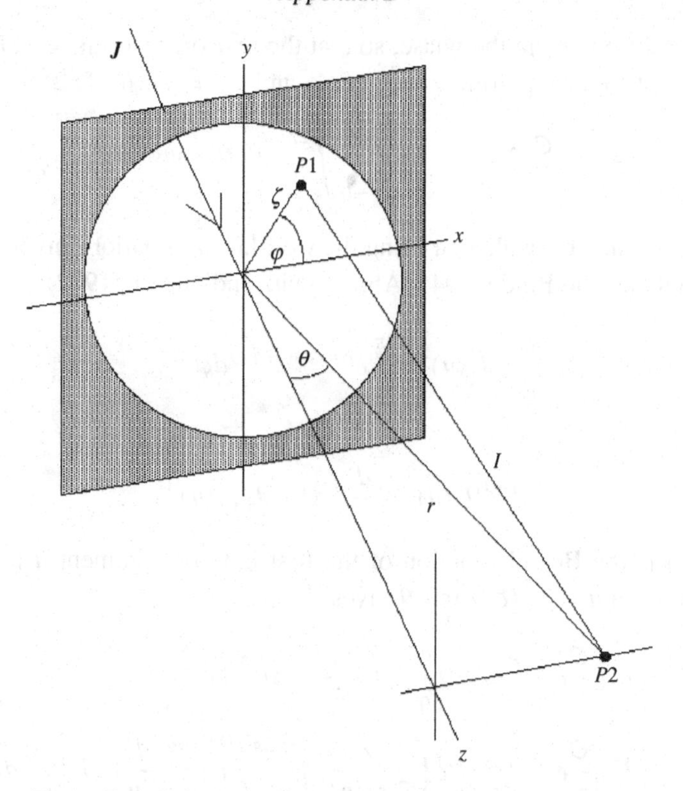

Figure D.1 Schematic diagram of the diffraction by a hole.

Then the total field at P_2 generated by all the areas in the hole is

$$E_e(x_2, y_2, z_2) = CE_{e0} \int_A \frac{e^{2\pi i(l/\lambda - vt)}}{l} dA_1. \tag{D.2}$$

Now

$$\begin{aligned} l &= [(x_1 - x_2)^2 + (y_1 - y_2)^2 + (z_1 - z_2)^2]^{1/2} \\ &= [(x_1^2 + y_1^2 + z_1^2) + 2(x_1x_2 + y_1y_2 + z_1z_2) + (x_2^2 + y_2^2 + z_2^2)]^{1/2} \end{aligned}$$

But $\zeta^2 = x_1^2 + y_1^2 + z_1^2$, $r^2 = x_2^2 + y_2^2 + z_2^2$, and $y_2 = z_1 = 0$, so

$$l = [r^2 + 2x_1x_2 + \zeta^2]^{1/2} = r[1 + 2x_1x_2/r^2 + \zeta^2/r^2]^{1/2}. \tag{D.3}$$

Because of the Fresnel condition the third term inside the brackets is negligible, and the second term is small. Thus (D.2) my be expanded to first order in a Taylor series,

$$l \approx r + x_1x_2/r.$$

Since l is not very different from r, it may be replaced by r in the denominator of the integral in (D.2) with little error. However, because $l >> \lambda$, small changes in l

make a large difference in the phase, so that the first-order terms must be retained in the exponent. Hence, putting $x_1 = \zeta \cos\varphi$, and $x_2 = r\sin\theta$, (D.2) becomes

$$E_e \simeq E_{e0}\frac{C}{r}e^{2\pi i(r/\lambda - \nu t)}\int_{\zeta=0}^{\alpha}\int_{\varphi=0}^{2\pi} e^{i(\zeta/\lambda)\sin\theta\cos\varphi}\zeta\, d\zeta\, d\varphi. \tag{D.4}$$

These integrals may be evaluated using the well-known relations involving Bessel functions (Jahnke and Emde, 1945; Abramowitz and Stegun, 1972):

$$J_j(u) = \int_0^{2\pi} e^{i(u\cos\varphi + j\varphi)}d\varphi \tag{D.4}$$

and

$$u^{j+1}J_j(u) = \frac{d}{du}\left[u^{j+1}J_{j+1}(u)\right], \tag{D.5}$$

where $J_j(u)$ is the Bessel function of the first kind of argument u and order j. Putting $j = 0$ and $u = 2\pi(\zeta/\lambda)\sin\theta$ gives

$$\begin{aligned} E_e &= E_{e0}\frac{C}{r}e^{2\pi i(r/\lambda - \nu t)}\int_0^{a} J_0\left(\frac{2\pi\zeta}{\lambda}\sin\theta\right)\zeta\, d\zeta \\ &= E_{e0}\frac{C}{r}e^{2\pi i(r/\lambda - \nu t)}\left(\frac{\lambda}{2\pi\sin\theta}\right)^2\int_0^{(2\pi a/\lambda)\sin\theta}\frac{d}{du}[uJ_1(u)]du \\ &= E_{e0}\frac{C}{r}e^{2\pi i(r/\lambda - \nu t)}\frac{a^2}{X\sin\theta}J_1(X\sin\theta), \end{aligned} \tag{D.6}$$

where $X = 2\pi a/\lambda$. Hence, the diffracted power per unit area at P_2 is

$$\frac{dP_d}{dA_2} = \frac{1}{2}\sqrt{\varepsilon_{e0}/\mu_{m0}}|E_e|^2 = \frac{1}{2}\sqrt{\varepsilon_{e0}/\mu_{m0}}\left[E_{e0}\frac{C}{r}\frac{a^2}{X\sin\theta}J_1(X\sin\theta)\right]^2. \tag{D.7}$$

Since the solid angle $d\Omega$ subtended at the center of the hole by an increment of area dA_2 at P_2 is $d\Omega = dA_2/r^2$, the diffracted power per unit solid angle is

$$\frac{dP_d}{d\Omega} = \frac{dP_d}{dA_2}\frac{dA_2}{d\Omega} = \frac{1}{2}\sqrt{\varepsilon_{e0}/\mu_{m0}}\left[E_{e0}C\frac{a^2}{X\sin\theta}J_1(X\sin\theta)\right]^2. \tag{D.8}$$

The quantity $2J_1(u)/u$ and its square are plotted in Figures 5.7a and 5.7b.

The constant C is found by requiring that the total power in the diffraction pattern be equal to the power coming through the hole, $J\pi a^2 = \pi a^2\sqrt{\varepsilon_{e0}/\mu_{m0}}E_{e0}^2/2 = \int_{2\pi}\frac{dP_d}{d\Omega}d\Omega$. Let $\upsilon = X\sin\theta$. Then $d\upsilon = X\cos\theta d\theta$. However, Figure 5.7 shows that the diffraction pattern is appreciable only for $X\sin\theta <\sim 4$. Since $X >> 1$, θ is small over this region, so $\cos\theta \approx 1$, and $d\upsilon = Xd\theta$ with little error. Putting

$d\Omega = 2\pi \sin\theta d\theta,$

$$J\pi a^2 = \frac{1}{2}\sqrt{\varepsilon_{e0}/\mu_{m0}}E_{e0}^2 C^2 a^4 \int_0^{\pi/2} \left[\frac{J_1(X\sin\theta)}{(X\sin\theta)}\right]^2 2\pi \sin\theta d\theta$$

$$= J\frac{2\pi C^2 a^4}{X^2}\int_0^X \frac{J_1^2(\upsilon)}{\upsilon} d\upsilon. \qquad \text{(D.12)}$$

Since $X >> 1$, X may be replaced by ∞ in the upper limit of the last integral.

The integral can be evaluated by replacing u by υ in equation (D.5), putting $j = 0$ and differentiating to obtain

$$J_1(\upsilon)/\upsilon = J_0(\upsilon) - dJ_1(\upsilon)/d\upsilon. \qquad \text{(D.13)}$$

Putting $j = -1$ in (D.5) and using the fact that $J_{-1}(\upsilon) = -J_1(\upsilon)$ gives

$$J_1^2(\upsilon)/\upsilon = J_1[J_0 - dJ_1/d\upsilon]$$

$$= -J_0 dJ_0/d\upsilon - J_1 dJ_1/d\upsilon = -\frac{1}{2}\frac{d}{d\upsilon}\left[J_1^2(\upsilon) + J_0^2(\upsilon)\right]. \qquad \text{(D.14)}$$

Now, $J_0(0) = 1$ and $J_1(0) = J_0(\infty) = J_1(\infty) = 0$, hence, the integral has the value $\frac{1}{2}$, and the total power in the diffraction pattern is $J\pi a^2 = J\pi a^2(Ca/X)^2$, so that $C = X/a$. Hence, (D.6) becomes

$$E_e(r,\theta) = E_{e0}a\frac{X}{r}e^{2\pi i(r/\lambda - \nu t)}\frac{J_1(X\sin\theta)}{X\sin\theta}, \qquad \text{(D.14)}$$

and the power per unit solid angle (D.8) is

$$\frac{dP_d}{d\Omega} = J\frac{\sigma}{4\pi}X^2\left[2\frac{J_1(X\sin\theta)}{X\sin\theta}\right]^2. \qquad \text{(D.15)}$$

Appendix E

Table of symbols

a	mean particle radius
a_E	mean extinction radius
A	volume average absorption coefficient
A	area
A_B	bolometric albedo
A_h	directional–hemispherical or hemispherical reflectance or plane albedo
A_P	physical or geometric albedo
A_S	bihemispherical or spherical reflectance or Bond albedo
A_{L}	Lambert albedo
A_{M}	Minnaert albedo
b	coefficient in particle phase function
B_{C0}	amplitude of CBOE
B_{S0}	amplituce of SHOE
$\boldsymbol{B}_m$	magnetic field
BRDF	bidirectional reflectance distribution function
c	coefficient in particle phase function
c_0	speed of light in vacuum
c_1	first radiation constant
c_2	second radiation constant
C	specific heat per unit mass
CBOE	coherent backscatter opposition effect
ΔC	relative band contrast
D	mean particle diameter, $D = 2a$
D_E	effective particle size
$\langle D \rangle$	mean ray path length through particle
$\boldsymbol{D}_e$	electric displacement
e	zenith emission or viewing angle

e_0	charge of electron
E	volume-average extinction coefficient
$\mathcal{E}v$	volume-average emissivity factor
$\boldsymbol{E}_e$	electric field
$\boldsymbol{E}_{loc}$	local electric field
f	photometric function
f	remission function
f_j	oscillator strength of jth transition
F	source function
F_T	thermal volume average source function
F_T	thermal volume average emission coefficient
g	phase angle
G	volume average scattering coefficient
G_1, G_2	factors in specular reflection coefficient
h_C	angular width parameter of CBOE
h_S	angular width paramter of SHOE
H	Ambartsumian–Chandrasekhar H function
$\boldsymbol{H}_m$	magnetic intensity
i	zenith angle of incidence, $\sqrt{-1}$
I	radiance or intensity or first component of Stokes vector
I_D	radiance at a detector viewing the surface of a scattering medium
I_1	upwelling intensity
I_2	downwelling intensity
I_{SS}	singly scattered radiance
I/F	radiance factor
j	subscript denoting different types
$\boldsymbol{j}_e$	electric current density
J	irradiance
$J_p(u)$	Bessel function of order p and argument u
k	solid state thermal conductivity
k_{rad}	radiative conductivity
k_0	Boltzmann's constant
K	porosity coefficient
$\mathcal{K}$	roughness correction factor to integral phase function
K_e	dielectric constant, $K_e = K_{er} + iK_{ei}$
l	distance
l_E	extinction length
$<l>$	average coating thickness

L	distance, thickness, average distance between particles
L	luminance or photometric latitude
L_j	factor in modified IMSA model
$L(\Omega\iota, \Omega)$	Ambartsumian reflectance quantity
M	bulk density of jth type of material
$\boldsymbol{M}$	Mueller or scattering matrix
$\boldsymbol{M}_m$	magnetization
MOPL	mean optical path length
n	refractive index, $n = n_r + in_i$
N	number of particles per unit volume
N_e	number of free electrons per unit volume
N_0	Avogadro's number
OMAT	optical maturity parameter
$p(g)$	volume-average single scattering function
$\boldsymbol{p}_e$	electric dipole moment
P	polarization, polarization ratio
P_A	absorbed power
P_d	power in diffraction pattern
P_D	power at detector
$\boldsymbol{P}_e$	dipose moment per unit volume
P_E	extinguished power
P_j	Legendre polynomial of order j
P_{jk}	associated Legendre polynomial
P_S	scattered power
P_{em}	emitted power
P_{su}	power reflected from surface of particle
q	phase integral, thermal conductivity parameter
q_e	electric charge
Q	second component of Stokes vector
Q_A	volume-average absorption efficiency
Q_E	volume-average extinction efficiency
Q_s	volume-average scattering efficiency excluding diffraction
Q_S	volume-average scattering efficiency including diffraction
Q_{SB}	backscattering efficiency
Q_{SF}	forward-scattering efficiency
r	bidirectional reflectance
$\boldsymbol{r}$	vector position
r_B	reflectance of bottom layer
r_h	directional–hemispherical or hemispherical reflectance
r_{hd}	hemispherical–directional reflectance
r_i	internal diffusive reflectance

r_L	Lambert reflectance
r_M	Minnaert reflectance
r_0	diffusive reflectance
r_r	reduced reflectance
r_R	bidirectional reflectance of medium with rough surface
r_{sd}	combined reflectance–diffuse reflectance
r_U	reflectance of upper layer
R_P	Fresnel coefficient of specular reflection
$R_{\parallel}$	parallel specular reflection coefficient
$R_{\perp}$	perpendicular specular reflection coefficient
RADF	radiance factor = I/F
REFF	reflectance factor
s	distance, position, line element, internal scattering coefficient
S	volume-average scattering coefficient
S_e	external surface reflection coefficient
S_i	internal surface reflection coefficient
S_0	solar constant
S	shadowing function
$\boldsymbol{S}$	Stokes vector
SHOE	shadow-hiding opposition effect
t	time
t_D	thermal diffusion time
T_e	transmission probability for emerging radiance
T_i	transmission probability for incident radiance
t_0	transmittance of layer
$\tan \Delta$	loss tangent
T	temperature
T_b	brightness temperature
T_{BB}	black-body radiative equilibrium temperature
T_S	transmission coefficient of coating on particle
$\boldsymbol{u}_n$ $\boldsymbol{u}_p$ $\boldsymbol{u}_x$ $\boldsymbol{u}_y$ $\boldsymbol{u}_z$	unit vectors
U	Planck black-body function, third component of Stokes vector
U	roughness correction factor to the physical albedo
U	electric potential
v	velocity
V	Stefan–Boltzmann integral black-body radiation function, fourth component of Stokes vector

w	volume-average single-scattering albedo
W	espat function (effective single-particle absorption thickness)
W	molecular weight
X	size parameter
Y	emergent radiant power per unit solid angle per unit area of surface of medium
z_p	thickness of optically active layer (radialith)
α	absorption coefficient
α_E	electric polarizability
β	hemispherical asymmetry factor
χ	average cosine of the surface tilt angle
χ_e	electric susceptibility
$\delta(\mathrm{x})$	Dirac delta function of argument x
Δa	sensitive area of detector
ΔA	increment of area
ΔP	increament of power
ΔQ_s	scattering efficiency difference
$\Delta\phi$	difference between upwhelling and downwhelling radiances
$\Delta\omega$	acceptance solid angle of detector
$\Delta\Omega$	increment of solid angle
∇^2	Laplacian operator, $\nabla^2 = \mathrm{div} \cdot \mathrm{grad}$
ε	particle emissivity
$\bar{\varepsilon}$	integral or wavelength averaged emissivity
ε_d	directional emissivity
ε_e	electric permittivity
ε_{e0}	permittivity of free space
ε_h	hemispherical emissivity
$\varepsilon_{\mathrm{rms}}$	root-mean-square residual
ϕ	filling factor = 1–porosity
Φ_p	integral phase function
$\emptyset$	phase
γ	albedo factor = $\sqrt{1-w}$
Γ	relative reflectance
Γ_{T}	thermal intertia
η	amplitude of complex specular reflection coefficient at normal incidence, thermal beaming factor
φ	directionally averaged radiance
θ	angle
θ_{Fe}	spectral Fe parameter
$\bar{\theta}$	mean roughness angle

Θ	particle internal transmission factor
ϑ	angle
ϑ_B	Brewster's angle
ϑ_C	critical angle
λ	wavelength
λ_C	Christiansen wavelength
Λ	luminance or photometric longitude
Λ_A	absorption mean free path
Λ_E	extinction mean free path
Λ_S	scattering mean free path
Λ_T	transport mean free path
μ	$\cos e$
μ_m	magnetic permeability
μ_{m0}	permeability of free space
μ_0	$\cos i$
ν	frequency, exponent of particle size distribution, Minnaert index
ν_p	plasma frequency
ν_0	central frequency of absorption band
π	pi
$\mathbf{\Pi}$	Poynting vector
$\Pi(g)$	angular scattering function of single particle
ρ	mass density
ρ_e	electric charge density
σ	average particle cross sectional area
σ_e	electrical conductivity
σ_0	Stefan–Boltzmann constant
Σ	summation operator
τ	optical depth
τ_0	optical thickness
Υ	angle of refracted ray in scattering function of sphere
ϖ	single-scattering albedo of single particle
Ω	direction
ξ	cosine asymmetry factor
Ξ	electron collision frequency
ψ	azimuth angle, phase difference
Ψ	particle scattering difference factor
Ψ_{jkl}	spherical harmonic function
ζ	$\sqrt{1-\beta w}$

Bibliography

Abeles, F. (1966). *Optical Properties and Electronic Structure of Metals and Alloys.* Amsterdam: North Holland.

Abramowitz, M., and Stegun, I. (1972). *Handbook of Mathematical Functions.* Washington, DC: U.S. Government Printing Office.

Adams, C., and Kattawar, G. (1978). Radiative transfer in spherical shell atmospheres. I. Rayleigh scattering. *Icarus*, **35**, 139–51.

Adams, J. (1968). Lunar and Marian surfaces: petrologic significance of absorption bands in the near-infrared, *Science*, **159**, 1453–5.

Adams, J. (1975). Interpretation of visible and near-infrared diffuse reflectance spectra of pyroxenes and other rock-forming minerals. In *Infrared and Raman Spectroscopy of Lunar and Terrestrial Minerals*, ed. C. Karr (pp. 91–116). New York: Academic Press.

Adams, J., and Felice, A. (1967). Spectral reflectance 0.4 to 2.0 microns of silicate rock powder. *J. Geophys. Res.*, **72**, 5705–15.

Akkermans, E., Wolf, P., and Maynard, R. (1986). Coherent backscattering of light by disordered media: analysis of the peak line shape. *Phys. Rev. Lett.*, **56**, 1471–4.

Akkermans, E., Wolf, P., Maynard, R., and Maret, G. (1988). Theoretical study of the coherent backscattering of light by disordered media. *J. Phys. France*, **49**, 77–98.

Allen, C. (1946). The spectrum of the corona at the eclipse of 1940 October 1. *Proc. Roy. Astron. Soc. London*, **106**, 137–50.

Altobelli, N., Spilker, L., Pilorz, S., *et al.* (2009), Thermal phase curves observed in Saturn's main rings by Cassini-CIRS: detection of an opposition effect? *Geophys. Res. Lett.*, **36**, L10105, doi:10.1029/2009GL038163.

Ambartsumian, V. (1958). The theory of radiative transfer in planetary atmospheres. In *Theoretical Astrophysics*, ed. V. Ambartsumian (pp. 550–64). New York: Pergamon.

Arfken, G., and Weber, H. (2005). *Mathematical Methods for Physicists*. Boston, MA: Elsevier.

Arnold, G., and Wagner, C. (1988). Grain size influence on the mid-infrared spectra of the minerals. *Earth, Moon and Plan.*, **41**, 163–72.

Aronson, J., and Emslie, A. (1973). Spectral reflectance and emittance of particulate materials. II. Application and results. *Appl. Opt.*, **12**, 2573–84.

Aronson, J., and Emslie, A. (1975). Applications of infrared spectroscopy and radiative transfer to earth sciences. In *Infrared and Raman Spectroscopy of Lunar and Terrestrial Minerals*, ed. C. Karr (pp. 143–64). New York: Academic Press.

Aronson, J., Emslie, A., Ruccia, F., *et al.* (1979). Infrared emittance of fibrous materials. *Appl. Opt.*, **18**, 2622–33.

Asano, S., and Yamamoto, G. (1975). Light scattering by a spheroidal particle. *Appl. Opt.*, **14**, 29–49.

Bandermann, L., Kemp, J., and Wolstencroft, R. (1972). Circular polarization of light scattered from rough surfaces. *Mon. Not. Roy. Astron. Soc.*, **158**, 291–304.

Barber, P., and Yeh, C. (1975). Scattering of electromagnetic waves by arbitrarily shaped dielectric bodies. *Appl. Opt.*, **14**, 2864–77.

Barkey, B., Bailey, M., Liou, K., and Hallett, J. (2002). Light scattering properties of plate and column ice crystals generated in a laboratory cold chamber. *Appl. Opt.*, **41**, 5792–6.

Beckmann, P. (1965). Shadowing of random rough surfaces. *IEEE Trans. Antennas Propag.*, **13**, 384–8.

Belskaya, I., and Shevchenko, V. (2000). Oppostion effect of asteroids. *Icarus*, **147**, 94–105.

Berreman, D. (1970). Resonant reflectance anomalies: effect of shapes of surface irregularities. *Phys. Rev.*, B1, 381–9.

Bevington, P. (1969). *Data Reduction and Error Analysis for the Physical Sciences.* New York: McGraw-Hill.

Blevin, W., and Brown, W. (1967). Effect of particle separation on the reflectance of semi-infinite diffusers. *J. Opt. Soc. Amer.*, **57**, 129–34.

Blewett, D., Lucey, P., and Hawke, B. (1997). Clementine images of the lunar sample-return stations: refinement of FeO and TiO_2 mapping techniques. *J. Geophys. Res.*, **102**, 16 319–25.

Bloss, F. (1961). *An Introduction to the Methods of Optical Crystallography*. Philadelphia, PA: Holt, Rinehart, & Winston.

Bobrov, M. (1962). Generalization of the theory of the shadow effect on Saturn's rings to the case of particles of unequal size. *Sov. Astron. Astrophys. J.*, **5**, 508–16.

Bohren, C. (1986). Applicability of effective-medium theories to problems of scattering and absorption by nonhomogeneous atmospheric particles. *J. Atmos. Sci.*, **43**, 468–75.

Bohren, C., and Huffman, D. (1983). *Absorption and Scattering of Light by Small Particles*. New York: John Wiley.

Borel, C., Gerstl, S., and Powers, B. (1991). The radiosity method in optical remote sensing of structured 3-D surfaces. *Rem. Sens. Environ.*, **36**, 13–44.

Born, M., and Wolf, E. (1980). *Principles of Optics*, 6th edn. New York: Pergamon.

Bottcher, C. (1952). *Theory of Electric Polarization*. Amsterdam: Elsevier.

Bowell, E., and Lumme, K. (1979). Polarimetry and magnitudes of asteroids. In *Asteroids*, ed. T. Gehrels (pp. 132–69). Tucson, AZ: University of Arizona Press.

Bowell, E., and Zellner, B. (1974). Polarizations of asteroids and satellites. In *Planets, Stars and Nebulae Studied with Photopolarimetry*, ed. T. Gehrels (pp. 381–404). Tucson, AZ: University of Arizona Press.

Bowell, E., Dollfus, A., Zellner, B., and Geake, J. (1973). Polarimetric properties of the lunar surface and its interpretation. VI. Albedo determinations from polarimetric measurements. In *Proc. 4th Lunar Sci. Conf.*, ed. W. Gose (pp. 3167–74). New York: Pergamon.

Bowell, E., Hapke, B., Domingue, D., *et al.* (1989). Applications of photometric models to asteroids. In *Asteroids II*, ed. R. Binzel, T. Gehrels, and M. Matthews (pp. 524–56). Tucson, AZ: University of Arizona Press.

Bracewell, R. (2000). *The Fourier Transform and its Applications*. New York: McGraw-Hill.

Browell, E., and Anderson, R. (1975). Ultraviolet optical constants of water and ammonia ices. *J. Opt. Soc. Amer.*, **65**, 919–26.

Brown, J., and Churchill, R. (1996). *Complex Variables and Applications*. New York: McGraw-Hill.

Brown, R., and Cruikshank, D. (1983). The Uranian satellites: surface compositions and opposition brightness surges. *Icarus*, **55**, 83–92.

Brown, R., and Matson, D. (1987). Thermal effects of insolation propagation in the regoliths of airless bodies. *Icarus*, **72**, 84–94.

Bruggeman, D. (1935). Berechnung verschiedener physikalischer Konstanten von heterogen Substanzen. I. Dielectrizitätskonstanten und Leifähigkeiten der Mischkorper aus isotropen Substanzen. *Ann. Phys. (Leipzig)*, **24**, 636–79.

Bruning, J., and Lo, Y. (1971a). Multiple scattering of EM waves by spheres. I. Multiple expansions and ray optics solutions. *IEEE Trans. Antennas Propag.*, AP-19, 378–90.

Bruning, J., and Lo, Y. (1971b). Multiple scattering of EM waves by spheres. II. Numerical and experimental results. *IEEE Trans. Antennas Propag.*, AP-19, 391–400.

Buratti, B. (1985). Application of a radiative transfer model to bright icy satellites. *Icarus*, **61**, 208–17.

Buratti, B., and Veverka, J. (1985). Photometry of rough planetary surfaces: the role of multiple scattering. *Icarus*, **64**, 320–8.

Buratti, B., Hillier, J., and Wang, M. (1996). The lunar opposition surge: observations by clementine. *Icarus*, **124**, 490–9.

Burns, R. (1970). *Mineralogical Applications of Crystal Field Theory*. Cambridge University Press.

Burns, R. (1993). Origin of electronic spectra of minerals in the visible to near-infrared region. In *Remote Geochemical Analysis*, ed. C. Pieters and P. Englert (pp. 3–29). New York: Cambridge University Press.

Burns, R., Nolet, D., Parkin, K., McCammon, C., and Schwartz, K. (1980). Mixed-valence minerals of iron and titanium: correlations of structural, Mossbauer and electronic spectral data. In *Mixed Valence Compounds*, ed. D. Brown (pp. 295–336). Boston, MA: Reidel.

Camillo, P. (1987). A canopy reflectance model based on an analytical solution to the multiple scattering equation. *Rem. Sens. Environ.*, **23**, 453–77.

Campbell, M., and Ulrichs, J. (1969). Electrical properties of rocks and their significance for lunar radar absorptions. *J. Geophys. Res.*, **74**, 5867–81.

Capaccioni, F., Cerroni, P., Barucci, M., and Fuilchignoni, M. (1990). Phase curves of meteorites and terrestrial rocks: laboratory measurements and applications to asteroids. *Icarus*, **83**, 325–48.

Carrier, W., Mitchell, J., and Mahmood, A. (1973). The relative density of lunar soil. *Proc. 4th Lunar Sci. Conf.*, ed. W. Gose (pp. 2403–11). New York: Pergamon.

Chamberlain, J., and Smith, G. (1970). Interpretation of the Venus CO_2 absorption bands. *Astrophys. J.*, **160**, 755–65.

Chandrasekhar, S. (1960). *Radiative Transfer*. New York: Dover.

Chapman, C. (1996). S-type asteroids, ordinary chondrites and space weathering: the evidence from *Galileo's* fly-bys of Gaspra and Ida. *Meteorit. Planet. Sci.*, **31**, 699–725.

Chapman, S. (1931). The absorption and dissociative or ionizing effect of monochromatic radiation in an atmosphere on a rotating Earth. II. Grazing incidence. *Proc. Phys. Soc.*, **43**, 483–501.

Chiappetta, P. (1980). A new model for scattering by irregular absorbing particles. *Astron. Astrophys.*, **83**, 348–53.

Chorlton, F. (1976). *Vector and Tensor Methods*. New York: John Wiley.

Christensen, P., Bandfield, J., Fergason, R., Hamilton, V., and Rogers, A. (2008a). The compositional diversity and physical properties mapped from the Mars Odyssey Thermal Emission Imaging System (THEMIS). In *The Martian Surface: Composition, Mineralogy, and Physical Properties*, ed. J. Bell, III (pp. 221–41). Cambridge University Press.

Christensen, P., Bandfield, J., Rogers, A., *et al.* (2008b). Global mineralogy mapped from the Mars Global Surveyor Thermal Emission Spectrometer. In *The Martian Surface: Composition, Mineralogy, and Physical Properties*, ed. J. Bell, III, (pp. 195–220). Cambridge University Press.

Churchill, R. (1944). *Modern Operational Mathematics in Engineering*. New York: McGraw-Hill.

Chylek, P., Grams, G., and Pinnick, R. (1976). Light scattering by irregular randomly oriented particles. *Science*, **193**, 480–2.

Clark, R. (1983). Spectral properties of mixtures of montmorillonite and dark carbon grains: implications for remote sensing minerals containing chemically and physically adsorbed water. *J. Geophys. Res.*, **88**, 10 635–44.

Clark, R., and Lucey, P. (1984). Spectral properties of ice-particulate mixtures and implications for remote sensing. I. Intimate mixtures. *J. Geophys. Res.*, **89**, 6341–8.

Clark, R., and Roush, T. (1984). Reflectance spectroscopy: quantitative analysis techniques for remote sensing applications. *J. Geophys. Res.*, **89**, 6329–40.

Clark, R., Kierein, K., and Swayze, G. (1993). Experimental verification of the Hapke reflectance theory. I. Computation of reflectance as a function of grain size and wavelength based on optical constants. Preprint.

Cohen, A., and Janezic, G. (1983). Relationships among trapped hole and trapped electron centers in oxidized soda-silica glasses of high purity. *Phys. Stat. Sol. (a)*, **77**, 619–24.

Conel, J. (1969). Infrared emissivities of silicates: experimental results and a cloudy atmosphere model of spectral emission from condensed particulate mediums. *J. Geophys. Res.*, **74**, 1614–34.

Cord, A., Pinet, P., Daydou, Y., and Chevrel, S. (2003). Planetary regolith surface analogs: optimized determination of Hapke parameters using multi-angular spectro-imaging laboratory data. *Icarus*, **165**, 414–27.

Coulson, K. (1971). The polarization of light in the environment. In *Planets, Stars, and Nebulae Studied with Photopolarimetry*, ed. T. Gehrels (pp. 444–71). Tucson, AZ: University of Arizona Press.

Cox, C., and Munk, W. (1954). Measurement of the roughness of the sea surface from photographs of the sun's glitter. *J. Opt. Soc. Amer.*, **44**, 838–50.

Crank, J. (1975). *The Mathematics of Diffusion*. Oxford University Press.

Danjon, A. (1949). Photometrie et colorimetrie des planets Mercure et Venus. *Bull. Astron.*, **14**, 315–17.

Dexter, D. (1956). Absorption of light by atoms in solids. *Phys. Rev.*, **101**, 48–55.

Dickinson, R., Pinty, B., and Verstraete, M. (1990). Relating surface albedos in GCM to remotely sensed data. *Agricult. Forest Meteorol.*, **52**, 109–31.

Dollfus, A. (1956). Polarisation de la lumière renvoyée par les corps solides et les nuages naturels. *Ann. Astrophys.*, **19**, 83–113.

Dollfus, A. (1961). Polarization studies of planets. In *Planets and Satellites*, ed. G. Kuiper and B. Middlehurst (pp. 343–99). Chicago, IL: University of Chicago Press.

Dollfus, A. (1962). The polarization of moonlight. In *Physics and Astronomy of the Moon*, ed. Z. Kopal (pp. 131–60). New York: Academic Press.

Dollfus, A. (1998). Lunar surface imaging polarimetry. I. Roughness and grain size. *Icarus*, **136**, 69–103.

Dollfus, A., and Bowell, E. (1971). Polarimetric properties of the lunar surface and its interpretation. I. Telescopic observations. *Astron. Astrophys.*, **10**, 29–53.

Dollfus, A., Wolff, M., Geake, J., Lupishko, D., and Dougherty, L. (1989). Photopolarimetry of asteroids. In *Asteroids II*, ed. R. Binzel, T. Gehrels, and M. Matthews (pp. 594–615). Tucson, AZ: University of Arizona Press.

Domingue, D., and Hapke, B. (1989). Fitting theoretical photometric functions to asteroid phase curves. *Icarus*, **74**, 330–6.

Domingue, D., and Verbiscer, A. (1997). Reanalysis of the solar phase curves of the icy Galilean satellites. *Icarus*, **128**, 49–74.

Domingue, D., Hapke, B., Lockwood, G., and Thompson, D. (1991). Europa's phase curve: implications for surface structure. *Icarus*, **90**, 30–42.

Draine, B. (1988). The discrete dipole approximation: its application to interstellar graphite grains. *Astrophys. J.*, **333**, 848–72.

Draine, B. (2000). The discrete dipole approximation for light scattering by irregular targets. In *Light Scattering by Nonspherical Particles*, ed. M. Mischenko, J. Hovenier, and L. Travis (pp. 131–45). New York: Academic Press.

Draine, B., and Flatau, P. (1994). Discrete dipole approximation for scattering calculations. *J. Opt. Soc. Amer.*, **411**, 1491–9.

Draine, B., and Goodman, J. (1993). Beyond Clausiul–Mossotti: wave propagation on a polorizable point lattice and the discrete polar approximation. *Astrophys. J.*, **405**, 685–97.

Drude, P. (1959). *Theory of Optics*. New York: Dover.

Dwight, H. (1947). *Tables of Integrals and Other Mathematical Data*. New York: Macmillan.

Egan, W. (1985). *Photometry and Polarization in Remote Sensing*. New York: Elsevier.

Egan, W., and Hilgeman, T. (1976). Retroreflectance measurements of photometric standards and coatings. *Appl. Opt.*, **15**, 1845–9.

Egan, W., and Hilgeman, T. (1978). Spectral reflectance of particulate materials: a Monte Carlo model including asperity scattering. *Appl.Opt.*, **17**, 245–52.

Egan, W., and Hilgeman, T. (1979). *Optical Properties of Inhomogeneous Materials*. New York: Academic Press.

Elliott, R. (1966). *Electromagnetics*. New York: Academic Press.

Emslie, A., and Aronson, J. (1973). Spectral reflectance and emittance of particulate materials. I. Theory. *Appl. Opt.*, **12**, 2563–72.

Esposito, L. (1979). Extensions to the classical calculation of the effect of mutual shadowing in diffuse reflection. *Icarus*, **39**, 69–80.

Evans, J. (1962). Radio echo studies of the moon. In *Physics and Astronomy of the Moon*, ed. Z. Kopal (pp. 429–80). New York: Academic Press.

Evans, J., and Hagfors, T. (1968). *Radar Astronomy*. New York: McGraw-Hill.

Evans, J., and Hagfors, T. (1971). Radar studies of the moon. In *Advances in Astronomy and Astrophysics, Vol. 8*, ed. Z. Kopal (pp. 29–107). New York: Academic Press.

Fairchild, M., and Daoust, D. (1988). Goniospectrophotometric analysis of pressed PTFE powder for use as a primary transfer standard. *Appl. Opt.*, **27**, 3392–6.

Fountain, J., and West, E. (1970). Thermal conductivity of particulate basalt as a function of density in simulated lunar and Martian environments. *J. Geophys. Res.*, **75**, 4063–70.

Fowler, W. (1968). *Physics of Color Centers*. New York: Academic Press.

Fredricksson, K., and Keil, K. (1963). The light–dark structures in the Pantar and Kapoeta stone meteorites. *Geochim. Cosmochim. Acta*, **27**, 717–39.

French, R., Verbescer, A., Salo, H., McGhee, C. and Dones, L. (2007). Saturn's rings at true opposition. *Pub. Astronom. Soc. Pacific*, **119**, 623–42.

Frohlich, H. (1958). *Theory of Dielectrics*, 2nd edn. London: Oxford University Press.

Fuller, K., and Kattawar, G. (1988a). Consumate solutions to the problem of classical electromagnetic scattering by ensembles of spheres. I. Linear chains. *Opt. Lett.*, **13**, 90–2.

Fuller, K., and Kattawar, G. (1988b). Consumate solutions to the problem of classical electromagnetic scattering by ensembles of spheres. II. Clusters of arbitrary configurations. *Opt. Lett.*, **13**, 1063–5.

Fung, A., and Ulaby, F. (1983). Matter–energy interactions in the microwave region. In *Manual of Remote Sensing*, ed. D. Simonett (pp. 115–64). Falls Church, VA: American Society of Photogrammetry.

Gaffey, M., Bell, J., and Cruikshank, D. (1989). Reflectance spectroscopy and asteroid surface mineralogy. In *Asteroids II*, ed. R. Binzel, T. Gehrels, and M. Matthews (pp. 98–127). Tucson, AZ: University of Arizona Press.

Galileo (1638). *Dialogue on the Great World Systems*, trans. G. De Santillana (1953). Chicago, IL: University of Chicago Press.

Garbuny, M. (1965). *Optical Physics*. New York: Academic Press.

Geake, J., and Dollfus, A. (1986). Planetary surface texture and albedo from parameter plots of optical polarization data. *Mon. Not. Roy. Astron. Soc.*, **218**, 75–91.

Geake, J., Geake, M., and Zellner, B. (1984). Experiments to test theoretical models of the polarization of light by rough surfaces. *Mon. Not. Roy. Astron. Soc.*, **210**, 89–112.

Gehrels, T. (1974). Introduction and overview. In *Planets, Stars and Nebulae Studied with Photopolarimetry*, ed. T. Gehrels (pp. 3–44). Tucson, AZ: University of Arizona Press.

Gehrels, T., and Teska, T. (1963). The wavelength dependence of polarization. *Appl. Opt.*, **2**, 67–77.

Gehrels, T., Coffeen, D., and Owings, D. (1964). Wavelength dependence of polarization. III. The lunar surface. *Astron. J.*, **69**, 826–52.

Gerstl, S., and Zardecki, A. (1985a). Discrete-ordinates finite-element method for atmospheric radiative transfer and remote sensing. *Appl. Opt.*, **24**, 81–93.

Gerstl, S., and Zardecki, A. (1985b). Coupled atmosphere/canopy model for remote sensing of plant reflectance features. *Appl. Opt.*, **24**, 94–103.

Goguen, J. (1981). A theoretical and experimental investigation of the photometric functions of particulate surfaces. Ph.D. thesis, Cornell University, Ithaca, NY.

Goody, R. (1964). *Atmospheric Radiation*, Vol. 1, *Theoretical Basis*. Oxford University Press.

Gradie, J., and Veverka, J. (1982). When are spectral reflectance curves comparable? *Icarus*, **49**, 109–19.

Greenberg, J. (1974). Some examples of exact and approximate solutions in small particle scattering: a progress report. In *Planets, Stars and Nebulae Studied with*

Photopolarimetry, ed. T. Gehrels (pp. 107–34). Tucson, AZ: University of Arizona Press.

Grum, F., and Luckey, G. (1968). Optical sphere paint and a working standard of reflectance. *Appl. Opt.*, **7**, 2289–94.

Gustafson, B. (2000). Microwave analog to light scattering measurements. In *Light Scattering by Nonspherical Particles*, ed. M. Mishchenko, J. Hovenier, and L. Travis (pp. 367–92). New York: Academic Press.

Hagfors, T. (1964). Backscatter from an undulating surface with applications to radar returns from the moon. *J. Geophys. Res.*, **69**, 3779–84.

Hagfors, T. (1968). Relations between rough surfaces and their scattering properties as applied to radar astronomy. In *Radar Astronomy*, ed. J. Evans and T. Gehrels (pp. 187–218). New York: McGraw-Hill.

Hale, A., and Hapke, B. (2002). A time-dependent model of radiative and conductive thermal energy transport in planetary regoliths with applications to the moon and Mercury. *Icarus*, **156**, 318–34.

Hameen-Anttila, K. (1967). Surface photometry of the planet Mercury. *Ann. Acad. Sci. Fenn., Ser. A6*, **252**, 1–19.

Hansen, J., and Arking, A. (1971). Clouds of Venus: evidence for their nature. *Science*, **171**, 669–72.

Hansen, J., and Travis, L. (1974). Light scattering in planetary atmospheres. *Space Sci. Rev.*, **16**, 527–610.

Hapke, B. (1963). A theoretical photometric function for the lunar surface. *J. Geophys. Res.*, **68**, 4571–86.

Hapke, B. (1968). On the particle size distribution of lunar soil. *Planet. Space Sci.*, **16**, 101–10.

Hapke, B. (1971). Optical properties of the lunar surface. In *Physics and Astronomy of the Moon*, ed. Z. Kopal (pp. 155–211). New York: Academic Press.

Hapke, B. (1981). Bidirectional reflectance spectroscopy. I. Theory. *J. Geophys. Res.*, **86**, 3039–54.

Hapke, B. (1984). Bidirectional reflectance spectroscopy. III. Correction for macroscopic roughness. *Icarus*, **59**, 41–59.

Hapke, B. (1986). Bidirectional reflectance spectroscopy. IV. Extinction and the opposition effect. *Icarus*, **67**, 264–80.

Hapke, B. (1990). Coherent backscatter and the radar characteristics of outer planet satellites. *Icarus*, **88**, 407–17.

Hapke, B. (1993). *Theory of Reflectance and Emittance Spectroscopy*. Cambridge University Press.

Hapke, B. (1996a). A model of radiative and conductive energy trasfer in planetary regoliths. *J. Geophys. Res.*, **101**, 16 817–31.

Hapke, B. (1996b). Applications of an energy transfer model to three problems in planetary regloliths: the solid-state greenhouse, thermal beaming and emittance spectra. *J. Geophys. Res.*, **101**, 16 833–40.

Hapke, B. (1999). Scattering and diffraction of light by particles in planetary regoliths. *J. Quant. Spectrosc. Radiat. Transf.*, **61**, 565–81.

Hapke, B. (2001). Space weathering from Mercury to the asteroid belt. *J. Geophys. Res*, **106**, 10 039–073.

Hapke, B. (2008). Bidirectional reflectance spectroscopy. VI. Effects of porosity. *Icarus*, **195**, 918–26.

Hapke, B., and Blewett, D. (1991). Coherent backscatter model for the unusual radar reflectivity of icy satellites. *Nature*, **352**, 46–7.

Hapke, B., and Nelson, R. (1975). Evidence for an elemental sulfur component of the clouds from Venus spectrophotometry. *J. Atmos. Res.*, **32**, 1211–18.

Hapke, B., and Van Horn, H. (1963). Photometric studies of complex surfaces with applications to the moon. *J. Geophys. Res.*, **68**, 4545–70.

Hapke, B., and Wells, E. (1981). Bidirectional reflectance spectroscopy. II. Experiments and observations. *J. Geophys. Res.*, **86**, 3055–60.

Hapke, B., and Williams, A. (1988). Search for anomalous opposition spike in crystalline powders. *Bull. Amer. Astron. Soc.*, **20**, 808.

Hapke, B., Cassidy, W., and Wells, E. (1975). Effects of vapor phase deposition processes on the optical, chemical and magnetic properties of the lunar regolith. *The Moon*, **13**, 339–53.

Hapke, B., DiMucci, D., Nelson, R., and Smythe, W. (1996). The cause of the hot spot in vegetation canopies and soils. *Rem. Sens. Environ.*, **58**, 63–8.

Hapke, B., Nelson, R., and Smythe, W. (1993). The opposition effect of the moon: the contribution of coherent backscattering. *Science*, **260**, 509–11.

Hapke, B., Shepard, M., Nelson, R., Smythe, W., and Piatek, J. (2009). A quantitative test of the ability of models based on the equation of radiative transfer to predict the bidirectional reflectance of a well-characterized medium. *Icarus*, **199**, 210–18.

Hapke, B., Wells, E., and Wagner, J. (1981). Far-UV, visible and near-IR reflectance spectra of frosts of H_2O, CO_2, NH_3 and SO_2. *Icarus*, **47**, 361–7.

Harris, D. (1957). Diffuse reflection from planetary atmospheres. *Astrophys. J.*, **126**, 408–12.

Hartman, B., and Domingue, D. (1998). Scattering of light by individual particles and the implications for models of planetary surfaces. *Icarus*, **131**, 421–48.

Helfenstein, P. (1986). Derivation and analysis of geological constraints on the emplacement and evolution of terrains on Ganymede from applied differential photometry. Ph.D. thesis, Brown University, Providence, R.I.

Helfenstein, P. (1988). The geological interpretation of photometric surface roughness. *Icarus*, **73**, 462–81.

Helfenstein, P., and Shepard, M. (1999). Submillimeter-scale topography of the lunar regolith. *Icarus*, **141**, 107–31.

Helfenstein, P., and Veverka, J. (1987). Photometric properties of lunar terrains derived from Hapke's equation. *Icarus*, **72**, 343–57.

Helfenstein, P., and Veverka, J. (1989). Physical characterization of asteroid surfaces from photometric analysis. In *Asteroids II*, ed. R. Binzel, T. Gehrels, and M. Matthews (pp. 557–93). Tucson, AZ: University of Arizona Press.

Helfenstein, P., Veverka, J., and Thomas, P. (1988). Uranus satellites: Hapke parameters from *Voyager* disk-integrated photometry. *Icarus*, **78**, 231–9.

Henyey, C., and Greenstein, J. (1941). Diffuse radiation in the galaxy. *Astrophys. J.*, **93**, 70–83.

Herbst, T., Skrutskie, M., and Nicholson, P. (1987). The phase curve of the Uranian rings. *Icarus*, **71**, 103–14.

Hillier, J., Buratti, B., and Hill, K. (1999). Multispectral photometry of the moon and absolute calibration of the Clementine UV/Vis camera. *Icarus*, **141**, 205–25.

Hisdal, B. (1965). Reflectance of perfect diffuse and specular samples in the integrating sphere. *J. Opt. Soc. Amer.*, **55**, 1122–8.

Hodkinson, J. (1963). Light scattering and extinction by irregular particles larger than the wavelength. In *Electromagnetic Scattering*, ed. M. Kerker (pp. 87–100). New York: Macmillan.

Hodkinson, J., and Greenleaves, I. (1963). Computations of light scattering and extinction by spheres according to diffraction and geometrical optics, and some comparisons with the Mie theory. *J. Opt. Soc. Amer.*, **53**, 577–88.

Holland, A., and Gagne, G. (1970). The scattering of polarized light by polydisperse systems of irregular particles. *Appl. Opt.*, **9**, 1113–21.

Hopfield, J. (1966). Mechanism of lunar polarization. *Science*, **151**, 1380–1.

Hovenier, J. (2000). Measuring scattering metrices of small particles at optical wavelengths. In *Light Scattering by Nonspherical Particles*, ed. M. Mishchenko, J. Hovenier, and L. Travis (pp. 355–66). New York: Academic Press.

Huguenin, R., and Jones, J. (1986). Intelligent information extraction from reflectance spectra: absorption band positions. *J. Geophys. Res.*, **91**, 9585–98.

Hunt, G. (1980). Electromagnetic radiation: the communication link in remote sensing. In *Remote Sensing in Geology*, ed. B. Siegal and A. Gillespie (pp. 5–46). New York: John Wiley.

Hunt, G., and Vincent, R. (1968). The behavior of spectral features in the infrared emission from particulate surfaces of various grain sizes. *J. Geophys. Res.*, **73**, 6039–46.

Irvine, W. (1965). Multiple scattering by large particles. *Astrophys. J.*, **142**, 1563–75.

Irvine, W. (1966). The shadowing effect in diffuse reflectance. *J. Geophys. Res.*, **71**, 2931–7.

Irvine, W., and Pollack, J. (1968). Infrared properties of water and ice spheres. *Icarus*, **8**, 324–60.

Ishimaru, A. (1978). *Wave Propagation and Scattering in Random Media*. New York: Academic Press.

Ishimaru, A., and Kuga, Y. (1982). Attenuation of a coherent field in a dense distribution of particles. *J. Opt. Soc. Amer.*, **72**, 1317–20.

Jackson, J. (1999). *Classical Electromagnetics*. New York: John Wiley.

Jahnke, E., and Emde, E. (1945). *Tables of Functions*. New York: Dover.

Jakowsky, B., Finiol, G., and Henderson, B. (1990). Directional variations in thermal emission from geologic surfaces. *Geophys. Res. Lett.*, **17**, 985–8.

Jenkins, F., and White, H. (1950). *Fundamentals of Optics*, 2nd edn. New York: McGraw-Hill.

Jenkins, P., Smith, M., and Adams, J. (1985). Quantitative analysis of planetary reflectance spectra with principal components analysis. In *Proc. 15th Lunar Planet. Sci. Conf.*, ed. G. Ryder and G. Schubert (pp. C805–10). Washington, DC: American Geophysical Union.

Johnson, J., Grundy, W., and Shepard, M. (2004). Visible/near-infrared spectrogoniometric observations and modeling of dust-coated rocks. *Icarus*, **171**, 546–56.

Johnson, P., Smith, M., Taylor-George, S., and Adams, J. (1983). A semiempirical method for analysis of the reflectance spectra of binary mineral mixtures. *J. Geophys. Res.*, **88**, 3557–61.

Johnson, R., Nelson, M., McCord, T., and Gradie, J. (1988). Analysis of *Voyager* images of Europa: plasma bombardment. *Icarus*, **75**, 423–36.

Joseph, J., Wiscombe, W., and Weinman, J. (1976). The delta-Eddington approximation for radiative flux transfer. *J. Atmos. Sci.*, **33**, 2452–9.

Kaasalainan, S. (2003). Laboratory photometry of planetary regolith analogs. I. Effects of grain and packing properties on opposition effect. *Astron. Astrophys.* **409**, 765–9.

Kaasalainen, S., Peltoniemi, J., Naranen, J., *et al.* (2005). Small angle goniometry for backscattering measurements in the broadband spectrum. *Appl. Opt.*, **44**, 1485–90.

Kattawar, G. (1975). A three parameter analytic phase function for multiple scattering calculations. *J. Quant. Spectrosc. Radiat. Transf.*, **15**, 839–49.

Kattawar, G. (1979). Radiative transfer in spherical shell atmospheres. III. Application to Venus. *Icarus*, **40**, 60–6.

Kattawar, G., and Eisner, M. (1970). Radiation from a homogeneous isothermal sphere. *Appl. Opt.*, **9**, 2685–90.

Kattawar, G., and Humphreys, T. (1980). Electromagnetic scattering from two identical pseudospheres. In *Light Scattering by Irregularly Shaped Particles*, ed. D. Schuerman (pp. 177–90). New York: Plenum.

Kemp, J. (1974). Circular polarization of planets. In *Planets, Stars and Nebulae Studied with Photopolarimetry*, ed. T. Gehrels (pp. 607–16). Tucson, AZ: University of Arizona Press.

KenKnight, C., Rosenberg, D., and Wehner, G. (1967). Parameters of the optical properties of the lunar surface powder in relation to solar wind bombardment. *J. Geophys. Res.*, **72**, 3105–29.

Kerker, M. (1969). *The Scattering of Light*. New York: Academic Press.

Kimes, D., and Kerchner, J. (1982). Irradiance measurement errors due to the assumption of a Lambertian reference panel. *Rem. Sens. Environ.*, **12**, 141–9.

Kittel, C. (1976). *Introduction to Solid State Physics*, 5th edn. New York: John Wiley.

Kocinski, J., and Wojtczak, L. (1978). *Critical Scattering Theory: An Introduction*. New York: Elsevier.

Kolokolova, L. (1985). On the influence of the structure of atmosphereless bodies' surfaces to the polarimetric characteristics of reflected light. *Solar Syst. Res.*, **19**, 165–73.

Kolokolova, L. (1990). Dependence of polarization on optical and structural properties of the surfaces of atmosphereless bodies. *Icarus*, **84**, 305–14.

Kolokolova, L., Kimura, H., Ziegler, K., and Mann, I. (2006). Light scattering properties of random oriented aggregates: do they represent the properties of an ensemble of aggreagates? *J. Quant. Spectrosc. Radiat. Transf.*, **100**, 199–206.

Kortum, G. (1969). *Reflectance Spectroscopy*. New York: Springer.

Kourganoff, V. (1963). *Basic Methods in Transfer Problems: Radiative Equilibrium and Neutron Diffusion*. New York: Dover.

Kubelka, P. (1948). New contributions to the optics of intensely light-scattering materials. I. *J. Opt. Soc. Amer.*, **38**, 448–57.

Kubelka, P. (1954). New contributions to the optics of intensely light-scattering materials. II. Nonhomogeneous layers. *J. Opt. Soc. Amer.*, **44**, 330–5.

Kubelka, P., and Munk, F. (1931). Ein Beitrag zur Optik der Farberntricke. *Z. Techn. Physik*, **12**, 593–601.

Kuga, Y., and Ishimaru, A. (1984). Retroreflection from a dense distribution of spherical particles. *J. Opt. Soc. Amer.*, **8**, 831–5.

Landau, L., and Lifschitz, E. (1975). *The Classical Theory of Fields*, 4th edn. New York: Pergamon.

Lass, H. (1950). *Vector and Tensor Analysis*. New York: McGraw-Hill.

Lax, M. (1954). The influence of lattice vibrations on electronic transitions in solids. In *Photoconductivity Conference*, ed. R. Breckenridge, B. Russell, and E. Hahn (pp. 111–45). New York: John Wiley.

Lebofsky, L., Sykes, M., Tedesco, E., *et al.* (1986). A refined "standard" thermal model for asteroids based on observations of 1 Ceres and 2 Pallas. *Icarus*, **68**, 239–51.

Leinert, C., Link, H., Pitz, E., and Giese, R. (1976). Interpretation of a rocket photometry of the inner zodiacal light. *Astron. Astrophys.*, **47**, 221–30.

Lenoble, J. (1985). *Radiative Transfer in Scattering and Absorbing Atmospheres*. Hampton, VA: Deepak Publishing.

Liang, C., and Lo, Y. (1967). Scattering by two spheres. *Radio Sci.*, **2**, 1481–95.

Liou, K., and Coleman, R. (1980). Light scattering by hexagonal columns and plates. In *Light Scattering by Irregularly Shaped Particles*, ed. D. Schuerman (pp. 207–18). New York: Plenum.

Liou, K., and Hansen, J. (1971). Intensity and polarization for single scattering by polydisperse spheres: a comparison of ray optics and Mie theory. *J. Atmos. Sci.*, **28**, 995–1004.

Liou, K., and Schotland, R. (1971). Multiple backscattering and depolarization from water clouds for a pulsed lidar system. *J. Atmos. Sci.*, **28**, 772–84.

Liou, K., Cai, Q., and Pollack, J. (1983). Light scattering by randomly oriented cubes and parallelepipeds. *Appl. Opt.*, **22**, 3001–8.

Logan, L., Hunt, G., Salisbury, J., and Balsamo, S. (1973). Compositional implications of Christiansen frequency maximums for infrared remote sensing applications. *J. Geophys. Res.*, **78**, 4983–5003.

Lorentz, H. (1952). *The Theory of Electrons*. New York: Dover.

Lucey, P., Blewett, D., and Hawke, B. (1998). Mapping the FeO and TiO_2 content of the lunar surface with multispectral imagery. *J. Geophys. Res.*, **103** (E2), 3679–99.

Lucey, P., Taylor, G., and Malaret, E. (1995). Abundance and distribution of iron on the Moon. *Science*, **268**, 1150–3.

Lucey, P., Blewett, D., Taylor, G., and Hawke, B. (2000). Imaging of lunar surface maturity. *J. Geophys. Res.*, **105** (E8), 20 377–86.

Lumme, K., and Bowell, E. (1981a). Radiative transfer in the surfaces of atmosphereless bodies. I. Theory. *Astron. J.*, **86**, 1694–704.

Lumme, K., and Bowell, E. (1981b). Radiative transfer in the surfaces of atmosphereless bodies. II. Interpretation of phase curves. *Astron. J.*, **86**, 1705–12.

Lumme, K., Rahola, J., and Hovenier, J. (1997). Light scattering by dense clusters of spheres. *Icarus*, **126**, 455–69.

Lyot, B. (1929). Recherches sur la polarisation de la lumière des planètes et de quelques substances terrestres. *Ann. Obs. Paris*, Vol. 8, Book 1 (translated as NASA Tech. Transl. TT-F-187, 1964).

McEwan, A. (1991). Photometric functions for photoclinometry and other applications. *Icarus*, **92**, 298–311.

McGuire, A., and Hapke, B. (1995). An experimental study of light scattering by large irregular particles. *Icarus*, **113**, 134–55.

McKay, D., Fruland, R., and Heiken, G. (1974). Grain size and the evolution of lunar soils. In *Proc. 5th Lunar Sci. Conf.*, ed. W. Gose (pp. 887–906). New York: Pergamon.

Macke, A. (2000). Monte Carlo calculations of light scattering by large particles with multiple internal inclusions. In *Light Scattering by Nonspherical Particles*, ed. M. Mishchenko, J. Hovenier, and L. Travis (pp. 300–22). New York: Academic Press.

MacKintosh, F., and John, S. (1988). Coherent backscattering of light in the presence of time-reversal, non-invariant and parity-violating media. *Phys. Rev.*, B37, 1884–97.

MacKintosh, F., Zhu, J., Pine, D., and Weitz, D. (1989). Polarization memory of multiply scattered light. *Phys. Revo*, B40, 9342–45.

Mackowski, D., and Mishchenko, M. (1996). Calculation of the T-matrix and the scattering matrix for ensembles of particles. *J. Opt. Soc. Amer.*, **13**, 2266–78.

Margenau, H., and Murphy, G. (1956). *The Mathematics of Physics and Chemistry*. New York: Van Nostrand.

Marion, J. (1965). *Classical Electromagnetic Radiation*. New York: Macmillan.

Matson, D., and Brown, R. (1989). Solid state greenhouses and their implications for icy satellites. *Icarus*, **77**, 67–81.

Maxwell-Garnett, J. (1904). Colours in metal glasses and in metallic films. *Phil. Trans. Roy. Soc. London*, A203, 385–420.

Melamed, N. (1963). Optical properties of powders. I. Optical absorption coefficients and the absolute value of the diffuse reflectance. II. Properties of luminescent powders. *J. Appl. Phys.*, **34**, 560–70.

Middleton, W., and Sanders, C. (1951). The absolute spectral diffuse reflectance of magnesium oxide. *J. Opt. Soc. Amer.*, **41**, 419–24.

Minnaert, M. (1941). The reciprocity principle in lunar photometry. *Astrophys. J.*, **93**, 403–10.

Minnaert, M. (1961). Photometry of the moon. In *Planets and Satellites*, ed. G. Kuiper and B. Middlehurst (pp. 213–45). Chicago, IL: University of Chicago Press.

Mishchenko, M. (1995). Coherent backscattering by a two-sphere cluster. *Opt. Lett.* **21**, 623–5.

Mishchenko, M. (2002). Vector radiative transfer equation for arbitraritly shaped and arbitrarily oriented particles: a microphysical derivation from statistical electromagnetics. *Appl. Opt.*, **41**, 7114–34.

Mishchenko, M. (2008). Multiple scattering, radiative transfer and weak localization in discrete random media: unified microphysical approach. *Rev. Geophys.*, **46**, RG2003, doi.10.1029/2007RG200230.

Mishchenko, M., and Liu, L. (2007). Weak localization of electromagnetic waves by densely packed many-particle groups: exact 3D results. *J. Quant. Spectrosc. Radiat. Transf.*, **106**, 616–21.

Mishchenko, M., and Macke, A. (1997). Asymmetry parameters for the phase function for isolated and densely packed spherical particles with multiple internal inclusions in the geometric optics limit. *J. Quant. Spectrosc. Radiat. Transf.*, **57**, 767–94.

Mishchenko, M., and Mackowski, D. (1996). Electromagnetic scattering by randomly oriented bispheres: comparison of theory and experiment and benchmark calculations. *J. Quant. Spectrosc. Radiat. Transf.*, **55**, 683–694.

Mishchenko, M., Dlugach, J., Yanovitskij, E., and Zakharova, N. (1999). Bidirectional reflectance of flat, optically thick particulate layers: an efficient radiatve transfer solution and applications to snow and soil surfaces. *J. Quant. Spectrosc. Radiat. Transf.*, **63**, 409–32.

Mishchenko, M., Hovenier, J., and Travis, L. (2000a). Concepts, terms and notation. In *Light Scattering by Nonspherical Particles*, ed. M. Mishchenko, J. Hovenier, and L. Travis (pp. 3–27). San Diego, CA: Academic Press.

Mishchenko, M., Liu, L., Mackowski, D., Cairns, B., and Videen, G. (2007). Multiple scattering by random particulate media: exact 3D results. *Opt. Expr.*, **15**, 2822–36.

Mishchenko, M., Luck, J., and Nieuwenhuizen, T. (2000b). Full angular profile of the coherent polarization opposition effect. *J. Opt. Soc Amer.*, A17, 888–91.

Mishchenko, M., Mackowski, D., and Travis, L. (1995). Scattering of light by bispheres with touching and separated components. *Appl. Opt.*, **34**, 4589–99.

Montgomery, W., and Kohl, R. (1980). Opposition effect experimentation. *Opt. Lett.*, **5**, 546–8.

Morris, R., Lauer, H., Lawson, C., *et al.* (1985). Spectral and other physicochemical properties of submicron powders of hematite, maghemite, magnetite, goethite and lepidocrocite. *J. Geophys. Res.*, **90**, 3126–44.

Morrison, D., and Lebofsky, L. (1979). Radiometry of asteroids. In *Asteroids*, ed T. Gehrels (pp. 184–205), Tucson, AZ: University of Arizona Press.

Morse, P., and Feshbach, H. (1953). *Methods of Theoretical Physics*. New York: McGraw-Hill.

Muhleman, D. (1964). Radar scattering from Venus and the moon. *Astron. J.*, **69**, 34–41.

Muinonen, K. (1990). Light scattering by inhomogeneous media: backward enhancement and reversal of linear polarization. Ph.D. thesis, University of Helsinki, Finland.

Muinonen, K. (2000). Light scattering by stochastically shaped particles. In *Light Scattering by Nonspherical Particles*, ed. M. Mishchenko, J. Hovenier, and L. Travis (pp. 323–54). New York: Academic Press.

Muinonen, K. (2004). Coherent backscattering of light by complex random media of spherical scatterers: numerical solution. *Waves Random Media*, **14**, 365–88.

Muinonen, K., Lumme, K., Peltoniemi, J., and Irvine, W. (1989). Light scattering by randomly oriented crystals. *Appl. Opt.*, **28**, 3051–60.

Muinonen, K., Zubko, E., Tyynela, J., Shkuratov, Y., and Videen G. (2007). Light scattering by Gaussian random particles with discrete-dipole approximation. *J. Quant. Spectrosc. Radiat. Transf.*, **106**, 360–77.

Mukai, S., Mukai, T., Giese, R., Weiss, K., and Zerull, R. (1982). Scattering of radiation by a large particle with a random rough surface. *Moon and Planets*, **26**, 197–208.

Munoz, O., Volten, H., deHan, J., Vassen, W., and Hovenier, J. (2000). Experimental determination of scattering matrices of olivine and Allende meteorite particles. *Astron. Astrophys.*, **360**, 777–88.

Munoz, O., Volten, H., Hovenier, J., *et al.* (2006). Experimental and computation study of light scattering by irregular particles with extreme refractive indices: hematite and rutile. *Astron. Astrophys.*, **446**, 525–35.

Mustard, J., and Pieters, C. (1987). Quantitative abundance estimates from bidirectional reflectance measurements. In *Proc. 17th Lunar Planet. Sci. Conf.*, ed. G. Ryder and G. Schubert (pp. E617–26). Washington, DC: American Geophysical Union.

Mustard, J., and Pieters, C. (1989). Photometric phase functions of common geologic minerals and applications to quantitative analysis of mineral mixture reflectance spectra. *J. Geophys. Res.*, **94**, 13 619–34.

Naranen, J., Kaasalainen, S., Peltoniemi, J., *et al.* (2004), Laboratory photometry of planetary regolith analogs. II. Surface roughness and extremes of packing density. *Astron. Astrophys.*, **426**, 1103–9.

Nash, D. (1983). Io's 4-μm band and the role of adsorbed SO_2. *Icarus*, **54**, 511–23.

Nash, D. (1986). Mid-infrared reflectance spectra (2.3–22 μm) of sulfur, gold, KBr, MgO and halon. *Appl. Opt.*, **25**, 2427–33.

Nash, D., and Conel, J. (1974). Spectral reflectance systematics for mixtures of powdered hypersthene, labradorite and ilmenite. *J. Geophys. Res.*, **79**, 1615–21.

Nelson, R., Hapke, B., Smythe, W., and Horn, L. (1998). Phase curves of selected particulate materials: the contribution of coherent backscattering to the opposition surge. *Icarus*, **131**, 223–30.

Nelson, R., Hapke, B., Smythe, W., and Spilker, L. (2000). The opposition effect in simulated planetary regolths: reflectance and circular polarization ratio changes at small phase angle. *Icarus*, **147**, 545–58.

Nelson, R., Hapke, B., Smythe, W., Hale, A., and Piatek, J. (2004). Planetary regolith microsctucture: an unexpected opposition effect result. *Lunar Planet. Sci. XXXV*, Lunar and Planetary Institute, Houston, TX, abstract 1089.

Nelson, R., Smythe, W., Hapke, B., and Hale, A. (2002) Low phase angle laboratory studies of the opposition effect: search for wavelength dependence. *Planet. Space Sci.*, **50**, 849–56.

Nicodemus, F. (1970). Reflectance nomenclature and directional reflectance and emissivity. *Appl Opt.*, **9**, 1474–5.

Nicodemus, F., Richmond, J., Hsia, J., Ginsberg, I., and Limperis, T. (1977). *Geometrical Considerations and Nomenclature for Reflectance*. National Bureau of Standards Monograph 160. Gaithersburg, MD: National Bureau of Standards.

Niklasson, G., Granqvist, C., and Hunderi, O. (1981). Effective medium models for the optical properties of inhomogeneous materials. *Appl. Opt.*, **20**, 26–30.

Nitsan, U., and Shankland, T. (1976). Optical properties and electronic structure of mantle silicates. *Geophys. J. Roy. Astron. Soc.*, **45**, 59–87.

O'Donnell, K., and Mendez, E. (1987). Experimental study of scattering from characterized random surfaces. *J. Opt. Soc. Amer.*, A4, 1194–205.

Oetking, P. (1966). Photometric studies of diffusely reflecting surfaces with applications to the brightness of the moon. *J. Geophys. Res.*, **71**, 2505–13.

Ohman, Y. (1955). A tentative explanation of the negative polarization in diffuse reflection. *Ann. Obs. Stockholm*, **18**(8), 1–10.

Ostro, S. (1982). Radar properties of Europa, Ganymede and Callisto. In *Satellites of Jupiter*, ed. D. Morrison (pp. 213–36). Tucson, AZ: University of Arizona Press.

Ostro, S., and Shoemaker, E. (1990). The extraordinary radar echoes from Europa, Ganymede and Callisto: a geological perspective. *Icarus*, **85**, 335–45.

Otterman, J. (1983). Absorption of insolation by land surfaces with sparse vertical protrusions. *Tellus*, B35, 309–18.

Ozrin, V. (1992). Exact solution for coherent backscattering of polarized light from a random medium of Rayleigh scatterers. *Waves Random Media*, **2**, 141–64.

Palik, E. (ed.) (1991). *Handbook of Optical Constants of Solids*. New York: Academic Press.

Paliouras, J. (1975). *Complex Variables for Scientists and Engineers*. New York: Macmillan.

Panofsky, W., and Phillips, M. (1962). *Classical Electricity and Magnetism*. Cambridge, MA: Addison-Wesley.

Pasrev, V., Ovcharenko, A., Shkuratov, Y., Belshaya, I., and Videen G. (2007). Photometry of particulate surfaces at extremely small phase angles. *J. Quant. Spectrosc. Radiat. Transf.*, **106**, 455–63.

Peltoniemi, J., Lumme, K., Muinonen, K., and Irvine, E. (1989). Scattering of light by stochastically rough particles. *Appl. Opt.*, **28**, 4088–95.

Perrin, J., and Lamy, P. (1983). Light scattering by large rough particles. *Optica Acta*, **30**, 1223–44.

Perry, R., Hunt, A., and Huffman, D. (1978). Experimental determinations of Mueller scattering matrices for non-spherical particles. *Appl. Opt.*, **17**, 2700–10.

Petrova, E., Tishkovets, V., and Jockers, K. (2007). Modeling of opposition effects with ensembles of clusters: interplay of various scattering mechanisms. *Icarus*, **186**, 233–45.

Piatek, J., Hapke, B., Nelson, R., Smythe, W., and Hale, A. (2004). Scattering properties of planetary regolith analogs. *Icarus*, **171**, 531–45.

Pinnick, R., Carroll, D., and Hofmann, D. (1976). Polarized light from monodisperse randomly oriented nonspherical aerosol particles: measurements. *Appl. Opt.*, **15**, 384–93.

Pinty, B., and Verstraete, M. (1991). Extracting information on surface properties from bidirectional reflectance measurements. *J. Geophys. Res.*, **96**, 2865–74.

Pinty, B., Verstraete, M., and Dickinson, R. (1989). A physical model for predicting bidirectional reflectances over bare soil. *Rem. Sens. Environ.*, **27**, 273–88.

Pinty, B., Verstraete, M., and Dickinson, R. (1990). A physical model of the bidirectional reflectance of vegetation canopies. II. Inversion and validation. *J. Geophys. Res.*, **95**, 11 767–75.

Pollack, J., and Cuzzi, J. (1980). Scattering by nonspherical particles of size comparable to a wavelength: a new semi-empirical theory and its application to tropospheric aerosols. *J. Atmos. Sci.*, **37**, 868–81.

Pollack, J., and Whitehill, L. (1972). A multiple scattering model of the diffuse component of the lunar radar echoes. *J. Geophys. Res.*, **77**, 4289–303.

Purcell, E. M., and Pennypacker, C. R. (1973). Scattering and absorption of light by nonspherical dielectic grains. *Astrophys. J.*, **186**, 705–14.

Ramsey, M., and Christensen, P. (1998). Mineral abundance determination: quantitative deconvolution of thermal emission spectra. *J. Geophys. Res.***103**, 577–96.

Rayleigh, Lord (1871). On the light from the sky, its polarization and colour. *Philos. Mag.*, **41**, 107–20, 274–9.

Reichman, J. (1973). Determination of absorption and scattering coefficients for nonhomogeneous media. I. Theory. *Appl. Opt.*, **12**, 1811–23.

Richter, N. (1962). The photometric properties of interplanetary matter. *Quart. J. Roy. Astron. Soc.*, **3**, 179–86.

Ross, J., and Marshak, A. (1984). Calculation of the canopy bidirectional reflectance using the Monte-Carlo method. *Rem. Sens. Environ.*, **24**, 213–25.

Rosenbush, V., and Kiselev, A. (2005). Polarization opposition effect for the Galilean satellites of Jupiter. *Icarus*, **179**, 490–6.

Rosenbush, V., Avramchuk, V., Rosenbush, A., and Mishchenko, M. (1997). Polarization properties of the Galilean satellites of Jupiter: observations and preliminary analysis. *Astrophys. J.*, **487**, 402–14.

Rozenberg, G. (1966). *Twilight.* New York: Plenum.

Russell, H. (1916). On the albedo of planets and their satellites. *Astrophys. J.*, **43**, 173–87.

Salisbury, J. (1993). Mid-infrared spectroscopy: laboratory data. In *Remote Geochemical Analysis*, ed. C. Pieters and P. Englert (pp. 79–98). Cambridge University Press.

Salisbury, J., and Eastes, J. (1985). The effect of particle size and porosity on spectral contrast in the mid-infrared. *Icarus*, **64**, 586–8.

Salisbury, J., and Wald, A. (1992). The role of volume scattering in reducing spectral contrast of restrahlen bands in spectra of powdered minerals. *Icarus*, **96**, 121–8.

Salisbury, J., and Walter, L. (1989). Thermal infrared (2.5–13.5 μm) spectroscopic remote sensing of igneous rock types on particulate planetary surfaces. *J. Geophys. Res.*, **94**, 9192–202.

Salisbury, J., Hapke, B., and Eastes, J. (1987). Usefulness of weak bands in midinfrared remote sensing of particulate planetary surfaces. *J. Geophys. Res.*, **92**, 702–10.

Saunders, P. (1967). Shadowing on the ocean and the existence of the horizon. *J. Geophys. Res.*, **72**, 4643–9.

Schaber, G., Berlin, G., and Brown, W., Jr. (1976). Variations in surface roughness within Death Valley, California: geologic evaluation of 25-cm wavelength radar images. *Geol. Soc. Amer. Bull.*, **87**, 29–41.

Schatz, E. (1966). Effect of pressure on the reflectance of compacted powders. *J. Opt. Soc. Amer.*, **56**, 389–94.

Schiffer, R., and Thielheim, K. (1982a). Light reflection from randomly oriented convex particles with rough surfaces. *J. Appl. Phys.*, **53**, 2825–30.

Schiffer, R., and Thielheim, K. (1982b). A scattering model for the zodiacal light particles. *Astron. Astrophys.*, **116**, 1–9.

Schlatter, T. (1972). The local surface energy balance and subsurface temperature regime in Antarctica. *J. Appl. Meteor.*, **11**, 1048–62.

Schönberg, E. (1929). Theoretische Photometrie. In *Handbuch der Astrophysik, Vol. 2*, ed. G. Eberhard, A. Kohlschutter, and H. Ludendorff (pp. 1–280). Berlin: Springer.

Schuerman, D. (1980). *Light Scattering by Irregularly Shaped Particles*. New York: Plenum.

Schuerman, D., Wang, R., Gustafson, B., and Schaefer, R. (1981). Systematic studies of light scattering. I. Particle shape. *Appl. Opt.*, **20**, 4039–50.

Schulman, J., and Compton, W. (1962). *Color Centers in Solids*. New York: Pergamon.

Schuster, A. (1905). Radiation through a foggy atmosphere. *Astrophys. J.*, **21**, 1–22.

Seeliger, H. (1887). Zur Theorie der Beleuchtung der grossen Planeten inbesondere des Saturn. *Abhandl. Bayer. Akad. Wiss. Math.-Naturw. Kl. II*, **16**, 405–516.

Seeliger, H. (1895). Theorie der Beleuchtung staubformiger kosmischen Masses insbesondere des Saturinges. *Abhandl. Bayer. Akad. Wiss. Math.-Naturw. Kl. II*, **18**, 1–72.

Shepard, M., and Arvidson, R. (1999). The opposition surge and photopolarimetry of fresh and coated basalts. *Icarus*, **141**, 172–8.

Shepard, M., and Campbell, R. (1998). Shadows on a planetary surface and implications for photometric roughness. *Icarus*, **134**, 279–91.

Shepard, M., and Helfenstein, P. (2007). A test of the Hapke photometric model. *J. Geophys. Res.*, **112**, E03001, doi:10.1029/2005JE0026252007.

Shkuratov, Y. (1982). A model for negative polarization of light by cosmic bodies without atmospheres. *Sov. Astron.*, **26**, 493–6.

Shkuratov, Y. (1988). Diffractional model of the brightness surge of complex structure surfaces. *Kin., Phys., Cel. Bodies*, **4**, 33–9.

Shkuratov, Y. (1989). New mechanism of formation of negative polarization of light scattered by the solid surfaces of cosmic bodies. *Solar Syst. Res.*, **23**, 111–13.

Shkuratov, Y., and Ovcharenko, A. (1998). Brightness opposition effect: a theoretical model and laboratory measurements. *Solar Syst. Res.*, **32**, 276–86.

Shkuratov, Y., Kreslavsky, M., Ovcharendo, A., *et al.* (1999a). Opposition effect from *Clementine* data and mechanisms of backscatter. *Icarus*, **141**, 132–51.

Shkuratov, Y., Starukhina, L., Hoffmann, H., and Arnold, G. (1999b). A model of spectral albedo of particulate surfaces: implications for optical properties of the Moon. *Icarus*, **137**, 235–46.

Shkuratov, Y., Kaldash, V., Kreslavsky, M., and Opanasenko, N. (2001). Absolute calibration of the Clementine UVVIS data: comparison with ground-based observation of the moon. *Solar Syst. Res.*, **35**, 29–34.

Shkuratov, Y., Opanasenko, N., and Kreslavsky, M. (1992a). Polarimetric and photometric properties of the moon: telescopic observations and laboratory simulations. I. The negative polarization. *Icarus*, **95**, 283–99.

Shkuratov, Y., Opanasenko, N., and Kreslavsky, M. (1992b). Polarimetric and photometric properties of the moon: telescopic observations and laboratory simulations. II. The positive polarization. *Icarus*, **99**, 468–84.

Shkuratov, Y., Opanasenko, N., Zubko, E., *et al.* (2007). Multispectral polarimetry as a tool to investigate texture and chemistry of lumar regolith particles. *Icarus*, **187**, 406–16.

Shkuratov, Y., Ovcharenko, A., Zubko, E., *et al.* (2002). The opposition effect and negative polarization of structural analogs for planetary regoliths. *Icarus*, **159**, 396–416.

Shkuratov, Y., Ovcharenko, A., Zubko, E., *et al.* (2004). The negative polarization of light scattered from particulate surfaces and of independently scattring particles. *J. Quant. Spectrosc. Radiat. Transf.*, **88**, 267–84.

Shkuratov, Y., Stankevich, D., Ovcharenko, A., and Korokhin, V. (1997). A study of light backscattering from planetary regolith type surfaces phase angles 0.2°–3.5°. *Solar Syst. Res.*, **31**, 56–63.

Shkuratov, Y., Stankevich, D. Petrov, D., *et al.* (2005). Interpreting photometry of regolith-like surfaces with different topographies: shadowing and multiple scattering. *Icarus*, **173**, 3–15.

Simonelli, D., and Veverka, J. (1987). Phase curves of minerals on Io: interpretation in terms of Hapke's function. *Icarus*, **68**, 503–21.

Simpson, R., and Tyler, G. (1982). Radar scattering laws for the lunar surface. *IEEE Trans. Antennas Propag.*, AP30, 438–49.

Skorobogatov, B., and Usoskin, A. (1982). Optical properties of ground surfaces of nonabsorbing materials. *Opt. Spectr.*, **52**, 310–13.

Smith, D. (1985). Dispersion theory, sum rules and their application to the analysis of optical data. In *Handbook of Optical Constants of Solids*, ed. E. Palik (pp. 35–154). New York: Academic Press.

Smith, J. (1983). Matter–energy interactions in the optical region. In *Manual of Remote Sensing*, 2nd edn., ed. R. Colwell (pp. 61–113). Falls Church, VA: American Society of Photogrammetry.

Smith, M., Johnson, P., and Adams, J. (1985). Quantitative determination of mineral types and abundances from reflectance spectra using principal components analysis. In *Proc. 15th Lunar Planet. Sci. Conf.*, ed. G. Ryder and G. Schubert (pp. C797–804). Washington, DC: American Geophysical Union.

Smythe, W. (1975). Spectra of hydrate frosts: their application to the outer solar system. *Icarus*, **24**, 421–7.

Sobolev, V. (1975). *Light Scattering in Planetary Atmospheres*. New York: Pergamon.

Sokolov, A. (1967). *Optical Properties of Metals*. New York: Elsevier.

Spencer, J. (1990). A rough-surface thermophysical model for airless planets. *Icarus*, **83**, 27–38.

Spitzer, W., and Kleinman, D. (1961). Infrared lattice bands of quartz. *Phys. Rev.*, **121**, 1324–35.

Sproull, R., and Phillips, W. (1980). *Modern Physics*, 3rd edn. New York: John Wiley.

Stamnes, K., Tsay, S., Wiscombe, W., and Jayaweeta, K. (1988). Numerically stable algorithm for discrete-ordinate method radiative transfer in multiple scattering and emitting layered media. *Appl. Opt.*, **27**, 2502–9.

Steigman, G. (1978). A polarimetric model for a dust covered planetary surface. *Mon. Not. Roy. Astron. Soc.*, **185**, 877–88.

Stratton, J. (1941). *Electromagnetic Theory*. New York: McGraw-Hill.

Stroud, D., and Pan, F. (1978). Self-consistent approach to electromagnetic wave propagation in composite media: application to model granular metals. *Phys. Rev.*, B17, 1602–10.

Suits, G. (1972). The calculation of the directional reflectance of a vegetative canopy. *Rem. Sens. Environ.*, **2**, 117–25.

Sung, C., Singer, R., Parkin, K., and Burns, R. (1977). Temperature dependence of Fe^{2+} crystal field spectra: implications to mineralogical mapping of planetary surfaces. In *Proc. 8th Lunar Sci. Conf.*, ed. R. Merrill (pp. 1063–79). New York: Pergamon.

Sunshine, J., and Pieters, C. (1993). Estimating modal abundances from the spectra of natural and laboratory pyroxene mixtures using the modified Gaussian model. *J. Geophys. Res.*, **98**, 9075–87.

Sunshine, J., Pieters, C., and Pratt, S. (1990). Deconvolution of mineral absorption bands: an improved approach. *J. Geophys. Res.*, **95**, 6955–66.

Tanashchuk, M., and Gilchuk, L. (1978). Experimental scattering matrices of ground glass surfaces. *Opt. Spectr.*, **45**, 658–62.

Thompson, T., Pollack, J., Campbell, M., and O'Leary, B. (1970). Radar maps of the moon at 70 cm wavelength and their interpretation. *Rad. Sci.*, **5**, 253–62.

Thorpe, T. (1973). *Mariner 9* photometric observations of Mars from November 1971 through March 1972. *Icarus*, **20**, 482–9.

Thorpe, T. (1978). *Viking* orbiter observations of the Mars opposition effect. *Icarus*, **36**, 204–15.

Tishkovets, V., Shkuratov, Y., and Litvinov, P. (1999). Comparison of collective effects of scattering by randomly oriented clusters of spherical particles. *J. Quant. Spectrosc. Radiat. Transf.*, **61**, 767–73.

Tishkovets, V., Petrova, E., and Jockers, K. (2004). Optical properties of aggregate particles comparable in size to the wavelength. *J. Quant. Spectrosc. Radiat. Transf.*, **86**, 241–65.

Torrance, K., and Sparrow, E. (1967). Theory for off-specular reflection from roughened surfaces. *J. Opt. Soc. Amer.*, **57**, 1105–14.

Trowbridge, T. (1978). Retroreflection from rough surfaces. *J. Opt. Soc. Amer.*, **68**, 1225–42.

Umov, N. (1905). Chromatische Depolarization durch Lichtzerstreuung. *Phys. Z.*, **6**, 674–6.

Ungut, A., Grehan, G., and Gouesbet, G. (1981). Comparisons between geometrical optics and Lorentz–Mie theory. *Appl. Opt.*, **20**, 2911–18.

Van Albada, M., Van der Mark, M., and Lagendijk, A. (1988). Polarization effects in weak localization of light. *J. Phys.*, D21, S28–S31.

Van Albada, M., Van der Mark, and Lagendijk, A. (1990). Experiments on weak localization of light and their interpretation. In *Scattering and Localization of Classical Waves in Random Media*, ed. P. Sheng (pp. 97–136), Teaneck, NJ: World Scientifc Publications.

Van de Hulst, H. (1957). *Light Scattering by Small Particles*. New York: John Wiley.

Van de Hulst, H. (1974). The spherical albedo of a planet covered with a homogeneous cloud layer. *Astron. Astrophys.*, **35**, 209–14.

Van de Hulst, H. (1980). *Multiple Light Scattering*. New York: Academic Press.

Van der Mark, M., Van Albada, M., and Lagendijk, A. (1988). Light scattering in strongly scattering media: multiple scattering and weak localization. *Phys. Rev.*, B37, 3575–92.

Van Diggelen, J. (1959). Photometric properties of lunar crater floors. *Rech. Obs. Utrecht*, **14**, 1–114.

Van Diggelen, J. (1965). The radiance of lunar objects near opposition. *Planet. Space Sci.*, **13**, 271–9.

Van Ginneken, B., Stavridi, M., and Koenderink, J. (1988). Diffuse and specular reflectance from rough surfaces. *Appl. Opt.*, **3**, 130–9.

Vanderbilt, V., Grant, L., Biehl, L., and Robinson, B. (1985). Specular, diffuse and polarized light scattered by two wheat canopies. *Appl. Opt.*, **24**, 2408–18.

Vaughan, D. (1990). Some contributions of spectral studies in the visible and near visible light region to mineralogy. In *Absorption Spectroscopy in Mineralogy*, ed. A. Mottana and T. Burragato (pp. 1–37). New York: Elsevier.

Verstraete, M., Pinty, B., and Dickinson, R. (1990). A physical model of the bidirectional reflectance of vegetation canopies. I. Theory. *J. Geophys. Res.*, **95**, 11 755–65.

Veverka, J., Goguen, J., Yang, S., and Elliot, J. (1978a). Near-opposition limb darkening of solids of planetary interest. *Icarus*, **33**, 368–79.

Veverka, J., Goguen, J., Yang, S., and Elliot, J. (1978b). How to compare the surface of Io to laboratory samples. *Icarus*, **34**, 63–7.

Veverka, J., Goguen, J., Yang, S., and Elliot, J. (1978c). Scattering of light from particulate surfaces. I. A laboratory assessment of multiple scattering effects. *Icarus*, **34**, 406–14.

Veverka, J., Helfenstein, P., Hapke, B., and Goguen, J. (1988). Photometry and polarimetry of Mercury. In *Mercury*, ed. F. Vilas and C. Chapman (pp. 37–58). Tucson, AZ: University of Arizona Press.

Videen, G., Muinonen, K., and Lumme, K. (2003). Coherence, power laws and the negative polarization surge. *Appl. Opt.*, **42**, 3647–52.

Vilaplana, R., Moreno, F., and Molina, A. (2006). Study of the sensitivity of size-averaged scattering matrix elements of nonspherical particles to changes in shape, porosity and refractive index. *J. Quant. Spectrosc. Radiat. Transf.*, **100**, 415–28.

Vincent, R., and Hunt, G. (1968). Infrared reflectance from mat surfaces. *Appl. Opt.*, **7**, 539.

Volten, O., Munoz, O., Rol. E., *et al.* (2001). Scattering matrices of mineral aerosol particles at 441.6 nm and 632.8 nm. *J. Geophys. Res.*, **106**, 17 375–401.

Wagner, J., Hapke, B.W., and Wells, E.N. (1987). Atlas of reflectance spectra of terrestrial, lunar, and meteoritic powders and frosts from 92 to 1800 nm. *Icarus,* **69**, 14–28.

Wagner, R. (1967). Shadowing of randomly rough surfaces. *J. Acoust. Soc. Amer.*, **41**, 138–47.

Wallach, D., and Hapke, B. (1985). Light scattering in a spherical exponential atmosphere, with applications to Venus. *Icarus*, **63**, 354–73.

Walter, L., and Salisbury, J. (1989). Spectral characterization of igneous rocks in the 8 to 12 μm region. *J. Geophys. Res.*, **94**, 9203–13.

Warren, S. (1982). Optical properties of snow. *Rev. Geophys. Space Phys.*, **20**, 67–89.

Waterman, T. (1965). Matrix formulation of electromagnetic scattering. *Proc. IEEE*, **53**, 805–12.

Waterman, T. (1979). Matrix methods in potential theory and electromagnetic scattering. *J. Appl. Phys.*, **50**, 455–66.

Watson, G. (1958). *A Treatise on the Theory of Bessel Functions*. Cambridge University Press.

Watson, K. (1969). Multiple scattering of electromagnetic waves in underdense plasma. *J. Mathemat. Phys.*, **16**, 688–702.

Weaver, R. (1993). Anomalous diffusivity and localization of classical waves in disordered media: the effect of dissipation. *Phys. Rev.*, B47, 1077–80.

Weidner, V., and Hsia, J. (1981). Reflection properties of pressed polytetrafluoroethylene powder. *J. Opt. Soc. Amer.*, **71**, 856–61.

Weidner, V., Hsia, J., and Adams, B. (1985). Laboratory intercomparison study of pressed polytetrafluoroethylene powder reflectance standards. *Appl. Opt.*, **24**, 2225–30.

Weiss-Wrana, K. (1983). Optical properties of interplanetary dust: comparison with light scattering by larger meteoritic and terrestrial grains. *Astron. Astrophys.*, **126**, 240–50.

Wells, E. (1977). Optical absorption bands in glasses of lunar composition. Ph.D. thesis, University of Pittsburgh, PA.

Wells, E., and Hapke, B. (1977). Lunar soil: iron and titanium bands in the glass fraction. *Science*, **195**, 977–9.

Wells, E., Veverka, J., and Thomas, P. (1984). Mars: experimental study of albedo changes caused by dust fallout. *Icarus*, **58**, 331–8.

Wendtland, W., and Hecht, H. (1966). *Reflectance Spectroscopy*. New York: Wiley-Interscience.

Wesselink, A. (1948). Heat conductivity and the nature of the lunar surface material. *Bull. Astron. Inst. Netherlands*, **66**, 3033–45.

Whitaker, E. (1969). An investigation of the lunar heiligenschein. In *Analysis of Apollo 8 Photography and Visual Observations* (pp. 38–9). NASA SP-201. Washington, DC: NASA.

White, W., and Keester, K. (1966). Optical absorption spectra of iron in the rock-forming silicates. *Amer. Min.*, **51**, 774–91.

Widorn, T. (1967). Zur photometrischen Bestimmung der Durchmesser derkleinen Planeten. *Ann. Univ. Sternw. Wien*, **27**, 112–19.

Wildey, R. (1978). The moon in heiligenschein. *Science*, **200**, 1265–7.

Woessner, P., and Hapke, B. (1987). Polarization of light scattered by clover. *Rem. Sens. Environ.*, **21**, 243–61.

Wolf, P., and Maret, G. (1985). Weak localization and coherent backscattering of photons in disordered media. *Phys. Rev. Lett.*, **55**, 2696–9.

Wolf, P., Maret, G., Akkermans, E., and Maynard, R. (1988). Optical coherent backscattering by random media: an experimental *study. J. Phys. France*, **49**, 63–75.

Wolff, M. (1975). Polarization of light reflected from rough planetary surface. *Appl. Opt.*, **14**, 1395–405.

Wolff, M. (1980). Theory and application of the polarization-albedo rules. *Icarus*, **44**, 780–92.

Wolff, M. (1981). Computing diffuse reflection from particulate planetary surface with a new function. *Appl. Opt.*, **20**, 2493–8.

Wooten, F. (1972). *Optical Properties of Solids*. New York: Academic Press.

Yolken, H., and Kruger, J. (1965). Optical constants of iron in the visible region. *J. Opt. Soc. Amer.*, **55**, 842–4.

Xu, Y. (1995). Electromagnetic scattering by a aggregate of spheres. *Appl. Opt.*, **34**, 4573–88.

Young, A. (1973). Are the clouds of Venus sulfuric acid? *Icarus*, **18**, 564–82.

Zellner, B., and Gradie, J. (1976). Polarimetric evidence for the albedos and compositions of 94 asteroids. *Astron. J.*, **81**, 262–80.

Zellner, B., Gehrels, T., and Gradie, J. (1974). Polarimetric diameters. *Astron. J.*, **79**, 1100–10.

Zerull, R. (1976). Scattering measurements of dielectric and absorbing non-spherical particles. *Contr. Atmos. Phys.*, **49**, 168–88.

Zerull, R., and Giese, R. (1974). Microwave analogue studies. In *Planets, Stars and Nebulae Studied with Photopolarimetry*, ed. T. Gehrels (pp. 901–15). Tucson, AZ: University of Arizona Press.

Zubko, E., Shkuratov, Y., Mishchenko, M., and Videen, G. (2008). Light scattering in a finite multi-particle system. *J. Quant. Spectrosc. Radiat. Transf.* **109**, 2195–206.

Index

absorbance 405
absorption
  band in emittance 428
  band in reflectance 378
  band shape 43, 382
  bands in specular reflection 62
  coefficient 21
  coefficient retrieval 395
  cross section, *see* cross section, absorption
  crystal field 39
  length 21
  of light 27
  mean free path 69
  mechanisms 34
  volume-average coefficient of 151
albedo
  bolometric 296
  Bond, *see* reflectance, spherical
  factor 152
  geometric 298
  hemispherical 297
  normal 296
  physical 298
  plane, *see* reflectance, hemispherical
  single-scattering: calculating from reflectance, 24; 375; of mixture 25; volume average 152; of particle 67
  spherical, *see* reflectance, spherical
Allen diffraction approximation 101
Ambartsumian–Chandrasekhar H functions 201
analytic function 5, 467
asymmetry factor
  cosine 71
  hemispherical 190
  volume average 151

Babinet's principle 83
band
  model of electrons in solid 36
  conduction 37
  valence 37
Bessel functions 8, 475
bidirectional–reflectance distribution function 263
black-body radiation 413
boundary conditions 45, 199
boundary-layer approximation 447
Brewster's angle 56
bright shadow, *see* opposition effect,
  brightness temperature 428
Buratti–Veverka model 337

Cauchy–Gorsat theorem 6, 468
Cauchy integral formula 6, 468
Cauchy–Riemann condition 5, 467
Cellini, Benvenuto 221
charge, electric 5
charge-transfer band 40
Christiansen
  feature 429
  frequency 62
Clausius–Mossotti relation 13
cloudbow 91
coherent backscatter 237
coherent effects 163
collision frequency 28
color center 38
complex variables 5, 467
conductivity
  electric 18
  optical 20
  radiative 440
constitutive equations 6
critical angle 57
cross section
  absorption 68
  extinction 68
  geometric 69, 101
  scattering 68
crossing symmetry 61, 469
crystal-field theory, *see* absorption, crystal field
curl 5
current, electric 5

delta function 67
depolarization ratio 252
dielectric constant 12
  complex 19
  of ice 34, 399
  of water 35
diffraction 79
  by disk 82
  Fraunhofer 82, 84, 478
  Fresnel 159, 478
  by hole 82
  by irregular particle 102, 109
  by isolated sphere 79
  by particles that are not isolated, 158
diffuse surface 188
diffusion time 459
diffusivity, thermal 459
dilution method 402
dipoles
  electric 12
  magnetic 15
  scattering by 110
discrete dipole approximation 111
discrete ordinates method, *see* multistream method
dispersion
  anomalous 30, 369
  normal 30
  quantum-mechanical 32
  relation 33, 139
displacement, electric 5
divergence 2, 5, 464
doubling method 170
Drude model 27
dust, lunar 297

Eddington approximation 171
effective-medium theory 146
effective particle size 398
effective single-particle absorption thickness, *see* espat function
efficiency
  absorption 69
  extinction 69
  scattering 69
eikonal approximation 113
electronic transitions 36
emergence, plane of 183
emission
  thermal 156, 412
  volume coefficient 156
emissivity 415
  directional 418
  effect of particle size on 421
  factor 418
  hemispherical 425
  integrated 200
  spectral 416
emittance 415
  and reflectance combined 427
  spectroscopy 428
equivalent-slab approximation,
  of irregular particle 101
  of sphere 95
error function 231
espat function 69, 136, 397, 428
Europa 326
evanescent wave 57
exciton 41
extended-boundary-condition method, *see* T-matrix method
extinction
  cross section, *see* cross section, extinction
  efficiency, *see* efficiency, extincion
  mean free path, *see* mean free path, extinction

fairy castle structures xiii, 224
field
  electric 5
  magnetic induction 5
filling factor 147
Fraunhofer diffraction, *see* diffraction, Fraunhofer
free carriers 42
Fresnel
  condition 159, 478
  diffraction, *see* diffraction, Fresnel
  equations 46

Galileo, Galilei 1
Gaussian quadratures method 173
Gauss's theorem 465
geometric-optics scattering 78
glory 91
Gold, T. xii
gold black 407
gradient 2, 464
gray body 416
greenhouse effect, solid state 457

H functions, *see* Ambartsumian–Chandrasekhar *H* functions
  approximations for 203
  moments of 203
Hankel functions 72, 475
heilegenschein, *see* opposition effect
helicity 23
  reversal by reflection 49
Helmholtz's equation 472
Henyey–Greenstein function 104
hot spot, *see* opposition effect
Huygens's principle 10

ice 399
IMSA, *see* isotropic multiple scattering approximtion
incidence, plane of 11
integral phase function 299
intensity 66

magnetic 5
specific 66
interference 22, 237, 359
internal scatterers 118
internal-transmission factor 96, 127
invariance, method of 174, 200, 419
irradiance 11
isotropic multiple scattering approximation 206

Jones matrix 340
Jones vector 24

Kirchhoff's law 416
Kramers–Kronig relations 7, 61, 469
Kubelka–Munk theory 400

Lambert
surface, *see* diffuse surface
sphere 107
Lambert's law 107, 187
Laplace's equation 472
Laplacian operator 7, 464
Laporte rule 39
lattice vibrations 35
acoustical branch 36
optical branch 36
layers
effect of, on absorption bands 392
reflectance of 272
leaf-area index 271
leaf-droop index 270
Legendre functions
addition theorem for 290, 474
associated 103, 473
polynomials 103, 473
Lommel–Seeliger law 107, 197
Lommel–Seeliger sphere 108
Lorentz–Lorenz relation, *see* Clausius–Mossotti relation
Lorentz model 29
loss tangent 19
luminance coordinates 184
Lumme–Bowell model 335

magnetization 15
Maxwell-Garnett model 146
difficulty with 147
role of density fluctuations in 148
Maxwell's field equations 5
mean free paths 152
mean optical path length 405
Melamed model 128
Mercury 326
Mie theory 72
Minnaert's law 188
mixtures 282
areal 282
espat function of 286
intimate 282
spectra of 388
modified Gaussian method 407
modified IMSA, *see* isotropic multiple scattering approximation
molecular rotation 34
Monte Carlo method 112, 170
Moon 1, 342
Mueller matrix 71, 340
multistream method 172
Murphy's law 4

near fields 73

Ohm's law 18
opposition effect 221
coherent backscatter 237
combined 260
shadow-hiding 224
thermal 435
optical
depth 157
flatness, Rayleigh criterion for 64
orthogonal set 476
oscillator strength 32

permeability
of free space 6
magnetic 16
permittivity
electric 12
of free space 7
phase
angle 70
function of a single particle 70
function, volume average 152
integral 302
of a wave 23
phonons 35
photometric
coordinates, *see* luminance coordinates
function 295
Planck function 413
plasma
frequency 28
oscillations 29
resonance 76
polarizability, electric 12
polarization
–albedo relation 351
and Brewster's angle 351
broad negative branch of 118, 354
circular 23, 341
elliptical 23
from irregular particles 118
linear 23, 340
negative ranch of 355
opposition effect 355

polarization (*cont.*)
of a particulate medium 339
phase curve 339
positive branch of 344
ratio 26, 339
of a sphere 95
transverse electric 52
transverse magnetic 52
of vegetation 351
of a wave 26
Pollack–Cuzzi model 109
porosity 147
coefficient 166
Poynting vector 11
principal plane 184

radialith 394
radiance 66
factor 264
radiative transfer, equation of 145
in a medium of arbitrary particle separation 158
in a medium of well-separated scatterers 148
methods of solution 169
radiosity method 170
radius, equivalent of irregular particle 101
rainbow 91
Rayleigh
absorber 76
criterion for optical flatness 64
region 75
scatterer 75
Rayleigh–Jeans law 414
reciprocity principle 264
reflectance
and absorption coefficient 378
biconical 181, 287
bidirectional 181, 197
bihemispherical 181, 301
coefficient, *see* reflectance factor
combined specular and diffuse 266
diffuse 188
diffusive 189
directional–hemispherical, *see* reflectance, hemispherical
factor 263
hemispherical 182, 288
hemispherical–directional 182
inversion 372
Lambert, *see* reflectance, diffuse
of layered media 273, 278
measurement of 370
of mixtures 282, 388
reduced 263
relative 264
spectroscopy 369
spherical, *see* reflectance, bihemispherical
of vegetation 270
reflection
dielectric 54
metals 59
normal 47
regular 51, 266
specular 45
total internal 57
refraction 47
index of 17, 20
regolith, lunar 234, 352, 391, 408
relaxation time 34
remission function 402
reststrahlen 64
roughness
effects of, on planetary photometry 323
effects of, on reflectance 303
Russell's rule 302

scattering
angle 69
by cylinder 95
by ellipsoid 95
by irregular particle 100
matrix 71
mean free path 152
by oriented particles 268
plane 69, 183
by sphere 72
volume coefficient 151
Schönberg function, *see* Lambert sphere
Schuster–Schwarzschild method, *see* two-stream method
separation-of-variables method 470
shadowing
function 309
interparticle 224
large-scale roughness 309
shadows
projected 311
tilt 311
Shkuratov albedo model 404
Shkuratov reflectance model 337
similarity relations 194, 205
size parameter 69
slope
–albedo relation in polarization 352
distribution function 304
mean angle of 305
Snell's law 51
source function 156
space weathering 408
specific inductive capacity, *see* dielectric constant
spectral-ratio–albedo diagram 409
spectroscopy
emittance 428
reflectance 369
spherical harmonic functions 476
spin-multiplicity rule 39

Stefan–Boltzmann law 415
Stokes theorem 466
Stokes vector 24
sum rule 32
scattering region
  strong surface 382
  volume 383
  weak surface 384
susceptibility
  electric 12
  magnetic 15

thermal
  beaming, *see* opposition effect, thernal
  black-body radiation 413
  emission, *see* emission, thermal
  inertia 435, 459
  infrared 412
  source function 417
  volume emission coefficient, 418
tilt, effective angles of 309
T-matrix method 112
transition minimum 385
transmission coefficient
  of dielectric 54
  normal 47
transmissivity 155, 163
transparency feature 431
transport
  coefficient 152
  mean free path 152
two-stream method, 173, 189

Umov effect, *see* polarization–albedo relation

vector
  calculus 463
  cross-product 463
  dot product 463
vegetation
  polarization by 351
  scattering by 269
Venus 215

water 34
wave equation 6, 8, 470
weak photon localization, *see* coherent backscatter
Wien displacement law 415

Printed in the United States
By Bookmasters